Next-Generation Materials for Energy Chemistry

Xiamen, China

27–29 October 2014

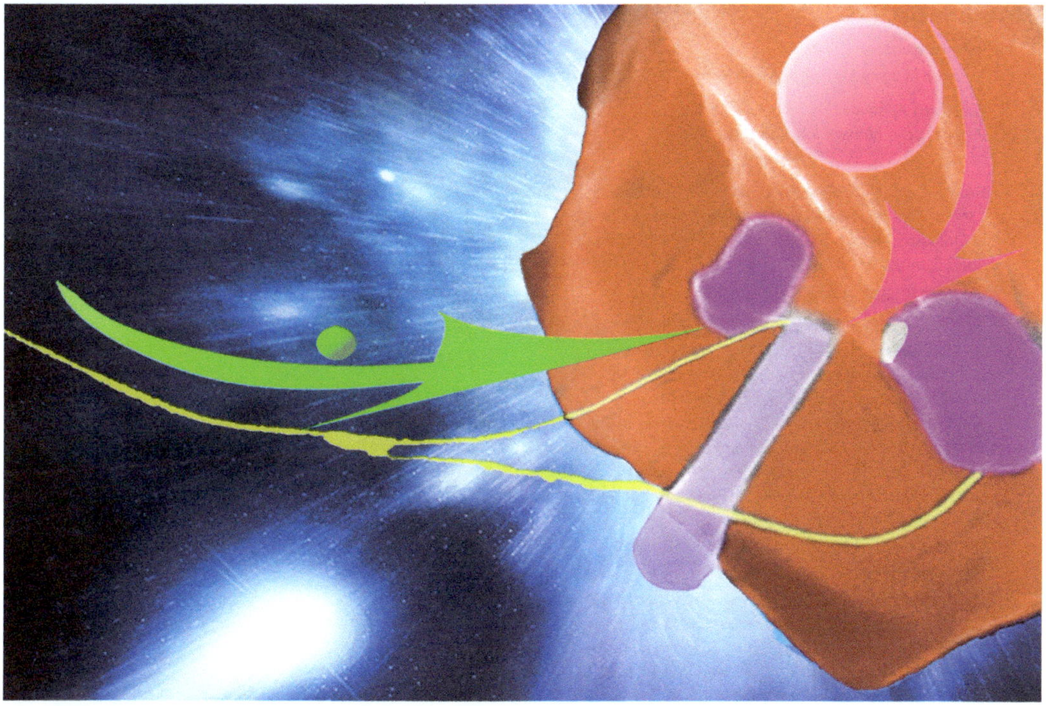

FARADAY DISCUSSIONS

Volume 176, 2014

ROYAL SOCIETY OF CHEMISTRY

The Faraday Division of the Royal Society of Chemistry, previously the Faraday Society, founded in 1903 to promote the study of sciences lying between Chemistry, Physics and Biology.

Faraday Discussions (Print ISSN 1359-6640, Electronic ISSN 1364-5498) is published 8 times a year by the Royal Society of Chemistry, Thomas Graham House, Science Park, Milton Road, Cambridge, UK CB4 0WF.

Volume 176 ISBN-13: 978-1-78262-044-0

2015 annual subscription price: print+electronic £877, US $1635; electronic only £833, US $1553. Customers in Canada will be subject to a surcharge to cover GST. Customers in the EU subscribing to the electronic version only will be charged VAT. All orders, with cheques made payable to the Royal Society of Chemistry, should be sent to RSC Order Department, Royal Society of Chemistry, Thomas Graham House, Science Park, Milton Road, Cambridge, CB4 0WF, UK.
Tel +44 (0) 1223 432398; E-mail www.orders@rsc.org

If you take an institutional subscription to any RSC journal you are entitled to free, site-wide web access to that journal. You can arrange access via Internet Protocol (IP) address at www.rsc.org/ip. Customers should make payments by cheque in sterling payable on a UK clearing bank or in US dollars payable on a US clearing bank.

Printed in the UK

MIX
Paper from responsible sources
FSC® C013604

Faraday Discussions documents a long-established series of Faraday Discussion meetings which provide a unique international forum for the exchange of views and newly acquired results in developing areas of physical chemistry,
biophysical chemistry and chemical physics.

Next-Generation Materials for Energy Chemistry

Faraday Discussions

www.rsc.org/faraday_d

A General Discussion on Next-Generation Materials for Energy Chemistry was held in Xiamen, China on the 27th, 28th and 29th of October 2014.

RSC Publishing is a not-for-profit publisher and a division of the Royal Society of Chemistry. Any surplus made is used to support charitable activities aimed at advancing the chemical sciences. Full details are available from www.rsc.org

CONTENTS

ISSN 1359-6640; ISBN 978-1-78262-044-0

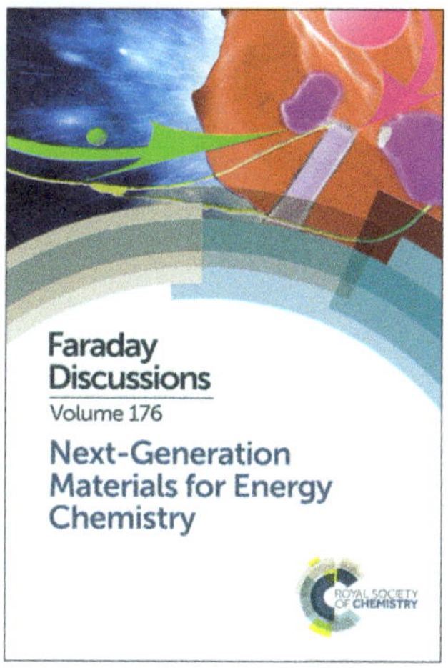

Cover
See Wenjun Li *et al.*, *Faraday Discuss.*, 2014, **176**, 109.

Growth of lithium dendrite in an all-solid-state battery has been investigated *in situ* in a scanning electronic microscope. After applying a −3 V bias voltage between the metal lithium anode and the carbon nanotube cathode, a rod-like lithium dendrite grows on the surface of metal lithium covered by a layer of Li_2O solid electrolyte. This image was captured after 1100 s.

Image courtesy of Wenjun Li, Hao Zheng, Geng Chu, Fei Luo, Jieyun Zheng, Dongdong Xiao, Xing Li, Lin Gu, Hong Li, Xianlong Wei, Qing Chen and Liquan Chen, Peking University, Beijing, China.

INTRODUCTORY LECTURE

PAPERS AND DISCUSSIONS

SUPPORTING ORGANISATIONS

elkington and fife LLP

PCOSS State Key Laboratory of Physical Chemistry of Solid Surfaces

固体表面物理化学国家重点实验室（厦门大学）

MIKROUNA

国家自然科学基金委员会 NSFC

SYN 新兴公司

超威 CHILWEE

Pen-Tung Sah Education and Scientific Research Foundation

SUO-JIN (XIAMEN) Chemical Technology Co.Ltd

PERSPECTIVE

CONCLUDING REMARKS

ADDITIONAL INFORMATION

Faraday Discussions

PAPER

Introductory lecture: Systems materials engineering approach for solar-to-chemical conversion

Chong Liu and Peidong Yang*

Received 15th December 2014, Accepted 17th December 2014
DOI: 10.1039/c4fd00264d

Solar-to-chemical (STC) production using a fully integrated system is an attractive goal, but to-date there has yet to be a system that can demonstrate the required efficiency or durability, or could be manufactured at a reasonable cost. One can learn a great deal from the natural photosynthesis where the conversion of carbon dioxide and water to carbohydrates is routinely carried out at a highly coordinated system level. There are several key features worth mentioning in these systems: spatial and directional arrangement of the light-harvesting components; charge separation and transport; as well as the desired chemical conversion at catalytic sites in compartmentalized spaces. In order to design an efficient artificial photosynthetic materials system, at the level of the individual components better catalysts need to be developed, new light-absorbing semiconductor materials will need to be discovered, architectures will need to be designed for effective capture and conversion of sunlight, and more importantly, processes need to be developed for the efficient coupling and integration of the components into a complete artificial photosynthetic system.

System-level planning is increasingly important for the development of modern materials science.[1] Materials science has evolved over the past decades so that there is now an increasing need for efforts from various subfields of materials chemistry and physics to come together to solve grand challenges in energy conversion and storage. Today, materials and system design for novel energy conversion and storage applications requires significant attention towards interfaces between different materials components,[1] as very often these interfaces are rate-limiting for energy transfer, and consequently limit the overall energy conversion efficiency.

Let's start the discussion with the following design of a solar-to-chemical (STC) generation system (Fig. 1).[1,2] For a typical STC generation process, there are a couple of fundamental reactions that one can think of. The simplest one is solar

Department of Chemistry and Department of Materials Science & Engineering, University of California, Materials Sciences Division, Lawrence Berkeley National Lab, Berkeley, CA 94720, USA. E-mail: p_yang@berkeley.edu

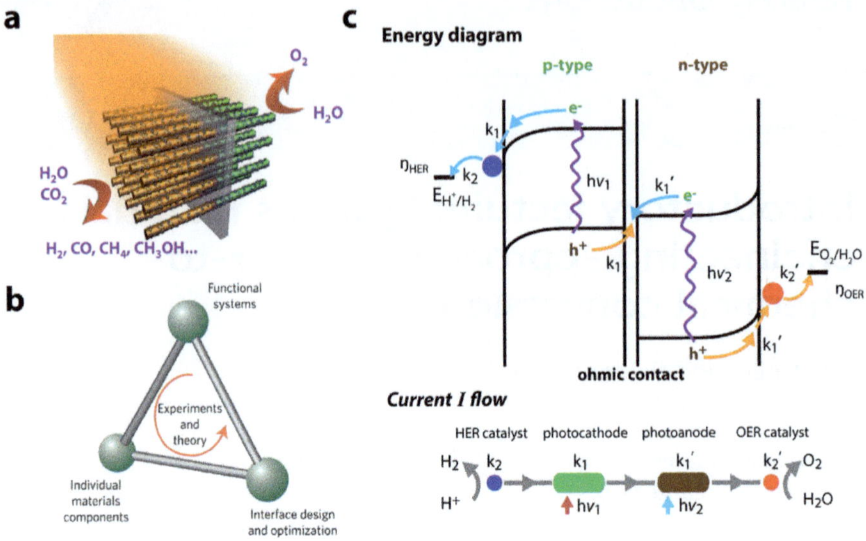

Fig. 1 Semiconductor nanowires for artificial photosynthesis of value added chemicals. A schematic standalone device based on nanowires is shown in (a), which captures many of the features in natural photosynthesis. Its energy diagram under the "Z-scheme" is depicted in (c), using water splitting as an example. In an ideal situation, the flux of charge carriers at different components, including the photocurrent from light absorbers (k_1, k_1') and the TOF from electrocatalysts (k_2, k_2'), should be comparable to ensure maximum efficiency. Because of the interrelated nature of the components in artificial photosynthesis, a system-level consideration is needed to consider the balances among the various components, and further device optimization should be based on such consideration (b). Panel (a) and (c) reproduced with permission from ref. 2, copyright © 2014 American Chemical Society; panel (b) reproduced with permission from ref. 1, copyright © 2012 Nature Publishing Group.

water splitting while CO_2 reduction can produce different final products depending on the number of electrons involved. To make these reactions happen efficiently, we will need to consider several processes synergistically.[2] One will need to worry about light-absorbing semiconductors and design of suitable heterojunctions for charge separation. It is important that the semiconductor electrodes have high minority carrier mobilities so that photo-excited charges can migrate efficiently towards catalyst/electrolyte interfaces where oxidation or reduction would happen. The length scale of the materials design spans from the atomic/molecular level, to the nanometer, micrometer, and all the way up to macroscopic device level. The time scales for many of the processes involved also cover many order of magnitudes, from femtoseconds for light harvesting/charge generation, to milliseconds or seconds for chemical reactions on the surface (Fig. 1). Such complexity in the materials design for this solar-to-chemical conversion process is indeed challenging!

First of all, there is something that we can learn from nature. Nature can do all of these with remarkable ease and provides an exquisite archetype. Natural photosynthesis is not only a source of oxygen, food and fossil fuels, but can capture roughly 100 out of the 178 000 terawatts solar energy striking on the earth daily with an overall efficiency typically less than 1%.[3,4] This accounts for approximately ten times the amount of energy that is produced and consumed by humans all over the globe. In plants, photosystems I and II are responsible for the two light-capturing events and cover essentially the entire visible part of solar spectrum. These two light absorption systems generate enough photovoltage or driving force to move the two reactions forward, one is water oxidation, and the other is CO_2 reduction. Even though the efficiency of the natural photosynthesis is not that high for its own good reason, the physics of coupling light absorbers with functional catalysts is quite useful when one tries to develop artificial materials and systems for solar water splitting or CO_2 reduction.

One basic concept, the photochemical diode, was proposed back in the late 70s.[5] It basically captures the essence of what is happening in natural photosynthetic systems. Within this original design, two small bandgap semiconductors were used, one is a *p*-type semiconductor serving as photocathode, the other side is an *n*-type semiconductor serving as photoanode. Additionally an ohmic contact exists in between these two semiconductors (Fig. 1). When a system of such two light-absorbers is in contact with the electrolyte, effectively two liquid junction photoelectrochemical cells are established.[6] Each of these two semiconductor/electrolyte junctions will generate their own photovoltages, and the combined photovoltage will provide sufficient driving force to carry out the proton/CO_2 reduction and water oxidation.

Theoretically, one would expect high energy conversion efficiency from such a dual light-absorber approach.[7,8] By using two small band gap semiconductors, photons from the whole solar spectrum can be captured and converted into photocurrent. The resultant theoretical photocurrent output readily approaches above 20 mA cm^{-2}. Assuming all of this photocurrent is used for chemcial reaction, a theoretical solar energy conversion efficiency of ~24% can be expected. This is one of the major advantages of using two semiconductor photoelectrodes within this photochemical diode concept. In reality, when we examine this fundamental design carefully, it turns out to be quite challenging to construct such a device with reasonable efficiency. In one aspect, light capture and charge separation in solid-state semiconductors typically happens on the femtosecond time scale; while the chemical reaction typically turns over in the order of magnitude of milliseconds or even seconds. Such a mismatch of time scales implies that the expected photocurrent (in the form of the minority carrier) from sunlight may not be converted efficiently into the flux of the chemical reaction. Subsequently, the matching between the solar flux and catalytic activity is a big challenge.[2]

Now, let's first take a look at the issue of matching solar flux to the catalytic activity. A commercially viable STC device should reasonably provide 10 mA cm^{-2} photocurrent,[8] which corresponds to delivering about 600 electrons onto a 1 nm^2 flat surface every second. This leads to an important interface issue, since these photo-generated charges have to be passed onto the catalysts on the surface so that energy can be stored in chemical bonds. In other words, the turnover frequency (TOF) of the catalysts has to match the output photocurrent from the photoelectrodes. With an expected current output of 10 mA cm^{-2} from the semiconductor, the catalysts located in a surface of 1 nm^2 area will, accordingly, need to be able to process 600 electrons every second, otherwise the overall efficiency of the STC device will suffer.

Such consideration leads us to revise the original concept of photochemical diodes[5] and attempt to remove this obstacle.[9] In the revised design, we have replaced the planar electrode with a high surface-area nanowire array (Fig. 1a), which effectively reduces the required flux of chemical reaction on the electrode surface. More importantly, using semiconductor nanowire array electrodes is beneficial due to their high carrier mobility and short distance that minority carriers have to travel in order to reach the electrolyte interface.[9] In this way the issue of matching solar flux with TOF can be removed or at least significantly alleviated. Assuming the nanowire array has 100 times more surface area to load the catalysts, effectively the 10 mA cm^{-2} photocurrent can now spread over a surface that is 100 times larger, resulting in lower effective operating current density and subsequently lower electrochemical overpotentials. Because of the unique capability of stacking the catalysts in the third dimension, the requirement of catalyst TOF for solar flux matching can be decreased by orders of magnitude, which will greatly facilitate the construction of a fully integrated photochemical diodes loaded with functional catalyst. As shown more quantitatively in Fig. 2, the increase of roughness factor (the ratio between surface area and the projected electrode area) leads to a reduced requirement of overpotential, which is certainly a significant improvement upon the planar design. While an increased surface area could be accomplished in many forms, the single-crystalline nanowire morphology provides a well-controlled platform for quantitative investigation of the catalyst

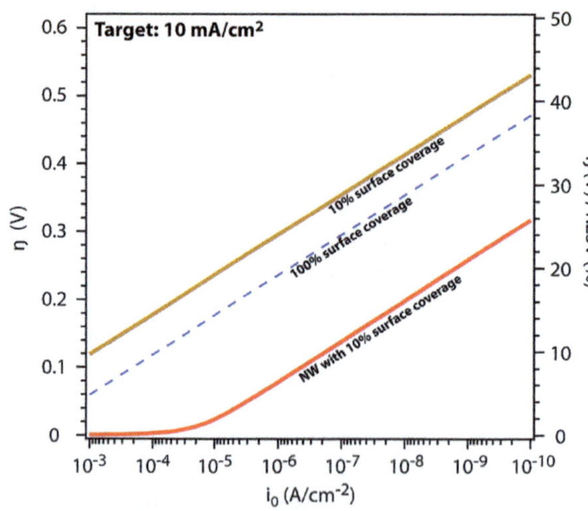

Fig. 2 An examination of the flux matching between electrocatalysts and light-absorbers. Here, the required overpotential η (left y-axis) to satisfy a charge flux of 10 mA cm^{-2} is plotted against the exchange current density i_0 (x-axis) of a planar electrode. The dashed blue curve represents a simulation for one electrode with 100% surface coverage of an electrocatalyst, assuming a 59 mV dec^{-1} Tafel slope in the Bulter–Volmer equation. The solid yellow curve represents the same electrode but with only 10% surface coverage. The benefit of nanowires is shown by the solid red curve, which is the simulation for one electrode with 10% electrocatalyst coverage, but with a roughness factor of 1000. The required η is significantly reduced, which is also illustrated by calculating the ratio between the overpotential η and 1.23 V, the thermodynamic energy required to split water (right y-axis). Assuming negligible catalytic activity from the semiconductor surface and negligible catalyst-support interaction, the loss of overpotential is not dependent on the semiconductor used. Reproduced with permission from ref. 2, copyright © 2014 American Chemical Society.

requirements, at the same time providing a direct current path for charge collection with high carrier mobility. This sheds new light on the fundamental properties of the interface between the electrocatalysts and semiconductor light absorbers.

With this concept in mind, we investigated quantitatively the required loading of Pt nanoclusters on a Si nanowire array photocathode for the hydrogen evolution reaction (HER).[10] As a typical model system, Pt is still the standard HER electrocatalyst used at the photocathode. It is worthwhile to investigate the lower limits of Pt loading on Si photocathodes under the desirable charge flux for practical applications. Taking advantage of the precise mass-loading control of atomic layer deposition (ALD), a highly conformal coating of Pt nanoparticles with tunable sizes ranging from 0.5–3 nm was deposited onto Si nanowire array photocathodes (Fig. 3). The electrochemical and photoelectrochemical performance were measured using as little as 1 cycle of ALD Pt, which corresponds to a surface mass loading of ~10 ng cm^{-2}, or an equivalent surface coverage of about 2.7% of a monolayer. Despite these low Pt loading quantities, Si nanowire photocathodes could still provide the necessary charge flux for STC conversion. Furthermore, the slope of the *J–V* curve could be directly controlled by varying the Pt mass loading in sub-monolayer quantities. By exploring quantitatively the lower limits of Pt co-catalyst loading on high surface area nanowire photoelectrodes, we can establish a general approach for minimizing the cost of precious metal co-catalysts for affordable, efficient solar-to-fuel applications.

With this revised overall design in mind, a system level of planning for materials synthesis and assembly is required.[1] First photocurrents from the photoanode and photocathode have to match, based on the effective circuit diagram (Fig. 1c). Second their combined output photovoltage has to be sufficient to drive the two half reactions. Last but not least, catalysts loaded onto these high surface area arrays must have a suitable TOF to match with the current generated by the nanowire photoelectrodes.

For the photocathode, Si[11] ($E_g = 1.1$ eV) and InP[12] ($E_g = 1.3$ eV) have been reported to generate high current densities with photovoltages of greater than 500 mV. Therefore, the development of photocathodes should focus on reducing the raw

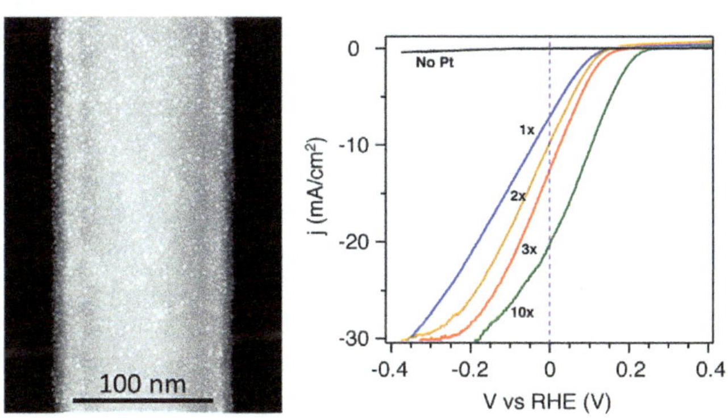

Fig. 3 ALD is a versatile deposition technique that can precisely deposit electrocatalysts on high-aspect-ratio surfaces in a controllable manner (left), which helps to explore the lower limits of Pt electrocatalyst loading on p-Si nanowire array photocathodes (right). The electrode is tested under 100 mW cm^{-2} simulated AM1.5G illumination, with 0.5 M H_2SO_4 electrolyte. Reprinted from ref. 10. Copyright 2013 American Chemical Society.

material usage and fabrication costs, for example *via* solution-phase synthesis of semiconductor nanomaterials.[13] Comparatively, the photoanode deserves more attention.[14] Although compound semiconductors with junction engineering and corrosion protection have shown promising progress,[15] a more cost-effective candidate for the photoanode, such as oxide or nitride based materials, should be considered. At present there is no oxide- or nitride-based photoanode device capable of providing a photocurrent of greater than 10 mA cm^{-2}, while at the same time providing adequate photovoltage to couple with the photocathode. Advanced materials engineering, including improving the material quality as well as designing an appropriate nanostructure, is expected to contribute to this exciting field.

In the above discussion, we considered individual components for artificial photosynthesis, with a focus on the constraints of current matching and the corresponding improvements needed. However, in the end, an integrated system with interconnected components is the key concept to be addressed.[16] It is critical to assemble all the pieces together and demonstrate the overall STC conversion under such an integrated system. It also acts as the final milestone for consideration of an STC application for commercialization. Furthermore, an integrated standalone system is fundamentally intriguing, in order to mimic the micro and nanoscopic spatial control of natural photosynthesis. In chloroplasts, the components of photosynthesis are carefully arranged:[3] photosystems I and II are arranged side-by-side on the thylakoid membrane with the electron transport chain between them for efficient charge transfer, while the reduction and oxidation catalytic centers are spatially separated to minimize the undesirable back-reaction of the photosynthetic products. Achieving such nanoscopic spatial control is a crucial aspect of our pursuit of artificial photosynthesis.

Using Si and TiO_2 nanowires as building blocks, a proof-of-concept STC conversion nanodevice was demonstrated for solar water splitting (Fig. 4).[16] Equipped with a knowledge of the various interfaces involved,[17,18] a tree-shaped nanowire-based heterojunction device was constructed and loaded selectively with HER and oxygen evolution reaction (OER) electrocatalysts. Compared with the Si nanowire photocathode, the smaller feature size of the TiO_2 nanowire photoanode is motivated by the smaller minority carrier diffusion length in the photoanode and the slower reaction kinetics of OER, in accordance with the design principles of nanowire photoelectrodes.[19] As discussed above, the bottleneck of the constructed system is the TiO_2 photoanode, whose band gap is in the UV region (3.0 eV for anatase). Despite this limitation, unassisted solar water splitting was observed with 0.12% solar-to-fuel efficiency, comparable with natural photosynthesis.[3]

This proof-of-concept device leaves much room for improvement and welcomes performance enhancements by updating individual components with the latest improvements, allowing for a modular design approach.[1] By implementing the advances discussed earlier, continued efficiency improvement is expected. Specifically, the balance between electron flux from the light-absorber and the TOF of the electrocatalyst requires development of electrocatalyst materials, especially for the OER and CO_2 reduction. Ideally, an overpotential of less than 100 mV at 10 mA cm^{-2} is desired. Nanowire electrodes can help accomplish this by providing a large surface area for catalyst loading. They can also provide a well-controlled system for fundamental studies such as tuning the interface between electrocatalyst and underlying semiconductor light-absorber. The

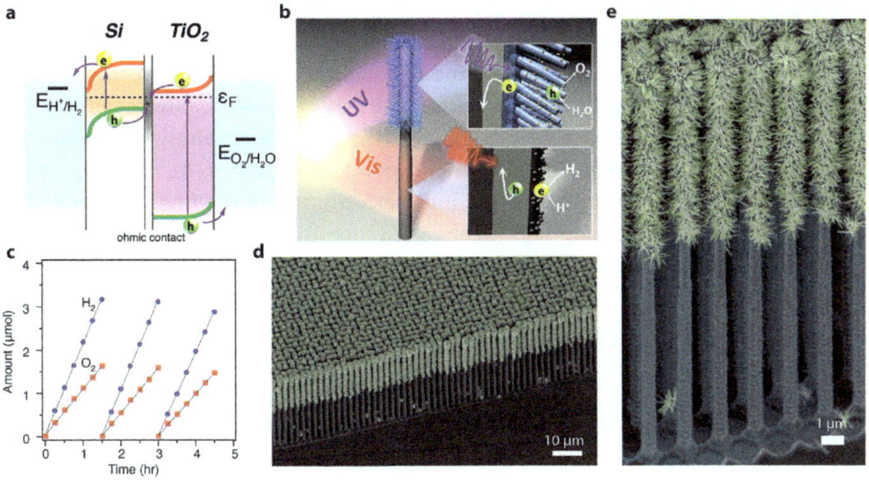

Fig. 4 A proof-of-concept fully integrated nanosystem for direct solar water splitting was demonstrated (a). The tree-shaped heterostructure is designed based on different material properties of Si and TiO_2 photoelectrodes (b), which leads to unassisted water splitting (c). The overall integrated structure contains spatial separation of the photocathode and photoanode (d and e), analogous with natural photosynthesis. Reproduced with permission from ref. 16, copyright © 2013 American Chemical Society.

requirements for high photocurrent and photovoltage output rely heavily on photoanode research, which can be further supplemented by improvement of photovoltage output in the photocathode. Nanowire structures provide opportunities for new material discovery, as well as improving the performance of existing materials. Based on these expectations, it is evident that semiconductor nanowires can function as essential building blocks and help to solve many of the issues in artificial photosynthesis.[9]

To wrap up on this discussion we think that, in the next 5–10 years, there will be two levels of research that we have to focus on. The first is to explore advanced components, including both light-absorbers and electrocatalysts, which are stable, scalable, and earth-abundant. The second is to achieve an efficient coupling and integration of these advanced components into a complete artificial photosynthetic system. Moreover, it is important to bear in mind that these two levels of researches are mutually related: the development of individual components requires a blueprint of the integrated system; and the assembled device not only relies on the performance of individual building blocks but also provides feedback for further optimization. When conducting this line of research, one will find that these seemingly disparate disciplines all become integral parts of the solution. There is no doubt that the ultimate solution of artificial photosynthesis or STC totally relies on insights not just from traditional materials science, but also from chemistry, physics, electrical engineering, mechanical engineering, and other research fields that may be beyond our imagination.

Now here is the take-home message from this solar-to-chemical conversion research. While the paradigm of the traditional materials-science tetrahedron is still highly relevant in many areas of materials research, it is now increasingly important to think and plan research at the systems level for both experimental and theoretical efforts (Fig. 1b). It is no surprise that we are on a similar path as how biology has evolved: with continued advances in the understanding of individual components in biology, the interaction and feedback between these components turned out to be critical and the discipline of systems biology was born. Similarly, systems materials engineering,[1] which emphasizes the study of the interactions between individual components of a system, and how these interactions give rise to the function and performance of the final system, is now emerging and will provide significant guidance in the advancement of energy storage and conversion research.

Acknowledgements

This work is supported by the Office of Science, Office of Basic Energy Sciences of the U. S. Department of Energy under Contract no. DE-AC02-05CH11231 (PChem).

References

1 P. Yang and J.-M. Tarascon, *Nat. Mater.*, 2012, **11**, 560.
2 C. Liu, N. P. Dasgupta and P. Yang, *Chem. Mater.*, 2014, **26**, 415.
3 D. Hall and K. Rao, *Photosynthesis*, Cambridge University Press, Cambridage, United Kingdom, 6th edn, 1999.
4 R. E. Blankenship, D. M. Tiede, J. Barber, G. W. Brudvig, G. Fleming, M. Ghirardi, M. R. Gunner, W. Junge, D. M. Kramer, A. Melis, T. A. Moore, C. C. Moser, D. G. Nocera, A. J. Nozik, D. R. Ort, W. W. Parson, R. C. Prince and R. T. Sayre, *Science*, 2011, **332**, 805.
5 A. J. Nozik, *Appl. Phys. Lett.*, 1977, **30**, 567.
6 M. Gratzel, *Nature*, 2001, **414**, 338.
7 J. R. Bolton, S. J. Strickler and J. S. Connolly, *Nature*, 1985, **316**, 495.
8 S. Hu, C. Xiang, S. Haussener, A. D. Berger and N. S. Lewis, *Energy Environ. Sci.*, 2013, **6**, 2984.
9 Semiconductor nanowire array: potential substrates for photocatalysis and photovoltaics, Y. Wu, H. Yan and P. Yang, *Top. Catal.*, 2002, **19**, 197.
10 N. P. Dasgupta, C. Liu, S. Andrews, F. B. Prinz and P. Yang, *J. Am. Chem. Soc.*, 2013, **135**, 12932.
11 S. W. Boettcher, E. L. Warren, M. C. Putnam, E. A. Santori, D. Turner-Evans, M. D. Kelzenberg, M. G. Walter, J. R. McKone, B. S. Brunschwig, H. A. Atwater and N. S. Lewis, *J. Am. Chem. Soc.*, 2011, **133**, 1216.
12 M. H. Lee, K. Takei, J. Zhang, R. Kapadia, M. Zheng, Y.-Z. Chen, J. Nah, T. S. Matthews, Y.-L. Chueh, J. W. Ager and A. Javey, *Angew. Chem., Int. Ed.*, 2012, **51**, 10760.
13 J. Sun, C. Liu and P. Yang, *J. Am. Chem. Soc.*, 2011, **133**, 19306.
14 J. Z. Zhang, *MRS Bull.*, 2011, **36**, 48.
15 S. Hu, M. R. Shaner, J. A. Beardslee, M. Lichterman, B. S. Brunschwig and N. S. Lewis, *Science*, 2014, **344**, 1005.
16 C. Liu, J. Tang, H. M. Chen, B. Liu and P. Yang, *Nano Lett.*, 2013, **13**, 2989.
17 C. Liu, Y. J. Hwang, H. E. Jeong and P. Yang, *Nano Lett.*, 2011, **11**, 3755.
18 Y. J. Hwang, A. Boukai and P. Yang, *Nano Lett.*, 2009, **9**, 410.
19 J. M. Foley, M. J. Price, J. I. Feldblyum and S. Maldonado, *Energy Environ. Sci.*, 2012, **5**, 5203.

Faraday Discussions

PAPER

Control parameters for electrochemically relevant materials: the significance of size and complexity†

J. Maier*

Received 18th June 2014, Accepted 18th August 2014
DOI: 10.1039/c4fd00135d

This contribution is concerned with the control parameters for arriving at defined, electrochemically relevant materials. The treatment is precise as far as the equilibrium situation of simple crystals is concerned, but becomes more and more qualitative if the distance from equilibrium or the (structural or compositional) complexity increases. It proves useful to distinguish between *in situ* parameters and *ex situ* parameters, the number ratio of which decreases with increasing distance from equilibrium. A particularly complex situation is met if not only size, shape and phase distribution are important, but even morphological details are of relevance, as it is the case for modern battery electrodes ("electrochemical integrated circuits"). For such cases archetypical examples along with their advantages or disadvantages for electrochemical storage properties are discussed. In this context, special emphasis is placed upon the dimensionality and distribution topology of building elements.

1. Introduction

As far as applications are concerned, materials under consideration may be as simple as single crystals but also as complex as multiphase composites. Most materials of technological interest contain frozen structure elements. In biology and biotechnology, stationary structures are typically out of equilibrium but not frozen, their stationarity being maintained by export of entropy; at present they do not play an important role in usual electrochemical applications. Here, we rather deal with structures that are stationary and non-dissipative. Table 1 gives an overview on morphologies with increasing deviation from equilibrium. In complete equilibrium not only all fluxes (J), but also all driving forces (F) vanish; while in the case of frozen morphologies some driving forces are non-zero, yet the corresponding fluxes disappear owing to kinetic reasons. Note that in all these cases the entropy production (P, given by the product of forces and fluxes) is zero. This is however not the case for the aforementioned dissipative structures; yet it

Max Planck Institute for Solid State Research, 70569 Stuttgart, Germany. E-mail: s.weiglein@fkf.mpg.de

† This work is dedicated to Professor Martin Jansen on the occasion of his 70th birthday.

Table 1 Classification and characteristics of (quasi-) stationary morphologies (P = entropy production, F = force, J = flux)

Classification	Thermodynamic characteristics	Characteristic properties	Examples
Equilibrium morphology	$P = F = J = 0$	Absolutely stable/ reproducible/typically boring/low information content	Wulff-shaped single crystals
Frozen morphology	$P = J = 0 \neq F$	Partially stable/ reproducible/most relevant for technology	Technologically relevant materials
Dissipative morphology	$\delta_F P = 0$ $F \neq 0 \neq J$	Living/self-healing/ sustained by outer fluxes/dissipative/ high information content	Biologically relevant systems

can be shown that for specified boundary conditions the variation of the entropy production, at least with respect to forces, nullifies.[1]

This paper considers the control parameter for materials definition for various degrees of complexity. The complexity is determined by the number of (practical) degrees of freedom and hence the information content. Complexity of a given material (Fig. 1) can be increased by varying composition (introduction of zero-dimensional defects), but also by varying morphology (introduction of higher-dimensional defects). As to the latter, nano-structuring has proved to be an efficient means of defining electrochemically relevant materials. Even though not always completely independent, Fig. 2 depicts the two coordinate-axes, size and complexity, along which many modern materials developments have been made possible.

Fig. 3 refers to two extreme cases: the uniform single crystal in equilibrium with only point defects present (top) and a multi component composite with many frozen elements (bottom). It illustrates the theme of this paper which investigates: (i) the impact of building elements defining a real electrochemically relevant material on the properties, whereby we mainly concentrate on

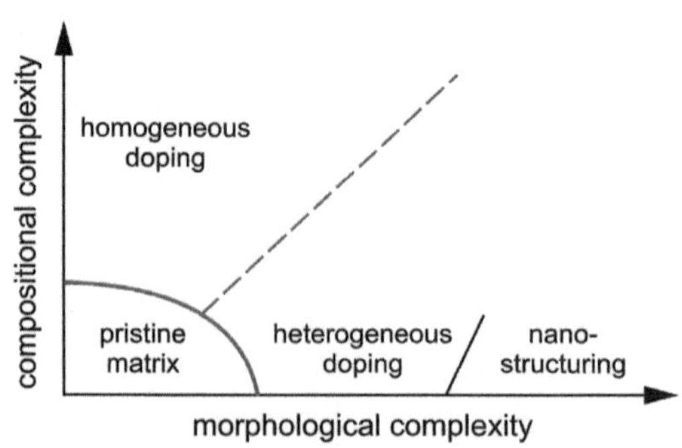

Fig. 1 Increase of materials complexity by introducing zero-dimensional defects (compositional complexity) and higher-dimensional defects (morphological complexity).

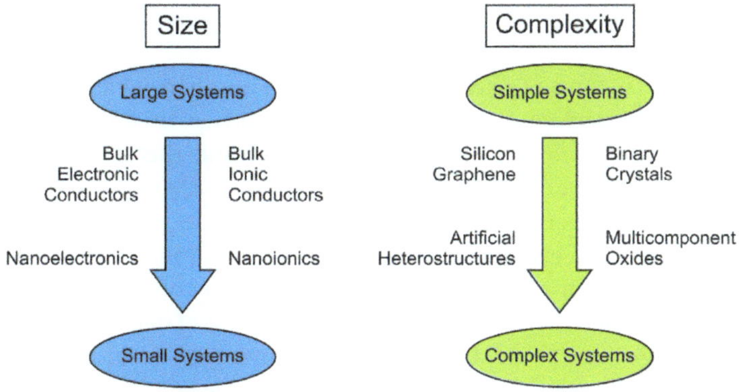

Fig. 2 Materials research *via* size reduction and complexity enhancement.

electrochemical storage, (ii) the parameters that are necessary to define the real material in terms of the constituent building elements.

2. Equilibrium

2.1. Point defect chemistry

The chemistry of real materials includes the chemistry of the perfect state as well as the defect chemistry to be superimposed. The simplest situation as far as charge carrier chemistry is concerned, is met for an infinite crystal in equilibrium. Then only point defects are of significance. They define the equilibrium charge carrier situation we are interested in. Under conditions at which point defect concentrations are small (Boltzmann conditions) and the complexity of charge carrier chemistry is moderate (Brouwer conditions, typically two majority carriers), ideal mass action laws (with mass action constants K_r) can be formulated for inner reactions and outer reactions connecting the inner situation with the component partial pressures (P_p) fixed by neighboring phases (*e.g.* gas phase).[2] For the concentration of any charge carrier (j), then it can be formulated that[3]

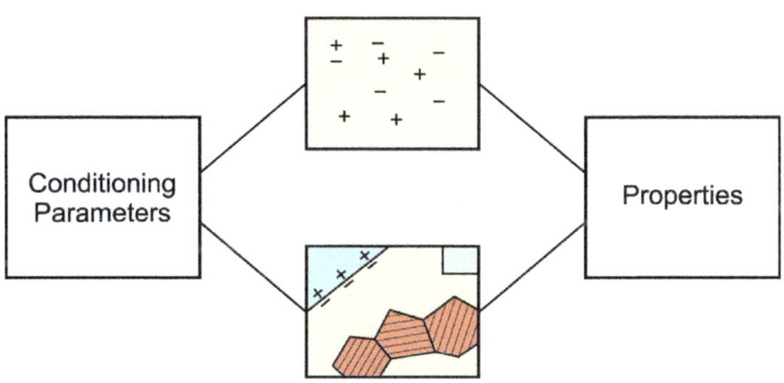

Fig. 3 The relation between properties and preparation parameters is comparatively simple and precise for equilibrium single crystals, yet only qualitative for morphologically and compositionally complex systems.

$$c_j(T, P_p, C) = \alpha_j\left(\Pi_p P_p^{N_{pj}}\right) C^{M_j}\left(\Pi_r K_r(T)^{\gamma_{rj}}\right) \tag{1}$$

where α_j, N_{pj}, M_j and γ_{rj} are simple rational numbers characteristic of the respective defect regime (*i.e.* majority charge carrier situation). The decisive control parameters are *in situ parameters* T, P_p's and the *ex situ parameter* C, denoting the effective doping content. While the *in situ* parameters can be reversibly varied under the conditions of interest, the variation of the *ex situ* parameters requires a change of the experimental conditions (*e.g.* going to higher temperatures) or even a new preparation. Accordingly one can formulate three rules for the characteristic exponents N, M, γ that are of decisive importance for materials design (see T-rule, P-rule and C-rule in ref. 3) as far as charge carrier chemistry is concerned. Here we only discuss the *rule of homogeneous doping* (C-rule) stating that

$$\frac{z_j \delta c_j}{z \delta C} < 0 \ \ i.e. \ \left(\frac{z_j}{z}\right) M_j < 0. \tag{2}$$

In words: on positive (negative) doping, *i.e.* an aliovalent substitution that leads to a fixed effective positive excess charge, the concentrations of all negative (positive) defects are increased and all positive (negative) defects that take part in the defect-chemical equilibrium are decreased (z_j: charge number of dopant defect). The non-trivial point here is that it affects all these charge carriers. This is not only a consequence of the electroneutrality condition but also of the mutual mass action laws. To give a well-known example: a partial substitution of Zr^{4+} by Y^{3+} does not only result in an increased oxygen vacancy ($V_O^{\cdot\cdot}$) concentration, also the hole ($h^{\cdot}$) concentration is increased while the conduction electron (e') concentration is decreased.

Eqn (2) ignores elastic effects of dopants. Such effects can even lead to homovalent dopant effects.[4]

Certainly in most cases the validity range of dilute conditions is exceeded. Then associates and activity coefficients are to be considered, that may introduce further materials constants (dielectric constants, mass action constants) but not further control parameters.

Taking account of an invariant dopant concentration already touches on the next paragraph and the next complication, namely the consideration of a (partially) *frozen-in* point defect chemistry.[5,6] Usually, at least at temperatures significantly below the melting point, not all components are mobile *via* point defects. Let us consider a binary oxide MO. At very high temperatures M- and O-defects may be mobile and in equilibrium (here creep can occur). On lowering the temperature, the O-lattice may stay mobile, while the M-defects now act as frozen dopants. When at even lower temperatures oxygen defects are also immobile, then we consider the situation typically assumed in semi-conductor physics (immobile ions, mobile electrons). Not only does the situation become more complex, but also the number of practical degrees of freedom increases since new parameters such as freezing-in temperature and partial pressures or even kinetic adjusting screws (cooling kinetics) come into play. Fig. 4 (we now refer to the concentrations not labelled by Q) shows how significantly the freezing procedure changes the defect chemistry.

Table 2 gives an overview of the electronic concentrations in an oxide in the different temperature regimes characterized by predominant ionic disorder. It assumes infinitely fast quenching which can be approximated by very long annealing at temperatures at which equilibrium takes a rather long time, followed by fast cooling. This temperature is denoted as T^Q. The concentrations labelled as Q, refer to the equilibrium values at T^Q (as discussed above). The other concentrations refer to situations where the ionic disorder is frozen. These concentrations are discussed as a function of P_{O_2} at T^Q. For details the reader is referred to ref. 6. The above introduced distinction between parameters that can be reversibly changed *in situ* (*in situ* parameters) and those the definition of which is requiring higher temperatures and then a new preparation (*ex situ* parameters) is very helpful in this context. Eqn (1) is still valid under these assumptions but freezing is the equivalent of losing *in situ* parameters (P's) to the benefit of the *ex situ* parameter C (*cf.* Table 2).

3. Excursion: control parameter and thermodynamic state variables

Here it is important to make some remarks on the terms "control parameters" used in this paper. With this term we designate the parameters available and necessary to define the sample.

In the simplest case of the infinite crystal in equilibrium, they are *in situ* parameters and are identical to the thermodynamic state parameters (*e.g.* open isobaric system ("intensive system"): T, p, μ's *i.e.* P's, μ denoting the chemical potential).

This is already different in the case of defining dopant effects or frozen-in defects. Most simply we may take their concentrations as given and consider these concentrations as control parameters which are in fact *ex situ* parameters.

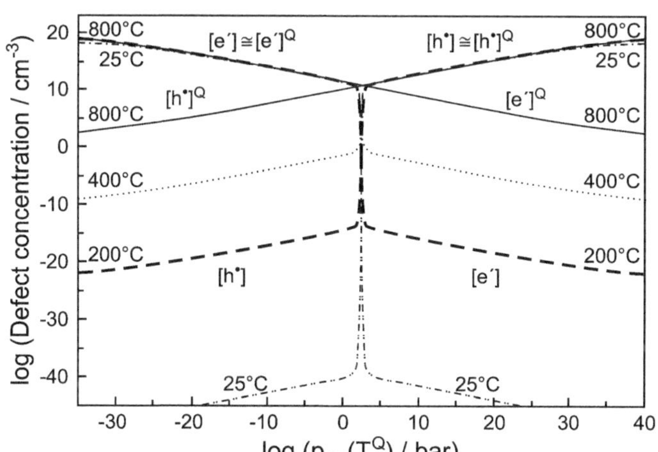

Fig. 4 Excess electron and electron hole concentrations as a function of oxygen partial pressure under the still reversible quenching conditions (labelled Q, here 800 °C) as well as for quenched state at lower temperatures. Ionic disorder is predominant and is frozen for $T < T^Q$. Reprinted with permission from ref. 6.

Table 2 Partially frozen-in defect chemistry of perovskite ABO_3[a]

Temperature range	Control parameters (total number = $F = 3$)		State parameters to be defined $\left(\text{total number} = \sum_{i=0}^{p}(F\text{-}1) \geq F\right)$
Very high	Reversible: T, P_{O2}, P_{AO}	*In situ*: 3	T, P_{O2}, P_{AO} (3)
	Fixed: ϕ	*Ex situ*: 0	
High	Reversible: T, P_{O2}	*In situ*: 2	T, P_{O_2}; T^{Q0}, $P_{O_2}^{Q0}$, P_{AO}^{Q0} (3 + 2)
	Fixed : $[V_A'']$	*Ex situ*: 1	
Low	Reversible: T	*In situ*: 1	T; T^{Q1}, $P_{O_2}^{Q1}$, T^{Q0}, $P_{O_2}^{Q0}$, P_{AO}^{Q0} (3 + 2 + 1)
	Fixed : $[V_O^{\cdot\cdot}]$, $[V_A'']$	*Ex situ*: 2	
(not realistic)	Reversible: ϕ	*In situ*: 0	T^{Q2}; T^{Q1}, $P_{O_2}^{Q1}$, T^{Q0}, $P_{O_2}^{Q0}$, P_{AO}^{Q0} (3 + 2 + 1 + 0)
	Fixed: $[h^{\cdot}]$,$[V_O^{\cdot\cdot}]$, $[V_A'']$	*Ex situ*: 3	

[a] The upper index Q labels quenching parameters (Q0 refers to quenching of A-sublattice, Q1 to quenching O-sublattice and Q2 to (unrealistically) quenching the electronic structure). The B-lattice is assumed to be frozen even at the very high temperatures (still far from melting point). Reprinted with permission from ref. 6.

Alternatively as shown in Section 2 one can prepare the situation under regard by freezing an equilibrium situation and instead use the thermodynamic state variables of the freezing conditions. This typically fails if higher-dimensional defects and formally morphological patterns are to be considered. It is for the vast majority of situations not possible to define the transformation from an equilibrium state to such a real situation. Thus, we simply take all the formal frozen structure elements (such as nature, density and orientation of interfaces; phase distribution, phase topology, size *etc.*) as given and the parameters necessary to describe them as *ex situ* control parameters.

4. Point defect chemistry in the vicinity of higher-dimensional defects

A further increase of complexity with respect to Section 1 is achieved if higher-dimensional defects are irreversibly introduced. Irrespective of varied mobilities within interfaces or dislocation networks, the defect concentrations near surfaces, interfaces or dislocations may be characteristically varied compared to the infinite bulk. This is particularly the case if we assume abrupt heterogeneities such as a mono-atomic interfacial core layer inserted with negligible structural relaxation. Owing to the novel boundary condition and the concomitant symmetry break, a point defect redistribution will occur such that an excess charge will establish, balanced by space charge zones with greatly varied carrier concentrations.

Owing to the enormous impact on *e.g.* conductivities and the similarity to homogeneous doping (= zero-dimensional doping), the technique of influencing carrier concentrations by implementing higher-dimensional defects has been termed *heterogeneous doping (= higher-dimensional doping)*. Even a similar rule (*rule of heterogeneous doping*[3]) can be formulated:

$$\frac{z_k \delta c_k}{\delta \Sigma} < 0. \tag{3}$$

Here we need to know the compensating interfacial charge density Σ (rather than the effective charge of an atomistic dopant) to predict how the individual concentrations will be varied. Again the rule includes the variation of minority species. The latter is of great significance (but is usually overlooked) for electronic conductors as often ionic point defects are in majority and the electronic effects may be dictated by the behavior of the ions (*fellow traveler effect*).[7]

5. Hetero-interfaces and phase combinations

Apart from grain or domain boundaries, interfaces separate different phases. The consequence of such hetero-interfaces on charge carrier redistribution is analogous to what was stated in the previous paragraph, up to one point: here a charge transfer from one phase to the other can occur and a bilateral space charge situation develops without even any excess charge in the interfacial core. (This however can be an additional effect.) Such effects depend critically on the phase combination and it is this circumstance[7] that provides a manifold of opportunities to purposefully vary defect chemistry and hence conductivities, reactivities, and related properties. If, *e.g.*, a combination of CaF_2 and BaF_2 phases result in a net transfer of F^- from BaF_2 to CaF_2 then $[V_F']$ in the space charge regions $BaF_2(CaF_2)$ is increased (decreased) while $[F_i^{\cdot}]$ shows the opposite behavior. The electronic minority species are changed accordingly: $[e']$ in $BaF_2(CaF_2)$ is increased (decreased) while $[h^{\cdot}]$ is oppositely varied.[8]

By selecting a suitable second phase contacted with the matrix, ion and electron conductivities in the matrix can be often varied by orders of magnitude, in particular if the density of interfaces is very high as it can be achieved in nanocomposites or heterolayers.

This heterogeneous doping effect has been predicted to be of great influence also for fields such as catalysis, superconductivity or thermoelectricity.[9]

6. Size effects

Size is an expression of the distance of neighboring surfaces. Many properties directly depend on it (*nano-ionics*).[9] As far as transport and storage parameters are concerned, in Fig. 5 L.H.S a distinction was made between explicit and implicit size dependencies.[10] The latter which concerns the intensive terms (materials constants), can then be divided into trivial and non-trivial size effects. Explicit size dependence (*cf.* L^n-term in Fig. 5 L.H.S) refers to the remaining geometry-relevant part of the size dependencies of resistances or capacitances and is often primarily responsible for the enormous influence of size on storage kinetics. R.H.S of Fig. 5 gives a collection of power laws for explicit size dependence for various mechanisms. As we refer to storage kinetics the reader's attention is to be drawn to the second line ($\tau \propto L^2$) highlighting the enormous influence of size on diffusion time. Implicit size dependencies refer, to give an example, to the size dependence of effective conductivities in inhomogeneous systems that may be caused by space charge effects at surfaces or internal boundaries. As long as neighboring space charge zones are well-separated, their contribution to the overall property scales simply with size, in contrast to situations where space charge layers overlap causing non-trivial size effects. Such a distinction is meaningless for homogeneous phenomena such as capillary pressure that can be of enormous consequences on phase stability, partial pressures of components, cell voltage, solubilities, defect concentrations and many others.

7. Shape and dimensionality in nano-sized systems

While in macroscopic crystals shape and dimensionality are only of influence if anisotropic or extremely surface-sensitive properties are concerned, these parameters are decisive for nano-sized systems.

For simplicity we refrain from distinguishing between different surface crystallographies and anisotropies caused by the perfect structure. Then we may be concerned with nano-particles, nano-grains in ceramics, nano-rods or nano-plates, the shapes of which are completely kinetically controlled or controlled by mechanical equilibrium (round nano-particles, nano-ceramics with equilibrium grain-angles[11] *etc.*).

It is important to note that having a stationary non-equilibrium structure or composition does not preclude reversible processes. In fact such situations are rather the rule than the exception as far as functional materials are concerned. Thus, oxygen ion conduction or even stoichiometric changes in the oxygen sub-lattice are possible in structures with frozen-in cation sub-lattices; equally, nano-sized or even amorphous electrodes can function as reversible anodes or cathodes in Li-based batteries. Only those processes that immediately affect the frozen-in structure elements cannot be reversible.

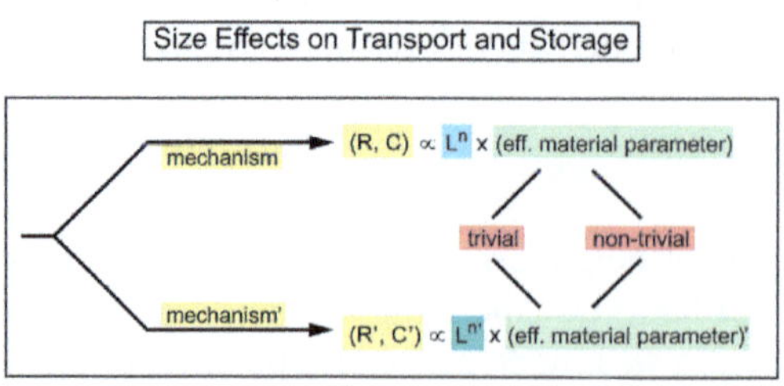

mechanism $\mathcal{E}$	$R_{\mathcal{E}}$	$C_{\mathcal{E}}$	$\tau_{\mathcal{E}}$
dielectric response	2-d	d-2	0
diffusion controlled bulk storage	2-d	d	2
interfacial rate controlled bulk storage	1-d	d	1
interfacial storage (semi-infinite)#	1-d	d-1	0

Fig. 5 L.H.S.: Implicit and explicit size effects on electrochemical, chemical or electrical resistors and capacitors. The first effects can be divided in trivial and non-trivial effects. Size effects may change R and C such that even another mechanism may become dominant. Reprinted with permission from ref. 10. R.H.S.: Dimensionality dependent exponents (n) of the explicit thickness (L) dependence of resistors, capacitors, relaxation times for various electrochemical mechanisms (ε) in nano-structured materials. The parameter d gives the number of nano-sized dimensions ($1 \leq d \leq 3$) of extension L (taken the same for simplicity). #In the finite size regime the situation also depends on the neighboring phases. Reprinted with permission from ref. 10.

Let us concentrate on conductivity and storage properties. As far as the equilibrium storage capacity of a component is concerned it is increasingly varied from nano-2 D (nano-plate) to nano-0 D (nano-dots) *via* nano-1 D (nano-rods) according to a modified chemical potential of this component. Note that *e.g.* the excess chemical potential of Li in a crystallite due to capillarity is given by $2(\gamma/r)\upsilon_{Li}$ (γ: effective surface tension, r: effective radius).[12,13] The partial molar volume of Li (υ_{Li}) in the compound under regard can be positive or negative depending on whether Li-addition leads to expansion or shrinkage of the crystallite. Even more important are kinetic issues that are dealt with below in the next section by means of a few notable examples.

In terms of transport, percolation efficiency is of great significance. While nano-0 D objects require agglomeration to higher-dimensional ones, nano-1 D objects are ideal as percolating objects provided their intrinsic transport properties are appreciable. If aligned properly, they are favorable solutions as far as covered transport length per mass is concerned. Equally good are nano-plate mixtures that connect two electrodes. Hierarchical geometries as well as fractal geometries can be very favorable, too, in order to achieve a good electrical connectivity.

The complexity is further increased by the presence a multitude of phases. A typical example is a composite electrode consisting of a mixture of an electroactive mass with a current collecting and electrolyte phase. Here the total morphology needs to be considered.

8. Nano-morphology and Li-storage

8.1. The necessity of electrochemical integrated circuits

As a rule an *efficient storage electrode consists of a bicontinuous network of electron collector and electrolyte which is heterogeneous on a scale comparable with the size of the electroactive particles, whereby this size is chosen such that diffusion problems become just negligible*. If the electroactive mass is identical to the current collector or the electrolyte, the morphology is significantly simplified.

The greatest simplification is achieved if the electroactive mass is simultaneously a good electron and Li^+ conductor. If then the related chemical diffusion coefficient is very high, the particle size can be very large. A wonderful but exceptional example is Ag_2S[14] in which Ag can be quickly stored, even if it is present as a macroscopic single crystal. There is no need of nano-structuring or of further network components, and the cell may be as simple as $Ag|AgI|Ag_2S|C$. But let us consider the frequently occurring case that the electroactive mass is electronically and ionically rather insulating.

Nano-0-D electro-active particles are then preferred owing to the fact that diffusion time scales with size squared. Size reduction from a 1 mm particle to an ensemble of 10 nm-size particles reduces the diffusional relaxation time by 10^{10}. On the other hand, the volume scales with size cubed meaning that we decompose the 1 mm-sized particle into 10^{15} 10 nm-sized particles, each of which needs to be sufficiently electronically and ionically connected, obviously requiring highly *sophisticated network solutions*. The use of liquid electrolyte wetting all the particles is the solution of choice for ion transport while adding carbon particles is the standard method to achieve sufficient electronic contact. On one hand the carbon volume fraction must be high enough to ensure percolation, on the other hand one has to think about connectivity down to nano-size which requires going beyond usual techniques. This is described in detail below.

Nano-1-D electroactive masses are providing special advantages (*e.g.* percolation efficiency) but may not be the best solution for high energy systems. An elegant example has been given in the literature[15] in the form of Si-nanowires growing on copper, where the nanobrush so-formed is embedded in liquid electrolyte. The rate performance is excellent yet the absolute energy stored small (but see 9.2.).

Nanoplates, or better thin films, are preferred in microbatteries as far as the electrolyte is concerned. Arranged parallel to the electrode they may also be useful as a protection layer or separation layer thin enough as not to cause too much resistance. Perpendicular arrangements would be necessary if connectivity is important. One might think of hetero-layered packages as electrolyte or electrode systems to exploit interfacial effects, yet for practical systems they might be too expensive.

Most successful proved composites that use various dimensionalities. Let us describe a few archetypes in the subsequent section and use the following notation: α:β denotes a two-phase system constituted by the phases α and β irrespective of phase distribution; α@β specifies that α and β are distributed more or less equitably (co-arranged rather than sub-ordinated). Characteristic examples are random two-phase systems or bi-continuous two-phase systems. The symbol $\alpha \subset \beta$ denotes a situation where α is completely embedded in a β-surrounding, characteristic examples being non-percolating particles or rods in a solid or liquid matrix or non-percolating particles in a tube. Naturally, this simple nomenclature cannot cover all topological cases. Moreover, the inclusion of the electrolyte phase (*i.e.* of pores) is also important requiring the description of a three-phase morphology. Whenever avoidable, we will refrain from using a too detailed notation. Note that we use the terms 0 D, 1 D, 2 D to denote extreme aspect ratios on the scale specified by the respective prefix (*e.g.* nano-1 D means that one dimension is nano-scale (but not atomistic) while the others are macroscopic).

8.2. Electroactive nanoparticles in electronically conducting nanotubes

8.2.1. Sn-nanoparticles in carbon nanotubes: (nano-0 D Sn) $\subset$ (nano-1 D carbon). The first sketch in Table 3 refers to Sn-nanoparticles dispersed in electrospun carbon-nanotubes embedded in liquid electrolyte.[16] The excellent performance of this morphology is due to a variety of reasons: (a) the Sn-particles are small enough to provide fast diffusion. (b) The Sn-particles are easily addressed by electrons *via* the electronically conducting C-nano fiber which (c) easily form percolating pathways. (d) The Sn-particles also have access to Li^+ stemming from electrolyte, owing to appreciate Li^+ diffusivity in C and the short distances involved. (e) Though perfectly electrochemically coupled, the Sn-particles are mechanically decoupled which allows for an easy buffering of volume changes that usually lead to the most important failure mode of Sn-electrodes. (f) The fibers act as binder and holder serving for a distinctly stationary morphology. (g) As Sn-atoms or Sn-ions are not mobile in the carbon matrix thus impeding chemical or electrochemical Ostwald-ripening.

8.2.2. MoS_2 nanodots in C-nanotubes: (nano-0 D MoS_2) $\subset$ (nano-1 D carbon). This example refers to ref. 17 (*cf.* again first sketch in Table 3). Here the size of electroactive particles (MoS_2) is extremely small so that we can truly speak of nanodots. In fact the particles are atomistically thin (0.4 nm corresponding to a single SMoS layer) and ultrasmall (4 nm). Apart from the above mentioned advantages, this morphology allows for a completely reversible conversion reaction.

Lithiation of macroscopic MoS_2 leads to conversion to separate Mo and Li_2S phases. Reversing the situation is not trivial as it assumes reversible nucleation and fast diffusion.

In the present case we refer to extreme reaction confinement and hence almost to a cluster reaction whereby the cluster is fixed in space. It can be assumed that the transfer of $4e^-$ corresponding to the reaction with 4Li does not lead to a significant spatial separation, rather the reaction will occur at the tiny clusters lowering if not nullifying transport and nucleation problems. Indeed striking reversibility could be achieved for at least thousand cycles. Simultaneously the cell voltage is lowered by the small size, which is favorable for application as an anode.

8.3. Nanocomposites of electroactive mass and electron conductors

8.3.1. Hierarchical nanoporous TiO_2 with RuO_2 wiring: (nano-0 D TiO_2)@(nano-0 D RuO_2). If the electroactive mass was both a perfect e^- and Li^+ conductor, even a single crystal could be used. If the electroactive mass is well electrochemically conducting such as carbon, the morphological problem reduces to a distribution of electroactive mass and electrolyte (see *e.g.* ref. 18). If a

Table 3 Two examples of electrochemical network topologies[a]

Topology	Examples	Literature	Sketch
$[\alpha \subset \beta]$ @γ	α = nano-0 D Sn β = nano-1 D C γ = liquid electrolyte	Ref. 16	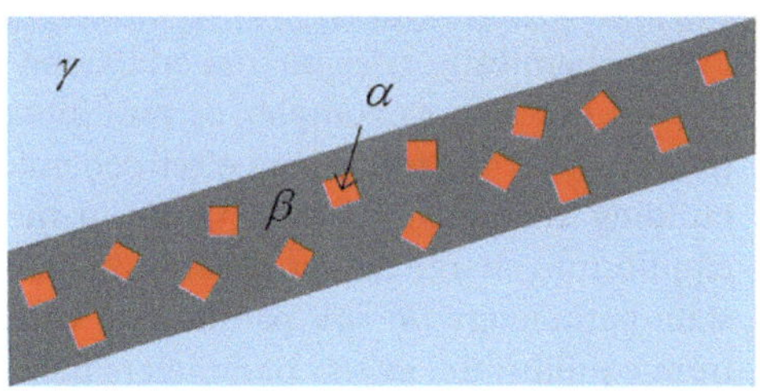
$[\alpha@\beta]$ @γ	α = nano-1 D TiO_2 β = nano-1 D C γ = liquid electrolyte	Ref. 20	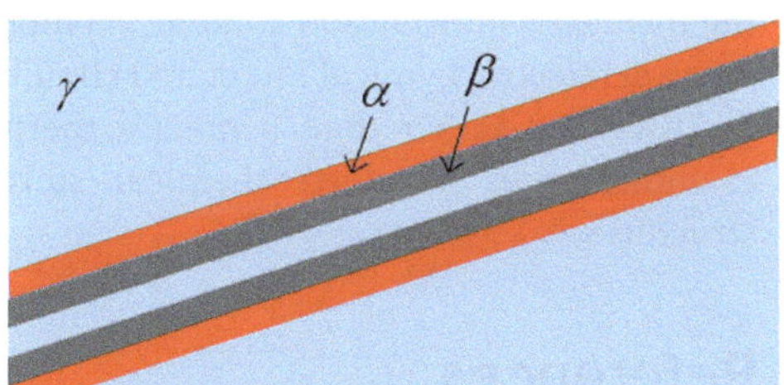

[a] α = electroactive phase. β = current collecting phase. γ = electrolyte phase.

liquid electrolyte is used then a hierarchical pore structure is best for rapid Li^+ transport to the particles.

If the electroactive mass is not a good electron conductor (such as TiO_2 or $LiFePO_4$) a hierarchically porous TiO_2 structure is not sufficient to achieve high rate performance; rather it is highly beneficial to incorporate an electronically conducting phase into the nanostructure. This has been achieved in ref. 19 with high capacities even at very high rates.

8.3.2. TiO_2/carbon-coaxial nanocables: (nano-1 D TiO_2)@(nano-1 D carbon). A further instructive structure is double-walled nanotubes.[20] The morphology refers to the second sketch in Table 3. Also here the performance is exceptional and even synergistic performance could be achieved, by virtue of carbon assisting with storage in TiO_2 while TiO_2 appears helpful for storage in carbon.

8.3.3. Carbon coating of $LiFePO_4$/carbon embedding: ((nano-0 D $LiFePO_4$) $\subset$ (nano-0 D carbon)) $\subset$ (porous-carbon). An effective principle is the carbon-coating of nanoparticles (see *e.g.* ref. 21). This carbon-coating delivers good contact to admixed carbon current collecting particles. If the coating is thin, the Li^+ can penetrate through the coating, otherwise it is important that the coating is porous allowing electrolyte access. A very efficient and simple technique is achieved with two organic precursors leading to carbon-coated particles embedded in a porous carbon matrix.[22] $LiFePO_4$ contacted in this way proved very efficient.

8.3.4. Carbon coating of $Na_3V_2(PO_4)_3$/carbon embedding: ((nano-0 D $Na_3V_2(PO_4)_3$) $\subset$ (nano-0 D carbon)) $\subset$ (porous-carbon). An even more exciting example of how very well this nanostructure can work is its application to $Na_3V_2(PO_4)_3$ (ref. 22) which outperforms Li-cathodes even per mass under high performance conditions. This extremely high rate performance also relies on the very high Na^+ conductivity in the NASICON structure of the phosphate and the low transfer resistance of Na^+ from electrolyte to electroactive mass.

9. Conclusions

Electrochemically relevant materials can be as simple as macroscopic single crystals but also as complex as multiphase nanostructured composites. In the present paper the focus is on electrochemical impact and on controllability of the building elements involved. While in the extreme case of a single crystal the equilibrium charge carrier situation can be fully defined by a few thermodynamic state parameters (*in situ* parameter), with increasing complexity and deviation from equilibrium, *ex situ* parameters become increasingly relevant. According to the increased information content, all morphological details may finally become relevant (electrochemical integrated circuit). Thus only archetypical examples can be discussed rather than a precise thermodynamical description given. Such examples refer to electrochemical storage in specially designed electrode structures.

References

1 P. Glansdorff and I. Prigogine, *Thermodynamic Theory of Structure, Stability and Fluctuations*, John Wiley & Sons, Ltd., New York, 1971.

2 G. Brouwer, *Philips Res. Rep.*, 1954, **9**, 366.

3 J. Maier, *Physical Chemistry of Ionic Materials. Ions and Electrons in Solids*, John Wiley & Sons, Ltd, Chichester, West Sussex UK, 2004.
4 D. Zahn, J. Maier and P. Heitjans, *Chem.–Eur. J.*, 2012, **18**, 6225.
5 F. A. Kroeger, *Chemistry of Imperfect Crystals*, North-Holland, 1964; R. Waser, *Ber. Bunsen-Ges.*, 1986, **90**, 1223; K. Sasaki and J. Maier, *J. Appl. Phys.*, 1999, **86**, 5422.
6 J. Maier, *Phys. Chem. Chem. Phys.*, 2003, **5**, 2164.
7 J. Maier, *Prog. Solid State Chem.*, 1995, **23**, 171; J. Maier, *Ber. Bunsen-Ges.*, 1989, **93**, 1468.
8 N. Sata, K. Eberman, K. Eberl and J. Maier, *Nature*, 2000, **408**, 946.
9 J. Maier, *Nat. Mater.*, 2005, **4**, 805; J. Maier, *Adv. Mater.*, 2009, **21**, 2571.
10 J. Maier, *Chem. Mater.*, 2014, **26**, 348.
11 W. D. Kingery, H. K. Bowen and D. R. Uhlmann, *An Introduction to Ceramics*, John Wiley & Sons, Ltd., New York, 1976.
12 R. Defay, I. Prigogine, A. Bellemans and H. Everett, *Surface Tension and Adsorption*, Longmans, Green & Co, Ltd, London, 1960.
13 J. Maier, *Angew. Chem., Int. Ed.*, 2013, **52**, 4998.
14 U. von Oehsen and H. Schmalzried, *Ber. Bunsen-Ges.*, 1981, **85**, 7.
15 C. K. Chan, H. Peng, G. Liu, K. McIlwrath, X. F. Zhang, R. A. Huggins and Y. Cui, *Nat. Nanotechnol.*, 2008, **3**, 31.
16 Y. Yu, L. Gu, C. Zhu, P. A. van Aken and J. Maier, *J. Am. Ceram. Soc.*, 2009, **131**, 15984.
17 C. Zhu, X. K. Mu, P. A. van Aken, Y. Yu and J. Maier, *Angew. Chem., Int. Ed.*, 2014, **53**, 2152.
18 Y.-S. Hu, P. Adelhelm, B. M. Smarsly, S. Hore, M. Antonietti and J. Maier, *Adv. Funct. Mater.*, 2007, **17**, 1873.
19 Y.-G. Guo, Y.-S. Hu, W. Sigle and J. Maier, *Adv. Mater.*, 2007, **19**, 2087.
20 F.-F. Cao, Y.-G. Guo, S.-F. Zheng, X.-L. Wu, L.-Y. Jiang, R.-R. Bi, L.-J. Wan and J. Maier, *Chem. Mater.*, 2010, **22**, 1908.
21 R. Dominko, M. Gaberscek, J. Drofenik, M. Bele, S. Pejovnik and J. Jamnik, *J. Power Sources*, 2003, **119–121**, 770.
22 C. Zhu, K.-P. Song, P. A. van Aken, Y. Yu and J. Maier, *Nano Lett.*, 2014, **14**, 2175.

Faraday Discussions

PAPER

Oxide diffusion in innovative SOFC cathode materials

Y. Hu,[a] V. Thoréton,[b] C. Pirovano,[b] E. Capoen,[b] C. Bogicevic,[a] N. Nuns,[b] A.-S. Mamede,[b] G. Dezanneau[a] and R. N. Vannier*[b]

Received 11th June 2014, Accepted 28th July 2014
DOI: 10.1039/c4fd00129j

Oxide diffusion was studied in two innovative SOFC cathode materials, $Ba_2Co_9O_{14}$ and $Ca_3Co_4O_{9+\delta}$ derivatives. Although oxygen diffusion was confirmed in the promising material $Ba_2Co_9O_{14}$, it was not possible to derive accurate transport parameters because of an oxidation process at the sample surface which has still to be clarified. In contrast, oxygen diffusion in the well-known $Ca_3Co_4O_{9+\delta}$ thermoelectric material was improved when calcium was partly substituted with strontium, likely due to an increase of the volume of the rock salt layers in which the conduction process takes place. Although the diffusion coefficient remains low, interestingly, fast kinetics towards the oxygen molecule dissociation reaction were shown with surface exchange coefficients higher than those reported for the best cathode materials in the field. They increased with the strontium content; the Sr atoms potentially play a key role in the mechanism of oxygen molecule dissociation at the solid surface.

Introduction

With the possibility of cogeneration of electricity and heat, solid oxide fuel cells are promising. The first units were commercialised in Japan in 2009 and demonstrators are running in Europe. Their principles are based on a dense ceramic electrolyte which separates two compartments, one containing hydrogen, and the other containing air. Due to a difference in oxygen chemical potential between the two compartments, oxygen molecules are reduced at the cathode into oxide ions, which migrate through the electrolyte to the anode to react with hydrogen molecules and produce water. Hydrogen is usually produced by methane reforming, however, internal reforming of methane is also envisaged. The typical temperatures used range between 700 and 1000 °C. The main electrolyte materials which are currently studied are Yttria Stabilised Zirconia (YSZ),

[a]*Laboratoire Structures, Propriétés et Modélisation des Solides (SPMS), Ecole Centrale Paris, Grande Voie des Vignes, 92295 Châtenay-Malabry, France*

[b]*Univ Lille Nord de France, CNRS UMR 8181, Unité de Catalyse et de Chimie du Solide – UCCS, ENSCL, Université Lille 1, CS 90108, 59652 Villeneuve d'Ascq Cedex, France. E-mail: rose-noelle.vannier@ensc-lille.fr; Fax: +33(0)320436814; Tel: +33(0)320436583*

Gadolinia Doped Ceria (GDC) and lanthanum gallates (LSGM). At the anode, a Ni-YSZ cermet containing a 40% volume of nickel is usually used, but it is prone to carbon deposition when used with hydrocarbon fuels. It is also sensitive to sulphur poisoning, and suffers from nickel agglomeration and redox stability with prolonged usage. Alternative materials, such as chromo-manganites $(La_{0.75}Sr_{0.25})Cr_{0.5}Mn_{0.5}O_3$ or titanates such as lanthanum-doped $SrTiO_3$ have been proposed. At the cathode, strontium-substituted lanthanum manganite ($La_{1-x}Sr_xMnO_{3-\delta}$, LSM) is conventionally used. However, it displays low oxide ion conductivity, and composites with YSZ have been developed in order to allow use at temperatures lower than 800 °C. To operate at lower temperatures, Mixed Ionic Electronic Conductors (MIEC) are usually preferred in order to allow the oxygen reduction reaction to be sprayed on the whole electrode volume, and not only at the electrode/electrolyte interface. As an alternative to LSM, the $La_{0.6}Sr_{0.4}Co_{0.2}Fe_{0.8}O_{3-\delta}$ (LSCF) perovskite has been widely studied.[1] When used with a GDC electrolyte and after optimisation of its microstructure, an area specific resistance (ASR) of only 0.13 Ω cm^2 was obtained at 603 °C for this compound,[2] the target for the cathode being 0.15 Ω cm^2 at 700 °C. This high performance is likely due to the presence of Co(III) in the structure, which is known to be particularly active for the reduction of oxygen molecules into oxide ions.[3] Indeed, cobaltites are known to display high oxide ion transport parameters. Among these, cobaltite with a composition of $La_{0.6}Sr_{0.4}CoO_{3-\delta}$ is one of the best catalysts for the oxygen molecule reduction reaction with a surface exchange coefficient (k^*, cm s^{-1}) of 1.1×10^{-6} cm s^{-1} associated with an oxygen self-diffusion tracer (D^*, cm^2 s^{-1}) coefficient of 2.0×10^{-8} cm^2 s^{-1} at 680 °C under 230 mbar.[4] One drawback of these materials is their high thermal expansion coefficient (TEC). The aforementioned composition has a value of 23×10^{-6} K^{-1}, which is twice the thermal expansion of YSZ (10×10^{-6} K^{-1})[5] and GDC (12×10^{-6} K^{-1}).[6] However, partial substitution of cobalt with iron in LSCF led to a good compromise with a TEC in the range of 14–15 $\times 10^{-6}$ K^{-1} from 500 to 700 °C.[7] More recently, promising performances as SOFC cathodes have been shown for two innovative cobaltites, $Ba_2Co_9O_{14}$ and $Ca_3Co_4O_{9+\delta}$. An ASR of 0.5 Ω cm^2 was first measured on a symmetric cell made of a 70 wt% $Ba_2Co_9O_{14}$–30 wt% GDC composite electrode deposited on a GDC electrolyte.[8] The good electrochemical performance of this compound was later confirmed by Li *et al.*, who obtained ASRs of only 0.133 Ω cm^2 and 0.068 Ω cm^2 at 750 and 850 °C, and powder densities of 450 mW cm^{-2} and 770 mW cm^{-2} on a cell made of a LSGM electrolyte with a $Ba_2Co_9O_{14}$–samarium-doped ceria composite cathode at 800 and 850 °C, respectively.[3] This compound exhibits a three dimensional structure based on $[Co_8O_8]$ blocks sandwiched in between $[Ba_2CoO_6]$ layers (Fig. 1a).[9,10] With cobalt atoms at different valences, it displays high electronic conductivity with a value of 240 S cm^{-1} at 650 °C, but as with other cobaltites, its thermal expansion coefficient is high (22×10^{-6} K^{-1}), which necessitates the use of a composite, at least for mechanical compatibility, between the electrode and the electrolyte. After optimisation of its composition and thickness, an ASR of only 0.08 Ω cm^2 at 700 °C was recently measured on a symmetric cell made of a 50 wt% $Ba_2Co_9O_{14}$–50 wt% GDC composite electrode with a GDC electrolyte.[11]

In contrast, a thermal expansion coefficient of only 10×10^{-6} K^{-1} was measured for the well-known $Ca_3Co_4O_{9+\delta}$ thermoelectric compound,[12] which has a misfit structure composed of CdI_2-type $[CoO_2]$ layers sandwiched in between

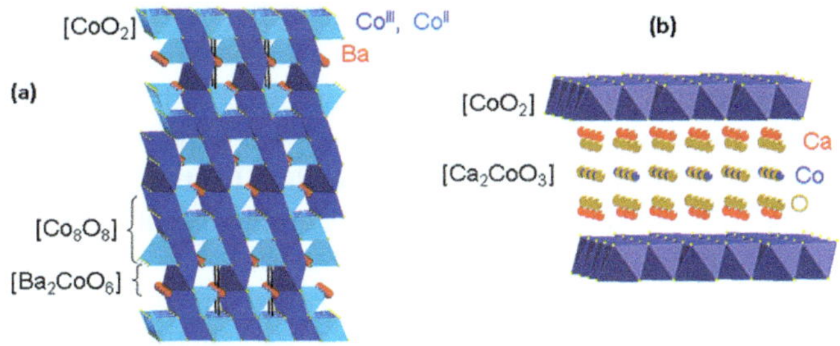

Fig. 1 Crystal structures of (a) $Ba_2Co_9O_{14}$ and (b) $Ca_3Co_4O_{9+\delta}$.

$[Ca_2CoO_{3-\delta}]$ rock salt slabs (Fig 1b).[13,14] With an electronic conductivity of 100 S cm^{-1} at 700 °C and oxygen vacancies in the rock salt layers, mixed ionic electronic conductivity was expected for this compound. When tested in a symmetrical cell with a GDC electrolyte, ASRs of 4.94 Ω cm^2 for the pure compound and 1 Ω cm^2 for a 70 wt% $Ca_3Co_4O_{9+\delta}$–30 wt% GDC composite electrode were first measured.[15,16] After optimisation of the composition and thickness, this was decreased to 0.5 Ω cm^2 at 700 °C for a 50 wt% $Ca_3Co_4O_{9+\delta}$–50 wt% GDC composite electrode with a thickness of 21 μm, and further decreased to 0.35 Ω cm^2 when calcium was partly substituted with 10% of strontium.[17] These good performances were confirmed by Samson *et al.*,[18] who reported a polarization resistance of 0.64 Ω cm^2 at 600 °C for a 50 vol% $Ca_3Co_4O_{9+\delta}$–50 vol% GDC composite electrode. Using a samarium-doped ceria electrolyte, Zou *et al.*[19] measured a maximum power density of 430 mW cm^{-2} at 700 °C for a 70 wt% $Ca_{2.9}Bi_{0.1}Co_4O_{9+\delta}$–30 wt% $Ce_{1.8}Sm_{0.2}O_{1.95}|Ce_{1.8}Sm_{0.2}O_{1.95}|Ce_{1.8}Sm_{0.2}O_{1.95}$–Ni button cell.

More recently, oxygen diffusion in this material was confirmed by combining $^{18}O/^{16}O$ isotope exchange, electrical conductivity relaxation and pulse isotopic exchange techniques.[20] Despite low diffusion with a coefficient of 2.7×10^{-10} cm^2 s^{-1} at 700 °C, this study showed high kinetics toward oxygen dissociation with a surface exchange coefficient as high as 1.6×10^{-7} cm s^{-1}.

In the present paper, oxygen diffusion in these two innovative cathodes was investigated, with emphasis placed on the impact of the partial substitution of calcium with strontium on the oxygen transport properties of $Ca_3Co_4O_{9+\delta}$.

Experimental

$Ba_2Co_9O_{14}$ and $(Ca_{1-x}Sr_x)_3Co_4O_{9+\delta}$ ($x = 0.1, 0.2$) powders were prepared by solid-state reaction from the corresponding carbonates and oxides $BaCO_3$ (Prolabo – 99.5%, dried at 100 °C), $CaCO_3$ (Prolabo – 99.5%, dried at 100 °C), $SrCO_3$ (Aldrich – 99.8%, dried at 100 °C) and Co_3O_4 (Alfa Aesar, 99.7%). Reactants were carefully ground in an agate mortar and then calcined repeatedly at 900 °C with intermediate grinding. The purity of the samples was checked by X-ray diffraction (D8 Advance Bruker AXS diffractometer, Cu Kα radiation, 1D LynxEye PSD detector). Fullprof software [v2.05, July 2011] was used for refinement of the unit cell parameters.[21] An Accupyc II 1340 micromeritic pychnometer was used to measure the powder density. The absolute oxygen stoichiometry of the sample was determined by iodometric titration.[22]

Thermal analyses were conducted under air, using a TG/DTA Setaram Setsys apparatus. 40 mg of the sample powder was introduced into a Pt crucible, and the experiment was carried out at a heating rate of 5 °C min^{-1} under flowing air (0.4 L h^{-1}) from room temperature to 950 °C. A blank experiment with an empty crucible was performed in order to correct the data for the effects of buoyancy. Thermogravimetry was also carried out for the evaluation of the oxygen stoichiometry of the sample as a function of pO_2. In this case, a Setaram 92-1750 apparatus was used. The experiment was carried out using 50 mg of powder in a Pt crucible under the flow of an O_2/N_2 mixture (5 L h^{-1}), with three successive oxygen partial pressures (100, 200 and 400 mbar), and the temperature ranging from 600 to 750 °C (50 °C step) for each pressure. For this set-up, the buoyancy was negligible and the data were directly used to derive the thermodynamic factor at each temperature.

Dense ceramics, 3 mm thick, were prepared by Spark Plasma Sintering at the IdF Platform in Thiais (France). A 20 mm (diameter) carbon die was used. The samples were sintered within 15 minutes with a 2 minute dwell at 850 °C, under a 50 MPa (8.8 kN) pressure. To remove carbon residues, the pellets were annealed for 12 h at 800 °C. A Hitachi S-3400N instrument was used to characterise the surface of the samples.

For isotopic exchange and electrochemical conductivity relaxation, sample surfaces were roughly polished in the first step, before they were polished down to a 1/4th of a micron with successive SiC papers and diamond pastes (Ra ~ 10–20 nm) in the second step. In the case of calcium cobaltites, to avoid the evolution of the surface roughness, samples were annealed for 72 h at 900 °C before the second polishing in order to release residual strains in the material.

The $^{18}O/^{16}O$ Isotope Exchange Depth Profiling technique was applied to all samples in order to derive the oxygen self-diffusion tracer (D^*, cm^2 s^{-1}) and surface exchange coefficients (k^*, cm s^{-1}). The diffusion time was chosen to get diffusion lengths significantly smaller than the sample dimensions to consider a 1D diffusion model perpendicular to the surface of the dense ceramic in a semi-infinite medium. In these conditions, assuming a first order surface exchange reaction and constant ^{18}O concentration during the exchanges, the solution for a semi-infinite solid to the Fick's 2nd law of diffusion is given by the following equation:[23]

$$C'(x,t) = \frac{C(x,t) - C_{\text{bg}}}{C_{\text{g}} - C_{\text{bg}}} = \text{erfc}\left[\left(\frac{x}{2 \times \sqrt{D^* \times t}}\right)\right] - \left[\exp\left(\frac{k^*x}{D^*} + \frac{k^{*2}t}{D^*}\right) \times \text{erfc}\left(\frac{x}{2 \times \sqrt{D^* \times t}} + k^*\sqrt{\frac{t}{D^*}}\right)\right] \quad (1)$$

where C_{bg} is the natural ^{18}O ratio, C_{g} is the ^{18}O ratio in the enriched exchange atmosphere, x is the distance from the surface of the specimen, and t is the time of the isotope exchange.

The exchange conditions are reported in Table 1. Experiments were carried out in between 650 and 750 °C under a pressure close to 210 mbar with pre-annealing

Table 1 Conditions for the $^{18}O/^{16}O$ isotope exchange

Temperature (°C)	Pressure (mbar)	Pre-annealing time (hours)	Annealing time (min)	^{18}O ratio in the gas phase
$Ba_2Co_9O_{14}$				
648	210	240	2860	0.79
700	210	96	2765	0.80
750	218	96	2790	0.80
$(Ca_{1-x}Sr_x)_3Co_4O_{9+\delta}$ $x = 0.10, 0.20$				
604	207	160	725	0.88
649	211	100	567	0.67
694	219	61	392	0.80
741	210	20	55	0.9

under normal oxygen (search grade oxygen 99.995% ALPHAGAZ) to reach thermodynamic equilibrium prior to the exchange under labelled oxygen. For both annealings, under the ^{18}O enriched atmosphere or pre-annealing, the samples were annealed for the required time, and quenched. To determine the ^{18}O content in the oxygen gas used for the exchange, a silicon wafer was oxidised overnight at 1000 °C under 200 mbar.

A ToF-SIMS[5] machine (ION-TOF GmbH, 25 keV Bi^+ LMIG analysis gun, DSCS sputter gun) was used to measure the ^{18}O diffusion profiles using the same experimental protocol as described in ref. 20. Both depth analysis from the sample surface (depth profile mode)[24] and edge analysis of the cut sample (line scan mode)[25] were performed and the profiles merged. The depth of the craters, after SIMS depth profiling, were measured with a KLA Tencor Alpha-step IQ profilometer. The same apparatus was used to measure the samples' roughness.

Electrical conductivity relaxation was also carried out on strontium-doped $Ca_3Co_4O_{9+\delta}$. For this purpose, plane-parallel samples (17 mm in length and 1 to 2.4 mm in width) were cut from the strontium substituted dense ceramics using a diamond saw. The electrical conductivity was measured using a 4 point configuration while changing the oxygen concentration of the surrounding atmosphere, as shown in ref. 20. A gold paste was used to paint gold contacts which were sintered at 750 °C for 1 h at a rate of 100 °C h^{-1} before the experiment. The distance between the two internal electrodes was about 14 mm.

Once the sample was at equilibrium at the given temperature, the pO_2 was changed step-wise and the conductivity was monitored as a function of time. The variation in pO_2 was performed by an abrupt switch between air and mixtures of O_2/Ar with oxygen partial pressures of 0.05 atm and 0.01 atm, respectively, at a 5 L h^{-1} flow rate. In addition, the temperature-dependent electrical conductivity was measured under air with a 5 L h^{-1} flow rate at a very slow speed of 25 °C h^{-1}.

To characterise the nature of the atoms at the uppermost surface of the sample, Low-Energy Ion Scattering (LEIS) was also applied to a dense ceramic $(Ca_{1-x}Sr_x)_3Co_4O_{9+\delta}$ with $x = 0.1$ composition which had been heated at 800 °C in air during the 12 hours prior to the analysis. The analysis was performed using a Qtac^{100} (ION-TOF GmbH) apparatus. Before analysis, the sample was subjected *in situ* to an atomic oxygen treatment (2–3 × 10^{-4} mbar) to remove possible organic surface contamination. The study was carried out with a $^4He^+$ (3 keV) beam on a 1 mm^2 analysed area.

Results and discussion

Characterization of powder and ceramics

The purity of the synthesised powders and dense ceramics was confirmed by X-ray diffraction. As shown in Fig. 2, good pattern matching was obtained for both type of ceramics.

Corresponding unit-cell parameters are given in Table 2. For $Ba_2Co_9O_{14}$, they are in good agreement with the parameters reported by Ehora *et al.* (5.6963(8) Å, 28.924(6) Å)[9] and Sun *et al.* (5.6958(4) Å, 28.909(4) Å).[10] The small differences in parameters we observed between the powders and the ceramics, on one hand, and these authors, on the other hand, are likely due to small differences in the oxygen content due to different thermal histories. For the refinement of $(Ca_{1-x}Sr_x)_3Co_4O_{9+\delta}$, the phases were considered as a mixture of two phases, one corresponding to the $[CoO_2]$ layers and the second one corresponding to the $[Ca_2CoO_{3-\delta}]$ rock salt layers, with common a, c and β cell parameters constrained

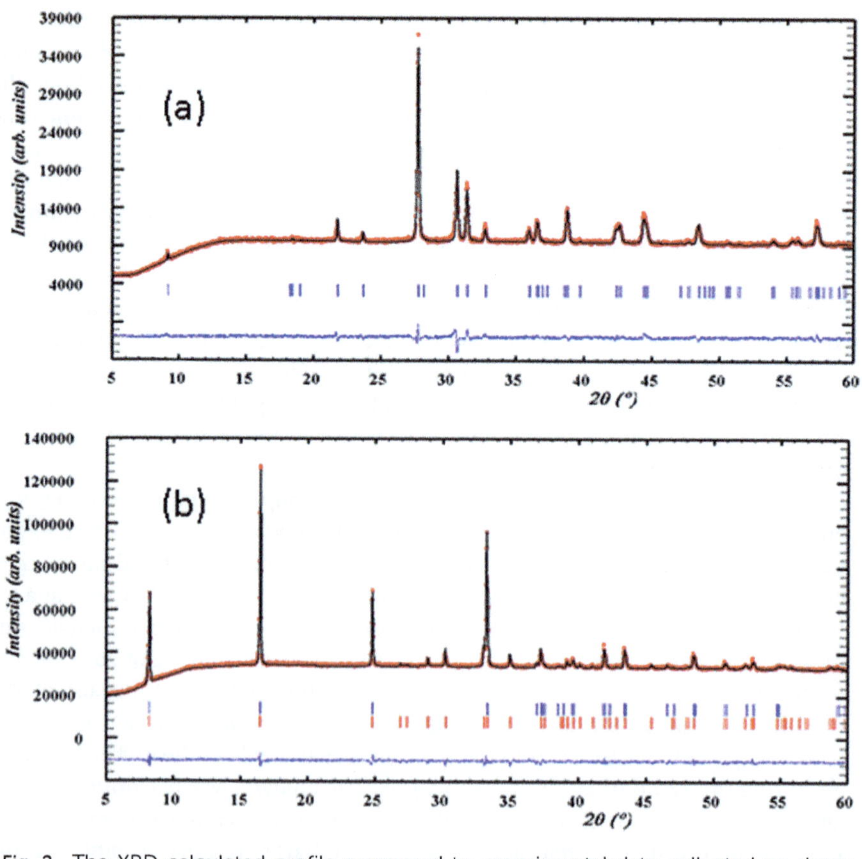

Fig. 2 The XRD calculated profile compared to experimental data collected on dense pellets of $Ba_2Co_9O_{14}$ (a), and $(Ca_{1-x}Sr_x)_3Co_4O_{9+\delta}$ with $x = 0.10$ (b).

Table 2 $Ba_2Co_9O_{14}$ and $(Ca_{1-x}Sr_x)_3Co_4O_{9+\delta}$ $x = 0.1, 0.2$ unit cell parameters from XRD on the powder and dense ceramics

	Powder	Dense ceramics
$Ba_2Co_9O_{14}$ ($R\bar{3}m$)		
a (Å)	5.6945(3)	5.700(1)
b (Å)	28.905(2)	28.926(6)
$(Ca_{1-x}Sr_x)_3Co_4O_{9+\delta}$ $x = 0.10$		
a (Å)	4.8429(3)	4.8433(9)
b_H (Å)	2.8255(3)	2.825(4)
b_{RS} (Å)	4.5968(6)	4.563(2)
c (Å)	10.8730(4)	10.881(2)
β (°)	98.137(9)	98.14(3)
$(Ca_{1-x}Sr_x)_3Co_4O_{9+\delta}$ $x = 0.20$		
A (Å)	4.8518(3)	4.8547(6)
b_H (Å)	2.8225(5)	2.821(1)
b_{RS} (Å)	4.6304(8)	4.6362(8)
c (Å)	10.8995(7)	10.904(2)
β (°)	98.123(7)	98.12(2)

to be the same, and different b_H and b_{RS}, refined independently. The evolution of the unit cell parameters with the strontium content is in good agreement with the results reported by Wang *et al.*,[26] who showed almost no variation of the b_H unit cell characterising the $[CoO_2]$ layer periodicity, and an increase of the b_{RS} unit cell with strontium content, in good agreement with the higher radius of Sr^{2+} (1.18 Å) compared to Ca^{2+} (1.00 Å).[27]

As shown by Ehora *et al.*,[9] $Ba_2Co_9O_{14}$ decomposed into CoO and a cubic $BaCoO_{\sim 2}$ form at 1030 °C. To check the stability of the calcium cobaltites, TGA and DTA analyses were carried out on the two strontium-doped compositions. Their thermograms are compared to the signals measured for the non-doped compound in Fig. 3. They exhibit similar behavior. The first transition, which is observed at 130 °C for the pure compound, disappears when strontium is introduced into the structure. The second transition at 550 °C, corresponding to the beginning of a mass loss, is maintained at the same temperature, but the intensity of the associated endothermic peak on heating decreases with the increase in strontium content. All compounds decomposed before 950 °C, and the decomposition temperature was slightly lowered when the strontium content was increased. This decomposition process is reversible, but it is worth noting that recombination happens at a lower temperature when the substitution rate increases.

Due to the lack of stability at high temperature, it was not possible to sinter these compounds using conventional sintering, but dense ceramics with relative densities higher than 96% were obtained using spark plasma sintering, as confirmed by SEM (see Fig. 4).

$^{18}O/^{16}O$ isotope exchange in $Ba_2Co_9O_{14}$

The normalised $^{18}O/(^{16}O + ^{18}O)$ isotope concentrations as a function of the penetration depth are given in Fig. 5 for the three studied temperatures. Both data

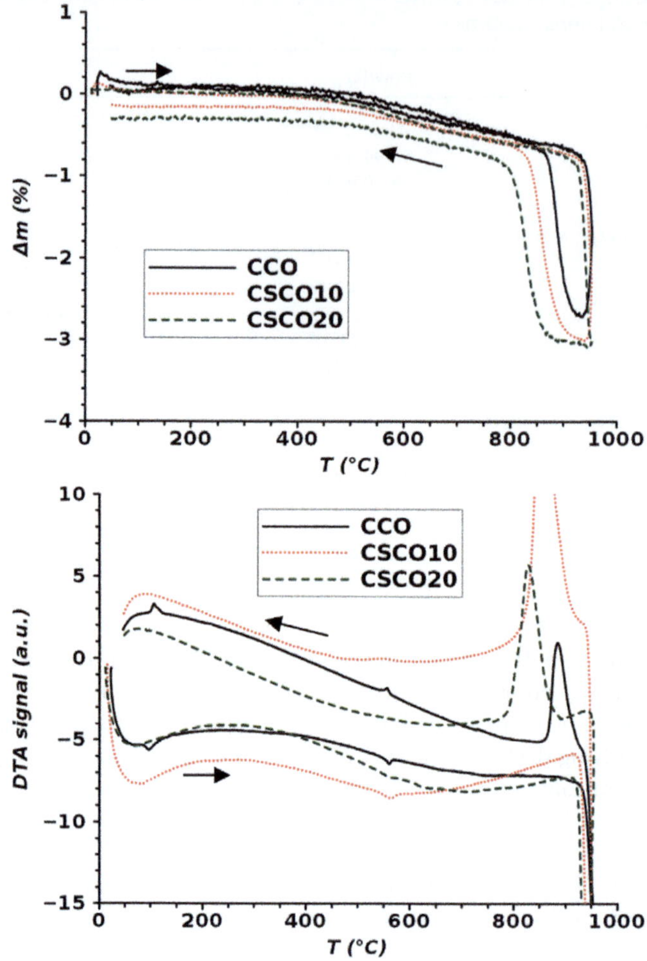

Fig. 3 Comparison of TGA and DTA analyses corresponding to the $Ca_3Co_4O_{9+\delta}$ parent compound (CCO, black solid line), $(Ca_{1-x}Sr_x)_3Co_4O_{9+\delta}$ with $x = 0.10$ (CSCO10, red line) and $(Ca_{1-x}Sr_x)_3Co_4O_{9+\delta}$ with $x = 0.20$ (CSCO20 black dotted line).

recorded in the depth-profile and line-scan modes are reported. A good agreement between the two sets of data was observed. However, whatever the temperature of exchange, a high oxygen content was measured at the surface of the samples within the first micron of the depth; the thickness of this high oxygen content layer increased with temperature. It was therefore not possible to fit these data to the diffusion equation (eqn (1)).

Taking into account only the data obtained far from the surface, the transport parameters given in Table 3 were derived.

Surprisingly, despite very good electrochemical performances, rather low oxygen diffusion and kinetics towards oxygen molecule dissociation were shown. However, one must be careful with these results. The increase in ^{18}O at the sample surface is not clearly understood. It is likely due to an oxidation

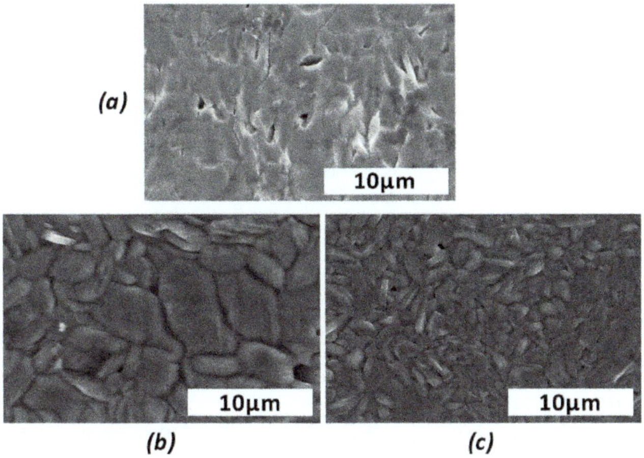

Fig. 4 SEM images of $Ba_2Co_9O_{14}$ (a), and $(Ca_{1-x}Sr_x)_3Co_4O_{9+\delta}$ with $x = 0.10$ (b) and 0.20 (c), showing the dense ceramics.

process which occurs when the sample is cooled, but this hypothesis still has to be confirmed.

Oxygen transport in strontium-doped $Ca_3Co_4O_9$

The oxygen transport properties in the two strontium-doped $Ca_3Co_4O_{9+\delta}$ materials were characterised by both electrical conductivity relaxation and the $^{18}O/^{16}O$ isotope depth profile technique. The electrical conductivity at the thermodynamic equilibrium corresponding to the $x = 0.10$ composition under different oxygen pressures is given in Fig. 6a. An increase in the electrical conductivity was observed when the oxygen partial pressure increased, confirming the p-type behaviour of these materials. As shown in Fig. 6b, the electrical conductivity slightly decreased when strontium was introduced into the structure. The evolution of the electrical conductivity corresponding to the

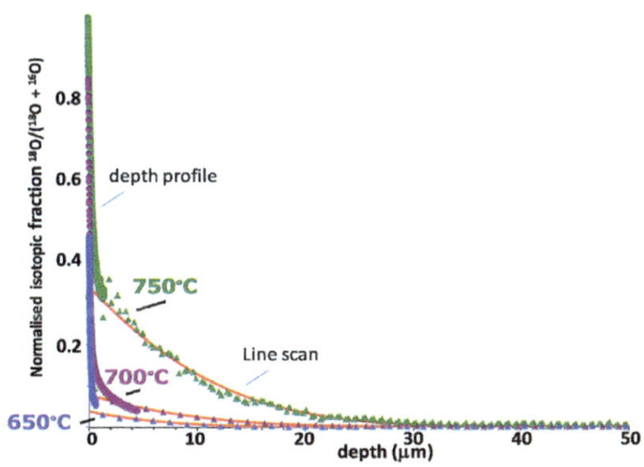

Fig. 5 Normalised ^{18}O diffusion profile measured on $Ba_2Co_9O_{14}$ dense ceramics after annealing under labelled oxygen at 650 °C, 700 °C and 750 °C for about 46 hours and corresponding fit (in red) taking into account the data obtained far from the surface.

Table 3 Tentative $Ba_2Co_9O_{14}$ transport parameters derived from normalised ^{18}O isotope profile far from the sample edges

T (°C)	k^* (cm s^{-1})	D^* (cm^2 s^{-1})
648	1.3×10^{-10}	2.5×10^{-12}
700	4.0×10^{-10}	5.0×10^{-12}
750	2.5×10^{-9}	6.0×10^{-12}

$(Ca_{1-x}Sr_x)_3Co_4O_{9+\delta}$ (x = 0.1) composition at 750 °C with the changes in surrounding atmosphere is shown Fig. 7. It is worth noting that oxidation and reduction are reversible processes.

Data obtained for both compositions in the 650–750 °C range were fitted to the theoretical model of diffusion. In these conditions, only the diffusion coefficient could be derived with precision, reflecting a D_{chem}/k_{chem} characteristic length that was much smaller than the size of the sample, and suggesting much higher surface exchange kinetics for these materials than for chemical diffusion.[28] Corresponding D_{chem} values are given in Table 4 and shown in Fig. 8. The D_{chem} values obtained when the air ↔ 0.05 atm and air ↔ 0.01 atm are close, and are similar to what was obtained for the pure material.[20] Moreover, the chemical diffusion coefficients are independent of whether the relaxation is a reduction or an oxidation process.

As shown in Fig. 8, as expected, the diffusion process is thermally activated, and activation energies of 1.03 ± 0.03 eV and 0.98 ± 0.04 eV were measured for the compositions with x = 0.1 and x = 0.2, respectively, which were slightly lower than the activation energy displayed by the pure compound (1.2 ± 0.02 eV), indicating that diffusion was easier for the strontium-doped compound, which also exhibits higher oxygen diffusion coefficients.

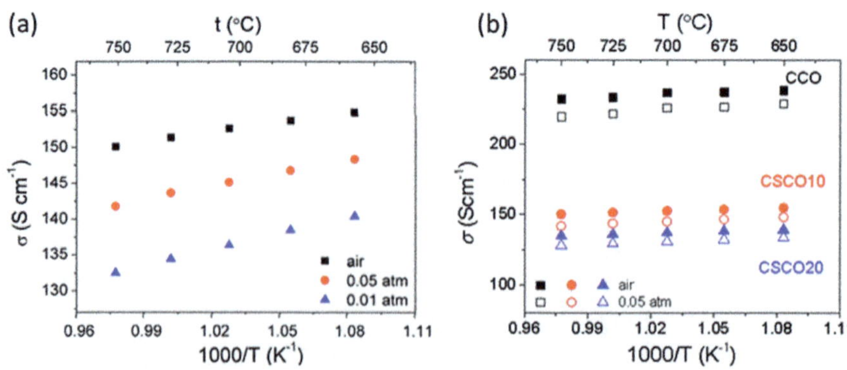

Fig. 6 Evolution of the electrical conductivity of $(Ca_{1-x}Sr_x)_3Co_4O_{9+\delta}$ with an x = 0.10 composition at the thermodynamic equilibrium under different oxygen pressures (a), and comparison of the electrical conductivities of the x = 0 (CCO in black), x = 0.10 (CSCO10 in red) and x = 0.20 (CSCO 20 in blue) compositions as a function of temperature under air and 0.05 atm (b).

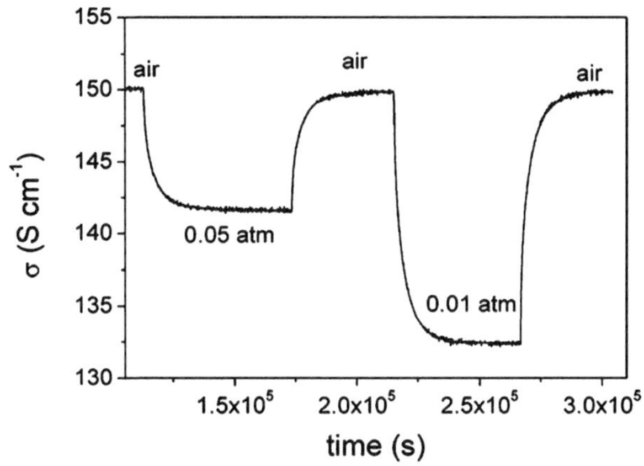

Fig. 7 Evolution of the electrical conductivity of $(Ca_{1-x}Sr_x)_3Co_4O_{9+\delta}$ with an $x = 0.10$ composition at 750 °C with surrounding atmosphere changes, showing the reversibility of the oxidation and reduction processes.

This increase in the oxygen diffusion coefficient with the substitution ratio was confirmed by $^{18}O/^{16}O$ isotope exchange. The normalized ^{18}O diffusion profile corresponding to the $(Ca_{1-x}Sr_x)_3Co_4O_{9+\delta}$ with $x = 0.1$ composition, exchanged at 649 °C over 567 minutes under 211 mbar dry oxygen, is shown in Fig. 9. Again, a good agreement was obtained between data collected in depth profile mode and line scan mode.

However, in this case, as shown for the parent compound,[20] a small decrease in the ^{18}O content was noticed at the sample surface, which is attributed to a possible carbonation or hydroxylation when back in the ambient atmosphere. In the inset, the same plot has been drawn on a decimal logarithmic scale to compare the experimental and theoretical profiles in depth. As previously noticed for the pure compound, a "tail" can be seen, most likely suggesting anisotropic

Table 4 D_{chem} ($cm^2\ s^{-1}$) values corresponding to $(Ca_{1-x}Sr_x)_3Co_4O_{9+\delta}$ with $x = 0.1$ and $x = 0.2$

T (°C)	Air-0.05 atm	0.05 atm-air	Air-0.01 atm	0.01 atm-air
$(Ca_{1-x}Sr_x)_3Co_4O_{9+\delta}$ $x = 0.10$				
650	1.37×10^{-7}	1.35×10^{-7}	1.34×10^{-7}	1.31×10^{-7}
675	1.92×10^{-7}	1.91×10^{-7}	1.85×10^{-7}	1.95×10^{-7}
700	2.58×10^{-7}	2.51×10^{-7}	2.52×10^{-7}	2.57×10^{-7}
725	3.47×10^{-7}	3.61×10^{-7}	3.44×10^{-7}	3.53×10^{-7}
750	4.68×10^{-7}	4.95×10^{-7}	4.73×10^{-7}	4.76×10^{-7}
$(Ca_{1-x}Sr_x)_3Co_4O_{9+\delta}$ $x = 0.20$				
650	3.09×10^{-7}	3.17×10^{-7}	3.04×10^{-7}	3.25×10^{-7}
675	4.62×10^{-7}	4.54×10^{-7}	4.57×10^{-7}	4.58×10^{-7}
700	5.99×10^{-7}	6.40×10^{-7}	5.87×10^{-7}	6.37×10^{-7}
725	8.41×10^{-7}	8.50×10^{-7}	8.00×10^{-7}	8.24×10^{-7}
750	1.04×10^{-6}	1.14×10^{-6}	0.96×10^{-6}	1.08×10^{-6}

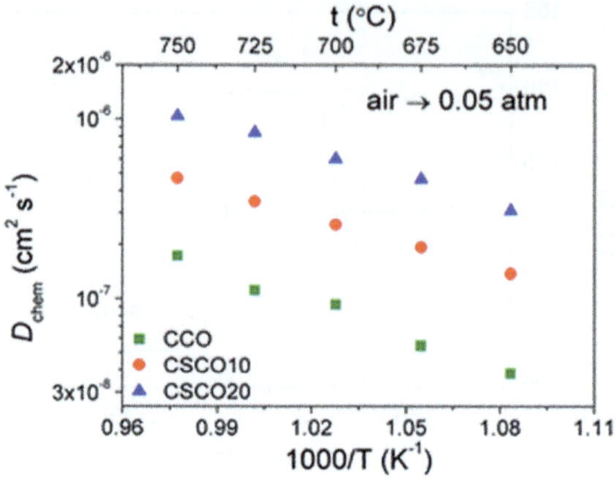

Fig. 8 Evolution of D_{chem} values as a function of temperature, measured on the parent compound $Ca_3Co_4O_{9+\delta}$ (CCO, green square), and $(Ca_{1-x}Sr_x)_3Co_4O_{9+\delta}$ with $x = 0.10$ (CSCO10, red dot) and $x = 0.20$ (CSCO20 blue triangle).

diffusion.[20] Nevertheless, as shown in Fig. 9, a good fit between the calculated and experimental data was obtained. The derived k^* and D^* parameters are given in Table 5. Their evolution as a function of temperature is shown on an Arrhenius plot in Fig. 10.

The tracer diffusion can be correlated to the chemical diffusion *via* the thermodynamic factor γ, according to the following equation:

$$D_{\text{chem}} = \gamma D^*$$

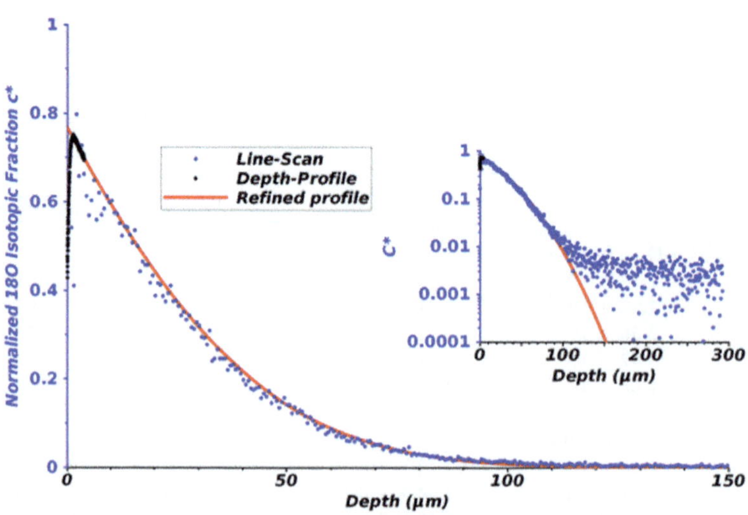

Fig. 9 Normalised ^{18}O diffusion profile measured on the $(Ca_{1-x}Sr_x)_3Co_4O_{9+\delta}$ with $x = 0.1$ composition, exchanged at 649 °C over 567 minutes under 211 mbar dry oxygen. The calculated profile is given in red.

$$\gamma = 1/2\frac{\text{d}\ln pO_2}{\text{d}\ln C_O} \approx 1/2\frac{\Delta\ln pO_2}{\Delta\ln C_O} \quad \text{when } \sigma_e \gg \sigma_i \tag{2}$$

where $\Delta \ln pO_2$ and $\Delta \ln C_O$ are the corresponding variations in the oxygen partial pressure and oxygen content, respectively, and σ_e and σ_i are the electronic and ionic conductivities, respectively.

For the sake of comparison, the evolution of $D^*_{ECR} = D_{chem}/\gamma$ is also reported in Fig. 10. The values are in good agreement. The slightly higher values observed for D^*_{ECR} compared to the tracer diffusion coefficient values can be explained by a lack of accuracy in the determination of the thermodynamic coefficient because of small oxygen stoichiometry variations (see the values of the thermodynamic coefficient in Table 6).

An increase in the diffusion coefficient with strontium content is confirmed, and may be explained by the increase in the volume of the rock salt layers, which may facilitate oxygen mobility within these layers. However, with a value in the range of 10^{-10} cm^2 s^{-1}, it is still low, which justifies the addition of gadolinia-doped ceria to improve the electrochemical performance.[15–19]

With surface exchange coefficients of 3.0×10^{-6} cm s^{-1} and 4.0×10^{-6} cm s^{-1} at 700 °C for both the $x = 0.1$ and $x = 0.2$ compositions, these materials exhibit very fast kinetics for the oxygen reduction reaction compared to the best cathode materials in the field, since values of only 1.3×10^{-7}, 1.1×10^{-6} and 1.3×10^{-7} cm s^{-1} were reported for $La_2NiO_{4+\delta}$ at 700 °C,[29] $La_{0.6}Sr_{0.4}CoO_{3-\delta}$ at 680 °C[4] and $GdBaCo_2O_{5+\delta}$ at 686 °C,[30] respectively.

To further understand oxygen transport in these materials, the $x = 0.1$ composition was characterised by LEIS. The initial spectrum, corresponding to the estimated removal of 0.2 monolayers, is compared to the one measured after the estimated removal of 3 monolayers in Fig. 11. Despite bulk contamination with sodium, which was also observed at the surface of the powders and ceramics of the parent compound,[20] mainly strontium and calcium were initially observed at the uppermost sample surface. No cobalt was initially observed, just a sub-surface signal was shown on the low energy side from the surface peak[31] in the background, indicating that cobalt atoms are beneath the first measured atomic layer. After ionic sputtering, cobalt was evidenced at the outermost surface, and an increase in calcium was also noticed. However, one

Table 5 D^* (cm^2 s^{-1}) and k^* (cm s^{-1}) values corresponding to $(Ca_{1-x}Sr_x)_3Co_4O_{9+\delta}$ with $x = 0.1$ and $x = 0.2$

T (°C)	604	649	694	741
$(Ca_{1-x}Sr_x)_3Co_4O_{9+\delta}$ $x = 0.10$				
k^*	1.4×10^{-8}	1.9×10^{-7}	3.0×10^{-6}	3.0×10^{-6}
D^*	1.1×10^{-10}	2.5×10^{-10}	4.1×10^{-10}	1.0×10^{-9}
$(Ca_{1-x}Sr_x)_3Co_4O_{9+\delta}$ $x = 0.20$				
k^*	6.7×10^{-8}	3.0×10^{-7}	4.0×10^{-6}	2.0×10^{-6}
D^*	3.0×10^{-10}	2.8×10^{-10}	6.1×10^{-10}	6.7×10^{-10}

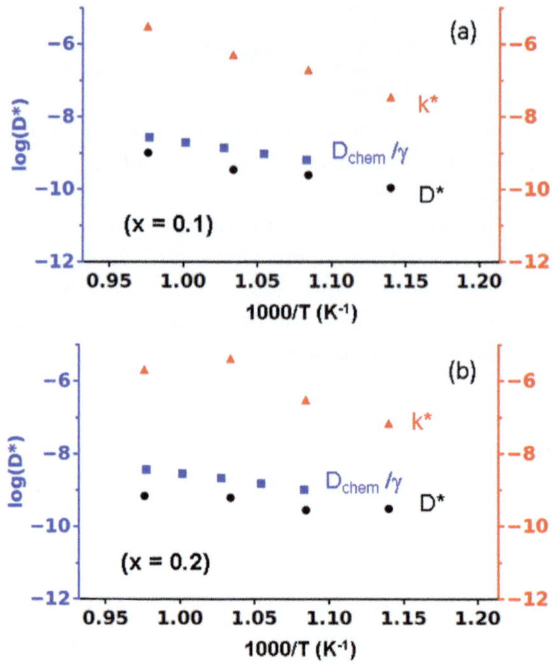

Fig. 10 Arrhenius plot of k^* and D^* values compared to D_{chem}/γ for the composition $(Ca_{1-x}Sr_x)_3Co_4O_{9+\delta}$ with $x = 0.1$ (a) and $x = 0.2$ (b).

Table 6 $(Ca_{1-x}Sr_x)_3Co_4O_{9+\delta}$ ($x = 0.1$, $x = 0.2$) thermodynamic coefficients derived from TGA analysis as a function of temperature and oxygen partial pressure

Temp. (°C)	$x = 0.1$	$x = 0.2$
650	211	313
675	200	305
700	189	297
725	183	299
750	178	302

must be careful with this technique, since strontium may also segregate at the sample surface. The same behaviour was observed for other compounds. For $SmCoO_3$, Fullarton *et al.* also showed that samarium was preferentially at the surface with only 5% of cobalt at the outermost surface, and segregation of strontium was evident at the surface for strontium-doped compounds.[32] Similarly, Viitanen *et al.* showed the absence of cobalt and iron in the outermost atomic layer of a $La_{0.6}Sr_{0.4}Co_{0.2}Fe_{0.8}O_3$ membrane.[33] We also observed calcium oxide at the uppermost surface of the $Ca_3Co_4O_9$ parent compound. Here, strontium seems to be mainly at the sample surface, and its presence may partly explain the increase in the kinetics towards the oxygen dissociation reaction when the strontium content increases.

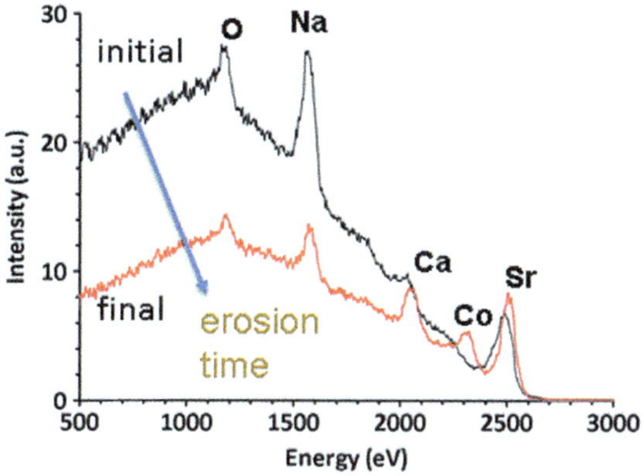

Fig. 11 $^4He^+$ (3 keV) LEIS spectra recorded on dense ceramics of $(Ca_{1-x}Sr_x)_3Co_4O_{9+\delta}$ with $x = 0.1$ after annealing at 800 °C for 12 hours prior to the analysis. The first spectrum (in black) corresponds to an estimated removal of 0.2 monolayers, and the last spectrum (in red) corresponds to an expected removal of around 3 monolayers.

Conclusions

Although oxygen diffusion was confirmed in the promising material $Ba_2Co_9O_{14}$, it was not possible to derive accurate transport parameters because of an oxidation process at the sample surface which still has to be clarified. In contrast, for $Ca_3Co_4O_{9+\delta}$ derivatives, it was shown that oxygen diffusion was improved when calcium was partly substituted with strontium, likely due to an increase in the volume of the rock salt layers of the structure, which may help oxygen mobility. Although the diffusion coefficient remains low, interestingly, fast kinetics towards the oxygen dissociation reaction were shown, with a surface exchange coefficient higher than those reported for the best cathode materials in the field. The surface exchange coefficient is increased when the strontium content increases, which may be correlated to the presence of a higher amount of strontium at the sample surface, which may play a key role in the mechanism of oxygen molecule dissociation at the surface of the solid.

Acknowledgements

The "Fonds Européen de Développement Régional (FEDER)", "CNRS", "Région Nord Pas-de-Calais" and French "Ministère de l'Education Nationale de l'Enseignement Supérieur et de la Recherche" are acknowledged for funding of X-ray and surface analysis facilities. V.T. is grateful to the University Lille 1 for funding his PhD grant. The authors are also grateful to the Agence Nationale de la Recherche for funding the ANR EVEREST (ANR-07-PANH-005), NanOxyDesign (ANR-10-BLAN-0814) and IDEA-MAT (ANR-12-BS08-0012-01) projects. Maxence Vandewalle, Nora Djelal and Laurence Burylo from UCCS are acknowledged for SEM, TG-DTA analyses and conventional XRD experiments, respectively.

References

1 *Functional materials for sustainable energy applications*, Woodhead publishing series in energy, ed. J. A. Kilner, S. J. Skinner, S. J. C. Irvine and P. P. Edwards, Woodhead Publishing Limited, 2012.

2 D. Marinha, J. Hayd, L. Dessemond, E. Evers-Tiffée and E. Djurado, *J. Power Sources*, 2011, **196**, 5084.

3 Y. Li, M. W. Xu and J. B. Goodenough, *J. Power Sources*, 2012, **209**, 40.

4 A. V. Berenov, A. Atkinson, J. A. Kilner, E. Bucher and W. Sitte, *Solid State Ionics*, 2010, **181**, 819.

5 O. Yamamoto, *Electrochim. Acta*, 2000, **45**, 2423.

6 K. Kordesch and G. Simader, *Fuel Cells and their Applications*, VCH Wiley, NY, Weinheim, 1996.

7 B. Fan, J. Yan and X. Yan, *Solid State Sci.*, 2011, **13**, 1835.

8 A. Rolle, N. Preux, G. Ehora, O. Mentré and S. Daviero-Minaud, *Solid State Ionics*, 2011, **184**, 31.

9 G. Ehora, S. Daviero-Minaud, M. Colmont, G. André and O. Mentré, *Chem. Mater.*, 2007, **19**, 2180.

10 J. Sun, M. Yang, G. Li, T. Yang, F. Liao, Y. Wang, M. Xiong and J. Lin, *Inorg. Chem.*, 2006, **45**, 9151.

11 I. Kehal, to be published.

12 K. Nagasawa, S. Daviero-Minaud, N. Preux, A. Rolle, P. Roussel, H. Nakatsugawa and O. Mentré, *Chem. Mater.*, 2009, **21**, 4738.

13 H. Muguerra, D. Grebille and F. Bourée, *Acta Crystallogr., Sect. B: Struct. Sci.*, 2008, **64**, 676.

14 C. D. Ling, K. Aivazian, S. Schmid and P. Jensen, *J. Solid State Chem.*, 2007, **180**, 1446.

15 A. Rolle, S. Boulfrad, K. Nagasawa, H. Nakatsugawa, O. Mentré, J. Irvine and S. Daviero-Minaud, *J. Power Sources*, 2011, **196**, 7328.

16 A. Rolle, V. Thoréton, P. Rozier, E. Capoen, O. Mentré, B. Boukamp and S. Daviero-Minaud, *Fuel Cells*, 2012, **12**(2), 288.

17 A. Rolle, to be published.

18 A. J. Samson, M. Søgaard, N. Van Nong, N. Pryds and N. J. Bonanos, *J. Power Sources*, 2011, **196**, 10606.

19 J. Zou, J. Park, H. Yoon, T. Kim and J. Chung, *Int. J. Hydrogen Energy*, 2012, **37**, 8592.

20 V. Thoréton, Y. Hu, C. Pirovano, E. Capoen, N. Nuns, A.-S. Mamede, G. Dezanneau, C. Y. Yoo, H. J. M. Bouwmeester and R. N. Vannier, *J. Mater. Chem. A*, 2014, **2**, 19717.

21 J. Rodriguez-Carvajal, *Phys. B*, 1993, **192**, 55.

22 L. Karvonen, H. Yamauchi and M. Karppinen, *Chem. Mater.*, 2008, **20**, 7143.

23 J. Crank, *The Mathematics of diffusion*, Clarendon Press, Oxford, 2nd edn, 1975.

24 J. A. Kilner, B. C. H. Steele and L. Ilkov, *Solid State Ionics*, 1984, **12**, 89.

25 R. J. Chater, S. Carter, J. A. Kilner and B. C. H. Steele, *Solid State Ionics*, 1992, **53–56**, 859.

26 L. B. Wang, A. Maignan, D. Pelloquin, S. Hébert and B. Raveau, *J. Appl. Phys.*, 2002, **92**, 124.

27 R. D. Shannon, *Acta Crystallogr., Sect. A: Cryst. Phys., Diffr., Theor. Gen. Crystallogr.*, 1976, **32**, 751.
28 M. W. den Otter, H. J. M. Bouwmeester, B. A. Boukamp and H. Verweij, *J. Electrochem. Soc.*, 2001, **148**(2), J1.
29 R. Sayers, R. A. De Souza, J. A. Kilner and S. J. Skinner, *Solid State Ionics*, 2010, **181**, 386.
30 A. Tarancón, S. J. Skinner, R. J. Chater, F. Hernández-Ramirez and J. A. Kilner, *J. Mater. Chem.*, 2007, **17**, 3175.
31 H. R. J. ter Veen, T. Kim, I. E. Wachs and H. H. Brongersma, *Catal. Today*, 2009, **140**, 197.
32 I. C. Fullarton, J. P. Jacobs, H. E. Benthem, J. A. Kilner, H. H. Brongersma, P. J. Scanlon and B. C. H. Steele, *Ionics*, 1995, **1**, 51.
33 M. M. Viitanen, R. G. Welzenis, H. H. Brongersma and F. P. F. van Berkel, *Solid State Ionics*, 2002, **150**, 223.

Faraday Discussions

PAPER

Ion counting in supercapacitor electrodes using NMR spectroscopy†

John M. Griffin,*[a] Alexander C. Forse,[a] Hao Wang,[a] Nicole M. Trease,[b] Pierre-Louis Taberna,[c] Patrice Simon[c] and Clare P. Grey[ab]

Received 27th June 2014, Accepted 5th August 2014
DOI: 10.1039/c4fd00138a

^{19}F NMR spectroscopy has been used to study the local environments of anions in supercapacitor electrodes and to quantify changes in the populations of adsorbed species during charging. In the absence of an applied potential, anionic species adsorbed within carbon micropores (in-pore) are distinguished from those in large mesopores and spaces between particles (ex-pore) by a characteristic nucleus-independent chemical shift (NICS). Adsorption experiments and two-dimensional exchange experiments confirm that anions are in dynamic equilibrium between the in- and ex-pore environments with an exchange rate in the order of tens of Hz. ^{19}F *in situ* NMR spectra recorded at different charge states reveal changes in the intensity and NICS of the in-pore resonances, which are interpreted in term of changes in the population and local environments of the adsorbed anions that arise due to the charge-storage process. A comparison of the results obtained for a range of electrolytes reveals that several factors influence the charging mechanism. For a tetraethylammonium tetrafluoroborate electrolyte, positive polarisation of the electrode is found to proceed by anion adsorption at a low concentration, whereas increased ion exchange plays a more important role for a high concentration electrolyte. In contrast, negative polarization of the electrode proceeds by cation adsorption for both concentrations. For a tetrabutylammonium tetrafluoroborate electrolyte, anion expulsion is observed in the negative charging regime; this is attributed to the reduced mobility and/or access of the larger cations inside the pores, which forces the expulsion of anions in order to build up ionic charge. Significant anion expulsion is also observed in the negative charging regime for alkali metal bis(trifluoromethane)sulfonimide electrolytes, suggesting that more subtle factors also affect the charging mechanism.

[a]*Department of Chemistry, University of Cambridge, Lensfield Road, Cambridge, CB2 1EW, UK. E-mail: jg670@cam.ac.uk; cpg27@cam.ac.uk*
[b]*Department of Chemistry, Stony Brook University, Stony Brook, New York, 11794-3400, USA*
[c]*Université Paul Sabatier Toulouse III, CIRIMAT, UMR-CNRS 5085, F-31062 Toulouse, France*

† Electronic supplementary information (ESI) available. See DOI: 10.1039/c4fd00138a

Introduction

Electrochemical capacitors, or 'supercapacitors', are currently receiving widespread interest as a promising energy storage technology for high-power applications.[1–3] These devices store charge through electrostatic attraction between ionic charges within an electrolyte and the electronic charge on the electrode surface. Since this mechanism is non-Faradaic, very high charge and discharge rates can be achieved. In addition, since no structural changes take place within the electrode, supercapacitors can offer essentially unlimited cycle lives. Supercapacitor electrolytes typically comprise organic salts such as tetraethylammonium tetrafluoroborate (NEt_4–BF_4) dissolved in acetonitrile (ACN).[3] Such electrolytes offer high ionic conductivity and higher breakdown voltages than aqueous electrolytes, enabling more energy to be stored. There has also been an increasing focus on the use of room-temperature ionic liquids as supercapacitor electrolytes, as they offer greater thermal stability than organic electrolytes and have an even wider voltage window.[4–6] Supercapacitor electrodes are generally constructed using porous carbon due to its high surface area and conductivity.[7] In particular, porous carbon derived from organic precursors such as coconut shells and wood are widely used due to their low cost and high availability. These materials are predominantly microporous with pore sizes in the range 1–2 nm, although they can also exhibit a range of pore sizes in the mesoporous and macroporous regimes. The importance of the relationship between pore size and ion size was highlighted by Chmiola *et al.*, who reported an anomalous increase in capacitance for microporous titanium carbide-derived carbons with pore sizes that were smaller than the solvated ion size.[8] This challenged the previously-held view that pores smaller than the solvated ion size do not contribute significantly to the capacitance. It has been hypothesized that ion desolvation results in a closer proximity of the charge centres to the electrode surface, resulting in higher capacitance.[9] However, experimental validation of this hypothesis is scarce, and many questions relating to the mechanism of supercapacitance in general remain unanswered. A better understanding of this important phenomenon will lead to the design of devices with improved properties.

To gain insight into the mechanism of supercapacitance, a large number of theoretical studies have been carried out. Huang *et al.* developed theoretical models based upon a so-called electric-double cylinder and electric wire-in-cylinder capacitors in order to correlate capacitance with properties such as pore size, specific surface area and ion size.[10] Kondrat *et al.* have used mean field theory models to propose that image charges arising from the charged micropore surfaces facilitate the packing of ions of the same polarity during charging, accounting for the large increases in capacitance experimentally measured for carbons with micropores smaller than the solvated electrolyte ions.[11,12] A large amount of theoretical insight has also been gained through the application of molecular dynamics simulations.[13–15] In particular, these studies have revealed the importance of the overscreening effects at planar electrode–electrolyte interfaces, which give rise to the layering of ions in an oscillatory pattern, whereby capacitance decays with increasing pore size.[16–19] It has been suggested that the absence of such effects in microporous electrodes allows high capacitances to develop.[20]

A range of experimental methods have also been employed in order to study the structure of the electrode–electrolyte interface in supercapacitors. Neutron diffraction has been used to observe and quantify ion diffusion and concentration variations as a function of the applied potential.[21,22] Changes in electrolyte absorbances measured in *in situ* infrared spectroscopy experiments have been attributed to the net movement of positive and negative ion pairs into carbon micropores upon charging.[23] Changes in the height of electrodes in electrochemical dilatometry experiments have been interpreted in terms of shrinkage and expansion of pores as ions fill and leave the electrode.[24] Recently, a number of studies have demonstrated the utility of *in situ* electrochemical quartz crystal microbalance methodologies to track changes in electrode mass during charging and discharging.[25–28] Deviations from theoretical mass changes based on Faraday's law have been interpreted in terms of distinct charging regimes, which are characterized by the compositional differences in the ionic part of the electric double layer. For aqueous and organic electrolytes, these studies have generally identified two distinct charging regimes for the systems studied; at low potentials charging was found to proceed *via* ion exchange, whereas at high potentials counter-ion adsorption plays a more dominant role. Recently, this approach has been extended to the study of ionic liquids where similar observations were reported for the positive polarisation of the electrode, whereas for negative polarisation the observed mass change was consistent with adsorption of counter-ions only across the voltage range studied.[28]

Despite the progress that has been made in both theoretical and experimental methods for the study of supercapacitance, further work is required to fully understand this complex process. In particular, the development of experimental methods that can selectively observe and quantify anions or cations will add significant insight. In this respect, nuclear magnetic resonance stands out as a highly localised and element-selective probe of chemical environments. A number of early studies by Harris *et al.* demonstrated the potential of NMR for the study of molecules adsorbed on porous carbon, where distinct resonances were observed for adsorbed and non-adsorbed species within activated carbons.[29–31] Subsequent studies on a range of different carbon morphologies have revealed similar trends, whereby resonances corresponding to species adsorbed on the carbon surface are always observed at a lower frequency than their non-adsorbed counterparts, regardless of the species studied.[32–39] This effect has also been investigated in theoretical studies, where it has been shown that the characteristic shift in frequency for adsorbed species arises due to the circulation of delocalised electrons in the carbon surface, which results in a locally reduced magnetic field.[40–43] This so-called nucleus-independent chemical shift (NICS) therefore provides a useful means by which to identify ions confined within carbon micropores and can in principle be used to obtain information about their local chemical environments such as pore diameter and local curvature of the carbon surface. Indeed, we have recently shown how NMR can identify adsorbed ions in model supercapacitor electrodes comprising pieces of microporous carbon film soaked with a NEt_4–BF_4/ACN electrolyte.[38,39] Furthermore, since NMR is in principal a quantitative technique (*i.e.*, the integrated spectral resonance intensity is directly proportional to the number of species in a particular chemical environment), it can be used to monitor changes in adsorbed ion populations during charging. Recent work has shown how magic-angle spinning (MAS) NMR experiments can identify differences in the number, and local environments, of ions in electrodes taken from supercapacitors that were disassembled at different charge states.[44] To obtain a direct insight into working supercapacitor devices under realistic charging conditions, *in situ* NMR methods have also been developed. These approaches involve the use of a specially-designed supercapacitor cell which is situated inside the NMR spectrometer, enabling the acquisition of NMR spectra under a range of charging conditions. We have recently demonstrated the utility of this approach whereby systematic changes were observed in ^{11}B NMR spectra of a supercapacitor comprising a NEt_4–BF_4/ACN electrolyte and activated carbon electrodes which were held at different applied potentials.[38] Through the use of ^{19}F NMR spectroscopy, this approach was further developed to enable the quantification of relative adsorbed anion populations at different charge states, and the observation of changes in anion environments under dynamic charging conditions.[45]

Having developed and refined the *in situ* NMR methodology for the study of supercapacitors, the focus of the current study is to apply this approach to build up an understanding of different factors that affect the charging mechanism. We first investigate the effects of electrolyte concentration on the adsorption of anions in model electrodes in the absence of an applied potential. We then measure changes in the number and local environment of the adsorbed anions when a potential is applied to a working supercapacitor device. This reveals important information about the charge storage process in each system, and by comparing a range of different electrolytes it is possible to correlate the observed differences with the properties of the electrolyte, such as concentration and relative ion size.

Experimental details

Materials

All carbon electrodes were prepared from YP-50F activated carbon (Kuraray Chemical, Japan). Free-standing carbon films were prepared in the standard way[46] by mixing carbon powder (95 wt%) with polytetrafluoroethylene (5 wt%) (Sigma Aldrich, 60 wt% dispersion in water). The films were rolled to an approximate thickness of 0.25 mm. Prior to sample preparation, carbon film pieces were dried under vacuum at 200 °C overnight and then transferred to an argon glove box with H_2O and O_2 levels less than 0.1 ppm. Inside the glove box, the film pieces were cut to the specified mass with the approximate dimensions 4 × 10 mm. The electrolyte salts used were tetraethylammonium tetrafluoroborate (NEt_4–BF_4, Sigma Aldrich, electrochemical grade >99.0%), tetrabutylammonium tetrafluoroborate (NBu_4–BF_4, Sigma Aldrich, >99.0%), lithium bis(trifluoromethane)sulfonimide (Li-TFSI, 3M HQ-115) and sodium bis(trifluoromethylsulfonyl)imide (Na-TFSI), synthesized by neutralizing bis(trifluoromethylsulfonyl)imide (Iolitec GmbH, >98%) with excess sodium carbonate, followed by drying *in vacuo* and filtration with dry ACN.

Sample preparation

All bag cell samples were prepared inside an argon glove box with H_2O and O_2 levels less than 0.1 ppm. Model electrodes were prepared by sealing a carbon film piece inside a plastic bag (3M packaging film) with the chosen volume of electrolyte, which was added using a microsyringe. All model supercapacitor cells in this study were also constructed using plastic bag cell designs with ‘overlaid’ electrode configurations as discussed in detail in ref. 45: carbon film pieces were pressed onto strips of carbon-coated aluminium mesh, and a Celgard 2500 (monolayer polypropylene, 25 μm thickness) separator was placed between them. These components were placed inside a plastic bag (3M packaging film), which was heat sealed on three sides. The cell was then filled with electrolyte before making the final seal to close the bag. For NMR measurements, the supercapacitor bag cells were held between two rigid plastic strips inside the NMR coil in order to provide rigidity and allow better control over the cell orientation.

NMR details

Adsorption studies and *in situ* NMR experiments were performed using a Bruker Avance spectrometer operating at a magnetic field strength of 7.05 T, corresponding to a ^{19}F Larmor frequency of 284.2 MHz. A Bruker HX double resonance static probe was used, with a 6.8 mm inner diameter solenoid coil. The “depth” pulse sequence[47] with a 90° pulse length of 5.5 μs was used in order to reduce the background signal from the PTFE probe-head housing. The total delay between excitation of transverse magnetization and acquisition of the free induction decay was 90 μs. A recycle interval of 30 s was used, which was sufficient for the spectra to be quantitative. ^{19}F exchange MAS NMR experiments were carried out using a Bruker Avance spectrometer operating at a magnetic field strength of 9.4 T, corresponding to a ^{19}F Larmor frequency of 376.6 MHz. A Bruker 2.5 mm probe was used, and all experiments were performed at a MAS rate of 5 kHz with a 90° pulse length of 3.5 μs. For each exchange experiment, either 8 transients (1.5 M electrolyte) or 16 transients (0.5 M electrolyte) were coadded for each of the 160 t_1 increments. A short recycle interval of 2 s was used in order to enable the acquisition of multiple two-dimensional spectra within a reasonable time. For each experiment, 3 μL of electrolyte was added to 4 mg of film inside the MAS rotor. All ^{19}F NMR spectra are referenced relative to neat hexafluorobenzene (C_6F_6) at −164.9 ppm.

In situ NMR methodology

In situ experiments were carried out by placing the pristine cell in the NMR coil in the vertical orientation[39] and charging sequentially to a series of different voltages (0, 0.25, 0.5, 0.75, 1, 1.25, 1.5, 0, −0.25, −0.5, −0.75, −1, −1.25 and −1.5 V). The applied potential was controlled using a Bio-logic cycler in a two-electrode configuration. The cells were held at each voltage until a constant current was obtained, and then the ^{19}F NMR spectra were acquired; in most cases this was 45–60 minutes, while for NBu_4–BF_4 90 minutes was required. For the discharge of the cell (1.5 → 0 V), current relaxations of double duration were used. Spectral fitting was carried out using DMfit software.[48] Deconvolutions were carried out as described previously in ref. 45. Some spectra required several lineshapes to describe the free electrolyte/ex-pore resonance. This is ascribed to local variations in the magnetic field across the bag cell due to the different susceptibilities of the cell components, which results in a range of different local fields and thus shifts for the same chemical species.[45] For this reason, it is not necessarily appropriate to assume purely Lorentzian lineshapes for the ex-pore feature (as would be expected for an isotropic liquid in a homogeneous magnetic field), and so multiple lineshapes with mixtures of Gaussian and Lorentzian components were used to accurately model this feature. Similarly, the lineshape of the in-pore feature is expected to be influenced by both local field variations, motional effects, and distributions of the chemical environments, and so the Gaussian/Lorentzian ratio for this feature was allowed to vary. In some cases, the fitting routine gave unrealistic values for the Gaussian/Lorentzian ratio for some components (*i.e.*, less than 0 or greater than 1); in these cases, the ratios were fixed to either 0 or 1 and the fit was repeated. For each electrolyte, the spectrum obtained at 0 V was fitted first, as this showed the best resolution of the in-pore resonance. The peak positions and intensities obtained were then used as a starting point to fit the spectrum at the next highest voltage. Fits were repeated two or three times for each series of data in order to estimate errors. Representative deconvolutions and deconvolution parameters are given in the ESI.†

Results

Adsorption of BF_4 anions at zero applied potential

We first investigate factors affecting anion adsorption in porous carbon in the absence of an applied potential. While ^1H NMR is an attractive option for the study of organic tetra-alkylammonium cations, the large number of plastic components used in the bag cell assembly results in a broad and intense background ^1H signal that precludes the observation of the relatively low intensity

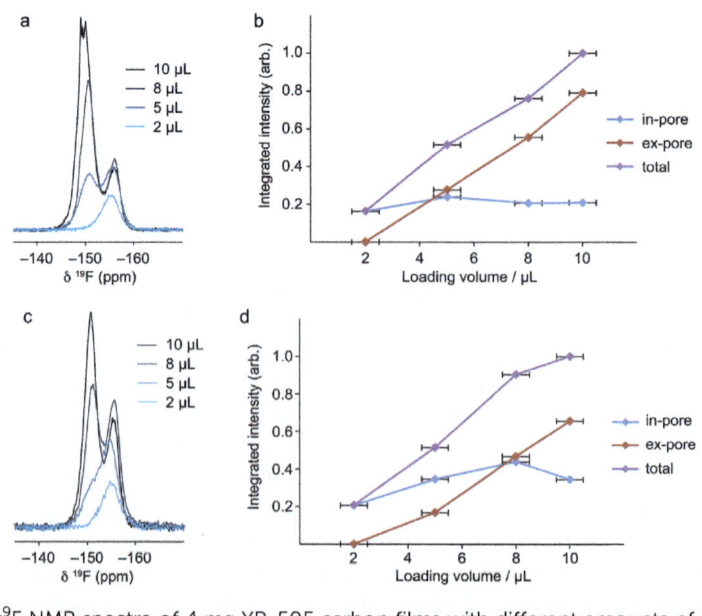

Fig. 1 ^{19}F NMR spectra of 4 mg YP-50F carbon films with different amounts of (a) 1.5 M NEt_4-BF_4/ACN and (c) 0.5 M NEt_4-BF_4/ACN electrolyte in plastic bags. Deconvoluted relative integrated intensities for the two main resonances are shown in (b) and (d).

electrolyte resonances. For this reason, we focus this study on ^{19}F NMR which provides a probe of the fluorine-containing anions.[38,45]

Fig. 1a shows ^{19}F NMR spectra of 4 mg YP-50F carbon films soaked with different amounts of 1.5 M NEt_4–BF_4/ACN electrolyte. Across the series of spectra, two main resonances are observed. Deconvoluted intensities determined for the two resonances at each loading volume are shown in Fig. 1b. For the lowest loading volume of 2 μL, the single resonance observed at −155 ppm corresponds to the BF_4 anions that are adsorbed to the carbon surface inside the micropores (referred to hereafter as 'in-pore'). As more electrolyte is added, the intensity of the in-pore resonance increases but then saturates around a loading volume of 5 μL. Simultaneously, a second resonance is observed at −150 ppm which corresponds to the BF_4 anions that are situated outside of the micropores (hereafter referred to as 'ex-pore'). As more electrolyte is added, the ex-pore resonance continues to grow, while there appears to be a small reduction in the in-pore resonance intensity. Fig. 1c and d show ^{19}F NMR spectra and integrated resonance intensities for films soaked in the same electrolyte at a lower concentration of 0.5 M. The spectra also exhibit in-pore and ex-pore resonances, although the relative build up rates of the two environments as a function of loading volume are slightly different. For this system, saturation of the in-pore resonance intensity occurs at around 8 μL, and there is a noticeable decrease in intensity of this resonance at the next loading volume of 10 μL.

Comparison of the adsorption behaviour for the two concentrations provides insight into the nature of the adsorption process. YP-50F has a specific surface area of 1600 m^2 g^{-1} and a total pore volume of approximately 0.7 cm^3 g^{-1}, according to data supplied by the manufacturer.[49] According to gas adsorption measurements, 46% of the pore volume is less than 1 nm in size, 46% is between 1–2 nm, and the rest is greater than 2 nm.[50] For NEt_4–BF_4, solvated ion diameters are 1.16 nm (BF_4^-) and 1.30 nm (NEt_4^+),[9] meaning that ions can easily access a large fraction of the pores. For a 4 mg piece of YP-50F film, the total pore volume is estimated to be between 2.7–3.3 μL. For the 1.5 M electrolyte at the highest loading volume of 10 μL, 16% of the anions are found to occupy the in-pore environment, corresponding to 3.1 μmol of anions. Assuming equal adsorption of anions and cations, the total volume occupied by in-pore electrolyte ions in this sample is estimated to be 3.67 μL on the basis of the solvated ion diameters. This value is slightly higher than the estimated total pore volume of the film piece, and may therefore indicate partial desolvation or overlap of anion and cation solvation shells inside the pores. Indeed, a recent electron radial distribution function study has indicated that electrolyte solvent molecules can become partially aligned when confined within sub-nanometre pores in activated carbon.[51] For the 0.5 M electrolyte at a loading volume of 10 μL, 34% of anions (corresponding to 1.7 μmol) occupy the in-pore environment. Again assuming equal adsorption of anions and cations, the total volume taken up by the solvated ions is estimated to be 2.01 μmol. The packing density of the ions inside the pores is therefore reduced for the lower concentration electrolyte, being approximately equal to 0.55 times that of the high concentration electrolyte. However, the packing density is reduced by less than a factor of 2/3, which may be expected based on the ratios of the concentrations, indicating that there is still a preference for anions to occupy the in-pore rather than the ex-pore environment. Overall, these results show that for electrolyte volumes smaller than the total pore volume, all of the electrolyte occupies the space within the micropores. When the loading volume exceeds the total pore volume, larger voids in the carbon become filled with electrolyte and ions start to be distributed between the in- and ex-pore environments. The observed decreases in the number of in-pore anions at high loading volumes indicate that the equilibrium between the occupation of the in- and ex-pore environments changes slightly as more ex-pore reservoirs become accessible. For the high concentration electrolyte, the ions are densely packed within the micropores at a high loading volume.

To further investigate the nature of the equilibrium between the occupation of the in-pore and ex-pore environments, ^{19}F two-dimensional exchange (2D-EXSY) NMR spectra were recorded. The ^{19}F 2D-EXSY NMR experiments measure the retained magnetization as the BF_4 anions exchange between in-pore and ex-pore environments during a mixing time, t_m. The spectra were recorded under magic-angle spinning (MAS) conditions in order to improve the resolution between the in-pore and ex-pore resonances. Representative spectra for YP-50F film soaked with 1.5 M NEt_4–BF_4/ACN are shown in Fig. 2a and b. Using a short t_m of 0.5 ms (Fig. 2a), only resonances on the autocorrelation diagonal are observed, showing that there is no movement of anions between the in-pore and ex-pore environments on this timescale. Using a much longer t_m of 50 ms (Fig. 2b), significant off-diagonal cross peak intensity is observed correlating to the in-pore and ex-pore resonances, showing that the exchange of anions between these environments takes place. By recording a set of 2D-EXSY NMR spectra with a range of t_m values, it is possible to estimate a rate constant, k, for a two-site exchange process by fitting integrated off cross-peak/diagonal-peak intensities to

$$\frac{a_{\mathrm{cross}}}{a_{\mathrm{diag}}} = \tanh(kt_{\mathrm{m}}) \tag{1}$$

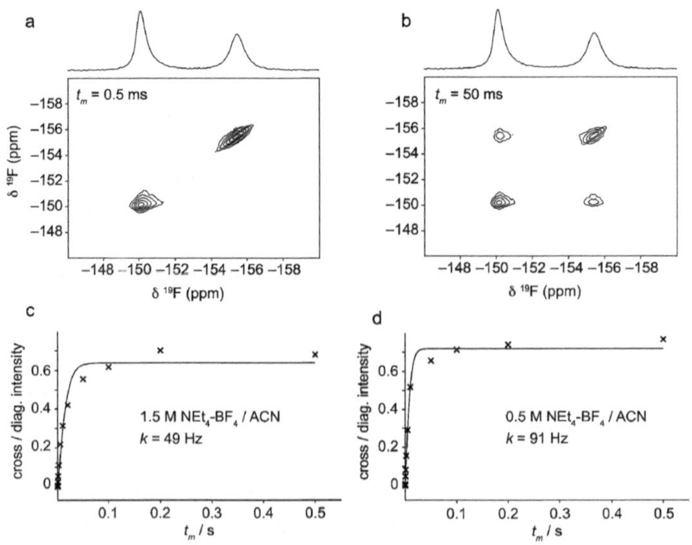

Fig. 2 ^{19}F 2D-EXSY NMR spectra of the YP-50F film soaked with 1.5 M NEt_4–BF_4/ACN electrolyte recorded with mixing times of (a) 5 ms and (b) 50 ms. Fits to the cross-peak/diagonal-peak intensity ratio as a function of t_m are shown for 1.5 M and 0.5 M NEt_4–BF_4/ACN electrolytes in (c) and (d), respectively.

where a_{cross} is the cross-peak intensity and a_{diag} is the diagonal-peak intensity.[52] Fits to the data obtained for the YP-50F film soaked with both 1.5 M and 0.5 M NEt_4–BF_4/ACN are shown in Fig. 2c and d, respectively. For 1.5 M NEt_4–BF_4/ACN, exchange between the in-pore and ex-pore environments is found to take place with a rate constant of 49 Hz, while for the lower concentration a higher rate constant of 91 Hz occurs. We note that the fits do not exactly reproduce the experimental data at long t_m values. This is likely to be due to the differences in longitudinal relaxation times between the in- and ex-pore environments,[38] and the fact that in reality the dynamic processes in these systems are taking place across a range of time- and length-scales. In-pore anions located near the edge of the microporous carbon particles will exchange with the ex-pore environment on a faster timescale than anions located near the centre of the carbon particles, which first need to diffuse to the edge. Since the NMR spectra do not distinguish ions near the edges and centres of the carbon particles, the observed exchange effects will be the superposition of the intraparticle diffusion process and the in-pore–ex-pore exchange process. However, the measured rate constants do provide an indication of the timescale of the in-pore–ex-pore exchange process in each system and show that a faster exchange takes place in the lower concentration electrolyte.

^{19}F *in situ* NMR of supercapacitor electrodes under applied potentials

To gain insight into the changes in the populations and local environments of the in-pore anions during charging, ^{19}F *in situ* NMR experiments were carried out on supercapacitor cells held at fixed voltages between −1.5 and 1.5 V. For each experiment, the cell was first held at 0 V, and the applied cell voltage was then increased in steps of 0.25 V to a maximum value of 1.5 V. The cell was then discharged to 0 V in a single step, before the applied voltage was changed in steps of −0.25 V to a maximum value of −1.5 V. At each voltage, a ^{19}F NMR spectrum was recorded after the cell current had fully relaxed. Gravimetric capacitances for each cell as determined from the integration of the total discharge current for the 1.5 → 0 V step are summarised in Table 1.

Fig. 3a and b show ^{19}F *in situ* NMR spectra recorded for a supercapacitor cell containing 1.5 M NEt_4–BF_4/ACN electrolyte. At 0 V, the in-pore resonance is observed at −155.0 ppm together with an intense feature centred around −148 ppm, which correspond to the ex-pore and free electrolyte anions. As the applied voltage is increased, the in-pore resonance moves to a higher frequency and appears to increase in intensity. The ex-pore/free electrolyte feature does not

Table 1 Gravimetric capacitances determined from the 1.5 → 0 V discharge for each system studied in this work

Electrolyte	Gravimetric capacitance/F g^{-1}
1.5 M NEt_4–BF_4/ACN	101.3
0.5 M NEt_4–BF_4/ACN	97.7
1.5 M NBu_4–BF_4/ACN	97.2
1.5 M Li-TFSI/ACN	95.1
1.5 M Na-TFSI/ACN	92.7

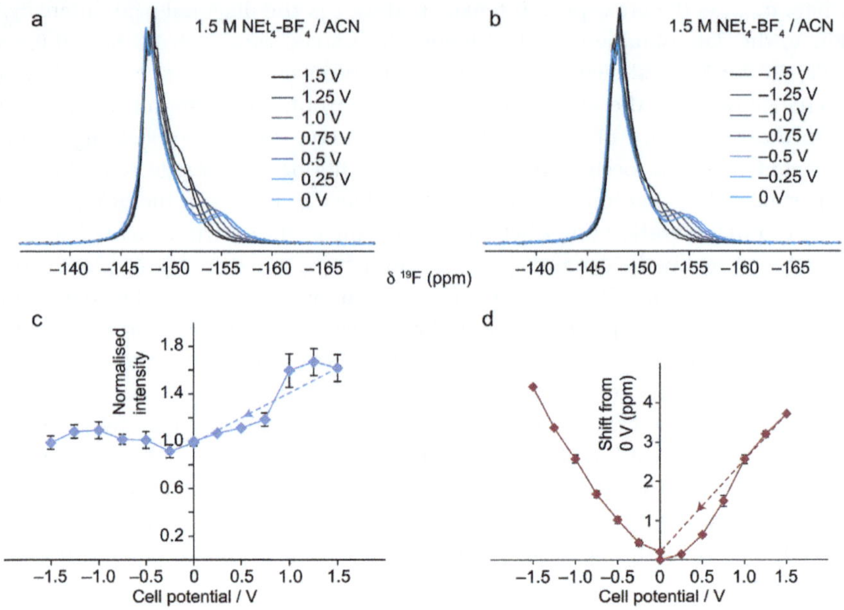

Fig. 3 ^{19}F *in situ* NMR spectra of the supercapacitor electrode containing 1.5 M NEt_4–BF_4/ACN electrolyte charged to potentials between (a) 0 and 1.5 V and (b) 0 and −1.5 V. Deconvoluted in-pore resonance intensities and shift differences are shown in (c) and (d), respectively. Lines are added as guides for the eye. Dotted arrows indicate the 1.5 V → 0 V discharge step.

display significant changes upon charging. After discharging to 0 V, the in-pore resonance returns to close to its original position at −154.8 ppm. As the applied voltage is then increased towards −1.5 V, the in-pore resonance again shifts to a high frequency; however, the changes in intensity are much less pronounced than for the positive charging regime. These observations are consistent with results reported previously for a similar system.[45] It was shown that changes in the intensity of the in-pore resonance correspond to changes in the in-pore anion population as anions are adsorbed or expelled from the electrode during charging. Changes in the frequency of the in-pore resonance arise due to changes in the electronic structure of the carbon as electrons are added to or removed from the electrode during charging. Interestingly, the in-pore resonance shifts in the same direction during charging and discharging. This is consistent with DFT calculations of NICS values for positively- and negatively-charged model carbon fragments.[45] Deconvoluted intensities and shifts for the in-pore resonance, shown in Fig. 3c and d, give a more precise picture of the changes in the in-pore anion population and local environment. In the positive charging regime the intensity of the in-pore resonance increases throughout the voltage range as anions are adsorbed inside the pores, although a 'jump' in intensity is observed at around 1 V. In contrast, the total charge stored in the supercapacitor, as determined from integration of the current relaxation, was found to vary approximately linearly with voltage step (see the ESI†). This was also observed previously,[45] and suggests a transition between two different charge storage processes. At a low voltage, the relatively small changes seen in the in-pore anion population indicate that charge is being stored at least in part through the expulsion of cations from the pores (which are not observed in the ^{19}F NMR spectra). At a high voltage, the abrupt increase in the in-pore anion resonance shows that charge storage is increasingly being achieved through the adsorption of anions. At 1.5 V, the in-pore anion population is approximately 1.6 times that at 0 V. When the capacitor is discharged to 0 V, the in-pore resonance returns to its original intensity showing that the additional adsorbed anions are released from the pores as the system returns to equilibrium. A small hysteresis in the shift of the in-pore resonance is observed, indicating that the charging process may result in small irreversible changes in the electrode (such as oxidation or reduction of the functional groups, or reactions involving electrolyte species). However, we note that the total charge and discharge currents are almost identical (Fig. S2†), showing that the extent of any irreversible processes must be small. Across the negative voltage range, very little overall change in the intensity of the adsorbed resonance is observed. However, the integration of the charging current (Fig. S2†) and the observed change in shift of the in-pore resonance confirm that a similar amount of charge is stored in this voltage range. This indicates that charge storage must take place through the adsorption of cation into the micropores, while the number of anions inside the pores remains relatively constant.

Fig. 4a and b show the ^{19}F *in situ* NMR spectra recorded for the analogous supercapacitor cell containing 0.5 M NEt_4–BF_4/ACN electrolyte. For this system, the in-pore resonance is again observed to change intensity and shift to a high frequency during charging. However, the deconvoluted resonance intensities shown in Fig. 4c reveal that the relative intensity changes in the positive voltage range are much more pronounced than for the higher concentration electrolyte. Between 0 and 1.5 V, the intensity of the in-pore resonance increases by a factor of approximately 3.2. At the same time, the total change in resonance frequency of 3.6 ppm over this range (Fig. 4d), and the gravimetric capacitance (Table 1), are similar to that observed for the higher concentration electrolyte, showing that the total amount of charge stored is not significantly affected. If it is assumed, on the basis of the NMR adsorption experiments, that the packing density of the ions inside the electrode for the 0.5 M electrolyte at 0 V is 0.55 times that of the 1.5 M electrolyte, then the relative increase factors of 3.2 (0.5 M electrolyte) and 1.6 (1.5 M electrolyte) indicate that approximately twice as many anions are adsorbed in the low concentration electrolyte as compared to the high concentration electrolyte. If at 0 V equal adsorption of anions and cations is assumed, the higher number of cations present at 0 V in the 1.5 M electrolyte means that some proportion must be expelled from the pores in order to build up a similar total ionic charge to the 0.5 M electrolyte. This suggests that for positive polarization of the electrode, anion adsorption plays a more important role in the charge storage process at low concentrations, while ion exchange takes place at higher concentrations. When the supercapacitor is discharged to 0 V, the intensity of the in-pore resonance remains slightly higher than for the pristine state and a small hysteresis in the shift is observed, again indicative of small irreversible changes in the electrode. As the cell is charged to −1.5 V, very little change is observed in the in-pore anion intensity, while the resonance frequency again changes by 3.7 ppm. This suggests that, as is the case for the higher concentration electrolyte, the

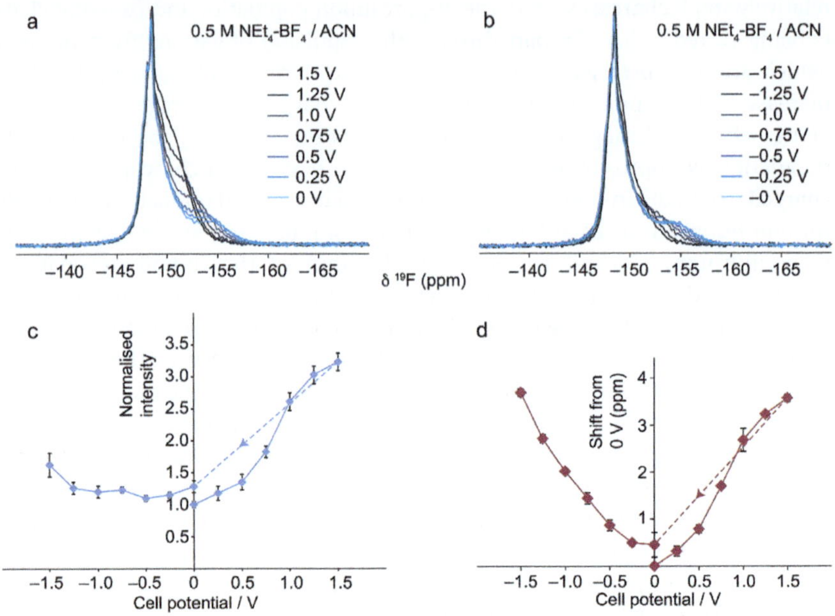

Fig. 4 ^{19}F *in situ* NMR spectra of a supercapacitor electrode containing 0.5 M NEt_4–BF_4/ACN electrolyte charged to potentials between (a) 0 and 1.5 V and (b) 0 and −1.5 V. Deconvoluted in-pore resonance intensities and shift differences are shown in (c) and (d), respectively. Lines are added as guides for the eye. Dotted arrows indicate the 1.5 V → 0 V discharge step.

charging storage mechanism in the negative voltage range largely involves cation adsorption only.

Fig. 5 shows the ^{19}F *in situ* NMR spectra and deconvoluted in-pore resonance intensities and shifts for the supercapacitor cell containing 1.5 M NBu_4–BF_4/ACN electrolyte. In this system, the same BF_4 anions are present but the NBu_4 cations are significantly larger, with an estimated desolvated diameter of 1.1 nm (based on an X-ray crystal structure for tetrabutylammonium hydrogen sulphate[53]), as compared to 0.67 nm for NEt_4^+.[9] Across the positive voltage range, a steady increase in the in-pore anion intensity is observed. The gravimetric capacitance (Table 1) and the total shift change of 3.1 ppm (Fig. 5d) are very similar to the NEt_4–BF_4 electrolytes. However, the factor of 1.3 increase in the in-pore anion population is slightly less than was observed for the 1.5 M NEt_4BF_4 electrolyte, suggesting that the degree of cation expulsion is slightly higher in this system. After discharging to 0 V, there is a noticeable reduction in the intensity of the in-pore anion resonance as compared to the pristine state, and a small hysteresis in the shift. However, the integrated charge and discharge currents over the positive charging range are again almost identical (Fig. S2†), showing that the charging process is largely reversible. It is possible that the bulky cations have difficulty re-entering the pores after charging, and so slightly more anions are expelled from the pores to maintain local charge neutrality. As the capacitor is charged to −1.5 V, a considerable reduction in the in-pore resonance intensity is observed, showing that for this electrolyte the anions are expelled from the electrode under

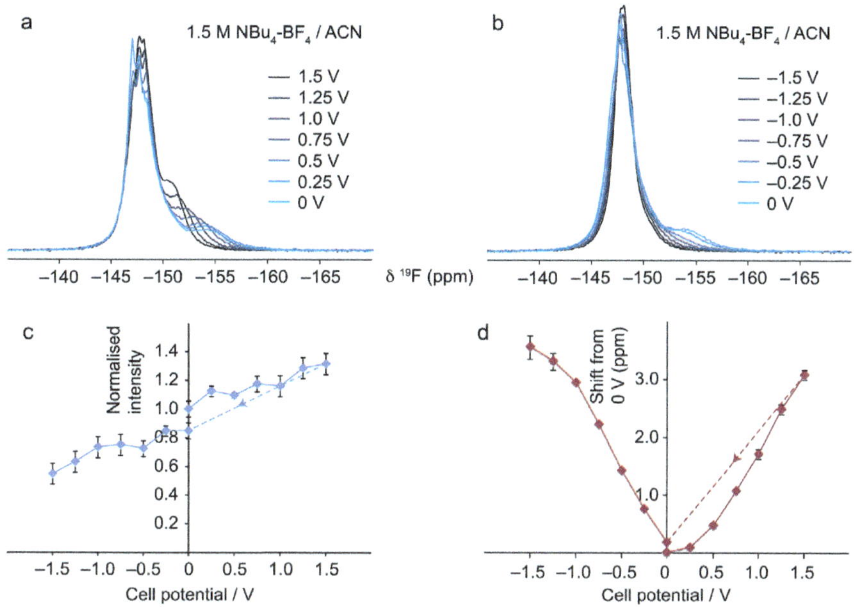

Fig. 5 ^{19}F *in situ* NMR spectra of a supercapacitor electrode containing 1.5 M NBu_4–BF_4/ACN electrolyte charged to potentials between (a) 0 and 1.5 V and (b) 0 and −1.5 V. Deconvoluted in-pore resonance intensities and shift differences are shown in (c) and (d), respectively. Lines are added as guides for the eye. Dotted arrows indicate the 1.5 V → 0 V discharge step.

negative polarisation. The total reduction in the in-pore anion population across this range is by a factor of approximately 0.5. Integration of the total current shows that a very similar amount of charge is stored to the positive charging range (Fig. S2†). The reduction in the in-pore anion population could reflect the fact that the large cations are unable to access the pores, or have reduced mobility inside a significant proportion of the pores, and so expulsion of the smaller and more mobile anions provides a larger contribution to the charge storage process.

To further investigate the factors affecting the charge storage process, experiments were carried out on cells containing Li-TFSI and Na-TFSI electrolytes. In these systems, the anions are significantly larger than the cations, with the longest dimension of the TFSI anion being equal to 0.79 nm, whereas the Li^+ and Na^+ cations have diameters of 0.09 and 0.12 nm, respectively. However, the cations are expected to be strongly solvated with resulting estimated solvated diameters of 0.65 and 1.06 nm (assuming tetrahedral coordination of Li^+ to four ACN molecules, and octahedral coordination of Na^+ to six ACN molecules[54]).

The ^{19}F *in situ* NMR spectra and deconvoluted in-pore resonance intensities and shifts for electrodes containing 1.5 M Li-TFSI and Na-TFSI/ACN electrolytes are shown in Fig. 6. At 0 V, the resonances corresponding to the in-pore TFSI anions are observed at −85.1 and −84.4 ppm for the Li- and Na-based electrolytes, respectively. As the cells are charged, characteristic changes in the position and intensity of the in-pore resonance are observed, while the ex-pore/free electrolyte feature remains relatively unchanged. For both systems, an increase in the in-pore

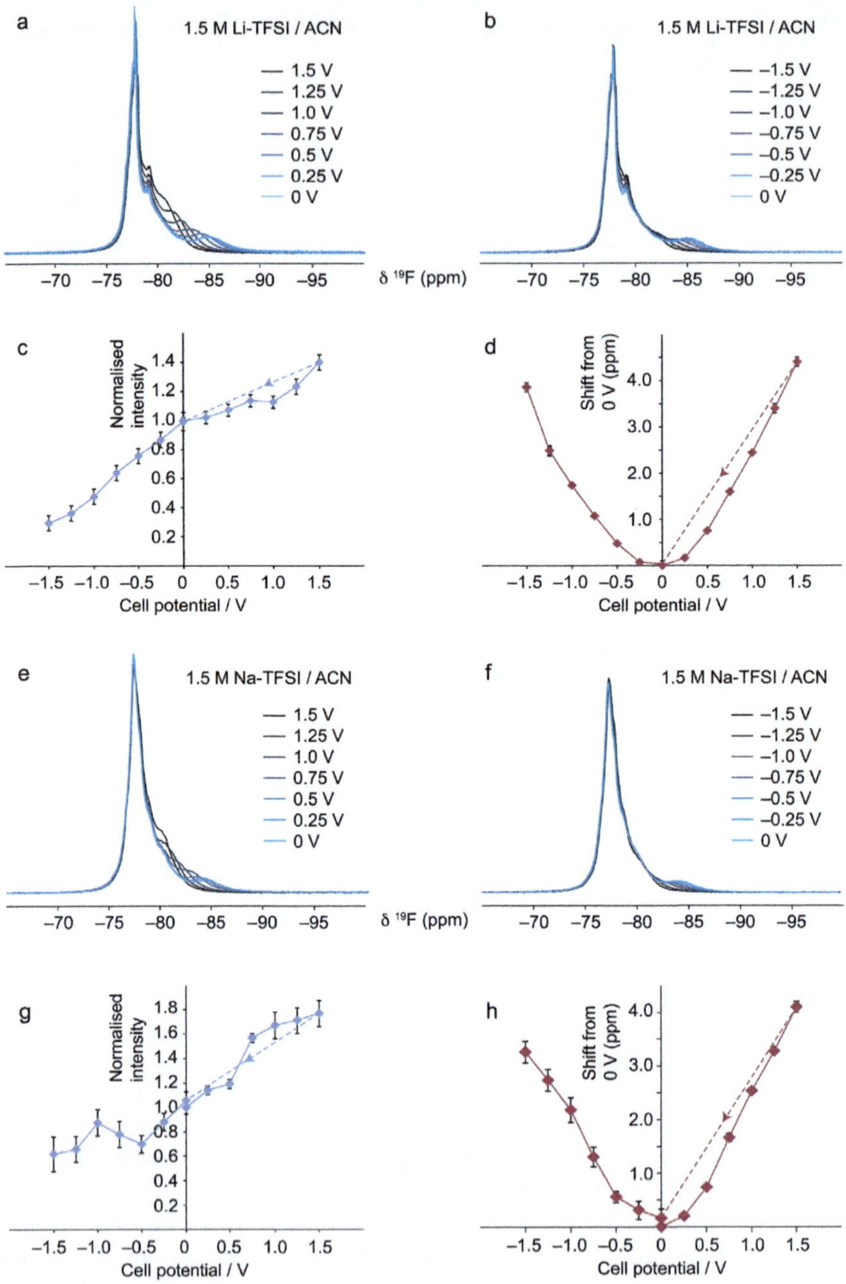

Fig. 6 ^{19}F *in situ* NMR spectra of supercapacitor electrodes containing (a) and (b) 1.5 M Li-TFSI/ACN and (e) and (f) 1.5 M Na-TFSI/ACN electrolytes charged to potentials between (a) and (e) 0 and 1.5 V and (b) and (f) 0 and −1.5 V. Deconvoluted in-pore resonance intensities and shift differences are shown for the LiTFSI electrolyte in (c) and (d), and for the Na-TFSI electrolyte in (g) and (h). Lines are added as guides for the eye. Dotted arrows indicate the 1.5 V → 0 V discharge step.

anion population is observed upon charging to 1.5 V. For Na-TFSI an abrupt increase is observed at 0.75 V, perhaps indicating a transition between two different charge storage processes as was observed for the 1.5 M NEt_4–BF_4/ACN electrolyte. At 1.5 V, the in-pore anion populations are increased by factors of 1.4 (Li-TFSI) and 1.8 (Na-TFSI) compared to at 0 V. The capacitances (Table 1) are very similar to the organic electrolytes, showing that a similar amount of charge is stored in both systems. Interestingly, the total shift change for the in-pore resonances over the positive voltage range is slightly larger for both of the TFSI-based electrolytes (4.1–4.4 ppm). This may indicate a chemical contribution to the change in NICS upon charging; it is possible that the anisotropic geometry of the TFSI anions could result in a different arrangement or proximity to the pore walls. Upon discharging to 0 V, the in-pore resonance intensity and shift returns to close to its original value for both systems. Over the negative voltage range, significant reductions in the in-pore resonance intensities are observed for both systems as anions are expelled from the pores. For the Li-TFSI electrolyte, a marked reduction is observed, and the in-pore anion population at −1.5 V is reduced to 0.3 times that at 0 V. For Na-TFSI, the reduction is less pronounced but still significant, with the anion population at −1.5 V being equal to 0.6 times the 0 V population, although the errors in this measurement are larger due to the low intensity of the in-pore resonance as compared to the ex-pore/free electrolyte feature. Negative polarization of the electrode in these systems clearly results in significant anion expulsion from the pores, in contrast to what was observed for the organic electrolytes. This indicates that there are additional factors that determine which ions undergo greater movement during charging, other than concentration and relative size effects.

Discussion

The results presented in this study offer a rare insight into one of the key questions regarding the mechanism of supercapacitance: what is the nature of the *ionic* charge that is developed within the micropores to balance the *electronic* charge developed on the electrode surface? The ^{19}F adsorption and 2D-EXSY measurements show that anions exist in equilibrium between the in-pore and ex-pore environments at zero applied potential. Assuming there is an equal number of in-pore anions and cations in order to maintain local charge neutrality at zero applied potential, this leads to a number of possible mechanisms for developing ionic charge within the micropores as the supercapacitor is charged, as shown schematically in Fig. 7. One possible mechanism is simply to expel co-ions, thereby leaving an excess of counter ions within the pores at a reduced overall packing density (Fig. 7a). Alternatively, it may be possible to adsorb counter-ions from outside the pores (Fig. 7b), thereby giving rise to a net ionic charge at an increased overall packing density, while the number of co-ions inside the pores remains unchanged. Finally, a third possible mechanism is ion exchange (Fig. 7c), whereby the combination of co-ion expulsion and counter-ion adsorption can give rise to a net ionic charge inside the pores, while the total ion population remains constant.

For all of the systems studied in this work, gravimetric capacitances of between 90–100 F g^{-1} were determined at 1.5 V, showing that the total amount of charge stored is very similar in each case. However, the *in situ* NMR results reveal that the

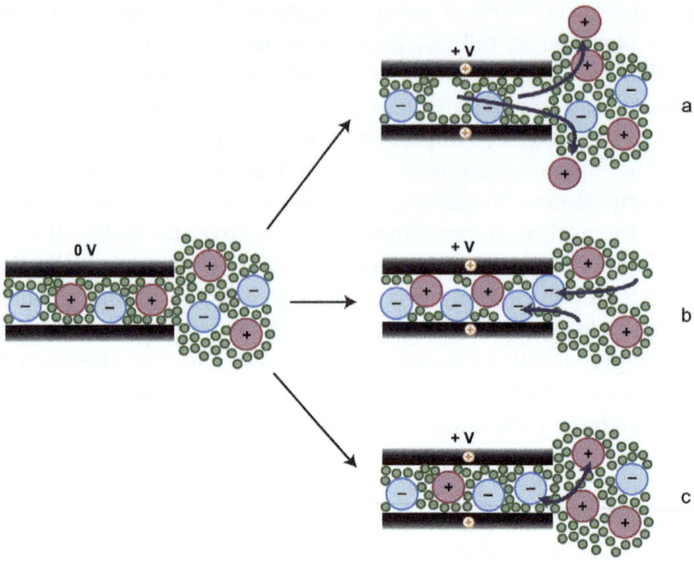

Fig. 7 Schematic illustrations of possible charge storage mechanisms for micropores that are wetted at zero potential. If the electrode surface is charged to a positive potential, an equal negative ionic charge can arise through either (a) the expulsion of cations from the pores, (b) adsorption of ex-pore anions into the pores, or (c) exchange of ex-pore anions for in-pore cations.

way in which the ionic charge is developed inside the micropores in the electrode as it charges differs depending on the properties of the electrolyte. The results obtained for the NEt_4–BF_4 electrolyte show that the ion concentration can significantly affect the way in which charge is stored when the electrode is positively polarized. The charge storage mechanism for the low concentration electrolyte seems to be dominated by anion adsorption, whereas for the high concentration electrolyte ion exchange takes place to a greater extent. In contrast, when the electrode is negatively polarized, the anion population was not found to change significantly for both concentrations, indicating that cation adsorption must be the main mechanism for charge storage. This cation adsorption-dominated mechanism in the negative charging regime will lead to an increased packing density within the pores. For the 1.5 M electrolyte, the NMR adsorption measurements show that solvated ions are already densely packed inside the pores at zero potential, taking up a volume that is comparable to the total pore volume. The adsorption of further ions into the pores upon charging will take up further space inside the pores. It therefore appears that desolvation must take place to some extent in order to accommodate the extra ions inside the pores.

The results obtained for the NBu_4–BF_4 electrolyte show that relative ion size also plays a role in determining the charge storage mechanism. In this case, the significantly larger NBu_4 cations seem to have difficulty accessing the smaller pores, forcing the expulsion of anions in the negative voltage range in order to balance the charge developed on the electrode surface. In the positive voltage range, the fact that large NBu_4 cations do not access the smaller, more confined environments within the electrode may also mean that they are more easily expelled from the larger spaces that they occupy, with the result that the in-pore anion population has to increase by a smaller fraction.

The ^{19}F *in situ* NMR experiments on the Li- and Na-TFSI electrolytes indicate that other factors may also govern the charge storage process. In these systems, significant movement of the TFSI anions was observed in the positive and negative charging regimes. This is not consistent with the ion-size effect observed for the NEt_4–BF_4 and NBu_4–BF_4 systems, since the solvated anion and cation sizes are small enough in each case that they should both be able to access a large proportion of the pores. It is possible that other factors such as anion–cation association may also alter the charge storage mechanism. In the TFSI electrolytes, the high charge density of the cations may have caused them to 'carry' strongly associated anions out of the pores as they are expelled during negative polarization of the electrode. How significant this effect is, and whether there are other factors that determine which ionic species are adsorbed or expelled from the pores, requires further investigation through the study of a wider range of electrolyte systems. However, the results presented in this work show that the nature of the charge storage process in supercapacitor electrodes is not straightforward and does not arise solely due to the adsorption of counter-ions at the electrode surface as is often assumed.

Conclusions

^{19}F NMR spectroscopy has been used to quantify anions in model supercapacitor electrodes, and to selectively observe changes in anion populations and local environments during charging. The results presented in this work provide a rare insight into the behaviour of electrolyte ions under such conditions. In particular, we are able to draw the following conclusions:

(1) For the electrolytes studied in this work, the observation of resonances corresponding to in-pore and ex-pore anions shows that they exist in equilibrium between these two environments in the electrode at zero applied potential. This is confirmed by ^{19}F 2D-EXSY experiments, which show that the exchange takes place between the in- and ex-pore environments on the timescale of tens of Hz.

(2) For the commonly-used NEt_4–BF_4 electrolyte, the charge storage mechanism for the negative polarization of a YP-50F electrode involves cation adsorption and no significant change in the in-pore anion population. The charge storage mechanism for positive polarization of the electrode involves a combination of anion adsorption and ion exchange, with anion adsorption favoured at low concentrations.

(3) The charge storage mechanism is altered by increasing the size of the cation, which forces the expulsion of anions upon negative polarization of the electrode.

In addition to concentration and ion size effects on the charging mechanism observed for the tetra-alkylammonium tetrafluoroborate electrolytes, experiments on the Li-TFSI and Na-TFSI electrolytes indicate that there are other factors governing the charging mechanism, which are not yet fully understood. Comparison of the behaviour of a wider range of electrolytes will enable a better understanding of the different factors governing the extent to which ions are adsorbed and expelled from the electrode under charging. Furthermore, the development of *in situ* NMR methods that can observe both anions and cations, and quantify them in absolute terms, will enable a much more detailed picture of the charging mechanism to be obtained in each system. Indeed, this work is ongoing in our laboratory and will be reported elsewhere.

Acknowledgements

A.C.F., J.M.G., H.W., and C.P.G. acknowledge the Sims Scholarship (A.C.F.), EPSRC (*via* the Supergen consortium, J.M.G. and H.W.), and the EU ERC (*via* an Advanced Fellowship to C.P.G.) for funding. N.M.T. was supported by the Northeastern Center for Chemical Energy Storage and by Energy Frontier Research Center funded by the U.S. DOE, BES under Award no. DE-SC0001294. P.S. acknowledges the support from the European Research Council (ERC, Advanced Grant, ERC-2011-AdG, Project 291543-IONACES). A.C.F. and J.M.G. thank the NanoDTC Cambridge for travel funding. We thank Dr. Paul Bayley for the synthesis of the TFSI electrolyte salts.

References

1 B. E. Conway, *Electrochemical Supercapacitors: Scientific Fundamentals and Technological Applications*, Springer, New York, 1st edn, 1999.
2 J. R. Miller and A. F. Burke, *Electrochem. Soc. Interface*, 2008, **17**, 53.
3 P. Simon and Y. Gogotsi, *Nat. Mater.*, 2008, **7**, 845.
4 M. T. Alam, J. Masud, M. M. Islam, T. Okajima and T. Ohsaka, *J. Phys. Chem. C*, 2011, **115**, 19797.
5 W. J. Cho, C. G. Yeom, B. C. Kim, K. M. Kim, J. M. Ko and K. H. Yu, *Electrochim. Acta*, 2013, **89**, 807.
6 S. Li, K. L. Van Aken, J. K. McDonough, G. Feng, Y. Gogotsi and P. T. Cummings, *J. Phys. Chem. C*, 2014, **118**, 3901.
7 Y. Zhai, Y. Dou, D. Zhao, P. F. Fulvio, R. T. Mayes and S. Dai, *Adv. Mater.*, 2011, **23**, 4828.
8 J. Chmiola, G. Yushin, Y. Gogotsi, C. Portet, P. Simon and P.-L. Taberna, *Science*, 2006, **313**, 1760.
9 J. Chmiola, C. Largeot, P.-L. Taberna, P. Simon and Y. Gogotsi, *Angew. Chem., Int. Ed.*, 2008, **47**, 3392.
10 J. Huang, B. G. Sumpter and V. Meunier, *Chem. - Eur. J.*, 2008, **14**, 6614.
11 S. Kondrat and A. A. Kornyshev, *J. Phys. Chem. C*, 2013, **117**, 12399.
12 S. Kondrat, P. Wu, R. Qiao and A. A. Kornyshev, *Nat. Mater.*, 2014, **13**, 387.
13 C. Merlet, B. Rotenburg, P. A. Madden and M. Salanne, *Phys. Chem. Chem. Phys.*, 2013, **15**, 15781.
14 R. Burt, G. Birkett and X. S. Shao, *Phys. Chem. Chem. Phys.*, 2014, **16**, 6519.
15 M. V. Fedorov and A. A. Kornyshev, *Chem. Rev.*, 2014, **114**, 2978.
16 G. Feng and P. T. Cummings, *J. Phys. Chem. Lett.*, 2011, **2**, 2859.
17 D. Jiang, Z. Jin and J. Wu, *Nano Lett.*, 2011, **11**, 5373.
18 G. Feng, R. Qiao, J. Huang, S. Dai, B. G. Sumpter and V. Meunier, *Phys. Chem. Chem. Phys.*, 2011, **13**, 1152.
19 D. Henderson, *J. Colloid Interface Sci.*, 2012, **374**, 345.
20 C. Merlet, B. Rotenberg, P. A. Madden, P.-L. Taberna, P. Simon, Y. Gogotsi and M. Salanne, *Nat. Mater.*, 2012, **11**, 306.

21 K. Sharma, H. Z. Bilheux, L. M. H. Walker, S. Voisin, R. T. Mayes, J. O. Kiggans, S. Yiacoumi, D. W. DePaoli, S. Dai and C. Tsouris, *Phys. Chem. Chem. Phys.*, 2013, **15**, 11740.
22 S. Boukhalfa, L. He, Y. B. Melnichenko and G. Yushin, *Angew. Chem., Int. Ed.*, 2013, **52**, 1.
23 F. W. Richey, B. Dyatkin, Y. Gogotsi and Y. A. Elbad, *J. Am. Chem. Soc.*, 2013, **135**, 12818.
24 P. Ruch, R. Kötz and A. Wokaun, *Electrochim. Acta*, 2009, **54**, 4451.
25 M. D. Levi, G. Salitra, N. Levy, D. Aurbach and J. Maier, *Nat. Mater.*, 2009, **8**, 872.
26 S. Sigalov, M. D. Levi, G. Salitra, D. Aurbach and J. Maier, *Electrochem. Commun.*, 2010, **12**, 1718.
27 M. D. Levi, S. Sigalov, D. Aurbach and L. Daikhin, *J. Phys. Chem. C*, 2013, **117**, 14876.
28 W.-Y. Tsai, P.-L. Taberna and P. Simon, *J. Am. Chem. Soc.*, 2014, **136**, 8722.
29 R. K. Harris, T. V. Thompson, P. R. Norman, C. Pottage and A. N. Tretheway, *J. Chem. Soc., Faraday Trans.*, 1995, **91**, 1795.
30 R. K. Harris, T. V. Thompson, P. R. Norman and C. Pottage, *J. Chem. Soc., Faraday Trans.*, 1996, **92**, 2615.
31 R. K. Harris, T. V. Thompson, P. R. Norman and C. Pottage, *Carbon*, 1999, **37**, 1425.
32 K. Shen and T. Pietrass, *J. Phys. Chem. B*, 2004, **108**, 9937.
33 W. Sekhaneh, M. Kotecha, U. Dettlaff-Weglikowska and W. S. Veeman, *Chem. Phys. Lett.*, 2006, **428**, 143.
34 Q. Chen, J. L. Herberg, G. Mogilevsky, H.-J. Wang, M. Stadermann, J. K. Holt and Y. Wu, *Nano Lett.*, 2008, **8**, 1902.
35 R. J. Anderson, T. P. McNicholas, A. Kleinhammes, A. Wang, J. Liu and Y. Wu, *J. Am. Chem. Soc.*, 2010, **132**, 8618.
36 S. Mitani, C. Murakami, Y. Korai, Y. Minato, S. Ishimoto, S. Suematsu and K. Tamamitsu, *Electrochim. Acta*, 2013, **94**, 30.
37 L. Borchardt, M. Oschatz, S. Paasch, S. Kaskel and E. Brunner, *Phys. Chem. Chem. Phys.*, 2013, **15**, 15177.
38 H. Wang, T. K.-J. Köster, N. M. Trease, J. Ségalini, P.-L. Taberna, P. Simon, Y. Gogotsi and C. P. Grey, *J. Am. Chem. Soc.*, 2011, **133**, 19270.
39 A. C. Forse, J. M. Griffin, H. Wang, N. M. Trease, V. Presser, Y. Gogotsi, P. Simon and C. P. Grey, *Phys. Chem. Chem. Phys.*, 2013, **15**, 7722.
40 D. Sebastiani, *ChemPhysChem*, 2006, **7**, 164.
41 M. Kibalchenko, M. C. Payne and J. R. Yates, *ACS Nano*, 2011, **5**, 537.
42 P. Ren, A. Zheng, X. Pan, X. Han and X. Bao, *J. Phys. Chem. C*, 2013, **117**, 23418.
43 A. C. Forse, J. M Griffin, V. Presser, Y. Gogotsi and C. P. Grey, *J. Phys. Chem. C*, 2014, **118**, 7508.
44 M. Deschamps, E. Gilbert, P. Azais, E. Raymundo-Piñero, M. R. Ammar, P. Simon, D. Massiot and F. Béguin, *Nat. Mater.*, 2013, **12**, 351.
45 H. Wang, A. C. Forse, J. M. Griffin, N. M. Trease, L. Trognko, P.-L. Taberna, P. Simon and C. P. Grey, *J. Am. Chem. Soc.*, 2013, **135**, 18968.
46 G. Laudisio, R. K. Dash, J. P. Singer, G. Yushin, Y. Gogotsi and J. E. Fischer, *Langmuir*, 2006, **22**, 8945.
47 D. G. Cory and W. M. Ritchey, *J. Magn. Reson.*, 1988, **80**, 128.

48 D. Massiot, F. Fayon, M. Capron, I. King, S. Le Calvé, B. Alonso, J.-O. Durand, B. Bujoli, Z. Gan and G. Hoatson, *Magn. Reson. Chem.*, 2002, **40**, 70.
49 Information available at http://www.kuraraychemical.com/products/sc/SCcarbon.htm.
50 Y. Gogotsi, P. Simon, C. Largeot, C. Portet, J. Chmiola and P.-L. Taberna, *US Pat.*, US0182000A1, 2011.
51 M. Fukano, T. Fujimori, J. Ségalini, E. Iwama, P.-L. Taberna, T. Iiyama, T. Ohba, H. Kanoh, Y. Gogotsi, P. Simon and K. Kaneko, *J. Phys. Chem. C*, 2013, **117**, 5752.
52 M. H. Levitt, *Spin Dynamics*, Wiley, Chichester, 1st edn, 2001.
53 M. E. Light, P. A. Gale and M. B. Hursthouse, *Acta Crystallogr.*, 2001, **E57**, o705.
54 D. Spångberg and K. Hermansson, *Chem. Phys.*, 2004, **300**, 165.

Faraday Discussions

PAPER

Effect of the alkali insertion ion on the electrochemical properties of nickel hexacyanoferrate electrodes

Hyun-Wook Lee,[a] Mauro Pasta,[a] Richard Y. Wang,[a] Riccardo Ruffo[b] and Yi Cui*[ac]

Received 8th July 2014, Accepted 13th August 2014
DOI: 10.1039/c4fd00147h

Nickel hexacyanoferrate (NiHCFe) is an attractive cathode material in both aqueous and organic electrolytes due to a low-cost synthesis using earth-abundant precursors and also due to its open framework, Prussian blue-like crystal structure that enables ultra-long cycle life, high energy efficiency, and high power capability. Herein, we explored the effect of different alkali ions on the insertion electrochemistry of NiHCFe in aqueous and propylene carbonate-based electrolytes. The large channel diameter of the structure offers fast solid-state diffusion of Li^+, Na^+, and K^+ ions in aqueous electrolytes. However, all alkali ions in organic electrolytes and Rb^+ and Cs^+ in aqueous electrolytes show a quasi-reversible electrochemical behavior that results in poor galvanostatic cycling performance. Kinetic regimes in aqueous electrolyte were also determined, highlighting the effect of the size of the alkali ion on the electrochemical properties.

Introduction

The demand for advanced energy storage devices has greatly increased recently, and current research efforts focus on solving scientific issues associated with this formidable challenge.[1,2] Depending on the targeted application, researchers in this area look for improvements in power output[3,4] and energy density[5,6] for portable applications and for cycle/calendar life and cost for grid-scale stationary energy storage systems (GSES).[7–9] Li-ion batteries (LIBs) based on conventional organic electrolytes have been widely used and studied for portable applications due to their high energy density (deriving from the high operating voltage and low electrochemical equivalent weight of lithium) and reasonable durability, which allow them to meet current market requirements.[10] Nevertheless, the constant

[a]*Department of Materials Science and Engineering, Stanford University, Stanford, California 94305, USA. E-mail: yicui@stanford.edu*

[b]*Dipartimento di Scienza dei Materiali, Università di Milano Bicocca, via Cozzi 55 20125, Milan, Italy*

[c]*Stanford Institute for Materials and Energy Science, SLAC National Accelerator Laboratory, Menlo Park, California 94025, USA*

increase in the demand for Li raw materials and their limited availability could lead to a considerable rise in the cost of LIBs in the near future.[11] In this regard, Na or K-ion based secondary batteries are often proposed as promising alternatives due to the ready and lower-cost availability of the raw materials.[12]

In the case of GSES applications, the key performance metrics are cost, environmental compatibility, cycle life, energy efficiency, and response time rather than energy density. In fact, the $ per kW h cost of the stored energy should be comparable to the price of the energy delivered by the primary source. Therefore, long calendar life and high energy efficiencies are required to reduce their lifetime cost.[13] Aqueous rechargeable batteries can meet these requirements, and previous research efforts in this area have been very promising.[14–16]

Open-framework Prussian blue analogues (PBAs) have recently received a great deal of attention as alkali-ion rechargeable battery electrodes due to their impressive electrochemical cyclability, their straightforward and low-cost synthetic processes, and the possibility of using them in both aqueous and organic electrolytes.[17–19] This class of materials contains large interstitial sites that allow for the insertion of alkali[20] and divalent[21] ions with very little crystallographic lattice strain. However, the insertion of different cations results in very different electrochemical behavior in both aqueous and organic electrolytes.[22] Hence, a deeper understanding of the effect of the nature of the inserting ion in aqueous and organic electrolytes on the electrochemical features of PBAs is required to achieve improved battery performance.

Herein, we investigated the electrochemical behavior of nickel hexacyanoferrate (NiHCFe) in different aqueous and organic electrolytes with particular emphasis on the role of the guest cation. Among different PBAs, we chose NiHCFe as a model system because of its chemical stability in both water and propylene carbonate and because of its redox behavior potential well within the electrochemical stability window of the investigated electrolytes.

Materials and methods

NiHCFe nanoparticles were prepared by a simple co-precipitation method[22,23] in which 120 mL of a 40 mM aqueous solution of $Ni(NO_3)_2$ and 120 mL of a 20 mM solution of $K_3Fe(CN)_6$ were simultaneously added dropwise into 60 mL of deionized water with vigorous stirring. The precipitation was carried out at a constant temperature of 70° C until a yellow–orange precipitate formed. The precipitate was then filtered, washed several times with deionized water, and dried under vacuum at 40° C for 12 h. High-resolution powder X-ray diffraction (XRD) patterns were obtained at 12 keV ($\lambda = 1.0332$ Å) at BL 2-1 at the Stanford Synchrotron Radiation Lightsource (SSRL) using zero-background quartz holders in reflection geometry with a monochromator and Soller slits. The structural analysis was performed on the XRD data with the Rietveld refinement method using the EXPGUI interface for the General Structure Analysis System (GSAS).[24–26] Morphology and particle size were investigated with transmission electron microscopy (TEM, Tecnai G^2 F20 X-Twin).

The electrochemical investigation was carried out on electrodes prepared by casting a slurry made of 80 wt% NiHCFe, 10 wt% carbon (Timcal Super P Li), and 10 wt% polyvinylidene fluoride (PVDF, Kynar HSV 900) dispersed in 1-methyl-2-pyrrolidone (NMP) on a carbon cloth (Fuel Cell Earth/Ballard AvCarb) current collector. The electrodes were then dried under vacuum at room temperature for 10 h, resulting in an active material loading of about 8–10 mg cm^{-2}. The electrodes were tested in a flooded, three-electrode electrochemical cell equipped with a commercial Ag/AgCl (4 M KCl) reference electrode and an activated carbon counter electrode. In-depth kinetics experiments were carried out in a Swagelok, three-electrode cell equipped with a pseudo-reference Ag/AgCl electrode, an activated carbon counter electrode, and an active electrode – prepared by coating the surface of a stainless steel current collector with the slurry previously described, achieving an active material loading of about 1 mg cm^{-2}. The Ag/AgCl pseudo-reference electrode was calibrated before and after each measurement in both aqueous and organic electrolyte using a commercial Ag/AgCl 3.5 M reference electrode and the Fc^+/Fc^0 redox couple,[27] respectively. All the potentials were converted relative to the standard hydrogen electrode (SHE). 1 M aqueous solutions of $LiNO_3$, $NaNO_3$, KNO_3, $RbNO_3$, $CsNO_3$ and 1 M $LiClO_4$, $NaClO_4$, and 0.8 M KPF_6 solutions in propylene carbonate were prepared and tested. The acidity of all aqueous electrolytes was increased to a pH of 2 by the addition of HNO_3 to take advantage of the lower solubility of PBAs in acidic conditions.[28]

Results and discussion

Physicochemical characterization

The as-synthesized NiHCFe has the general perovskite-type open framework crystal structure, shown in Fig. 1a, in which hexacyanoferrate complexes and nickel ions bound by cyanide linkages form a face-centered cubic structure. In the framework, guest cations can occupy large interstitial A sites around the center of each cube, as confirmed by high-resolution synchrotron powder diffraction measurements (Fig. 1b). Sharp peaks indicate the high crystallinity of the powder, and the absence of secondary peaks verifies the presence of a single phase and the absence of impurities. Rietveld refinement of the XRD data confirms that NiHCFe has a perovskite structure with a face-centered cubic phase ($fm\bar{3}m$, $a_0 = 10.211$ Å) that matches closely the crystal structure of Prussian blue (JCPDS no. 73-0687). The TEM image in Fig. 1c shows spherical particles with an average diameter of about 20–50 nm, which is beneficial for increasing the surface area (better contact with the electrolyte and the conductive additive) and reducing diffusion lengths of the charge carriers (both ions and electrons). In Fig. 1d and e, the HR-TEM image and its corresponding inverse fast Fourier transform (FFT) show an interplanar distance of about 2.076 Å, which well matches the (422) crystal plane.

Effect of insertion ion and solvent on the electrochemical properties of NiHCFe

The effect of the insertion ion on the insertion potential of NiHCFe bulk electrodes in aqueous electrolytes was investigated in a previous study by our research group.[22,23] In that study, increases in the insertion potential were observed for ions further down Group I on the periodic table. This behavior has been previously interpreted by Scholz *et al.*[29–31] They suggested that the change in Gibbs free energy due to the alkali ion insertion ΔG_i (corresponding to the insertion potential: $\Delta G_i = -nF\Delta E_i$) is equal to the difference between the Gibbs free energy of "solvation" of the alkali ion in the NiHCFe lattice (ΔG_l) and the Gibbs free energy of solvation of the alkali ion in the solvent (ΔG_s):[29]

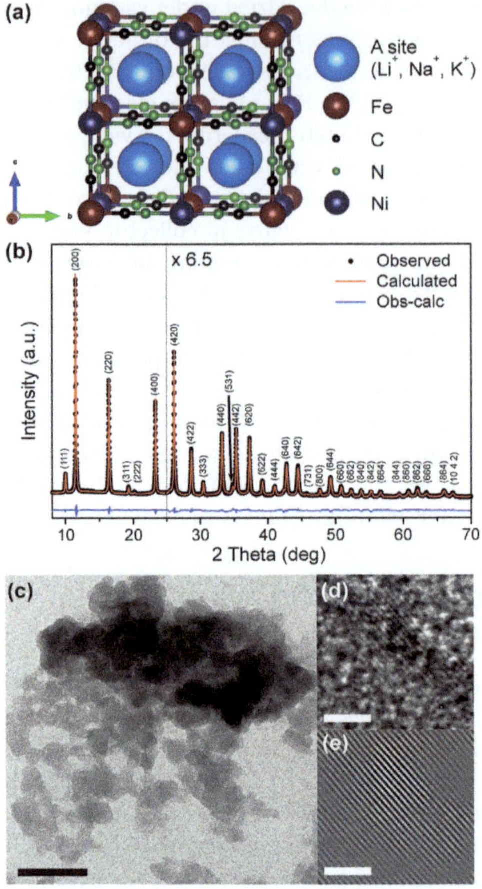

Fig. 1 (a) Schematic illustration of the unit cell of the open-framework NiHCFe structure with carbon-coordinated Fe ions and nitrogen-coordinated Ni ions. (b) Rietveld-refined XRD data of the as-synthesized NiHCFe. Black dots and red lines indicate the observed and the calculated patterns, respectively. The blue line at the bottom indicates the difference between observed and calculated patterns. The magnified patterns (6.5×) are shown for clarity above 25°. (c) Transmission electron microscope (TEM) images (black scale bar, 100 nm) and (d) high-resolution TEM images of NiHCFe nanopowders. (e) The inverse fast Fourier transform (FFT) of (d) (white scale bar, 1 nm).

$$\Delta G_{\mathrm{i}} = -nF\Delta E_{\mathrm{i}} = \Delta G_{\mathrm{l}} - \Delta G_{\mathrm{s}}$$

ΔG_{l} corresponds to the free energy of any residual hydration of the alkali ion after insertion. The magnitude of this value is smaller than the free energy of solvation in solution (ΔG_{s}) because insertion ions are mostly dehydrated as they go into the structure. Fully hydrated ions cannot fit in the interstitial sites, which are about 3.2 Å in diameter, because they are much larger in diameter.[32] Thus, the variation in insertion potential for different insertion ions is mainly a function of just the Gibbs free energy of solvation (hydration in the case of water) of the insertion ion. ΔG_{i} decreases as the ionic radius of the ion increases further down Group I of the periodic table since larger ions have lower hydration energies in solution.[33] In the present study, the same behavior is observed (Fig. 2a and Table 1).

It is worth noticing the presence of multiple electrochemical processes between 0.3 and 0.8 V *versus* SHE in the case of Li^+ at pH 6 (Fig. 2b, blue line). Since the concentration of H^+ is so low in this experiment, all the peaks can be attributed to the insertion of Li^+. The peaks are at different potentials because of the varying levels of hydration of Li^+ when in the structure. This is only observed for Li^+ because its small ionic radius allows for a significant degree of hydration even when in the structure. In other words, ΔG_l takes on varying values for Li^+ and has a more significant impact on the change in Gibbs free energy for ion insertion. The reduction peak is located at the lowest potential because the ions are almost fully dehydrated as they diffuse through the channels in the framework, but once they insert into interstitial sites, they may be partially rehydrated by mobile water molecules in the structure.[29] At pH 2 (Fig. 2b, red line), we see an additional electrochemical process developing at 0.6 V that is ascribed to the insertion of hydronium ions, as confirmed by CV cycling in an acidified lithium-free cell at pH 3 (Fig. 2b, black line). Data obtained at pH 6 (Fig. 2c) clearly show the capacity fading upon cycling caused by the higher solubility of NiHCFe.[28] The solubility of NiHCFe at pH 2 and 1 M salt solution is infinitesimal and definitely not enough to affect the measurements. In the case of K^+ and Na^+ insertion, the H_3O^+ peak is not observed because of the higher insertion potentials of these ions,

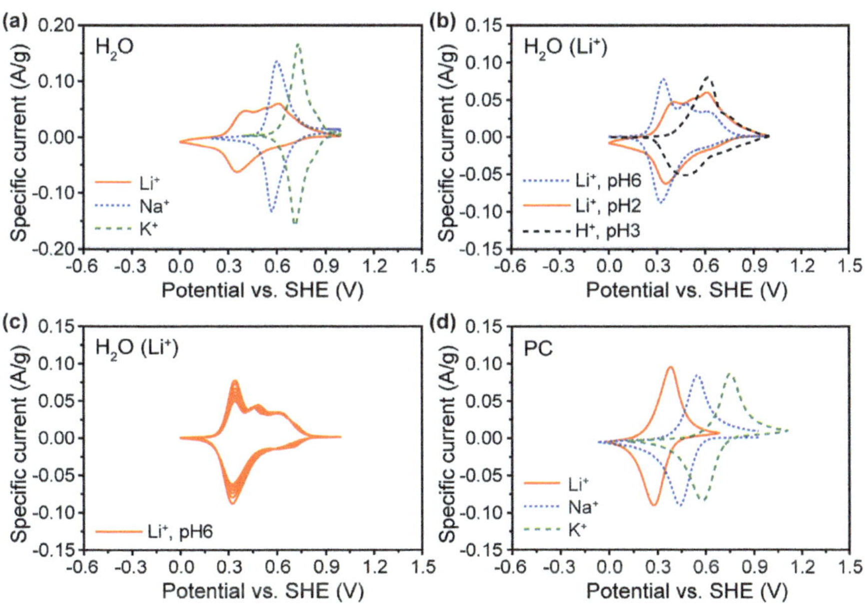

Fig. 2 Cyclic voltammetry (CV) curves of NiHCFe. The scan rate is 0.1 mV s^{-1}. (a) CV curves using aqueous electrolytes of Li^+, Na^+, and K^+ ions. (b) CV curves (in blue, red, and black) are measured by using 1 M $LiNO_3$ in H_2O (pH 6, controlled by HNO_3), 1 M $LiNO_3$ in H_2O (pH 2, controlled by HNO_3), and 1 mM HNO_3 in H_2O electrolytes, respectively. (c) 7 cycles of CV in pH 6, 1 M $LiNO_3$ in H_2O electrolyte. The oxidation and reduction peaks of NiHCFe electrode decrease with cycling. (d) Propylene carbonate (PC) is used as the electrolyte solvent. 1 M $LiClO_4$, 1 M $NaClO_4$, and 0.8 M KPF_6 are used as salts in PC.

Table 1 The redox potentials (E_{half}), the potential difference between the oxidation and reduction reactions (ΔE) recorded in cyclic voltammograms, and the specific capacities from galvanostatic cycling of NiHCFe electrodes in propylene carbonate and aqueous electrolytes[a]

	PC			Aqueous		
	E_{half}/V	ΔE/V	Specific capacity/mA h g^{-1}	E_{half}/V	ΔE/V	Specific capacity/mA h g^{-1}
Li^+	0.327	0.107	53.4	0.379	0.049	57.8
Na^+	0.492	0.111	49.6	0.583	0.032	57.4
K^+	0.665	0.17	43.7	0.721	0.02	57.7

[a] The CV data were measured at 0.1 mV s^{-1} and the all capacities in galvanostatic curves were characterized at 1 C.

which indicates that K^+ and Na^+ insertion are more thermodynamically favored processes.

In organic propylene carbonate-based electrolytes (PC), the same trend is observed (Fig. 2d). The insertion potential differences between the different cations in aqueous and organic electrolytes are similar due to the similarities in the solvation energies of the ions in water and PC.[30,34] There is only a single redox peak for each ion insertion process due to the lack of H_3O^+ insertion in PC and also due to the fact that ions cannot be resolvated in the structure because there are no mobile solvent molecules within the structure when in PC.

The cyclic voltammetry (CV) profiles suggest very different kinetics in aqueous and organic electrolyte. The CVs in aqueous conditions show symmetric anodic/cathodic peaks that demonstrate the reversibility and fast kinetics of aqueous ion insertion. In propylene carbonate, the peaks show asymmetry and a quasi-Nernstian behavior (the difference in potential between the anodic and cathodic peaks is larger than 100 mV) due to slower kinetics of ion insertion relative to the aqueous system. This behavior can be ascribed to different ion diffusion mechanisms within the structure in aqueous and organic electrolytes.[35] A Grotthus-like mechanism may enable the extremely rapid kinetics of ion insertion in water; this has previously been suggested for driving proton conductivity in hexacyanochromates.[35] Zeolitic water molecules exist throughout the structure, and ligand water molecules bond to the N-coordinated metal ions adjacent to hexacyanometallate vacancy sites. The N-coordinated metal ion acts as a Lewis acid, and protons are carried through the 3D hydrogen-bonding network, which is composed of zeolitic water and ligand water molecules. These different properties result in very different battery performance as described in the following section.

NiHCFe's battery performance comparison in aqueous and organic electrolyte

Galvanostatic charge–discharge curves are reported in Fig. 3. In all aqueous electrolytes, specific capacities of about 60 mA h g^{-1} were obtained at 1 C. The voltage profiles reflect the properties already observed during the CV experiments. In the presence of Li^+ ions (Fig. 3a), the potential spans from 0.3 to 0.7 V *versus* SHE. In particular, the potential hysteresis associated with Li^+ insertion is significantly greater than that associated with Na^+ or K^+ because of the partial

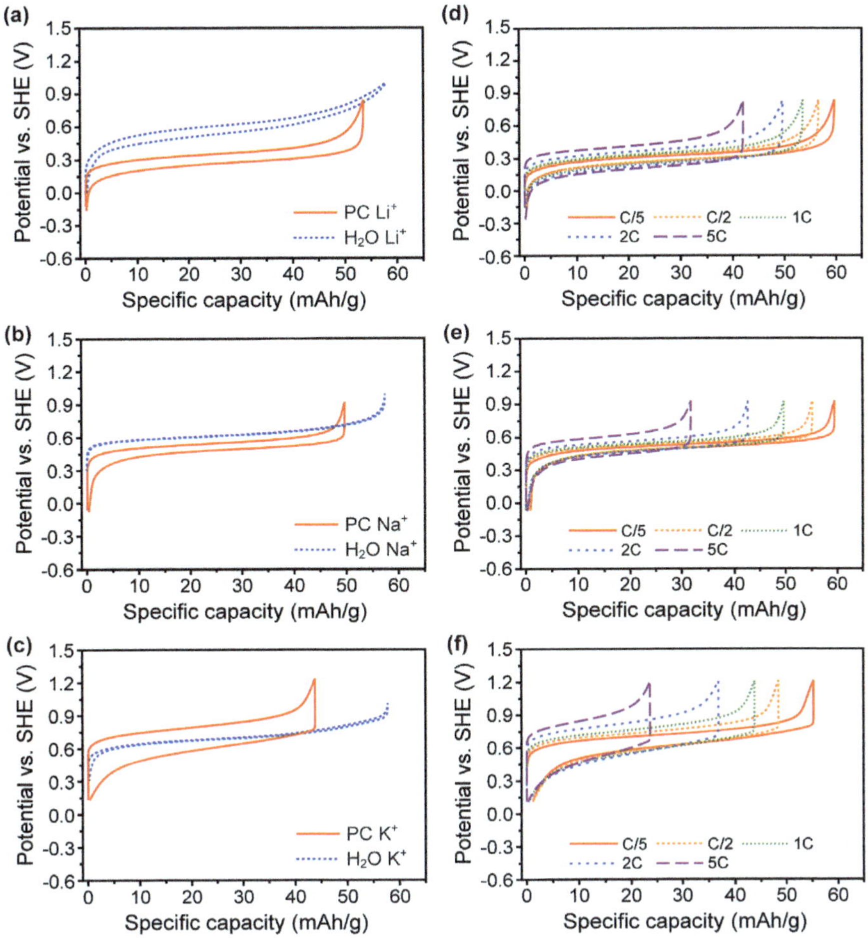

Fig. 3 Comparison of hysteresis in galvanostatic potential profiles at 1 C between aqueous and PC electrolytes. (a) Li^+, (b) Na^+, and (c) K^+ ion systems. Typical charge and discharge potential profiles of NiHCFe electrodes at different C rates (0.2, 0.5, 1, 2, and 5 C) with (d) Li^+, (e) Na^+, and (f) K^+ ions in PC.

rehydration of Li^+ after insertion as previously described. Curves obtained in the presence of Na^+ (Fig. 3b) and K^+ (Fig. 3c) show remarkably low potential hysteresis (high energy efficiency) and a higher reaction potential in the presence of K^+ relative to that of Na^+. The rate capability of the material in the presence of Li^+, Na^+, and K^+/PC solutions is reported in Fig. 3d–f, respectively, and the comparison between electrolytes is summarized in Fig. 4. The results show that the kinetics are much better in aqueous than in organic electrolytes.

Effect of the inserting ion on the kinetic features of NiHCFe's electrochemistry in aqueous electrolyte

The kinetic properties are not only affected by the nature of the solvent but also by the nature of the guest ion. To explore this effect, we have performed a set of measurements using the Swagelok experimental setup described in the

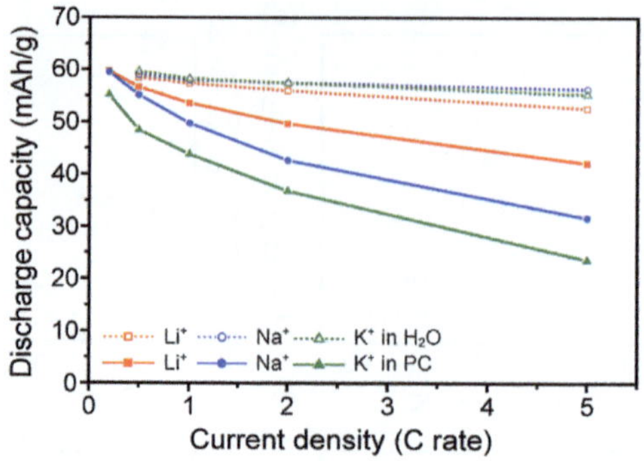

Fig. 4 Discharge capacity retention in aqueous and PC electrolytes with different insertion ions at varying C rates.

Experimental section. The Swagelok setup, besides allowing us to perform electrochemical experiments in a convenient three-electrode setup configuration, offers a consistent distance between the positive and negative electrode, which minimizes the varying effects of electrolyte resistance on kinetics experiments. Due to the use of flat, smooth stainless steel current collectors instead of porous carbon cloth, the active material loading is lower (about 1 mg cm^{-2}) but still many orders of magnitude higher than that in thin film electrodes.[36] Therefore, these electrodes in the Swagelok configuration can still be considered as bulk systems.

To highlight the kinetic properties of NiHCFe in different electrolytes, we performed CV scans at different scan rates (Fig. 5). We focused our attention on aqueous electrolytes because of the better electrochemical properties and the possibility of including Rb^+ and Cs^+ in the comparison since their nitrate salts have the required solubility (1 M). Fig. 5a shows the CV profiles obtained in 1 M solutions at a 1 mV s^{-1} scan rate. The peak positions are in perfect agreement with those obtained in the flooded configuration. The redox potentials corresponding to the different electrolytes are summarized in Table 2, which also reports the separation between the reduction and the oxidation peaks (ΔE). In a previous study,[29] the electrochemical reactions of hexacyanometallates were thought to be diffusion-limited, but our results suggest a different kinetic limitation for NiHCFe. The peak separation is too low for a one-electron reaction under diffusion-controlled kinetic conditions.

The kinetic regime can be better understood by analyzing the relationship between the CV scan rate and the corresponding peak current. The analysis was carried out at different scan rates from 0.25 to 5 mV s^{-1} for different guest ions (Fig. 5b–f). In the presence of small alkali cations (Li^+, Na^+, and K^+), very small potential shifts are observed for the redox processes, even at the highest scan rate. The corresponding peak currents show a linear relationship *versus* the scan rate (see Fig. 6a–c), which highlights the diffusionless behavior wherein the maximum current is only limited by the number of electroactive species at/in the electrode. This kind of behavior is typical of thin film electrodes,[37,38] and, to our knowledge, NiHCFe is the first case of diffusionless behavior in bulk systems. In the thin layer electrode treatment, the linear relationship between the peak current and the scan rate is ruled by:[37]

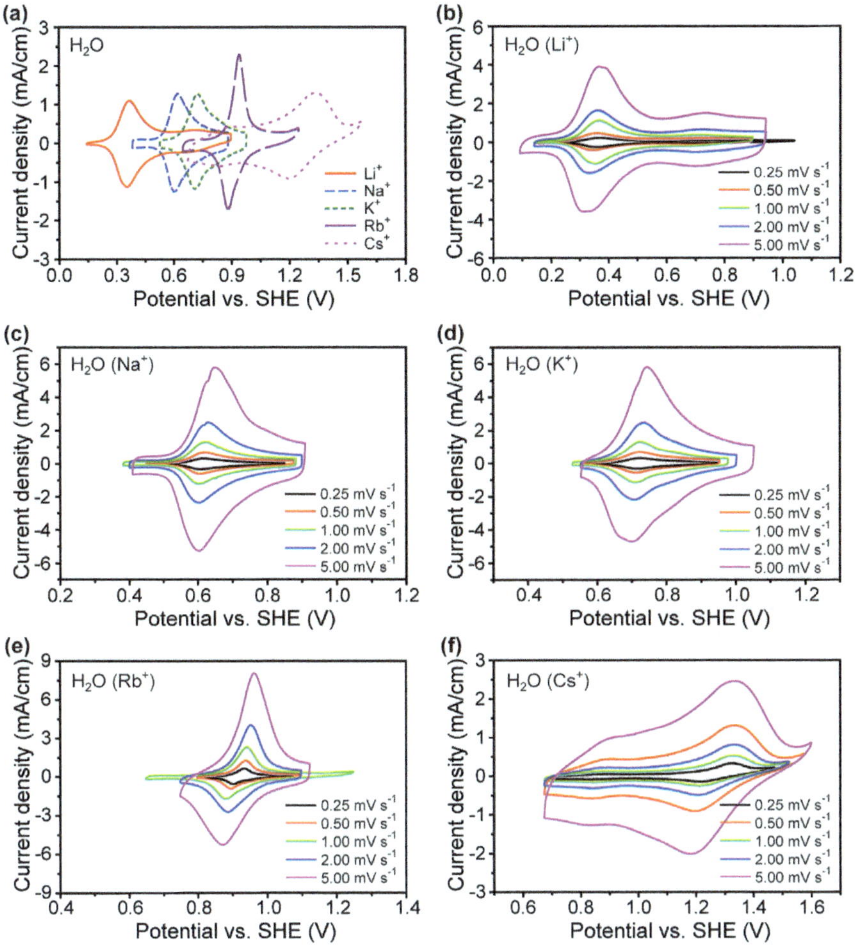

Fig. 5 CV curves of NiHCFe using the three-electrode Swagelok cell configuration in aqueous electrolytes. The scan rate is 1 mV s^{-1}. (a) CV curves of Li^+, Na^+, K^+, Rb^+, and Cs^+ ions. CV curves of (b) Li^+, (c) Na^+, (d) K^+, (e) Rb^+, and (f) Cs^+ ions at different scan rates from 0.25 to 5.00 mV s^{-1}.

Table 2 The redox potentials in different aqueous electrolytes as measured in the three-electrode Swagelok configuration

	E_{half}/V	ΔE/V
$LiNO_3$	0.350	0.015
$NaNO_3$	0.615	0.01
KNO_3	0.715	0.01
$RbNO_3$	0.920	0.06
$CsNO_3$	1.275 (0.870[a])	0.12 (0.05[a])

[a] The CV data were measured at 0.1 mV s^{-1} and stopped at 1.12 V.

$$i_{\text{peak}} = \frac{n^2F^2}{4RT} \times A \times \Gamma_0^* \times V_{\text{rate}}$$

where n is the number of electrons involved in the reaction, F the Faraday constant (C mol^{-1}), R the gas constant (J mol^{-1} K^{-1}), T the temperature (K), A the surface area (cm^2), Γ_0^* the concentration of active sites (mol cm^{-2}), and ν_{rate} the scan rate (mV s^{-1}). The Γ_0^* parameters obtained from the slopes of the straight lines in Fig. 6a–c are used to calculate the electrode specific capacity reported in Table 2, which are in good agreement with the values obtained in Fig. 4. The reduction and reoxidation processes in the presence of both Rb^+ and Cs^+ ions show asymmetrically shaped peaks. However, the kinetic properties differ greatly between the two guest ions. In the case of Rb^+, the ΔE is small (around 60 mV) and comparable with the value in the diffusion-controlled kinetic regime. In fact, the peak current depends linearly on the square root of the scan rate (Fig. 6d), whose dependency is given by:[37]

$$i_{\text{peak}} = 0.4463 \times \sqrt{\frac{n^3F^3}{RT}} \times \sqrt{D} \times C_0 \times \sqrt{V_{\text{rate}}}$$

where D is the average diffusion coefficient (cm^2 s^{-1}) in the whole composition range and C_0 the concentration of active sites (mol cm^{-3}). The slope can thus be used to determine the diffusion coefficient once C_0 and n are known. If we consider the reaction as monoelectronic and estimate C_0 from the specific capacity (60 mA h g^{-1}), we can calculate the diffusion coefficient D to be $5.7 \pm 1.0 \times 10^{-6}$ cm^2 s^{-1}, which is two times lower than that of Rb^+ in aqueous RbCl solution (2.06×10^{-5} cm^2 s^{-1})[39] but still remarkably high for solid-state diffusion.

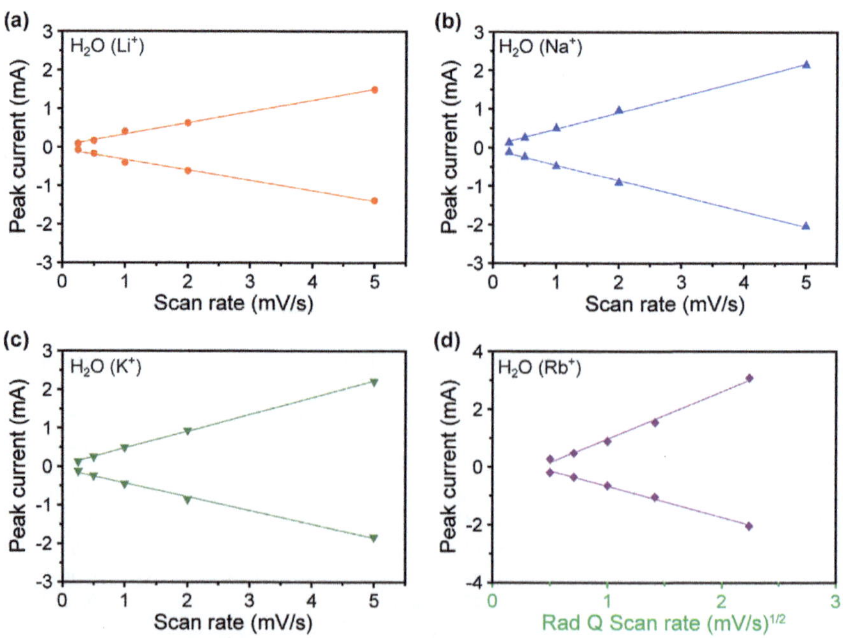

Fig. 6 The relationship between the peak currents and corresponding scan rates of (a) Li^+, (b) Na^+, (c) K^+, and (d) Rb^+ ions in aqueous electrolytes.

On the other hand, in the presence of Cs^+ ions the reaction is quasi-reversible, and no kinetic analysis can be carried out. The high hysteresis is a result of the size of Cs^+, which is larger than the channel size and smaller than the size of A-site, likely resulting in complete dehydration of the ion before insertion. The dehydration of the Cs^+ ion in this case is similar to the results in organic electrolytes as shown in Fig. 2d. In organic electrolytes, guest ions must also be fully desolvated before insertion, which results in high charge transfer energy.

Conclusions

The effect of the inserting alkali ion on the electrochemical properties of NiHCFe, which has the characteristic open framework crystal structure of Prussian blue analogues, was investigated in both aqueous and organic (propylene carbonate) electrolytes. A linear relationship between the insertion potential and the size of the alkali ion has been highlighted in both electrolyte systems. In particular, a smaller insertion ion has a higher solvation energy that needs to be overcome before insertion in the host structure and therefore a lower insertion potential.

On the other hand, profound kinetic dissimilarities between the two electrolytes systems were demonstrated. CVs in aqueous conditions show symmetric anodic/cathodic peaks that demonstrate the reversibility and fast kinetics of aqueous ion insertion. Moreover, smaller alkali ions like Li, Na and K show a diffusionless behavior wherein the maximum current is only limited by the number of electroactive species at/in the electrode. This enables the minimal hysteresis in the galvanostatic charge–discharge curves (extremely high energy efficiency) and the exceptionally long cycle life exhibited by nickel hexacyanoferrate and PBAs in general.

In propylene carbonate, the CV peaks show asymmetry and a quasi-Nernstian behavior that results in poor galvanostatic cycling performance. We believe that the different behavior could be ascribed to a different diffusion mechanism in the open framework crystal structure. Therefore, an aqueous electrolyte system is needed – to guarantee the extremely long cycle life and energy efficiencies required by grid scale energy storage applications.[13]

Acknowledgements

This work as well as use of the Stanford Synchrotron Radiation Lightsource, SLAC National Accelerator Laboratory, is supported by the U. S. Department of Energy, Office of Science, Office of Basic Energy Sciences under Contract no. DE-AC02-76SF00515 through the SLAC National Accelerator Laboratory LDRD project and the Assistant Secretary for Energy Efficiency and Renewable Energy. Further support comes from the Office of Vehicle Technologies of the U. S. Department of Energy under Contract no. DE-AC02-05CH11231, Subcontract no. 6951379, under the Batteries for Advanced Transportation Technologies (BATT) Program. H. W. L. acknowledges support from the Basic Science Research Program through the National Research Foundation of Korea (NRF) funded by the Ministry of Education, Science and Technology under Contract no. 2012038593. M.P. acknowledges support from the Fondazione Oronzio e Niccolò De Nora. R. Y. W. acknowledges support from the National Science Foundation Graduate Research Fellowship and the National Defense Science & Engineering Graduate Fellowship. We thank Dr

Johanna N. Weker, Dr Badri Shyam, and Dr Kevin H. Stone for help with synchrotron X-ray diffraction.

Notes and references

1 J.-M. Tarascon and M. Armand, *Nature*, 2001, **414**, 359–367.
2 M. Armand and J.-M. Tarascon, *Nature*, 2008, **451**, 652–657.
3 K. Kang, Y. S. Meng, J. Bréger, C. P. Grey and G. Ceder, *Science*, 2006, **311**, 977–980.
4 H. Lee, P. Muralidharan, R. Ruffo, C. M. Mari, Y. Cui and D. K. Kim, *Nano Lett.*, 2010, **10**, 3852–3856.
5 C. Chan, H. Peng, G. Liu and K. McIlwrath, *Nat. Nanotechnol.*, 2008, **3**, 31–35.
6 P. G. Bruce, S. A. Freunberger, L. J. Hardwick and J.-M. Tarascon, *Nat. Mater.*, 2012, **11**, 19–29.
7 B. Dunn, H. Kamath and J.-M. Tarascon, *Science*, 2011, **334**, 928–935.
8 Z. Yang, J. Zhang, M. C. W. Kintner-Meyer, X. Lu, D. Choi, J. P. Lemmon and J. Liu, *Chem. Rev.*, 2011, **111**, 3577–3613.
9 G. L. Soloveichik, *Annu. Rev. Chem. Biomol. Eng.*, 2011, **2**, 503–527.
10 M. S. Whittingham, *Chem. Rev.*, 2004, **104**, 4271–4301.
11 J.-M. Tarascon, *Nat. Chem.*, 2010, **2**, 510.
12 S.-W. Kim, D.-H. Seo, X. Ma, G. Ceder and K. Kang, *Adv. Energy Mater.*, 2012, **2**, 710–721.
13 C. Barnhart and S. Benson, *Energy Environ. Sci.*, 2013, **6**, 1083–1092.
14 W. Li, J. Dahn and D. Wainwright, *Science*, 1994, **264**, 1115–1118.
15 R. Ruffo, C. Wessells, R. A. Huggins and Y. Cui, *Electrochem. Commun.*, 2009, **11**, 247–249.
16 J. Luo, W. Cui, P. He and Y. Xia, *Nat. Chem.*, 2010, **2**, 760–765.
17 Y. Lu, L. Wang, J. Cheng and J. B. Goodenough, *Chem. Commun.*, 2012, **48**, 6544–6546.
18 L. Wang, Y. Lu, J. Liu, M. Xu, J. Cheng, D. Zhang and J. B. Goodenough, *Angew. Chem., Int. Ed.*, 2013, **52**, 1964–1967.
19 T. Matsuda, M. Takachi and Y. Moritomo, *Chem. Commun.*, 2013, **49**, 2750–2752.
20 C. D. Wessells, R. A. Huggins and Y. Cui, *Nat. Commun.*, 2011, **2**, 550.
21 R. Y. Wang, C. D. Wessells, R. A. Huggins and Y. Cui, *Nano Lett.*, 2013, **13**, 5748–5752.
22 C. D. Wessells, S. V. Peddada, M. T. McDowell, R. a. Huggins and Y. Cui, *J. Electrochem. Soc.*, 2012, **159**, A98–A103.
23 C. D. Wessells, S. V Peddada, R. A. Huggins and Y. Cui, *Nano Lett.*, 2011, **11**, 5421–5425.
24 H. M. Rietveld, *J. Appl. Crystallogr.*, 1969, **2**, 65–71.
25 A. C. Larson and R. B. Von Dreele, *Los Alamos Natl. Lab. Rep.*, **LAUR 86–74**, 1994.
26 B. H. Toby, *J. Appl. Crystallogr.*, 2001, **34**, 210–213.
27 G. Gritzner, *Pure Appl. Chem.*, 1990, **62**, 1839–1858.
28 Y. Yang, C. Brownell and N. Sadrieh, *Clin. Toxicol.*, 2007, **45**, 776–781.
29 F. Scholz and A. Dostal, *Angew. Chem., Int. Ed. Engl.*, 1996, **34**, 2685–2687.
30 A. Domenech, N. Montoya and F. Scholz, *J. Electroanal. Chem.*, 2011, **657**, 117–122.

31 M. Tissandier and K. Cowen, *J. Phys. Chem. A*, 1998, **5639**, 7787–7794.
32 J. Mähler and I. Persson, *Inorg. Chem.*, 2012, **51**, 425–438.
33 Y. Marcus, *J. Chem. Soc., Faraday Trans.*, 1991, **87**, 2995–2999.
34 M. Salomon, *J. Phys. Chem.*, 1970, **74**, 2519–2524.
35 S.-I. Ohkoshi, K. Nakagawa, K. Tomono, K. Imoto, Y. Tsunobuchi and H. Tokoro, *J. Am. Chem. Soc.*, 2010, **132**, 6620–6621.
36 A. Eftekhari, *J. Power Sources*, 2004, **132**, 291–295.
37 A. J. Bard and L. R. Faulkner, *Electrochemical methods: fundamentals and applications*, John Wiley, New York, 2nd edn, 1980.
38 J. Heinze, B. Frontana-Uribe and S. Ludwigs, *Chem. Rev.*, 2010, **110**, 4724–4771.
39 H. Harned and M. Blander, *J. Am. Chem. Soc.*, 1953, **159**, 1951–1953.

Faraday Discussions

PAPER

Synthesis, structure, and ionic conductivity of solid solution, $Li_{10+\delta}M_{1+\delta}P_{2-\delta}S_{12}$ (M = Si, Sn)

Satoshi Hori,[a] Kota Suzuki,[a] Masaaki Hirayama,[a] Yuki Kato,[b] Toshiya Saito,[b] Masao Yonemura[c] and Ryoji Kanno*[a]

Received 1st July 2014, Accepted 18th August 2014
DOI: 10.1039/c4fd00143e

Solid solutions of the silicon and tin analogous phases of the superionic conductor $Li_{10}MP_2S_{12}$ (M = Si, Sn) were synthesized by a conventional solid-state reaction in an evacuated silica tube at 823 K. The ranges of the solid solutions were determined to be $0.20 < \delta < 0.43$ and $-0.25 < \delta < -0.01$ in $Li_{10+\delta}M_{1+\delta}P_{2-\delta}S_{12}$ ($0.525 \leq k \leq 0.60$ and $0.67 \leq k \leq 0.75$ in $Li_{4-k}M_{1-k}P_kS_4$) for the Si and Sn systems, respectively. The ionic conductivity of these systems varied as a function of the changing M ions: the Si and Sn systems showed lower conductivity than the Ge system, $Li_{10+\delta}Ge_{1+\delta}P_{2-\delta}S_{12}$. The conductivity change for different elements might be due to the lattice size and lithium content affecting the ionic conduction. The relationship between ionic conduction, structure, and lithium concentration is discussed based on the structural and electrochemical information for the silicon, germanium, and tin systems.

Introduction

Solid lithium ionic conductors are of technological importance for future applications as solid electrolytes for all-solid-state batteries, which are expected to solve the safety issues in current lithium batteries by replacing flammable liquid electrolytes with solid electrolytes.[1] The conditions required for high lithium ionic conductors for battery applications are considered to be as follows: (i) a high lithium ionic conductivity ($\geq 1.0 \times 10^{-3}$ S cm^{-1}) at room temperature with negligible electronic conduction, (ii) high ionic transport number, (iii) electrochemical stability, and (iv) low-cost elements.[2] Recently, a lithium containing sulphide with a composition of $Li_{10}GeP_2S_{12}$ (LGPS) was reported to exhibit a wide electrochemical window of up to 5 V and an extremely

[a]Department of Electronic Chemistry, Tokyo Institute of Technology, 4259 Nagatsuta-cho, Midori-ku, Yokohama, 226-8502, Japan. E-mail: kanno@echem.titech.ac.jp; Fax: +81 045 924 5570

[b]Battery Research Division, Higashifuji Technical Center, Toyota Motor Corporation 1200 Mishuku, Susono, Shizuoka 410-1193, Japan

[c]Neutron Science Laboratory (KENS), Institute of Materials Structure Science, High Energy Accelerator Research Organization (KEK), 1-1 Oho, Tsukuba, Ibaraki 305-0801, Japan

high conductivity value of 1.2×10^{-2} S cm^{-1}, which almost equals the conductivity values of liquid electrolytes.[3] The discovery of LGPS opened the door to a new era of materials research, targeting new crystalline solid electrolytes with the same structure as LGPS. Its structure, shown in Fig. 1, consists of one-dimensional chains formed by LiS_6 octahedra and $(Ge_{0.5}/P_{0.5})S_4$ tetrahedra, which are connected by a common edge, and these chains are connected at a corner with PS_4 tetrahedra.

The search for new LGPS-type compounds is of practical importance for battery applications, because the use of rare and expensive germanium increases the cost of the electrolyte. The substitution of Ge^{4+} with Sn^{4+} yields $Li_{10}SnP_2S_{12}$,[4] which shows a lower conductivity value of 4.0×10^{-3} S cm^{-1} at room temperature; the Si^{4+}-substituted material, $Li_{11}Si_2PS_{12}$, was also synthesized under high-pressure conditions.[5] On the other hand, the ionic conduction mechanism of these phases was studied based on information on the crystal structures,[6] dynamics of ionic motion, and theoretical analysis.[7,8] In particular, $Li_{10}SiP_2S_{12}$ has been predicted to have a higher bulk conductivity value than the LGPS phase, from first principle calculations.[9] Materials synthesis and an understanding of the conduction mechanism in these LGPS-type analogous phases are still issues to be considered.

In the present study, the LGPS-related phases, $Li_{10+\delta}M_{1+\delta}P_{2-\delta}S_{12}$ where $M = $ Si and Sn, were synthesized by a conventional solid-state reaction in a sealed and evacuated silica tube. The ranges of the solid solutions were determined, and the conductivity and electrochemical stabilities were clarified. The relationships between ionic conduction, metal ions, lithium content, lattice volume, and local environment of lithium along the conduction pathway were studied on the basis of the crystal structures and conductivities of these solid solutions. The conduction mechanism is discussed for the $Li_{10+\delta}M_{1+\delta}P_{2-\delta}S_{12}$ ($M = $ Si, Ge, Sn) systems with the LGPS-type structure in order to develop further materials design of these solid lithium ion conductors.

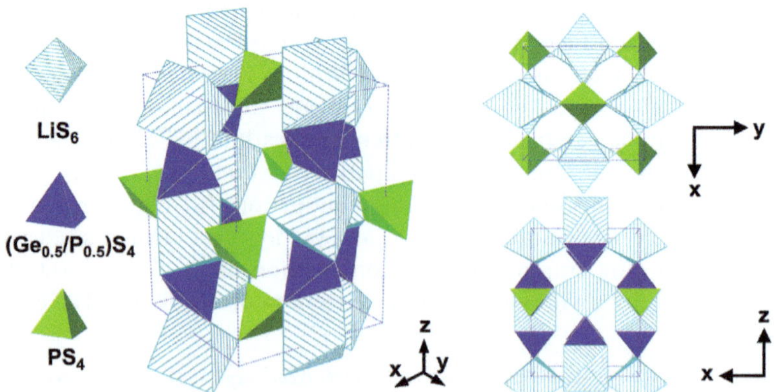

Fig. 1 Framework structure of the $Li_{10}GeP_2S_{12}$ (LGPS) crystal. Hatched blue octahedra, dark purple tetrahedra, and light green tetrahedra represent LiS_6, $(Ge_{0.5}/P_{0.5})S_4$, and PS_4. The dotted lines represent the unit cell. Lithium atoms participating in ionic conduction are not drawn for simplicity.

Experimental

The starting materials in this work were Li_2S (Idemitsu Kosan, >99.9% purity), P_2S_5 (Aldrich, >99.9% purity), SiS_2 (Mitsuwa Chemical, >99% purity), and SnS_2 (Kojundo Chemical Laboratory, >99% purity). These were weighed in appropriate molar ratios in an Ar-filled glove box, and mixed by planetary ball-milling for 40 h. The specimens were then pressed into pellets, sealed in a carbon-coated quartz tube at 10 Pa, and heated at a reaction temperature of 823 K for 3 days. After heating, the tube was slowly cooled to room temperature. The synthesized materials were characterized by an X-ray diffractometer (Rigaku SmartLab) with CuKα radiation. Diffraction data were collected at each 0.01° step width over the 2θ range from 10° to 35°. Synchrotron X-ray diffraction data for structural analysis were obtained using a high-flux synchrotron X-ray source at the BL19B2 beamline at SPring-8, operating at room temperature with a wavelength of 0.5 Å or 0.6 Å. A Debye–Scherrer diffraction camera was used for the measurements, and the specimens were sealed under vacuum in a Lindemann glass capillary (about 0.5 mm in diameter). The structural and profile parameters were refined by Rietveld analysis with the refinement program RIETAN-FP.[10] Neutron diffraction data were obtained using a special environmental neutron powder diffractometer (SPICA) installed on the BL09 beamline at the Japan Proton Accelerator Research Complex in Tokai, Japan. The specimen was sealed in a vanadium cell (about 6 mm diameter) with an indium ring. Rietveld refinement of the data was performed using the Z-Rietveld program.[11]

The electrochemical window of the solid electrolyte (SE) was evaluated by cyclic voltammetry. An Au electrode and a lithium electrode were used as a working, and a counter and reference electrode, respectively. The asymmetric Li/SE/Au cell was examined with a scan rate of 1 mV s^{-1} and scan range between -0.5 V and 5 V using a Solartron 1287 electrochemical interface. Ionic conductivity was measured by the ac impedance method in an Ar gas flow over a 253 K $\leq T \leq$ 373 K temperature range with applied frequencies in the range between 100 Hz and 15 MHz using a Solartron 1260 frequency response analyser. The sample was pressed into a pellet (diameter 11–12 mm; thickness 1–2 mm) and heated at 823 K in vacuum. The pellet was then coated with Au to form electrodes.

Results and discussion

Materials synthesis

Solid solutions of the germanium system, $Li_{10+\delta}Ge_{1+\delta}P_{2-\delta}S_{12}$, were synthesized with a composition range of $0 \leq \delta \leq 0.35$ at 823 K by the conventional solid-state reaction.[12] Similar solid solutions can be expected for silicon- and tin-based LGPS-type phases. As superionic conducting phases of the LGPS-type structure exist in the binary Li_4MS_4–Li_3PS_4 system (M = Si, Ge, Sn), the composition of the phase could be described by the formula, $Li_{4-k}M_{1-k}P_kS_4$, where the ratio, $1 - k : k$, corresponds to the molar ratio of $M^{4+} : P^{5+}$. Based on the original LGPS phase, the solid solution is described by $Li_{10+\delta}M_{1+\delta}P_{2-\delta}S_{12}$, where the original composition of $Li_{10}GeP_2S_{12}$ ($\delta = 0.0$ in $Li_{10+\delta}M_{1+\delta}P_{2-\delta}S_{12}$) corresponds to $k = 0.67$ in $Li_{4-k}M_{1-k}P_kS_4$. In the present study, the compositions of the solid solutions were described mainly by the k value in $Li_{4-k}M_{1-k}P_kS_4$.

Silicon system

A previous material search on the pseudo binary Li_4SiS_4–Li_3PS_4 system indicated the existence of the lithium conducting phase, $Li_{3.4}Si_{0.4}P_{0.6}S_4$, which was regarded as a member of the thio-LISICON family with a conductivity value of 6.4×10^{-4} S cm^{-1} at room temperature.[13] The $Li_{3.4}Si_{0.4}P_{0.6}S_4$ phase is composed of a three-dimensional framework structure formed by isolated PS_4 tetrahedral units, which is different from the LGPS-type structure. The solid-state synthesis of the binary system in this study confirmed the existence of the LGPS-type phase with the nominal composition of $Li_{3.45}Si_{0.45}P_{0.55}S_4$ and $Li_{10.35}Si_{1.35}P_{1.65}S_{12}$ ($k = 0.55$ in $Li_{4-k}Si_{1-k}P_kS_4$ and $\delta = 0.35$ in $Li_{10+\delta}Si_{1+\delta}P_{2-\delta}S_{12}$), which was close to that of the thio-LISICON phase, $Li_{3.4}Si_{0.4}P_{0.6}S_4$ ($k = 0.6$ in $Li_{4-k}Si_{1-k}P_kS_4$).

Fig. 2(a) shows synchrotron X-ray diffraction patterns of $Li_{3.45}Si_{0.45}P_{0.55}S_4$. All the diffraction peaks were indexed by the same space group as LGPS, $P4_2/nmc$ (137), although small diffraction peaks indexed by Li_3PO_4, which was present in the raw starting materials, were also observed. The amount of the impurity phase was calculated to be *ca.* 3.8 wt% by Rietveld refinements. The refined lattice parameters, $a = 8.67088(8)$ Å and $c = 12.53955(19)$ Å, are smaller than those of LGPS ($a = 8.71771(5)$ Å and $c = 12.63452(10)$ Å),[3] indicating that Ge^{4+} (ionic radii $r = 0.39$ Å) ions are substituted by Si^{4+} ($r = 0.26$ Å).[13] The structure parameters were refined by Rietveld analysis using neutron diffraction data at room temperature. The composition was fixed to $Li_{3.45}Si_{0.45}P_{0.55}S_4$ and multiphase models were employed for the small amount of impurity phase throughout the refinements. The structure model of the LGPS was used for the initial refinement. Four lithium sites reported recently,[9,12] the 16*h*, 4*d*, 8*f*, and 4*c* sites, were employed as initial positions. Initial atomic positions for other elements were as follows: S at 8*g* sites, P at 2*b* sites and 4*d* sites, and Si at 4*d* sites. Isotropic Debye–Waller factors were used for all atoms at the initial stage of the refinements, and then anisotropic Debye–Waller factors were applied to the four Li atoms. Several structure models were examined to determine the Si position; the model of Si in the 4*d* site provided better fitting results than the structure model with partial disordering of Si in the 2*b* site. Lithium was found to be distributed over four sites: Li(1) on the 16*h* site (site occupation parameter, $g = 0.488(4)$), Li(2) on the 4*d* site ($g = 1.0$), Li(3) on the 8*f* site ($g = 0.716(8)$), and Li(4) on the 4*c* site ($g = 0.791(3)$). These results confirmed

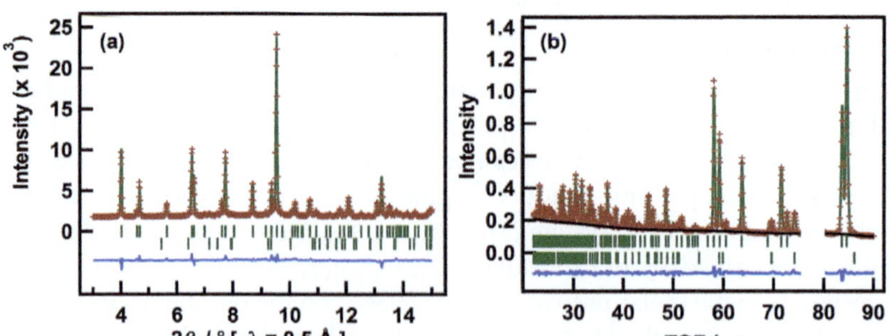

Fig. 2 (a) Synchrotron X-ray and (b) neutron Rietveld refinement patterns for $Li_{3.55}Si_{0.45}P_{0.55}S_4$ at 298 K. Bragg positions for secondary Li_3PO_4 phases are represented by the lower vertical marks.

Table 1 Rietveld refinement results for $Li_{3.45}Si_{0.45}P_{0.55}S_4$ at 298 K[a]

Atom	Site	g	x	y	z	U/Å^2
Li(1)	16h	0.488(4)	0.2657(5)	0.2641(5)	0.1946(3)	—
Li(2)	4d	1	0	0.5	0.94524(13)	—
Li(3)	8f	0.716(8)	0.24717(13)	0.24717(13)	0	—
Li(4)	4c	0.791(3)	0	0	0.24973(19)	—
Si(1)	4d	0.675	0	0.5	0.69145(6)	0.02252(18)
P(1)	4d	0.325	0	0.5	0.69145(6)	0.02252(18)
P(2)	2b	1	0	0	0.5	0.0322(2)
S(1)	8g	1	0	0.18779(7)	0.40795(5)	0.0290(2)
S(2)	8g	1	0	0.29962(8)	0.09879(7)	0.0310(2)
S(3)	8g	1	0	0.69561(9)	0.79026(6)	0.0262(2)

Atom	U_{11}[b]/Å^2	U_{22}/Å^2	U_{33}/Å^2	U_{12}/Å^2	U_{13}/Å^2	U_{23}/Å^2
Li(1)	0.126(4)	0.007(3)	0.329(14)	0.0261(18)	0.061(4)	−0.029(4)
Li(2)	0.0760(17)	0.0952(17)	0.0075(14)	0	0	0
Li(3)	0.075(2)	0.075(2)	0.352(9)	−0.0508(18)	0.141(2)	−0.141(2)
Li(4)	0.097(3)	0.207(4)	0.060(3)	0	0	0

[a] Note: space group $P4_2/nmc$ (137), $a = 8.669578(18)$ Å, $c = 12.536058(48)$ Å, $R_{wp} = 2.46\%$, $R_p = 1.92\%$, $S = R_{wp}/R_e = 5.66$, $R_B = 2.26\%$, $R_F = 3.30\%$. [b] The form of the anisotropic temperature factor is $\exp[-2\pi^2(h^2a^{*2}U_{11} + k^2b^{*2}U_{22} + l^2c^{*2}U_{33} + 2hka^*b^*U_{11} + 2hla^*c^*U_{13} + 2klb^*c^*U_{23})]$.

that the $Li_{3.45}Si_{0.45}P_{0.55}S_4$ phase is iso-structural to the LGPS phase. Fig. 2(b) shows the neutron Rietveld refinement pattern and Table 1 lists the final structure parameters for $Li_{3.45}Si_{0.45}P_{0.55}S_4$.

Silicon solid-solution system

Solid solutions were synthesized according to the composition $Li_{4-k}Si_{1-k}P_kS_4$. Fig. 3(a) shows X-ray diffraction patterns for the samples with $0.45 \le k \le 0.65$ in $Li_{4-k}Si_{1-k}P_kS_4$, together with the diffraction data for Li_4SiS_4 and β-Li_3PS_4,[13,14] which corresponds to the composition, $k = 0$ and 1.0, respectively. The figure shows select diffraction lines, indicating a continuous peak shift in the composition range of $0.525 \le k \le 0.60$. The samples for $k = 0.45$ and 0.50 show additional diffraction lines due to the phase with a similar structure to Li_4SiS_4. On the other hand, the sample with $k = 0.65$ shows diffraction lines due to the presence of β-Li_3PS_4. The lattice parameters, a and c, and volume V were calculated based on the LGPS-type structure. Fig. 3(b) shows the lattice parameters and volume of $Li_{4-k}Si_{1-k}P_kS_4$. A continuous decrease in the lattice parameters was observed with the variation of k from 0.525 to 0.60, which corresponds to the range of the solid solution.

The decrease in the lattice parameter is due to a decrease in the silicon content (Si^{4+}: $r = 0.26$ Å; P^{5+}: $r = 0.17$ Å), and also due to a decrease in the lithium content in the structure.

Tin solid-solution system

The solid solution of the Sn-substituted system was synthesized at 823 K by the solid-state reaction based on the composition, $0.50 \le k \le 0.75$ in $Li_{4-k}Sn_{1-k}P_kS_4$

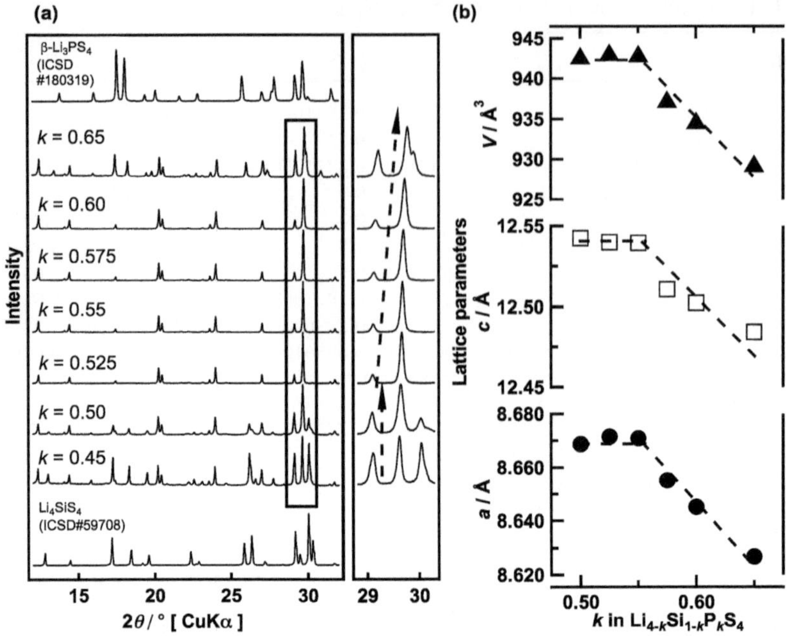

Fig. 3 (a) XRD patterns (left) and peak shifts (right) for the sample, $Li_{4-k}Si_{1-k}P_kS_4$ ($0.45 \leq k \leq 0.65$). XRD patterns for Li_4SiS_4 and β-Li_3PS_4 are also shown. (b) Composition dependence of the lattice parameters in LGPS-type phase with k in $Li_{4-k}Si_{1-k}P_kS_4$.

($-0.25 \leq \delta \leq 0.50$ in $Li_{10+\delta}Sn_{1+\delta}P_{2-\delta}S_{12}$), which is near the original LGPS composition of $Li_{10}SnP_2S_{12}$. Fig. 4(a) shows X-ray diffraction patterns for the sample with $Li_{4-k}Sn_{1-k}P_kS_4$ ($0.50 \leq k \leq 0.75$). The mono-phasic region of the LGPS-type structure was obtained near the composition $k = 0.73$, and the samples with the compositions $0.50 \leq k \leq 0.65$ and $k = 0.75$ showed additional diffraction peaks due to the Li_4SnS_4-type phase and the β-Li_3PS_4-type phase, respectively.[14,15] On the other hand, the sample with $0.67 \leq k \leq 0.72$ showed diffraction peaks from a small amount of impurity phases, and the samples with $k = 0.73$ and 0.74 were found to be mono-phasic in character. Fig. 4(b) shows the lattice parameters and volume of the $Li_{4-k}Sn_{1-k}P_kS_4$ phase obtained in the present study. The lattice parameters decreased from $k = 0.67$ to 0.75, indicating the formation of the solid solution. The decrease in the lattice parameter is due to a decrease in the amount of Sn^{4+} ions that have a larger ionic radius ($r = 0.55$ Å) than the P^{5+} ion ($r = 0.17$ Å), and also due to a decrease in the lithium content in the structure.

Solid solutions of the LGPS-type phases

LGPS-type phases were synthesized in the Si (ref. 5, present study), Ge (ref. 12), and Sn (ref. 4, present study) systems, and all the LGPS-type phases form a solid solution with the composition $Li_{4-k}M_{1-k}P_kS_4$ (M = Si, Ge, Sn). Fig. 5(a) summarizes the lattice volume data of the LGPS-type solid solutions as a function of k. Each system has a different range of the solid solutions: $0.525 \leq k \leq 0.60$, $0.50 \leq k \leq 0.67$, and $0.67 \leq k \leq 0.75$ in $Li_{4-k}M_{1-k}P_kS_4$ for the Si, Ge, and Sn systems,

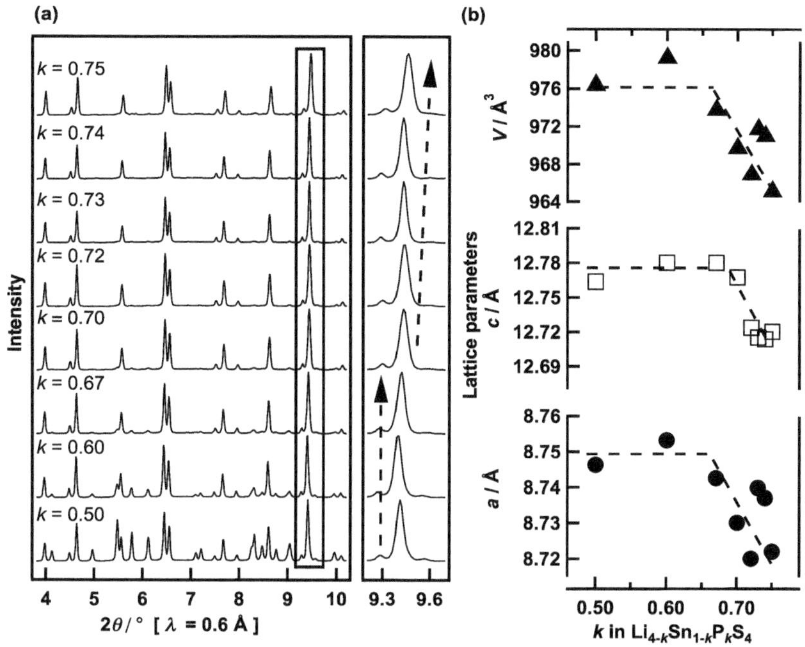

Fig. 4 (a) XRD patterns (left) and peak shifts (right) for the sample, $Li_{4-k}Sn_{1-k}P_kS_4$ ($0.50 \leq k \leq 0.75$). (b) Composition dependence of the lattice parameters in the LGPS-type phase as a function of k in $Li_{4-k}Sn_{1-k}P_kS_4$.

respectively. The differences in the single-phase region are discussed below in relation to the size of the cations (Si^{4+}, Ge^{4+}, and Sn^{4+}), which are present in the tetrahedral unit, $(M/P)S_4$ on the 4*d* sites. Fig. 5(b) shows the volume of the $(M/P)S_4$ unit as a function of the lattice volume together with other polyhedral units consisting of the framework structure, LiS_6 octahedra, and PS_4 tetrahedra. The volume of the $(M/P)S_4$ unit became larger depending on the cation size in the order of $M = Si^{4+}$, Ge^{4+}, and Sn^{4+}. This indicates that the M ions occupy the $(M/P)S_4$ tetrahedra. In contrast to the case of the $(M/P)S_4$ unit, no significant volume changes were observed for the PS_4 tetrahedra, which connect four LiS_6 octahedra in the one-dimensional framework chain by sharing S(1) as shown in Fig. 6(a). This is consistent with the structure analysis results that no metal ions are situated at the P (2*b*) sites. The dependence on the size of the M^{4+} ions was also observed in the volume of the LiS_6 unit. The volume of the octahedra is affected by the size of the $(M/P)S_4$ tetrahedra, because the octahedra share their S(2)–S(2) and S(3)–S(3) edges with the $(M/P)S_4$ tetrahedra.

Fig. 6(b) shows the atomic distances between sulphide atoms at LiS_6 octahedra as a function of k in $Li_{4-k}M_{1-k}P_kS_4$ (M = Si, Ge, Sn). The S(1)–S(1) distance was maintained constant with the variation of M and k, because no changes in the volume were observed for the PS_4 unit, which connects LiS_6 octahedra by sharing S(1). On the other hand, the S(2)–S(2) and S(3)–S(3) distances of all samples changed in the range of 3.47–3.70 and 3.35–3.63 Å, respectively. The ranges of the distances might be related to the composition range of the solid solution; the

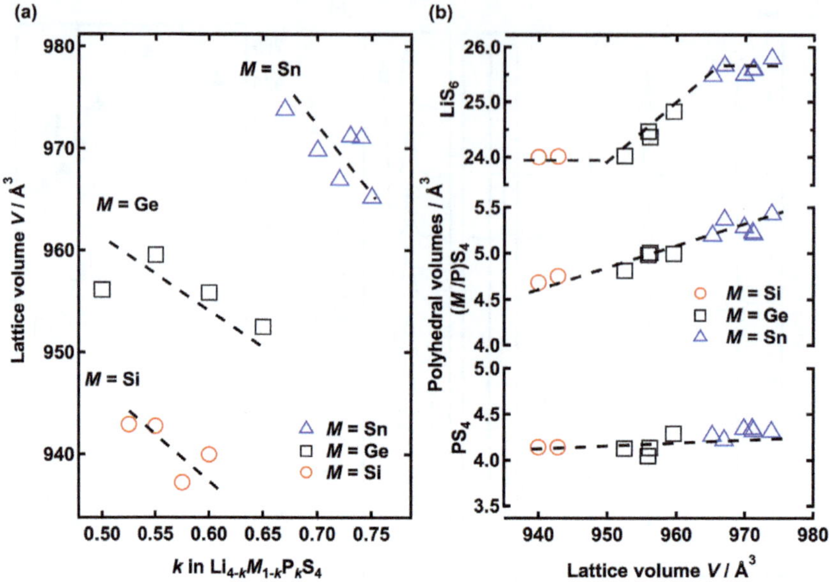

Fig. 5 (a) Composition dependence of the lattice volume in monophasic of LGPS-type phase with k in $Li_{4-k}M_{1-k}P_kS_4$ (M = Si, Ge, Sn). (b) Variation in polyhedral volumes of monophasic LGPS-type phase as a function of lattice volume V. The values were calculated based on Rietveld analysis of synchrotron X-ray diffraction data for M = Si and Sn. For M = Ge, the values were calculated based on a previous study.[12]

smallest and largest values of the S–S bond distances correspond to the limit of the solid solution.

The compositions of the mono-phasic region shift to larger k values in the order of Si^{4+}, Ge^{4+}, Sn^{4+}, which corresponds to the change in the ionic radius of $Si^{4+} < Ge^{4+} < Sn^{4+}$. The single-phase region in the Si system ranges from $k = 0.525$ to $k = 0.60$, while the region in the Ge system expands to $k = 0.67$. The mono-phasic region in the Sn system shifts to a higher k range compared to the Ge and Si systems.

Electrochemical properties

The electrochemical stability of the solid electrolytes synthesized in the present study was measured by cyclic voltammetry using Li and Au as electrodes. Fig. 7(a) shows the cyclic voltammograms of solid electrolytes with compositions of $Li_{3.45}Si_{0.45}P_{0.55}S_4$ ($k = 0.55$ in $Li_{4-k}Si_{1-k}P_kS_4$) and $Li_{3.27}Sn_{0.27}P_{0.73}S_4$ ($k = 0.73$ in $Li_{4-k}Sn_{1-k}P_kS_4$). For both electrolytes, oxidization and reduction peaks are observed near 0 V *vs.* Li^+/Li. On the other hand, no significant reactions were observed up to 5 V *vs.* Li^+/Li. These results indicate that both electrolytes have a wide electrochemical window.

Ionic conductivity was measured by the ac impedance method over the temperature range of 253 K ≤ T ≤ 373 K with applied frequencies in the range between 100 Hz and 15 MHz. The conductivity values were determined from impedance plots of the data.

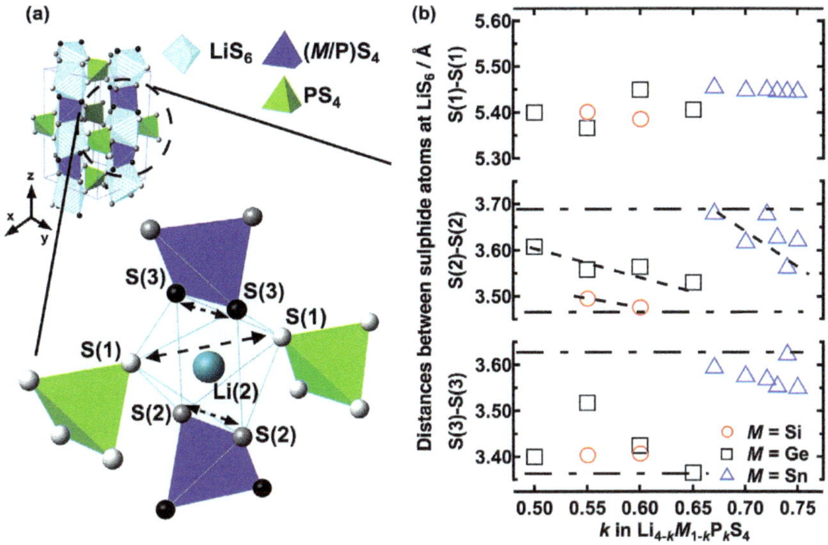

Fig. 6 (a) Coordination of sulphide atoms at LiS_6 octahedra. The large blue sphere represents Li(2), which is situated in the LiS_6 octahedra. Small spheres represent sulphide atoms. White spheres: S(1); grey spheres: S(2); black spheres: S(3). (b) Composition dependence of distances between sulphide atoms in the LiS_6 unit of the LGPS-type phase as a function of k in $Li_{4-k}M_{1-k}P_kS_4$. The values were calculated based on Rietveld analysis of synchrotron X-ray diffraction data for M = Si and Sn. For M = Ge, the values were calculated based on a previous study.[12]

Fig. 7(b) shows the impedance and Arrhenius plots of $Li_{3.45}Si_{0.45}P_{0.55}S_4$ ($k = 0.55$ in $Li_{4-k}Si_{1-k}P_kS_4$). The impedance plots consist of a semicircle and a spike, which respectively correspond to contributions from the bulk/grain boundary and the electrode. The total conductivity at each temperature was

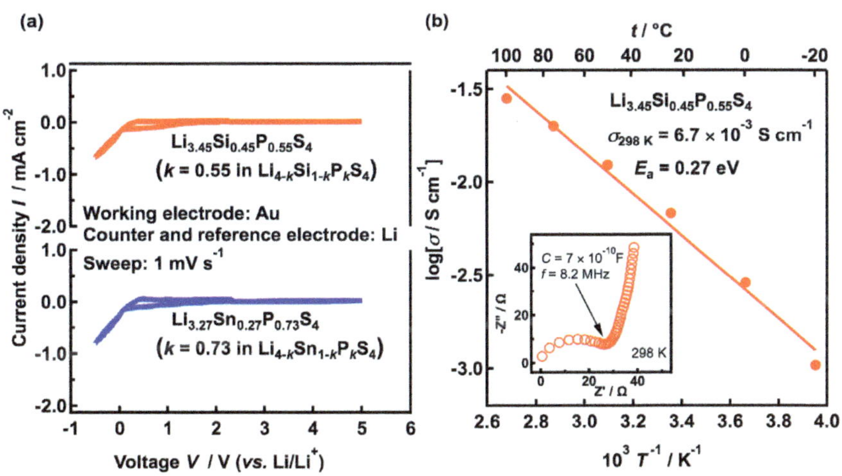

Fig. 7 (a) Cyclic voltammograms for LGPS type phases, $Li_{3.45}Si_{0.45}P_{0.55}S_4$ (upper) and $Li_{3.27}Sn_{0.27}P_{0.73}S_4$ (lower). (b) Arrhenius plots and cole–cole plots for $Li_{3.45}Si_{0.45}P_{0.55}S_4$.

obtained as a sum of the bulk and grain boundary resistances. The activation energy for ionic conduction was calculated to be 0.27 eV, which is a typical value for superionic conductors. The total conductivity value of 6.7 × 10^{-3} S cm^{-1} at room temperature was slightly lower than that of the Ge system, LGPS (1.2 × 10^{-2} S cm^{-1}), and was higher than that of $Li_{3.4}Si_{0.4}P_{0.6}S_4$ with the thio-LISICON phase (6.4 × 10^{-4} S cm^{-1}). Although these two compositions are close to each other ($Li_{3.45}Si_{0.45}P_{0.55}S_4$ with the LGPS-type and $Li_{3.4}Si_{0.4}P_{0.6}S_4$ with the thio-LISICON-type), the LGPS structure provided higher lithium ionic conduction.

Fig. 8(a) shows the total ionic conductivities at room temperature for the mono-phasic composition ranges for all the LGPS-type phases as a function of k in $Li_{4-k}M_{1-k}P_kS_4$ (M = Si, Ge, Sn). For the Si system, the maximum conductivity was obtained at $k = 0.55$, smaller than that observed for the Ge system at $k = 0.55$. For M = Sn, the conductivity of the sample with $k = 0.67$ (the nominal composition of $Li_{10}SnP_2S_{12}$, which is out of the solid-solution range) was close to the reported values (3.7 × 10^{-3} S cm^{-1} in this study and 4.0 × 10^{-3} S cm^{-1} in a previous study[4]). However, the maximum value of 5.0 × 10^{-3} S cm^{-1} was obtained at $k = 0.73$, which indicates that the single-phase of the LGPS-type has higher conductivity. Fig. 8(b) illustrates the relationships between the ionic conductivity, lattice volume, and composition parameter k in $Li_{4-k}M_{1-k}P_kS_4$ (M = Si, Ge, Sn). The lattice volume plays a key role for the Si and Ge systems; the compositions of the larger volume show higher ionic conductivity values. On the other hand, the k value also affects ionic conductivity for the Sn system. Although a large lattice volume was obtained for the Sn system, lower ionic conductivity was observed. This can be explained by larger k values of $Li_{4-k}M_{1-k}P_kS_4$ for the Sn system, which corresponds to smaller amounts of Li ions in the unit cell. Low concentration of lithium for conduction might result in lower ionic conductivity compared to the Si and Ge systems.

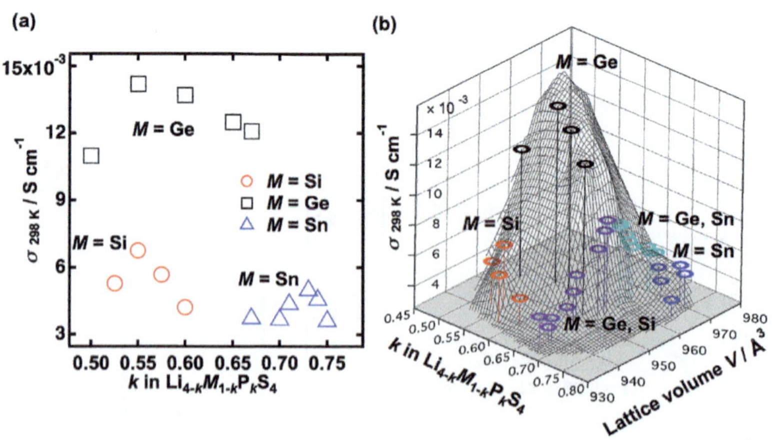

Fig. 8 (a) Composition dependence of the ionic conductivity at room temperature in the LGPS-type phase with k in $Li_{4-k}M_{1-k}P_kS_4$ (M = Si, Ge, and Sn). (b) The relationships between ionic conductivity, lattice volume and composition parameter in monophasic LGPS-type phases. For germanium related systems (M = Ge; M = Ge, Si; M = Ge, Sn), the values are cited from previous studies.[12,16]

The ionic conductivity of the LGPS-type phase might be understood by the lattice volume and the lithium concentration that participates in the ionic conduction.

(i) A large lattice volume yields high ionic conductivity. To increase the lattice volume, the choice of the M atoms is important. The M atoms situate in the $(M/P)S_4$ tetrahedra which form a one-dimensional framework chain and this expands the cell volume according to the ionic radii: $Si^{4+} < Ge^{4+} < Sn^{4+}$.

(ii) The change in the lattice volume is related to the structure restriction. The change in the size of the one-dimensional framework polyhedra (LiS_6 octahedra and $(M/P)S_4$ tetrahedra) might be restricted by the S(2)–S(2) and S(3)–S(3) distances, which were found to determine the single phase region.

(iii) As the aliovalent M cation changes the lithium composition, the lithium concentration that participates in the ionic conduction is determined by the higher and lower limits of the lattice volume. Therefore, the range of the solid solution changed with the lattice volume; $0.525 \le k \le 0.60$ for $M = Si$, $0.50 \le k \le 0.67$ for $M = Ge$, and $0.67 \le k \le 0.75$ for $M = Sn$ in the composition, $Li_{4-k}M_{1-k}P_kS_4$ ($M = Si, Ge, Sn$). Since the lithium concentration in the Sn system is lower than in the other two systems, the conductivity of the Sn system is lower than those of the Ge and Si systems.

Conclusion

The solid solutions of LGPS-type phases have been investigated for the composition of $Li_{4-k}M_{1-k}P_kS_4$ with $M = Si$, Ge, and Sn. The LGPS-type phase of the Si system was successfully synthesized by a solid-state reaction in an evacuated sealed tube. All the systems showed a certain range of solid solution, and the composition range in each system depended on the ionic radii of the substituted cation (M^{4+}). The ionic conductivity changed with the size of the M cations and Li compositions. Although the highest ionic conduction was proposed by first principle calculation, the Si system showed lower ionic conductivity than the Ge system.

Acknowledgements

This study was supported by the Post-LiEAD project of the New Energy and Industry Technology Development Organization (NEDO), Japan. The synchrotron radiation experiments were carried out as projects approved by the Japan Synchrotron Radiation Institute (JASRI) (proposal no. 2013A1768 and no. 2013B1807). The neutron radiation experiments were performed at the Japan Proton Accelerator Research Complex (J-PARC) (proposal no. 2009S10) using the facilities developed by the RISING project of the New Energy and Industrial Technology Development Organization (NEDO), Japan.

References

1 V. Thangadurai, S. Narayanan and D. Pinzaru, *Chem. Soc. Rev.*, 2014, **43**, 4714–4727.

2 A. D. Robertson, A. R. West and A. G. Ritchie, *Solid State Ionics*, 1997, **104**, 1–11.

3 N. Kamaya, K. Homma, Y. Yamakawa, M. Hirayama, R. Kanno, M. Yonemura, T. Kamiyama, Y. Kato, S. Hama, K. Kawamoto and A. Mitsui, *Nat. Mater.*, 2011, **10**, 682–686.

4 P. Bron, S. Johansson, K. Zick, J. Schmedt auf der Günne, S. Dehnen and B. Roling, *J. Am. Chem. Soc.*, 2013, **135**, 15694–15697.

5 A. Kuhn, O. Gerbig, C. Zhu, F. Falkenberg, J. Maier and B. V. Lotsch, *Phys. Chem. Chem. Phys.*, 2014, **16**, 14669–14674.

6 A. Kuhn, J. Kohler and B. V. Lotsch, *Phys. Chem. Chem. Phys.*, 2013, **15**, 11620–11622.

7 A. Kuhn, V. Duppel and B. V. Lotsch, *Energy Environ. Sci.*, 2013, **6**, 3548–3552.

8 C. H. Hu, Z. Q. Wang, Z. Y. Sun and C. Y. Ouyang, *Chem. Phys. Lett.*, 2014, **591**, 16–20.

9 S. P. Ong, Y. Mo, W. D. Richards, L. Miara, H. S. Lee and G. Ceder, *Energy Environ. Sci.*, 2013, **6**, 148–156.

10 F. Izumi and M. Koichi, *Solid State Phenom.*, 2007, **130**, 15–20.

11 R. Oishi, M. Yonemura, Y. Nishimaki, S. Torii, A. Hoshikawa, T. Ishigaki, T. Morishima, K. Mori and T. Kamiyama, *Nucl. Instrum. Methods Phys. Res.*, 2009, **600**, 94–96.

12 O. Kwon, M. Hirayama, K. Suzuki, Y. Kato, T. Saito, M. Yonemura, T. Kamiyama and R. Kanno, *J. Mater. Chem. A*, 2015, **3**, 438–446.

13 M. Murayama, R. Kanno, M. Irie, S. Ito, T. Hata, N. Sonoyama and Y. Kawamoto, *J. Solid State Chem.*, 2002, **168**, 140–148.

14 K. Homma, M. Yonemura, T. Kobayashi, M. Nagao, M. Hirayama and R. Kanno, *Solid State Ionics*, 2011, **182**, 53–58.

15 T. Kaib, S. Haddadpour, M. Kapitein, P. Bron, C. Schröder, H. Eckert, B. Roling and S. Dehnen, *Chem. Mater.*, 2012, **24**, 2211–2219.

16 Y. Kato, R. Saito, M. Sakano, A. Mitsui, M. Hirayama and R. Kanno, *J. Power Sources*, 2014, **271**, 60–64.

Faraday Discussions

PAPER

In situ TEM study of the Li–Au reaction in an electrochemical liquid cell

Zhiyuan Zeng,[a] Wen-I. Liang,[ab] Ying-Hao Chu[b] and Haimei Zheng*[ac]

Received 1st July 2014, Accepted 24th July 2014
DOI: 10.1039/c4fd00145a

We study the lithiation of a Au electrode in an electrochemical liquid cell using transmission electron microscopy (TEM). The commercial liquid electrolyte for lithium ion batteries (1 M lithium hexafluorophosphate $LiPF_6$ dissolved in 1 : 1 (v/v) ethylene carbonate (EC) and diethyl carbonate (DEC)) was used. Three distinct types of morphology change during the reaction, including gradual dissolution, explosive reaction and local expansion/shrinkage, are observed. It is expected that significant stress is generated from lattice expansion during lithium–gold alloy formation. There is vigorous bubble formation from electrolyte decomposition, likely due to the catalytic effect of Au, while the bubble generation is less severe with titanium electrodes. There is an increase of current in response to electron beam irradiation, and electron beam effects on the observed electrochemical reaction are discussed.

Introduction

In commercial lithium-ion batteries, the most commonly used anode materials are graphitic carbon or other carbonaceous materials, since such materials are cheap, have good cyclability and the working voltage is low.[1,2] However, the theoretical capacity of a carbon electrode is only 372 mA h g^{-1}. To decrease the reduction of the cell voltage and thereby increase the energy density, anode materials with extremely negative potentials and high capacities have been investigated. Metals or other materials including Sn, Cu, Sb and Si that can be alloyed with lithium are promising candidates due to their high theoretical capacities.[3–10] Gold (Au) reacts with lithium at low voltage forming several lithium–gold alloy phases, including ones with a very high percentage of Li (up to Au_4Li_{15}).[11] A Au electrode can be used as the anode (although it is not cost effective) since it has a wide range of stable cell voltages and the lowest possible alloying/de-alloying potential relative to the Li/Li^+ couple. It is known that any deviation from the potential of the Li/Li^+ couple will be at the expense of the

[a]Materials Sciences Division, Lawrence Berkeley National Laboratory, Berkeley, CA 94720, USA. E-mail: hmzheng@lbl.gov

[b]Department of Materials Science and Engineering, National Chiao Tung University, Hsinchu, 30010, Taiwan

[c]Department of Materials Science and Engineering, University of California, Berkeley, California 94720, USA

overall cell voltage, and the net effect will be a reduction in the energy density of the battery. The major problem encountered in gold, as well as other materials, is the large volume change due to structural changes accompanying lithiation and delithiation. The mechanical stress generated during these lithium uptake and release processes will cause pulverization and quick capacity fading,[7] which makes these materials not very practical for long term usage. Current strategies to overcome the pulverization of electrodes mainly focus on a reduction of the particle size, the use of composite materials, and optimization of the binder material. However, the reaction mechanisms of these materials with lithium are still far from being completely understood.

In situ liquid cell TEM allows the imaging of chemical reactions in liquids with high spatial resolution,[12–19] which provides the opportunity to address important questions regarding the electrode–electrolyte interfaces in a battery cell. We study the Au electrode reaction with lithium during charging and discharging using an electrochemical liquid cell under a transmission electron microscope. In previous reports using gold as an anode material for lithium ion batteries, although both *ex situ* and *in situ* studies were applied and different Au–Li alloy phases have been achieved,[4,11,20–22] the Au–Li reaction has never been observed directly. Recently, utilizing our newly designed electrochemical liquid TEM cells, we have captured the dynamic processes of inhomogeneous lithiation, lithium metal dendritic growth, electrolyte decomposition and solid–electrolyte interface formation.[23] Here, using the same electrochemical liquid cell, we captured different types of lithiation reaction for a Au electrode in a commercial battery electrolyte by real time imaging. The catalytic role of Au in electrolyte decomposition and beam effects are discussed based on control experiments or theoretical calculations.

Experimental

The electrochemical cells were fabricated following a procedure similar to that reported previously.[23] We used ultra-thin silicon wafers (200 μm, 4 inches, p-doped) purchased from Virginia Semiconductor (Fredericksburg, VA). A 25 nm thick low stress silicon nitride film was evaporated on the silicon wafer as a membrane of the viewing window. The viewing windows and two reservoirs were created by photolithographic patterning and etching with KOH solution (with water to KOH ratio of 2 : 1). The dimensions of the windows were 25 μm × 6 μm. Two 120 nm-thick gold electrodes were deposited on the bottom chips with a face-to-face distance of 20 μm. The bottom and top chips were bonded together with a 150 nm thick sputtered indium spacer. The commercial electrolyte for lithium ion batteries, *i.e.* 1 M lithium hexafluorophosphate $LiPF_6$ dissolved in 1 : 1 (v/v) ethylene carbonate (EC) and diethyl carbonate (DEC), was loaded into one of the reservoirs with a syringe. We sealed the cell, including the reservoirs, using Cu foil and epoxy. Since the window gap is less than 150 nm, no contamination from the epoxy was observed during the electrochemical experiments. The dimensions of the biasing cell were ~3 mm × 3 mm in a square and ~400 μm thick. Both of the working and counter electrodes were extended to two gold pads in two reservoirs. Gold wires were bonded onto each gold pad and they were connected to the two copper pads on the TEM holder tip, which allow an electric bias to be applied. The cells were fabricated at the Marvell Nanofabrication Laboratory of the University of California at Berkeley. The TEM holder fits a JEOL 2100 TEM operated at 200 kV.

Real-time videos of the electrochemical experiments were recorded at 2–15 frames per second. Faster recording, up to 400 frames per second, was also incorporated using a high speed camera provided by Direct Electron, LP (San Diego, CA) with a model number DE-12. The electrochemical process was controlled by an electrochemical workstation (Model 660D series made by CH Instruments) connected to the TEM holder. It was used to perform *in situ* cyclic voltammetry measurements.

For *ex situ* experiments, we first put a droplet of Au nanowire solution (used as purchased from Nanopartz Inc.) onto the lacey carbon Cu grids and let the Cu grids dry out so that lots of Au nanowires were attached to the carbon film. Then, the TEM grids were immersed into the lithium electrolyte with a short face-to-face distance. The two Cu grids were connected to an electrochemical working station and cyclic voltammetry with a voltage range of 0 to −3 V and a scan rate of 0.1 V s^{-1} was applied. The morphology changes and lithiated alloy phases were examined in great detail under TEM as a comparison to the *in situ* results.

Results and discussion

A scheme of how the gold electrode reacts with the lithium ion battery electrolyte in the electrochemical liquid cell set up is shown in Fig. 1. The bottom left of Fig. 1 shows a real electrochemical liquid cell with Au wires bonded onto the Au pads on the bottom chip in the reservoir area. In the home-made TEM holder, a cell pocket is designed to fit and hold the electrochemical liquid cell. Currently, only two electrodes are used. In the future, a reference electrode can be added. When the electrolyte $LiPF_6$/EC/DEC was loaded into one of the two reservoirs, the electrolyte flowed into the cell by capillary force and a thin liquid layer (~120 nm) was formed at the observatory window and everywhere inside the cell. We applied cyclic voltammetry and monitored the electrode–electrolyte interface under TEM. The dynamic reaction of the Au electrode with Li was captured in real time.

The phase diagram of the Au–Li binary system is shown in Fig. 2.[24] The alloys of Au_5Li_4, AuLi, $AuLi_3$ and Au_4Li_{15} are expected to form during the lithiation of

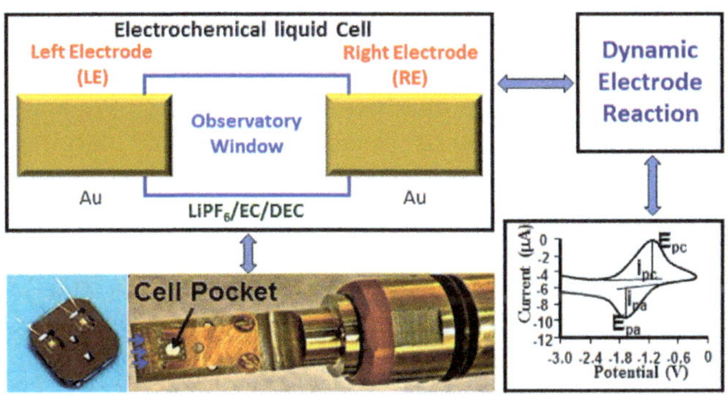

Fig. 1 A scheme of the *in situ* TEM set up for the observation of the electrochemical reaction using an electrochemical liquid cell. 120 nm thick gold (or Ti) electrodes are used. The dynamic electrochemical reaction on the electrode corresponding to the applied cyclic voltammetry is captured. The bottom left photographs show an electrochemical liquid cell and a home-made TEM holder.

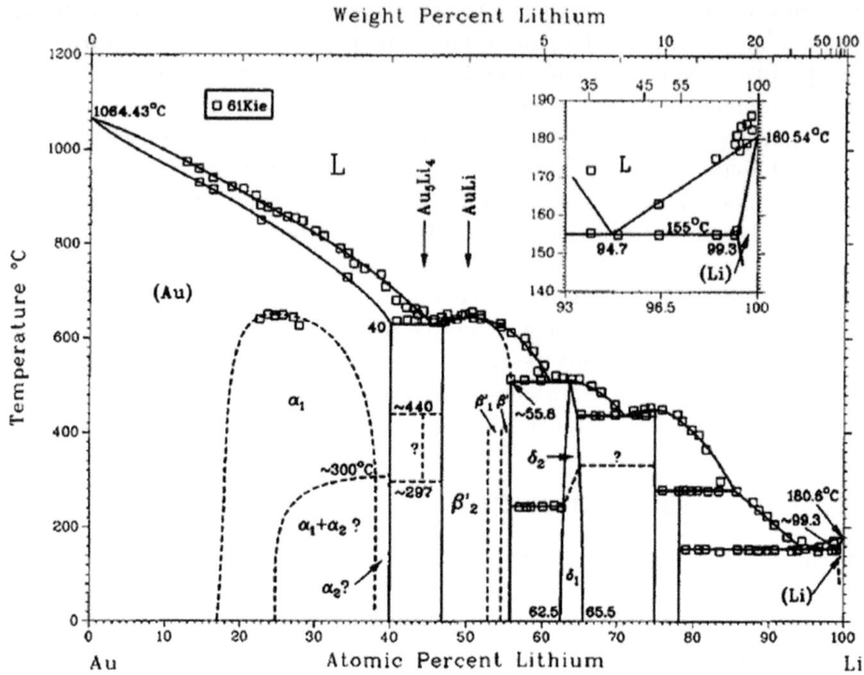

Fig. 2 Au–Li phase diagram.[24] With kind permission from Springer Science and Business Media.

gold at equilibrium conditions. However, these crystalline (Li, Au) compounds may not form directly during the electrochemical-driven alloying of Au with Li, despite the fact that the formation energies of the crystalline (Li, Au) compounds are lower than those of the corresponding amorphous phases, similar to the lithiation reaction of Si.[25] A recent study of lithium alloying with a gold electrode, conducted by *in situ* XRD, revealed that during the electrochemical cycling of the gold electrode two different crystalline Au–Li alloy phases were observed. For instance, $AuLi_1$ was formed during charging of the gold electrode. During discharging, $AuLi_1$ converted into $AuLi_2$ and subsequently dissolved, leaving an amorphous phase. Both the $AuLi_1$ and $AuLi_2$ phases are not shown in the Au–Li phase diagram, which suggests that they are metastable.[22] We aim to image in detail the phases and phase transitions accompanying the morphology changes during the Au–Li reaction. Here, we report the morphology changes and reaction kinetics during lithiation of the gold electrode from *in situ* observation. From our *ex situ* studies, only the Au_3Li phase was achieved, which is reasonable since only stable phases are expected after the reaction. A detailed *in situ* study of the phases and phase transitions during the reaction is in progress and the results will be reported in a future publication.

Fig. 3A–I show sequential images of the electrolyte decomposition, lithiation of the gold electrode and subsequent dissolution of the gold electrode during a discharge–charge cycle at an early stage. At the beginning, a sharp interface between the Au electrode and $LiPF_6$/EC/DEC is observed indicating that no reaction occurred. When the cyclic voltammetry is ramped down to

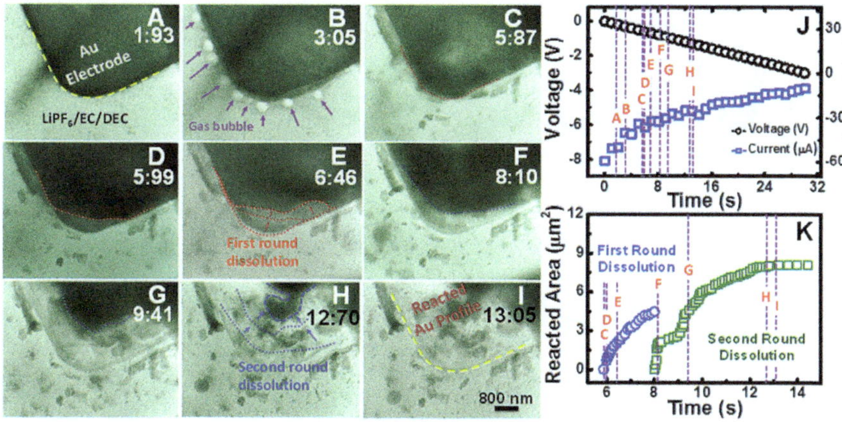

Fig. 3 (A)–(I) Time evolution of the dissolution of the Au electrode in the $LiPF_6$/EC/DEC electrolyte; (J) the corresponding applied electric potential and measured electric current from frame A to frame I; (K) the evolution of the dissolved area of the Au electrode as a function of time during cyclic voltammetry with the voltage range of 0 to −3 V and a scan rate of 0.1 V s^{-1}.

−0.3 V, many bubbles are formed, which suggests gaseous products are generated from electrolyte decomposition.[23] In the meantime, a thin Li–Au layer is formed because of the interfacial reaction between the Au electrode and the lithium ion battery electrolyte. Subsequently, the Au electrode reacts vigorously with the electrolyte, resulting in the Au layer stripping and dissolving from the outer layer to the inner part of the electrode. The starting etching point is the corner of the Au electrode, then the reaction front propagates along the diagonal direction (Fig. 3C–E). A large part of the Au corner is etched away within 2 s, defined as the "first round dissolution". The contrast changes from dark to light due to the surface layer of the Au electrode reacting and dissolving in the electrolyte. After the surface layer is peeled off, the exposed new surface area of the Au electrode reacts with the electrolyte resulting in "second round dissolution". As shown in Fig. 3F–H, the reaction front moves promptly in a diagonal direction from 8.10 s to 12.70 s, and almost all the Au electrode vanishes. Fig. 3J shows the corresponding applied electrical potential and the measured electrical current from frames (A)–(I), where the applied voltage is ramping from 0 V to −3 V and the measured electrical current decreases. It is noted that the measured electric current under TEM is higher than that from the same electrochemical liquid cell without an electron beam, and beam effects will be discussed in a later section. The reaction kinetics are measured by the dissolution area of the Au electrode as a function of time (Fig. 3K). In the first round dissolution, an area of 4.5 μm^2 is etched away within 2 s under −0.6 V to −0.8 V, thus the etching speed of the Au electrode is 2.25 $\mu m^2 s^{-1}$. The Au etching speed of the second round dissolution in the voltage range of −0.8 V to −1.4 V is 1.33 $\mu m^2 s^{-1}$. Intensive Au dissolution mostly occurs in the voltage range of −0.6 V to −1.4 V. According to Bach *et al.*,[22] metastable $AuLi_2$ can be formed and then dissolved during cyclic voltammetry. We captured the signal of dissolved gold in the area surrounding the electrode (Fig. 5).

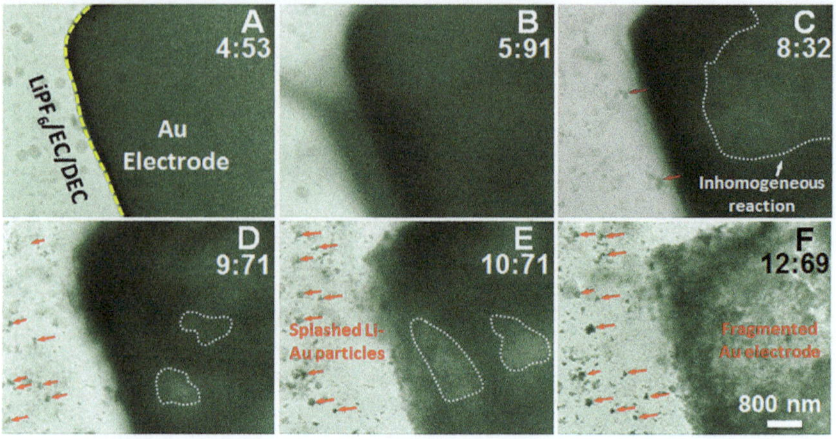

Fig. 4 Time series of TEM images showing the inhomogeneous reaction and fragmentation of the Au electrode during the Li–Au reaction. During this explosive reaction, a large number of nanoparticles splash from the Au electrode to the area surrounding the electrode.

Another type of Li–Au reaction that was observed is the explosive reaction of the Au electrode. In this case, during lithiation of the Au electrode, electrolyte decomposition is still expected but no gaseous products (bubbles) are released. It is likely that the gaseous products are absorbed and accumulated on the surface of the electrode, which triggers the explosion of the Au electrode at a later stage. This explosion generates a large number of Au or Li–Au nanoparticles, since Au or

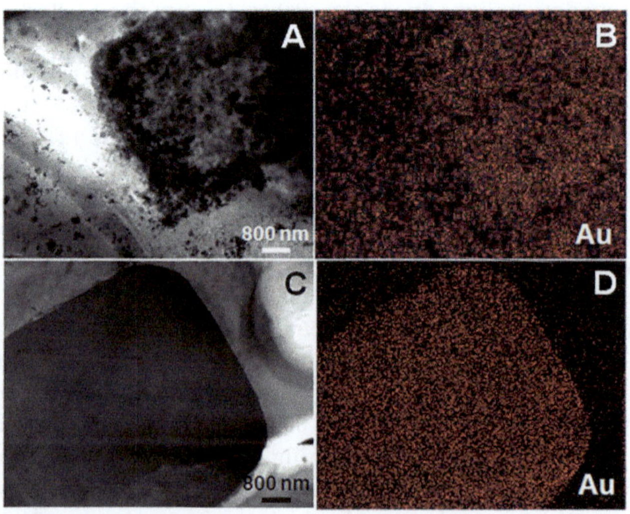

Fig. 5 (A) HAADF image of the right Au electrode that was electrochemically reacted with the lithium electrolyte (also shown in Fig. 4F). (B) The corresponding EDS map of (A). (C) The left side of the Au electrode. (D) The corresponding EDS map of the Au electrode in (C).

lithiated Au can peel off from the Au electrode during the explosion. In Fig. 4A–F, the sequential images show such an explosive Au lithiation reaction. Initially, the Au electrode is intact with a clean, sharp interface with the electrolyte. When the cyclic voltammetry is ramped down in the negative direction, an inhomogeneous reaction occurs as marked with a white dash line in Fig. 4C. Subsequently, the splashing of nanoparticles accompanied by erosion of the electrode takes place. The number of splashed particles dramatically increases within 3 s, and they reside in the area surrounding the fragmented electrode.

We performed energy dispersive X-ray spectroscopy (EDS) mapping of the reacted Au electrode, with the fragmented Au electrode and many small nanoparticles in the surrounding area, as well as the other side of the electrode. Fig. 5B and D show the elemental distribution profiles of Au in both the reacted and unreacted electrodes. In the reacted Au electrode residual, there is a higher concentration of Au at the edge than the inner part, indicating an inhomogeneous Li–Au reaction. The front area of the Au electrode has a higher intensity of Au, which corresponds to a stronger electric field during the reaction. It is clear that the particles splashed from the electrode during Au lithiation contribute to the measured Au signal in the EDS map. It also shows that there is Au in the "clean" area between the particles, suggesting a fraction of Au is dissolved in the electrolyte. On the other side of the Au electrode, we observe a sharp edge with Au uniformly distributed across the electrode.

In addition to the two types of reactions above, expansion and shrinkage of the Au electrode during discharging and charging are observed. This type of reaction is classical and frequently reported in the literature,[20] which is probably because such a reaction can easily be confirmed by *ex situ* characterization techniques, although detailed reaction dynamics are lacking. The sequential images in Fig. 6A–F show the morphology changes of the Au electrode during lithiation and delithiation. When the voltage is ramped down to −1.7 V, electrolyte bubbles start to form (Fig. 6A). This is followed by an expansion in the front part of the Au electrode (Fig. 6B), indicating that the Au is lithiated. The intensive expansion of

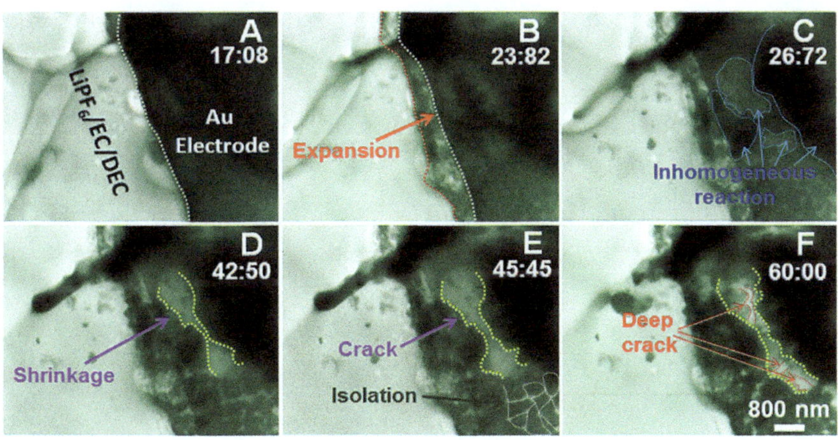

Fig. 6 Evolution of the Au electrode when electrochemically reacted with the lithium electrolyte. Expansion, cracking and partial shrinkage are observed.

the electrode induces cracks. The inhomogeneous reaction becomes more and more obvious and some heavy lithiated regions are distributed in the Au electrode. The electrode area with darker contrast represents light lithiation, while the brighter contrast indicates heavy lithiation, as marked by the blue dash line in Fig. 6C. During the charging process, shrinkage of the electrode is observed as indicated by the contour changes of the lithiated regions.

Au is likely catalyzing the electrolyte decomposition during the charge cycles. We saw lots of bubbles in Fig. 3 in the voltage range of 0 to −3 V. There are several factors influencing the electrolyte decomposition, including the applied voltage and irradiation by the electron beam. Our control experiments suggest the generation of gas bubbles is primarily due to electrolysis of the electrolyte. In a previous paper,[23] we discussed that the beam effect is secondary to electrolysis in electrolyte decomposition, since there is no bubble formation without an electric bias. Here, we found that Au may catalyze the electrolysis of the electrolyte. This is not surprising since Au is catalytically active in many chemical reactions,[26–29] including those in CO oxidation, Li–O_2 batteries and so forth. Fig. 7A–C (as well as Fig. 3) show that the Au electrode is clear with no reactions at the beginning (Fig. 7A). When the voltage is ramped down in the negative direction, many bubbles appear on the front of the electrode. Small bubbles can accumulate and merge into larger ones of up to several micrometers, as highlighted in Fig. 7B and C. At room temperature, an equilibrium exists:[23,30]

$$LiPF_6\ (s) \leftrightarrow LiF\ (s) + PF_5\ (g) \tag{1}$$

In the presence of non-aqueous solvents, PF_5 tends to cause the decomposition of carbonates:[30]

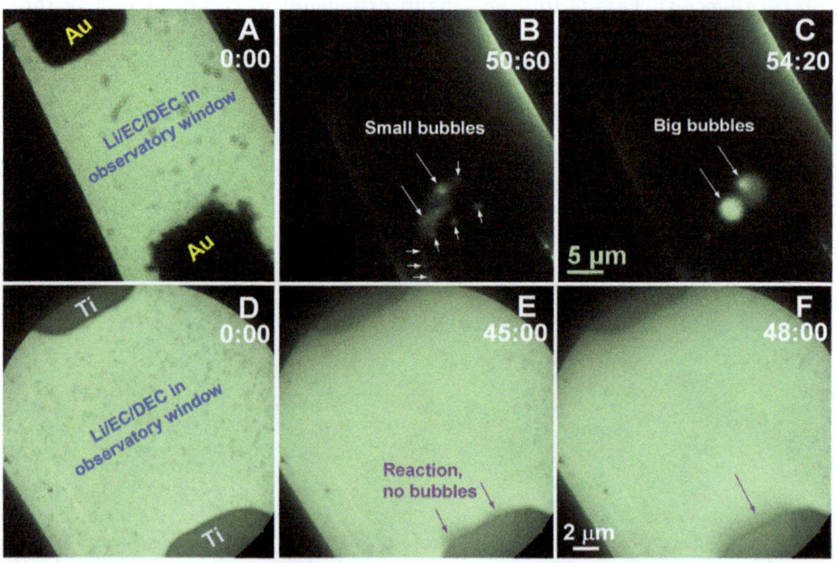

Fig. 7 Sequential images show the electrode–electrolyte interface reaction on both sides of the electrodes in two electrochemical cells. Au electrodes are used in (A)–(C) and Ti electrodes are used in (D)–(F).

(2)

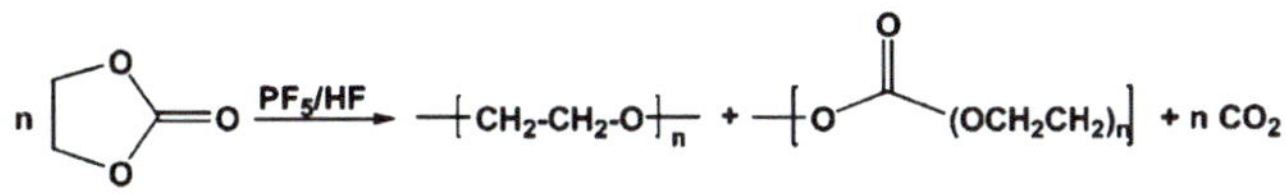

(3)

It is also known that moisture can induce the hydrolysis of $LiPF_6$ salts:[30]

$$LiPF_6\ (sol.) + H_2O \leftrightarrow LiF\ (s) + 2HF\ (sol.) + POF_3\ (sol.) \quad (4)$$

$$PF_5\ (sol.) + H_2O \leftrightarrow 2HF\ (sol.) + POF_3\ (sol.) \quad (5)$$

In our experiments, we loaded the electrolyte in a glove-box so that the moisture effects were minimized. It is likely that Au is catalyzing the above reactions 1–3. For our control experiments, we used Ti electrodes to replace the Au electrodes. The results show that, when cyclic voltammetry is applied in the voltage range of 0 to −3 V, no severe bubble formation is observed (Fig. 7E and F). This supports Au catalyzed bubble formation from electrolyte decomposition.

The electron beam has a strong effect on the current measured from the electrochemical liquid cell under TEM. Previously, we have shown that there is a drastic increase of the current when the cell is exposed to the electron beam.[31] Here, we show the cyclic voltammetry curves measured in an electrochemical cell with/without the electron beam. Fig. 8A is the CV curve without the electron beam passing through. It shows that when the voltage is ramped down from 0 to −3 V, there is a reduction peak located at −2.1 V, corresponding to the lithiation reaction of the Au electrode. When the voltage is swept back from −3 V to 0 V, there is an oxidation peak located at −1.45 V, which is attributed to the delithiation reaction of the Au electrode. It is noted that the reduction and oxidation peaks of −2.1 V and −1.45 V are not the exact lithiation/delithiation voltages of Au electrodes in lithium ion batteries, since there is no lithium reference electrode in the set up. However, it is clear that the CV plot without the electron beam show

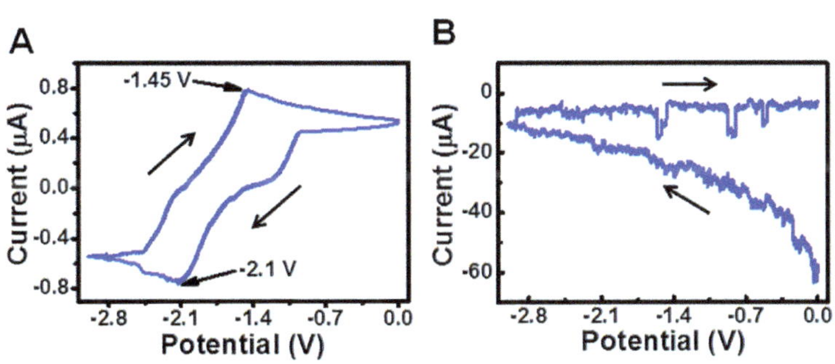

Fig. 8 Cyclic voltammetry without (A) or with (B) the electron beam passing through the SiN_x membrane in the same electrochemical liquid cell.

reduction/oxidation peaks corresponding to lithiation/delithiation of the Au electrode. When the electron beam passes through the electrochemical liquid cell, the CV plot changes drastically with strong current increases (Fig. 8B). As the voltage ramps from 0 to -3 V, the current decreases from -63 to -10 μA. Then, when the voltage is swept back to the positive direction, the current maintains a steady state in the -4 to -6 μA range, where the peak current without the electron beam is only 0.77 μA. By comparing these two CV curves, we conclude that in the current experiments, the major component of the measured current is from the electron beam.

The current changes due to the electron beam can be calculated based on the TEM imaging conditions and the electrochemical cell set up. We used an electron current density (J) of 2.5×10^3 electrons $Å^{-2}$ s^{-1} during imaging and the area (S) of the Au electrode exposed to the electron beam is $S = 2Lw$, where L is the length ($L = 10$ μm) and w is the width ($w = 6$ μm) of the Au electrode exposed to the electron beam. The current is measured in Amps (charge per second) and 1 A = -6.241×10^{18} e s^{-1}, therefore, the current (I) generated from the electron beam can be calculated: $I = JS = 2JLw = -5$ μA. The calculated value is comparable with the current measured by the electrochemical working station under the cyclic voltammetry program. This further confirms that the current increase is due to electrons collected directly by the electrodes from the electron beam. How such electrons affect the electrochemical reaction, including electrolyte decomposition, needs to be further studied. The thickness of the Au electrode in the current electrochemical liquid cells is about 120 nm. We expect that when the thickness of the electrode is reduced, the current increase from electron beam irradiation can be reduced.

Although the morphology changes of the Au electrode during the reaction with Li have been observed in great detail as discussed above, it is a great challenge to characterize the structure of the intermediate Li–Au alloy phases, due to the fast reaction kinetics and the amorphous-like reaction products dissolving at the end. We did parallel *ex situ* experiments to show the composition and crystal structure of the lithiated Au. As shown in Fig. 9,[23] when a single crystal Au nanowire is lithiated, both the morphology and crystal structure are changed. The electron diffraction pattern confirms the FCC structure of the Au crystal (space group $Fm\bar{3}m$) (inset of Fig. 9A). After cyclic voltammetry, the initially straight Au nanowire becomes twisted with a meandering morphology (Fig. 9B). Using selected area electron diffraction, we further identified the phases of the Li–Au section. The $Li_{15}Au_4$ phase is marked by the red circles (inset of Fig. 9B).[20] The blues circles are attributed to the original cubic structure of Au (space group $Fm\bar{3}m$). The atomic resolution of Au and Li ordering in an electrochemically treated Au nanowire reveals a L12 (Au_3Li) Li–Au structure in co-existence with the pure Au lattice (111) lattice planes[4,32] (Fig. 9C). From spatially resolved electron energy loss spectroscopy (EELS) spectra (Fig. 9D and E), the Au core with a Li–Au alloy shell can be achieved. The Li concentration profile can be extracted by integrating the near edge signal from 56–65 eV, as shown in Fig. 9E. The Au concentration is almost the same with no obvious deviation along the red dash line of the cross section of the Au nanowire. However, the Li concentration is lower in the center of the Au wire, indicating that the electrochemical lithiation starts from the surface of the Au wire, and Li propagates and diffuses towards the core.[23] Such *ex situ*

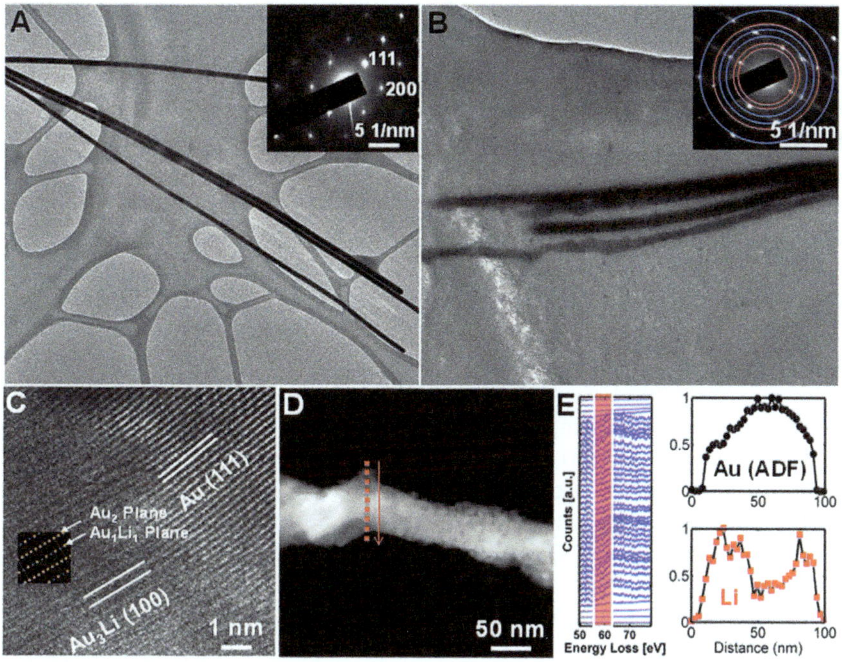

Fig. 9 *Ex situ* TEM studies of gold nanowires before and after the charge cycles. (A) TEM image of gold nanowires before lithiation, and the corresponding selected area electron diffraction pattern (SAED; inset). (B) TEM image of lithiated Au nanowires, and the corresponding SAED (inset). (C) High resolution TEM image of a lithiated Au nanowire with a Au_3Li structure. The alternating planes of Au_2 and Au_1Li_1 are shown in the model structure (Au in yellow; Li in blue). (D) HAADF image of a lithiated Au nanowire. The electron energy loss spectroscopy (EELS) line profile is along the dash line. (E) Extracted lithium and gold concentration along the line profile in (D) with the maximum intensity normalized to 1. The lithium concentration was integrated from the Li K-edge (56–65 eV). (Au O edge contribution in this energy range is negligible. Gold concentration was estimated from the ADF signals.)[23] Reprinted with permission from Z. Y. Zeng, W. I. Liang, H. G. Liao, H. L. L. Xin, Y. H. Chu and H. M. Zheng, *Nano Lett.*, 2014, **14**, 1745–1750. Copyright 2014 American Chemical Society.

study provide some useful information, although intermediate phases are missing.

Conclusions

In this paper, we demonstrated Au lithiation *in situ* by utilizing an electrochemical liquid cell. The observed Li–Au reaction indicated that there are three types of morphology change during Au lithiation: dissolution, explosive reaction and expansion/shrinkage. The possible factors that induce these kinds of Li–Au reactions are discussed. Besides the Li–Au reactions, we also studied bubble formation from electrolyte decomposition and the electron beam effect. The catalytic effects of Au in electrolyte decomposition are observed. More detailed characterization of the intermediate structure during the lithiation of Au is needed by *in situ* study in future work.

Acknowledgements

We used both the MSD TEM facility and TEAM0.5 and CM200 microscopes at the National Center for Electron Microscopy (NCEM) of Lawrence Berkeley National Laboratory (LBNL), which is supported by the U.S. Department of Energy (DOE) under contract # DE-AC02-05CH11231. W.L. is supported by the National Science Council in Taiwan under contract number NSC102-2911-I-009-502. H.Z. thanks the support of the DOE Office of Science Early Career Research Program. We thank Direct Electron, LP (San Diego, CA) for providing the high speed direct electron camera model DE-12 for movie capture.

References

1 W. Xu, J. Wang, F. Ding, X. Chen, E. Nasybulin, Y. Zhang and J.-G. Zhang, *Energy Environ. Sci.*, 2014, **7**, 513–537.

2 M. Gu, Z. Wang, J. G. Connell, D. E. Perea, L. J. Lauhon, F. Gao and C. Wang, *ACS Nano*, 2013, **7**, 6303–6309.

3 J. T. Yin, M. Wada, S. Yoshida, K. Ishihara, S. Tanase and T. Sakai, *J. Electrochem. Soc.*, 2003, **150**, A1129–A1135.

4 Y. J. Lee, Y. Lee, D. Oh, T. Chen, G. Ceder and A. M. Belcher, *Nano Lett.*, 2010, **10**, 2433–2440.

5 H. C. Shin and M. Liu, *Adv. Funct. Mater.*, 2005, **15**, 582–586.

6 W. Luo, S. Lorger, B. Wang, C. Bommier and X. L. Ji, *Chem. Commun.*, 2014, **50**, 5435–5437.

7 Z. Y. Zeng, J. P. Tu, Y. Z. Yang, J. Y. Xiang, X. H. Huang, F. Mao and M. Ma, *Electrochim. Acta*, 2008, **53**, 2724–2728.

8 Z. Y. Zeng, J. P. Tu, X. L. Wang and X. B. Zhao, *J. Electroanal. Chem.*, 2008, **616**, 7–13.

9 Z. Y. Zeng, J. P. Tu, X. H. Huang, X. L. Wang, X. B. Zhao and K. F. Li, *Electrochem. Solid-State Lett.*, 2008, **11**, A105–A107.

10 Z. Y. Zeng, J. P. Tu, X. H. Huang, X. L. Wang and J. Y. Xiang, *Thin Solid Films*, 2009, **517**, 4767–4771.

11 G. Taillades, N. Benjelloun, J. Sarradin and M. Ribes, *Solid State Ionics*, 2002, **152**, 119–124.

12 H. Zheng, R. K. Smith, Y.-w. Jun, C. Kisielowski, U. Dahmen and A. P. Alivisatos, *Science*, 2009, **324**, 1309–1312.

13 H.-G. Liao, L. Cui, S. Whitelam and H. Zheng, *Science*, 2012, **336**, 1011–1014.

14 K. W. Noh and S. J. Dillon, *Scr. Mater.*, 2013, **69**, 658–661.

15 M. Gu, L. R. Parent, B. L. Mehdi, R. R. Unocic, M. T. McDowell, R. L. Sacci, W. Xu, J. G. Connell, P. Xu, P. Abellan, X. Chen, Y. Zhang, D. E. Perea, J. E. Evans, L. J. Lauhon, J.-G. Zhang, J. Liu, N. D. Browning, Y. Cui, I. Arslan and C.-M. Wang, *Nano Lett.*, 2013, **13**, 6106–6112.

16 M. E. Holtz, Y. Yu, D. Gunceler, J. Gao, R. Sundararaman, K. A. Schwarz, T. A. Arias, H. D. Abruña and D. A. Muller, *Nano Lett.*, 2014, **14**, 1453–1459.

17 P. Abellan, B. L. Mehdi, L. R. Parent, M. Gu, C. Park, W. Xu, Y. Zhang, I. Arslan, J.-G. Zhang, C.-M. Wang, J. E. Evans and N. D. Browning, *Nano Lett.*, 2014, **14**, 1293–1299.

18 R. L. Sacci, N. J. Dudney, K. L. More, L. R. Parent, I. Arslan, N. D. Browning and R. R. Unocic, *Chem. Commun.*, 2014, **50**, 2104–2107.

19 C.-M. Wang, X. Li, Z. Wang, W. Xu, J. Liu, F. Gao, L. Kovarik, J.-G. Zhang, J. Howe, D. J. Burton, Z. Liu, X. Xiao, S. Thevuthasan and D. R. Baer, *Nano Lett.*, 2012, **12**, 1624–1632.
20 L. Yuan, H. K. Liu, A. Maaroof, K. Konstantinov, J. Liu and M. Cortie, *J. New Mater. Electrochem. Syst.*, 2007, **10**, 95–99.
21 B. Laik, L. Eude, J. P. Pereira-Ramos, C. S. Cojocaru, D. Pribat and E. Rouviere, *Electrochim. Acta*, 2008, **53**, 5528–5532.
22 P. Bach, A. Seemayer, U. Ruett, O. Gutowski and F. Renner, *DESY Photon Science Annual Reports*, 2012.
23 Z. Y. Zeng, W. I. Liang, H. G. Liao, H. L. L. Xin, Y. H. Chu and H. M. Zheng, *Nano Lett.*, 2014, **14**, 1745–1750.
24 A. D. Pelton, *Bull. Alloy Phase Diagrams*, 1986, **7**, 228–231.
25 P. Limthongkul, Y. I. Jang, N. J. Dudney and Y. M. Chiang, *Acta Mater.*, 2003, **51**, 1103–1113.
26 B. Hvolbæk, T. V. W. Janssens, B. S. Clausen, H. Falsig, C. H. Christensen and J. K. Nørskov, *Nano Today*, 2007, **2**, 14–18.
27 T. V. W. Janssens, A. Carlsson, A. Puig-Molina and B. S. Clausen, *J. Catal.*, 2006, **240**, 108–113.
28 S. H. Overbury, V. Schwartz, D. R. Mullins, W. Yan and S. Dai, *J. Catal.*, 2006, **241**, 56–65.
29 Z. Peng, S. A. Freunberger, Y. Chen and P. G. Bruce, *Science*, 2012, **337**, 563–566.
30 K. Xu, *Chem. Rev.*, 2004, **104**, 4303–4417.
31 M. Sun, H.-G. Liao, K. Niu and H. Zheng, *Sci. Rep.*, 2013, **3**, 3227.
32 H. L. Xin, J. A. Mundy, Z. Y. Liu, R. Cabezas, R. Hovden, L. F. Kourkoutis, J. L. Zhang, N. P. Subramanian, R. Makharia, F. T. Wagner and D. A. Muller, *Nano Lett.*, 2012, **12**, 490–497.

Faraday Discussions

PAPER

Effect of electrochemical dissolution and deposition order on lithium dendrite formation: a top view investigation

Wenjun Li,†[a] Hao Zheng,†[a] Geng Chu,[a] Fei Luo,[a] Jieyun Zheng,[a] Dongdong Xiao,[a] Xing Li,[b] Lin Gu,*[a] Hong Li,*[a] Xianlong Wei,[b] Qing Chen[b] and Liquan Chen[a]

Received 4th June 2014, Accepted 14th August 2014
DOI: 10.1039/c4fd00124a

Rechargeable metallic lithium batteries are the ultimate solution to electrochemical storage due to their high theoretical energy densities. One of the key technological challenges is to control the morphology of metallic lithium electrode during electrochemical dissolution and deposition. Here we have investigated the morphology change of metallic lithium electrode after charging and discharging in nonaqueous batteries by *ex situ* SEM techniques from a top view. Formation of the hole structure after lithium dissolution and the filling of dendrite-like lithium into the holes has been observed for the first time. In addition, an *in situ* SEM investigation using an all-solid Li/Li_2O/super aligned carbon nanotube set-up indicates that lithium ions could diffuse across through the surface oxide layer and grow lithium dendrites after applying an external electric field. The growth of lithium dendrites can be guided by electron flow when the formed lithium dendrite touches the carbon nanotube.

Introduction

High energy density rechargeable batteries are the key technology to support rapid developments of many emerging technologies, such as advanced portable and wearable electronic devices, electrical vehicles, smart grids, tools and toys, robots, and back-up batteries for various applications. According to thermodynamic calculations, the Li–O_2 battery has the highest theoretical gravimetric energy density among chemical energy storage devices.[1] For similar chemical reactions, batteries with a lithium anode have the highest energy density among other batteries with different metal anodes, such as Mg, Al, Na and Zn. In addition, rechargeable metallic lithium batteries (RMLB) have much higher energy

[a]*Beijing National Laboratory for Condensed Matter Physics, Institute of Physics, Chinese Academy of Sciences, Beijing, 100190, P. R. China. E-mail: l.gu@iphy.ac.cn; hli@iphy.ac.cn*

[b]*Key Laboratory for the Physics and Chemistry of Nanodevices, Department of Electronics, Peking University, Beijing 100871, P. R. China*

† Wenjun Li and Hao Zheng contributed equally.

densities than lithium ion batteries (LIB) with the same cathode.[1] Recently, the developing of RMLB has attracted increasing attention after the first wave in the 1970s, especially on Li–S, Li–air, and all solid lithium batteries (ASLB) using polymer or ceramic as electrolytes.[2–6] However, periodic dissolution and deposition of metallic lithium during cycling leads to the formation of microstructured or nanostructured lithium with different morphologies, such as needle-like, dendrite, granular and mossy lithium. The newly formed structured lithium reacts significantly with nonaqueous electrolyte electrochemically and chemically.[7,8] The products are electronically insulating and ionically conducting solid electrolyte interphase (SEI) films. The formation of SEI is unavoidable in nonaqueous lithium batteries, leading to the irreversible consumption of electrolyte and isolation of active lithium. Consequently, the coulombic efficiency of the battery is decreased and the cyclic performance is poor. In addition, structured lithium becomes more dangerous and could cause an internal short circuit, rapid heating and explosion of the battery.[7–9]

Several strategies have been attempted to control the morphology of metallic lithium electrodes during cycling, including tuning the solvent and salt, ionic liquids, functional additives to modify SEI, suppress lithium dendrites by tuning the local electrical field, lithium surface coating, solid electrolytes, ceramic separators, 3D current collectors, using lithium powder instead of lithium foil, and mechanical pressing.[8] In spite of great efforts in improving the performance of metallic lithium electrodes, investigation of the morphology and surface composition are quite challenging due to the soft and air-sensitive nature of the lithium electrode. *Ex situ* and *in situ* scanning electronic microscopy (SEM) has been used widely to investigate lithium dendrites in lithium nonaqueous batteries,[10–12] and lithium polymer batteries.[13–15] Atomic force microscopy,[16,17] optical spectroscopy,[18] and magnetic resonance imaging (MRI) technology[19] have also been used to investigate the morphology change of lithium electrodes. *In situ* NMR was also developed successfully to quantify the amount of dendritic/mossy Li by analyzing the change in the NMR signal intensity of the Li metal during cycling.[20] Fourier transmission infrared spectroscopy (FTIR),[11,21,22] X-ray photoelectronic spectroscopy (XPS),[22–24] Raman spectroscopy,[25,26] and secondary ion mass spectroscopy (SIMS)[22] have been used to identify the surface composition of reacted lithium electrodes. Electrochemical impedance spectroscopy (EIS) and cyclic voltammetry (CV) have been performed to analyze the electrical response of the lithium electrode.[11,16]

For the rechargeable lithium battery with a cathode containing the lithium source, such as a Li/$LiCoO_2$ cell, the deposition of lithium occurs firstly during the first charging then follows the dissolution of lithium during discharging. In the case of rechargeable lithium batteries with a lithium anode as the lithium source, such as Li–S, Li–air batteries, the dissolution of lithium occurs firstly during the first discharge then follows the deposition. There are few reports to distinguish the effect of the order of dissolution and deposition on the morphology of the lithium electrode. In this work, we have investigated the order effect of dissolution and deposition by comparing the morphology change of the metallic lithium electrodes in Li/Si cells, and Li/$LiCoO_2$ and Li/$LiFePO_4$ cells respectively. It is obvious that the top view is favorable for the investigation of the morphology change of the lithium electrode. There are very few top-view SEM images of reacted lithium since the lithium electrode is soft and the sample transfer could change the morphology completely. In order to solve this problem, a special stainless steel sample holder was designed. The soft lithium sheet can be embedded into the hole of the SS substrate and the surface of the lithium can be flattened by mechanical pressing. Transfer of the lithium electrode was performed by contact with the hard edge of the SS sample holder. In addition, our Hitachi SEM instrument has a specially designed sealed sample transfer box, which allows the transfer of the air-sensitive sample from the Ar-filled glove box to the vacuum chamber of the SEM microscope without exposure to air.

Experimental

Flatten lithium sheet for this study

The lithium sheet (4 mm in diameter, 0.45 mm in thickness) in the SS sample holder was pressed by a precisely polished stainless steel pillar to flatten its surface. The roughness of the lithium foil before and after pressing was measured by a MultiMode 8 (BRUKER) scanning probe microscope (SPM) equipped in an Ar-filled glove box.[27] Atomic force microscope (AFM) images are shown in Fig. 1. The surface roughness is defined by root-mean-squared (RMS) values calculated from the AFM images.[27] As shown in Fig. 1 and Table 1, the surface roughness of the pressed lithium foil is much smaller than the unpressed one. Therefore, the pressed lithium sheet is suitable for top-view investigation.

Electrochemical cell

A flooded two-electrode nonaqueous cell and a specially designed sample holder were used and are shown in Fig. 2, these can be easily assembled and disassembled and can avoid direct contact with the surface of the reacted lithium electrode. Our new design is especially suitable for sample transfer from the glove box for *ex situ* SEM measurements. Three types of cells were assembled with a surface flattened lithium sheet electrode. The cathode was a silicon/carbon (Si/C) composite electrode, $LiCoO_2$ electrode and $LiFePO_4$ electrode, respectively. All cathodes were composed of the active material, the conductive additive (carbon black) and the polyvinylidene fluoride (PVDF) binder at weight ratio of 92 : 3 : 5. These electrodes were prepared by spreading the active material slurry on the

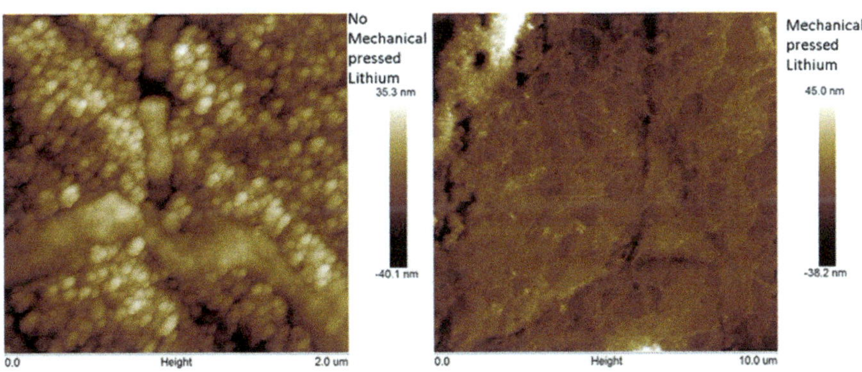

Fig. 1 The roughness of the lithium foil before and after mechanical pressing, measured by AFM.

Table 1 Lithium roughness parameter before and after mechanical pressing

Parameter	R_a/nm	R_q/nm	Roughness R_{max}/nm
Before press	8.02	10.0	72.7
After press	3.15	3.99	34.57

current collector (copper foils for the Si/C composites, Al foils for $LiCoO_2$ and $LiFePO_4$). Commercial electrolytes of 1 M $LiPF_6$ dissolved in ethylene carbonate–dimethyl carbonate (EC–DMC) (1 : 1 in v/v) without any additives were used in all cells.

Ex situ SEM investigation

As the cell was assembled, it was discharged or charged. After the discharge or charge, the cell was disassembled in the glove box and the reacted lithium foil was taken out from the cell and washed with dimethyl carbonate (DMC). The washed lithium electrode was dried in the vacuum chamber of the glove box for at least 6 hours before it was transferred to the SEM chamber of a Hitachi S4800 microscope. The special designed transfer box was used to transfer the sample, which avoid exposure of the sample to the air. In addition, a special SEM sample holder was designed to fix the installed position of the sample, which was suitable for finding the same position of the same reacted lithium electrode during each *ex situ* SEM investigation. After the SEM measurements, the lithium electrode was assembled into the cell again to be discharged or charged to next state. Such a procedure was repeated, as shown in Fig. 3.

Order of discharging and charging of lithium electrode

The Li/Si cell was discharged first. Consequently, dissolution of lithium from the lithium electrode will occur first. The Li/$LiCoO_2$ and Li/$LiFePO_4$ cell were charged first. Therefore, deposition of lithium on the counter lithium electrode will occur

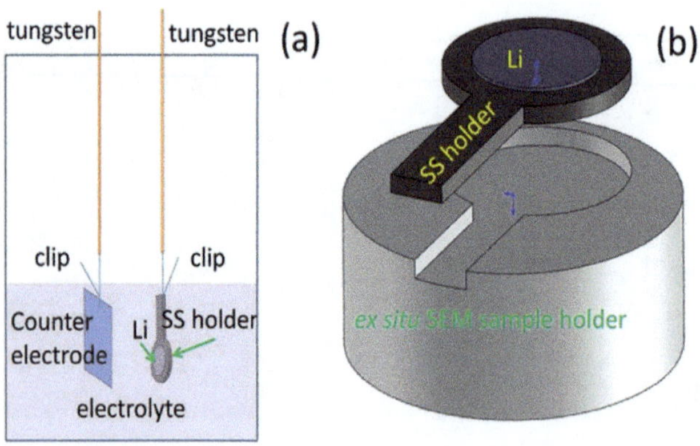

Fig. 2 Sketch of (a) a flooded two-electrode nonaqueous cell (b) sample holder for SEM investigation.

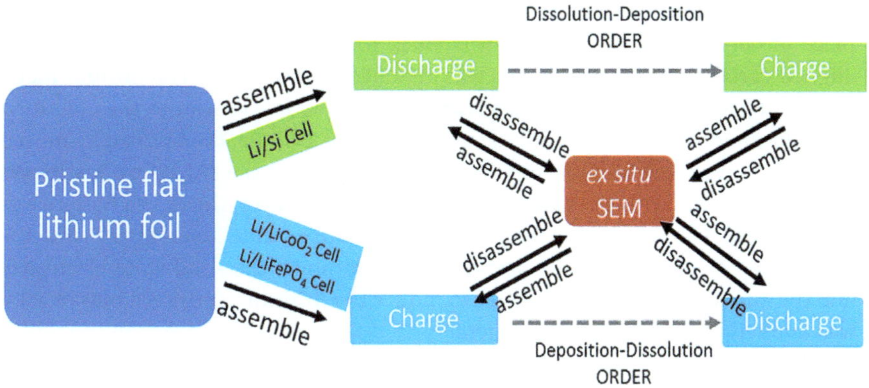

Fig. 3 The experimental design for the dissolution–deposition order and the deposition–dissolution order investigation.

first. Accordingly, we could investigate the effect of changing the discharging/charging order on the morphology change of metal lithium electrodes. The corresponding experimental details are listed in Table 2.

Second ion mass spectroscopy (SIMS) experimental

All mass spectroscopy and elemental SIMS mapping (ESM) were recorded by the Hiden SIMS Workstation to analyze the composition of the lithium electrode surface after cycling. An Ar ion gun was used. Full mass spectra were recorded using 1 nA ion current and the depth profiles were recorded using 100 nA ion current. ESM was recorded using 10 nA ion current. Two kinds of lithium foils, *i.e.* the pristine lithium foil, the lithium foil from the Li/$LiFePO_4$ battery at the second and 5th discharge state, were chosen for surface composition analysis. The sample transfer process was protected by a glove bag filled with argon.

In situ SEM investigation

An all-solid-state cell was used and investigated in a FEI Quanta 600 microscope, as shown in Fig. 4. The formation and growth of the lithium dendrite were investigated. In order to construct the *in situ* micro-scale all-solid-state battery setup, two tips including one Au tip and one tungsten tip were installed on the

Table 2 The sample details of the *ex situ* SEM measurements

Cell	Charge/discharge state	Charge/discharge capacity/ mAh	SEM images
Li/Si	Dis. to 20 mV	1.33	Fig. 6(a)
	Dis. to 5 mV	1.47	Fig. 6(b)
	Charge to 1 V	0.42	Fig. 6(c)
	Dis. to 5 mV	1.40	Fig. 6(d)
Li/$LiCoO_2$	Charge to 4.2 V	3.6	Fig. 8(c) and (d)
	Dis. to 3 V	1.78	Fig. 8(e) and (f)
Li/$LiFePO_4$	Charge to 4 V	0.26	Fig. 9(c) and (d)
	Dis. to 2.8 V	0.15	Fig. 9(e) and (f)

probe workstation (Kleindiek) as shown in Fig. 4. Before closing the chamber, a piece of commercial lithium foil was stuck to the tungsten tip, and the super-aligned carbon nanotube (SACNT) was attached to the Au tip. The lithium metal on the tungsten tip was oxidized naturally in air before transfer to the chamber. After the chamber is closed and evacuated to about 10^{-2} Pa, the tips can be moved in three dimensions by handling the manipulator; once the SACNT on Au tip was moved to contact with the surface of the native oxide layer, the *in situ* battery was constructed. The SACNT on the Au tip, the native oxide layer Li_2O[28] and the lithium metal on tungsten tip worked as the cathode, the solid state electrolyte and the anode, respectively. In order to induce the lithium dendrite, a bias of −3 V was applied to the SACNT *vs.* Li metal.

Results and discussions

SEM image of original lithium sheet electrode

As shown in Fig. 5, it is quite clear that lithium sheet after flattening treatment does have flat surface and the defect structure is decreased, which is suitable for further SEM and AFM investigation. In this work, all lithium sheet electrodes used for AFM and SEM studies are flattened lithium sheet electrodes embedded into the SS sample holder shown in Fig. 2. For simplicity, the term "lithium electrode" will be used in later sections without extra description.

Morphology change of Li electrode in Li/Si cell

Fig. 6(a) shows a typical SEM image of the lithium electrode after the first discharge to 20 mV in the Li/Si cell. Many holes with different size and depth can be observed. In this image, the diameter of the biggest hole is about 100 μm, and the depth of the hole cannot be measured. The appearance of such a large hole indicates that the electrochemical dissolution of lithium occurs more easily at the etched area. It is reasonable since the surface of the lithium electrode is covered unevenly by a layer of insulating SEI film once the lithium electrode is soaked and discharged in the electrolyte. Once the lithium electrode starts to dissolve in a certain area after discharging, some areas will be etched and the SEI surface layer covered on those areas can also be peeled off. The exposed fresh areas of the lithium electrode have a lower resistance compared to unreacted areas and are

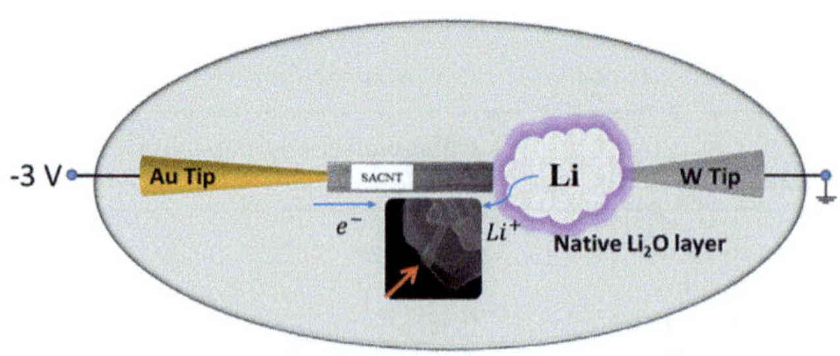

Fig. 4 The schematic illustration of the *in situ* SEM setup of all-solid-state battery. The SACNT, native Li_2O layer and Li metal worked as the cathode, solid-electrolyte and anode respectively. −3 V was applied on SACNT *vs.* Li metal to induce the formation of the lithium dendrite.

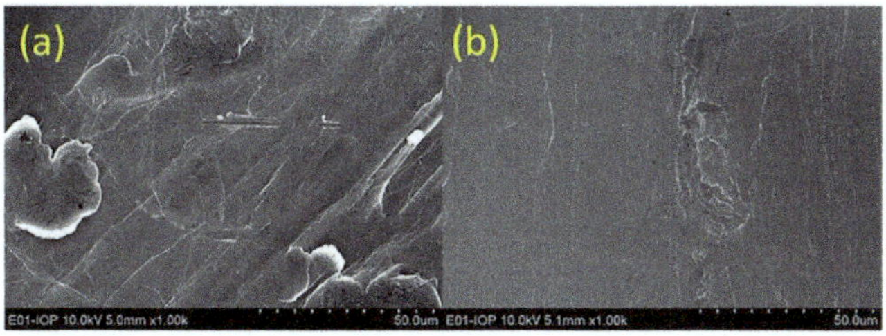

Fig. 5 (a) Original lithium foil, (b) flatten lithium foil.

can more easily be etched further. It is noted that further discharge to 5 mV seems not change too much the morphology of lithium electrode, compared to the lithium electrode discharged to 20 mV, as shown in Fig. 6(b). Some small holes can be seen clearly at the bottom inside the biggest hole and in some other regions as well.

Fig. 6(c) shows the morphology of the lithium electrode after the first charge to 1.0 V. It can be seen that the mossy-like lithium deposits back into the biggest hole of the lithium electrode which was formed mainly during discharging. We

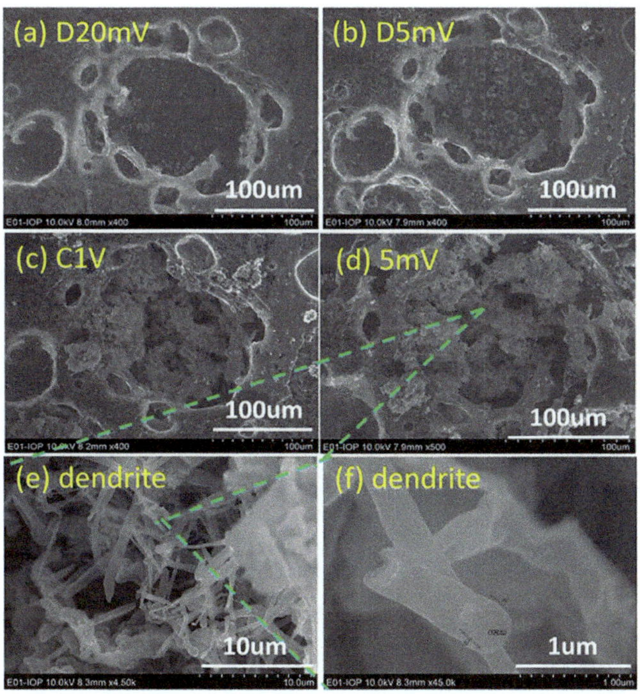

Fig. 6 Morphology evolution of the lithium electrode during lithium dissolution and deposition in the Li/Si cell. The current density for charging and discharging was 6.4 mA cm^{-2}. (a) 1st discharge to 20 mV; (b) 1st discharge to 5 mV; (c) 1st charge to 1 V; (d) 2nd discharge to 5 mV; (e and f) SEM images of the dendrite-like lithium which deposited in the hole.

did not observe that lithium is deposited on the flat surface of the lithium electrode during charging. This indicates that once the holes are formed, that area becomes more electrochemically active than the unreacted flattened areas.

Fig. 6(d) shows the morphology of the lithium electrode after the second discharge to 5 mV. The morphology does not change too much and the filled mossy-like lithium in the big hole still exists but the amount decreases. The edge part of the big hole is destroyed slightly. At high magnification in Fig. 6(e) and (f), it can be observed clearly that the mossy-like lithium has the dendrite appearance. The diameter is about 400 nm in this case.

To get a whole impression, low resolution images of the lithium electrode in the Li/Si cell at different states are shown in Fig. 7. In the center area of the lithium electrode, many holes can be seen clearly after the first discharge to 20 mV. The distribution of the holes is random. The hole size is about 50–100 μm. The density and size of the holes increases after further discharging to 5 mV. Some holes have merged together. At the edge part, it is noted that dendrite-like lithium is observed, as shown in Fig. 7b and c. After the second discharge, the surface becomes significantly rougher. From Fig. 7d, it is quite obvious, that the many deposited mossy-like lithium structures still remain on the surface of the lithium electrode; it is quite plausible that the remaining deposited lithium after the second discharge becomes "dead" lithium.

The morphology evolution of the lithium electrode in the Li/$LiCoO_2$ cell is shown in Fig. 8. The pristine lithium electrode has a flat surface and is free of any deposited lithium.

In the charged state, the surface of the lithium electrode is covered by granular species, as shown in Fig. 8c and d. The distribution of the granules can be regarded as random, although parallel stripes can be seen. At the discharge state, most of the granular species remain. Note that the coulombic efficiency of the Li/$LiCoO_2$ cell is 49% (calculated from the Table 2, 1.78/3.6 = 49%), which means that only 49% deposited lithium dissolves from the lithium electrode during

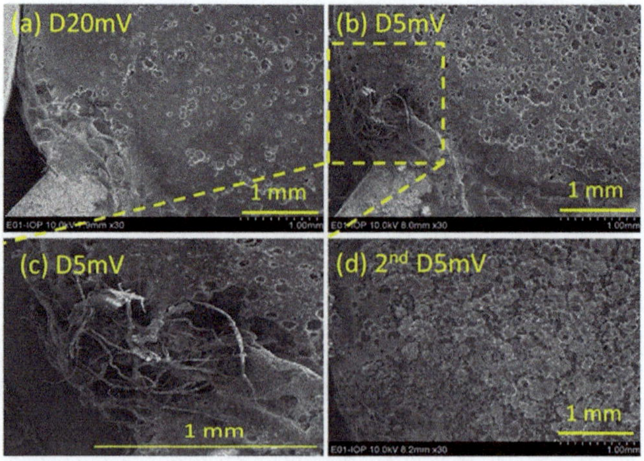

Fig. 7 Evolution of lithium electrode in Li/Si system at different states: (a) 1st discharge to 20 mV; (b) 1st discharge to 5 mV; (c) 1st discharge to 5 mV, edge part; (d) 2nd discharge to 5 mV.

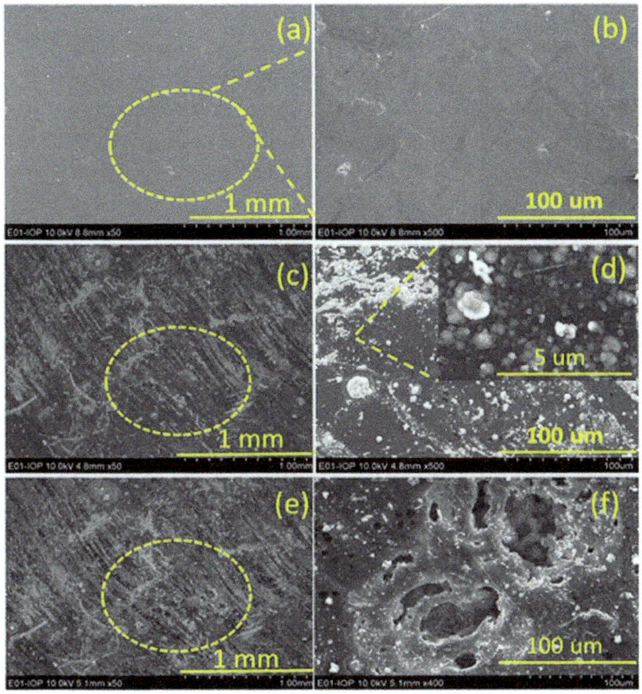

Fig. 8 Morphology evolution of the lithium electrode during lithium dissolution and deposition in the Li/$LiCoO_2$ cell. The current density for charging and discharging was 6.4 mA cm^{-2}. (a) pristine lithium; (b) zoomed image of the marked area of pristine lithium; (c) 1st charge to 4.2 V; (d) zoomed image of the marked area in the image (c); (e) 1st discharge to 3 V; (f) zoomed image of the marked area in the image (e).

discharging. This calculation is consistent with the phenomena that the granular species still remain after discharge. It is also noted that many holes appear, like in Fig. 6 and 7. This indicates further that the hole structure is more electrochemically active than the flat surface and deposited granular lithium on the surface of the lithium electrode.

In both cases of Li/Si cell and Li/$LiCoO_2$ cell, the current density during discharging and charging was 6.4 mA cm^{-2} and the corresponding capacity for the dissolution and deposition of lithium is about 1–3 mA h. Such capacity and current density are comparable values for the discharging and charging of commercial lithium batteries. In order to study the effect of the current density and capacity on the morphology of the lithium electrode, another Li/$LiFePO_4$ cell was tested with one tenth of the current density and the capacity of the Li/$LiCoO_2$ cell. As shown in Fig. 9c, a few granular particles can be seen on the surface of the lithium electrode after charging. After discharging many holes with a size of 10 μm appeared, accompanied with some smaller holes. These phenomena are similar to the case of Li/$LiCoO_2$ cell, but quite possibly in the early stages.

Since the lithium electrode has excellent electronic conductivity and ionic conductivity, the electrolyte has also high ionic conductivity and high concentration, therefore, it is not likely that the inhomogeneity of the transport of electron and lithium cations in the lithium electrode and inhomogeneous ionic flow leads to the formation of an inhomogeneous hole structure and dendrite/granular/mossy-like lithium. We suspect that the SEI on the lithium electrode plays a key role to influence the morphology of the lithium electrode during electrochemical dissolution and deposition in nonaqueous lithium cells. However, it is not very easy to detect the homogeneity of the SEI on the lithium electrode. We performed SIMS to investigate this issue indirectly.

SIMS investigation on surface compositions

With the sample holder, it is convenient to scan the negative ion full mass spectrum of the pristine and reacted lithium electrode by the SIMS technique. The lithium electrode after five cycles in a Li/$LiFePO_4$ cell is detected and the full negative ion mass spectrum is shown in Fig. 10.

According to the negative ion spectroscopy, the four elements (namely H, C, O, F) on the surface can be identified. Li, C and N can be identified from the positive ion spectrum. Accordingly, the SIMS results are consistent with previous investigations, which demonstrated that the SEI on the surface of lithium electrode contains Li_2CO_3, LiF, Li_2O.[29–31]

With the cycling scanning, the depth distribution of every element can be also detected. The scanning mode using a small ion beam current (1 nA) and in spot

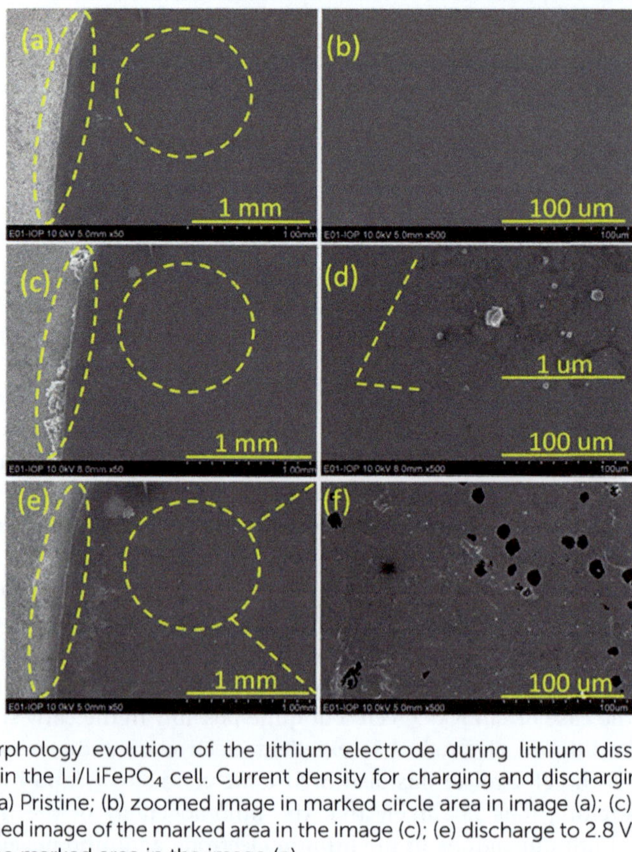

Fig. 9 Morphology evolution of the lithium electrode during lithium dissolution and deposition in the Li/$LiFePO_4$ cell. Current density for charging and discharging was 0.64 mA cm^{-2}. (a) Pristine; (b) zoomed image in marked circle area in image (a); (c) charge to 4 V; (d) zoomed image of the marked area in the image (c); (e) discharge to 2.8 V; (f) zoomed image of the marked area in the image (e).

mode can only get depth information from the surface of the sample. In this work, a raster mode provided by the Hiden SIMS was used, which applied a bigger ion beam current (100 nA) and can reflect the depth profile more accurately. Fig. 11 shows the SIMS signals from the strongest two peaks of F and O. It can be seen that concentration of fluorine on the surface is increased significantly after cycling and the concentration of oxygen decreases. This is reasonable, as LiF covers gradually the pristine oxide layer.

Elemental SIMS mapping was also performed and is shown in Fig. 12. The element distribution on the surface of the lithium electrode is inhomogeneous. The element mapping consists of the depth profiles: the concentration of fluorine is increased and the oxygen is decreased after cycling. Accordingly, the coverage of the SEI on the lithium electrode is inhomogeneous and the outer layer is enriched with fluorine.

In situ SEM investigation on the growth of lithium dendrite

Furthermore, lithium dendrite formation was investigated by the means of *in situ* SEM. Currently, we cannot perform *in situ* SEM investigation on a liquid nonaqueous cell. We built an all-solid-state battery in the SEM chamber (see Fig. 4), with the SACNT, the native oxide layer on lithium surface, the Li metal as the cathode, solid-electrolyte and anode, respectively.[32] As shown in Fig. 13, −3 V was applied to the SACNT *vs.* Li metal, to discharge the all-solid-state battery. It is observed that the lithium dendrite is formed after applying an external electrical field for 700 s (Fig. 13b), and it grew along a random direction until it overlapped the SACNT (Fig. 13a–c). Once the dendrite and SACNT overlapped each other, the dendrite started to grow along the SACNT, rather than along the previous growth direction. The growth of the lithium dendrite has to be accompanied with the transportation of lithium ions and electrons, which may be due to the high electronic conductivity of CNT. This phenomenon means that after growing to a certain length, the electron flow could also guide the formation of lithium dendrite.

Actually, the whole process of the dendrite growth was followed and captured; Fig. 13(a)–(d) were just selected for demonstration. In addition, the size evolution of the dendrite was extracted from the *in situ* captured images, as plotted in

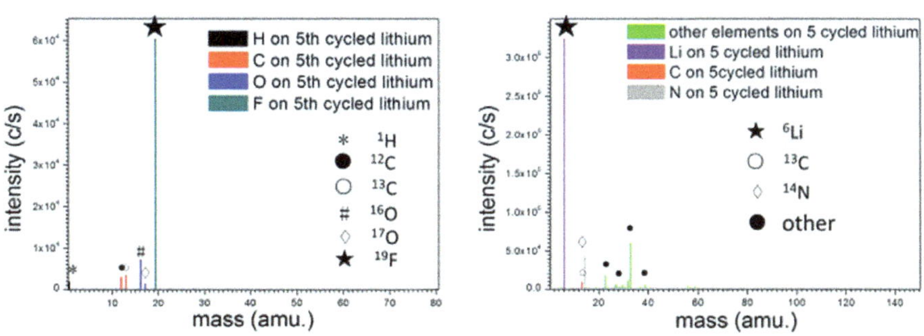

Fig. 10 SIMS full negative ion mass scanning spectrum (left) and positive ion spectrum (right) of the lithium electrode after 5 cycles. The current density was 0.64 mA cm^{-2}.

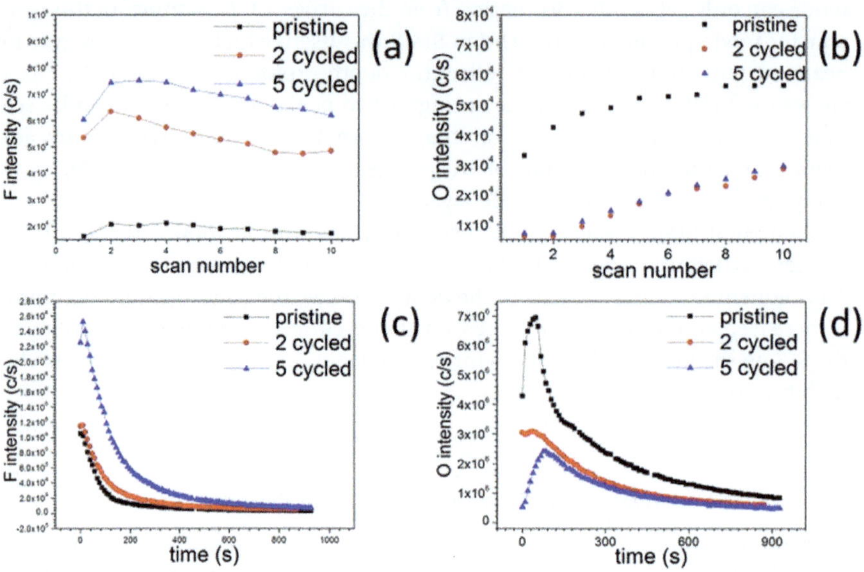

Fig. 11 The surface depth distribution of (a) fluorine and (b) oxygen, and the element depth profile of (c) fluorine and (d) oxygen in a pristine lithium electrode, after the second cycle, and the fifth cycle.

Fig. 13(e). It is demonstrated that the growth of the dendrite was a diffusion-controlled process.

The formation of the dendrite lithium in the *in situ* SEM experiment means that the lithium ions can migrate across the surface Li_2O layer, nucleate on the top of the Li_2O covered lithium electrode and grow into the dendrite. These results are consistent with the appearance of the lithium dendrite of the Li/Si cell during discharging (lithium extraction) shown in Fig. 7(c). It is also interesting to notice that as the diameter of the lithium dendrite remains the same (about 470 nm), the length of the dendrite increases continuously to as long as 1.72 μm (Fig. 13(d)). Further finite element simulation is being carried out.

As investigated above, in both *ex situ* SEM and *in situ* SEM, the formation of dendrite-like lithium can occur in both dissolution and deposition. In the case of dissolving first, large holes are formed randomly after delithiation and dendrite-like and mossy-like lithium preferentially fill the holes after lithiation. In the case of depositing first, granular-like lithium will deposit on the surface of the lithium electrode randomly after lithiation. The hole-like structure forms again on the surface in spite of covered by granular-like lithium. Since the surface of lithium electrode is covered unevenly by a layer of SEI, the dissolution and deposition of lithium are not uniform. Whatever the charging/discharging (dissolution–deposition) order, mossy-like or granular-like lithium cannot disappear. These phenomena will definitely lead to the formation of "dead" lithium, which is not favorable for achieving good cyclic performance and high coulombic efficiency for rechargeable lithium batteries.

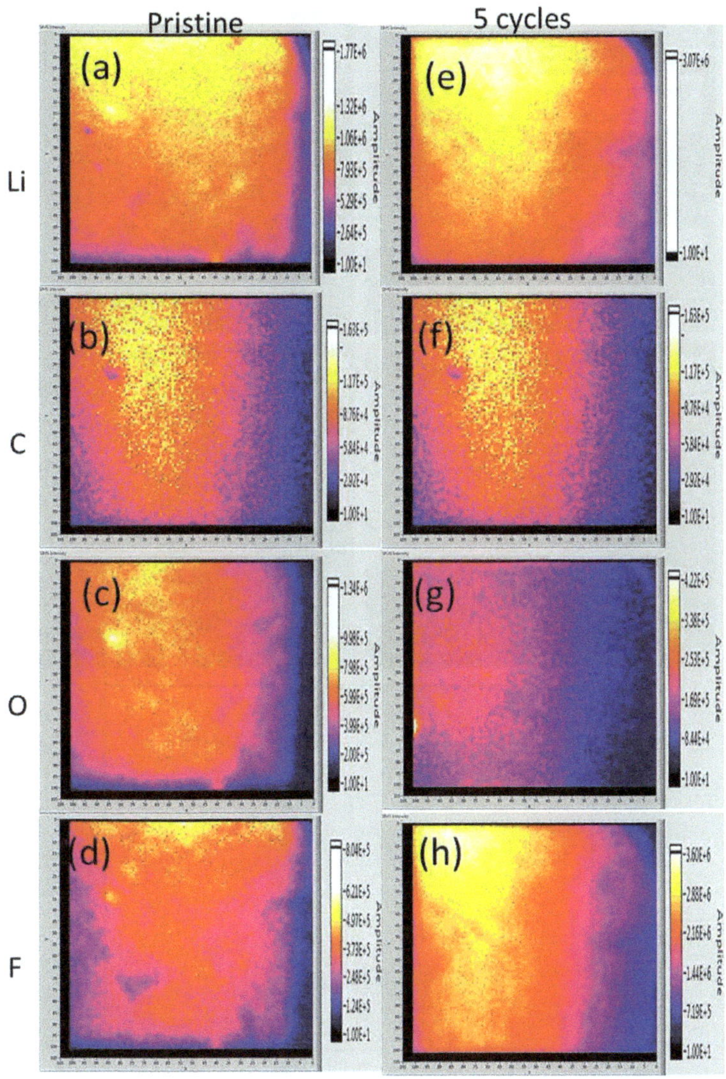

Fig. 12 ESM of Li,F, C, O: (a), (c), (e) and (g) in the pristine lithium and (b), (d), (f) and (h) in the 5-cycle lithium electrode.

Our report shows that top-view direct observation of the morphology of the lithium electrode is now technologically feasible. Combining with AFM, conducting AFM, XPS and SIMS, it is possible to understand clearly the evolution of the structure and chemical composition of the lithium electrode in rechargeable lithium batteries. Further investigations on the morphology of the lithium electrode in different lithium batteries, nonaqueous electrolytes with functional additives, surface coatings with solid electrolytes and different atmospheres are being carried out. Those investigations should be essential for developing rechargeable lithium batteries.

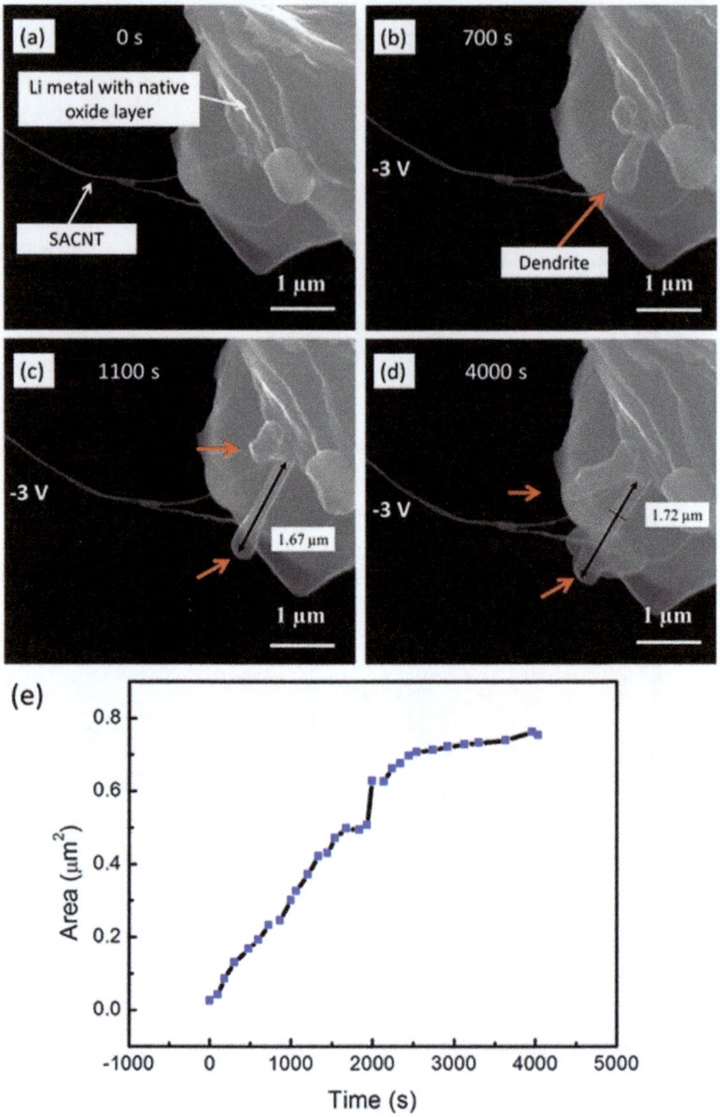

Fig. 13 The morphology evolution of the lithium dendrite in an *in situ* all-solid-state battery; the lithium dendrites were indicated by red arrows. (a–d) show the selected morphology of the lithium dendrite, with the electrical field applied for 0 s, 700 s, 1000 s, 4000 s respectively. The increase of the cross sectional area of the lithium dendrite is plotted in (e).

Conclusions

A flat surface of lithium electrode with a hard stainless steel sample holder is designed for *ex situ* SEM investigation. In the case of the Li/Si cell, a large amount of hole-like structures with 100 μm hole size and a small amount of lithium dendrites have been observed after discharging. Mossy-like and dendrite-like lithium are found mainly in the holes formed during discharging. Some of them do not disappear after the second discharge, indicating the existence of "dead lithium". In the case of Li/$LiCoO_2$ and Li/$LiFePO_4$ cells, granular like lithium is found on the surface of the lithium electrode after the first charging. Quite a lot of granular-like lithium remains after the first discharging. In addition, hole-like structures appear again. *In situ* SEM demonstrates clearly that the lithium dendrite could grow on the surface oxide (Li_2O) covered lithium during discharging of the Li/Li_2O/SACNT cell. Inhomogeneous deposition and dissolution of dendrite/mossy/granular like lithium and hole-like structures is believed to be related to inhomogeneous coverage of the SEI film, not likely to be related to the homogeneity of the electronic or ionic conductivity of the lithium electrode.

Acknowledgements

Financial supports from the NSFC project (51325206), Beijing S&T Project (Z13111000340000), and "Strategic Priority Research Program" of the Chinese Academy of Sciences, Grant no. (XDA0901010000) are appreciated. We thank Prof. Kaili Jiang's group from Tsinghua University, they provided very nice SACNTs in the *in situ* SEM experiment.

Notes and references

1 C. X. Zu and H. Li, *Energy Environ. Sci.*, 2011, **4**, 2614; J. Y. Peng, C. X. Zu and H. Li, *Energy Storage Sci. Technol.*, 2013, **2**, 55.
2 M. Winter, J. O. Besenhard, M. E. Spahr and P. Novak, *Adv. Mater.*, 1998, **10**, 725.
3 J. M. Tarascon and M. Armand, *Nature*, 2001, **414**, 359.
4 J. B. Goodenough and Y. Kim, *Chem. Mater.*, 2010, **22**, 587.
5 B. Scrosati and J. Garche, *J. Power Sources*, 2010, **195**, 2419.
6 B. Dunn, H. Kamath and J. M. Tarascon, *Science*, 2011, **334**, 928.
7 D. Aurbach, *Nonoaqueous Electrochemistry*, Marcel Dekker, New York, 1999.
8 H. Kim, G. Jeong, Y. U. Kim, C. J. H. Kim, C. M. Park and H. J. Sohn, *Chem. Soc. Rev.*, 2013, **42**, 9011.
9 K. Xu, *Chem. Rev.*, 2004, **104**, 4303.
10 Z. I. Takehara and K. I. Kanamura, *Electrochim. Acta*, 1993, **38**, 1169.
11 D. Aurbach, B. Markovsky, A. Shechter and Y. Ein-E1i, *J. Electrochem. Soc.*, 1996, **143**, 3809.
12 S. Shiraishi, K. Kanamura and Z. Takehara, *Langmuir*, 1997, **13**, 3542.
13 F. Orsini, A. Du Pasquier, B. Beaudoin, J. M. Tarascon, M. Trentin, N. Langenhuizen, E. De Beer and P. Notten, *J. Power Sources*, 1998, **76**, 19–29.
14 C. Brissot, M. Rosso, J. N. Chazalviel and S. Lascaud, *J. Power Sources*, 1999, **81–82**, 925.
15 M. Dollé, L. Sannier, B. Beaudoin, M. Trentin and J. M. Tarascon, *Electrochem. Solid-State Lett.*, 2002, **5**, A286.
16 D. Aurbach, B. Markovsky, M. D. Levi, E. Levi, A. Schechter, M. Moshkovich and Y. Cohen, *J. Power Sources*, 1999, **81–82**, 95.
17 K. Morigaki, *J. Power Sources*, 2002, **104**, 13.
18 W. S. Kim and W. Y. Yoon, *Electrochim. Acta*, 2004, **50**, 541.

19 S. Chandrashekar, N. M. Trease, H. J. Chang, L.-S. Du, C. P. Grey and A. Jerschow, *Nat. Mater.*, 2012, **11**, 311.
20 R. Bhattacharyya, B. Key, H. Chen, A. S. Best, A. F. Hollenkamp and C. P. Grey, *Nat. Mater.*, 2010, **9**, 504.
21 D. Aurbach, Y. Ein-Eli and A. Zaban, *J. Electrochem. Soc.*, 1994, **141**, L1.
22 D. Aurbach, M. L. Daroux, P. W. Faguy and E. Yeager, *J. Electrochem. Soc.*, 1988, **135**, 1863.
23 K. Kanamura, S. Shiraishi and Z.-I. Takehara, *J. Electrochem. Soc.*, 1994, **141**, L108.
24 D. Aurbach, I. Weissman, A. Schechter and H. Cohen, *Langmuir*, 1996, **12**, 3991.
25 M. Odziemkowski, M. Krell and D. E. Irish, *J. Electrochem. Soc.*, 1992, **139**, 3052.
26 P. C. Howlett, D. R. MacFarlane and A. F. Hollenkamp, *J. Power Sources*, 2003, **114**, 277.
27 J. Y. Zheng, H. Zheng, R. Wang, L. B. Ben, W. Lu, L. W. Chen, L. Q. Chen and H. Li, *Phys. Chem. Chem. Phys.*, 2014, **16**, 13229.
28 X. H. Liu and J. Y. Huang, *Energy Environ. Sci.*, 2011, **4**, 3844.
29 J. B. Bates, *US Pat.*, 5,314,765, 1994, pp. 5–24.
30 X. Wang, Y. Hou, Y. Zhu, Y. Wu and R. Holze, *Sci. Rep.*, 2013, **3**, 1401.
31 S. H. Lee, J. R. Harding, D. S. Liu, J. M. D'Arcy, Y. Shao-Horn and P. T. Hammond, *Chem. Mater.*, 2014, **26**, 2579–2585.
32 N. Choi, *Solid State Ionics*, 2004, **172**, 19–24.

Faraday Discussions

PAPER

Activity–stability relationship in the surface electrochemistry of the oxygen evolution reaction

Seo Hyoung Chang, Justin G. Connell, Nemanja Danilovic, Ram Subbaraman, Kee-Chul Chang, Vojislav R. Stamenkovic and Nenad M. Markovic*

Received 18th June 2014, Accepted 25th July 2014

DOI: 10.1039/c4fd00134f

Understanding the functional links between the stability and reactivity of oxide materials during the oxygen evolution reaction (OER) is one key to enabling a vibrant hydrogen economy capable of competing with fossil fuel-based technologies. In this work, by focusing on the surface chemistry of monometallic Ru oxide in acidic and alkaline environments, we found that the kinetics of the OER are almost entirely controlled by the stability of the Ru surface atoms. The same activity–stability relationship was found for more complex, polycrystalline and single-crystalline $SrRuO_3$ thin films in alkaline solutions. We propose that the electrochemical transformation of either water (acidic solutions) or hydroxyl ions (alkaline solutions) to di-oxygen molecules takes place at defect sites that are inherently present on every electrode surface. During the OER, surface defects are also created by the corrosion of the Ru ions. The dissolution is triggered by the potential-dependent change in the valence state (*n*) of Ru: from stable but inactive Ru^{4+} to unstable but active $Ru^{n>4+}$. We conclude that if the oxide is stable then it is completely inactive for the OER. A practical consequence is that the best materials for the OER should balance stability and activity in such a way that the dissolution rate of the oxide is neither too fast nor too slow.

Introduction

The development of true energy security requires the expansion of renewable energy sources as viable alternatives to fossil fuel-based technologies. In many of the concepts currently under consideration, hydrogen is the key energy carrier and, despite several hurdles that still need to be overcome, it seems that hydrogen is indeed the most likely fuel of the future.[1-3] Interestingly, the realization of a viable hydrogen economy is closely tied to developing greater understanding of the factors that influence the efficiency of di-oxygen electrochemistry, in

Materials Science Division, Argonne National Laboratory, 9700 S Cass Ave, Argonne, IL 60439, USA. E-mail: nmmarkovic@anl.gov

particular, designing active and stable anode materials for the oxygen evolution reaction (OER: $2H_2O \leftrightarrow O_2 + 4H^+ + 4e^-$ for acid and $4OH^- \leftrightarrow O_2 + 2H_2O + 4e^-$ for alkaline).[1,4] The OER generally suffers from significant kinetic limitations, so the reaction takes place well above the reversible potential for the OER of 1.23 V.[1] The kinetics of the OER have, for the most part, been closely tied to the concept of the volcano plot, which generally expresses the rate of the OER as a function of more fundamental properties of the oxide materials, known as descriptors.[4–7] These analyses showed that the interaction between the substrate and the reactants, intermediates, and the products has to be optimized for the reaction to proceed efficiently.[8–11] Although concepts resulting from volcano plot analyses have led to the establishment of important catalytic trends, many fundamental questions still remain open. One key question is what relationship exists between the kinetics of the OER and the stability of the oxide materials. The lack of understanding of such stability–activity relationship derives mainly from the fact that research directed at the development of anode materials for the OER has been strongly "activity-centric", while almost completely ignoring the stability of the active components *during* the OER.

Here, we show that this disparity in focus has masked the inherently close ties that exist between the stability and activity of monometallic and complex oxide Ru-based catalysts. By studying the stability–activity relationship of well-characterized Ru oxide and $SrRuO_3$ single crystal thin films in aqueous environments, we demonstrate that there is a fundamental link between the stability of catalysts and their reactivity in the OER. This trend is observed for Ru metal electrodes in both acidic and alkaline electrolytes, indicating that the stability–activity relationship is independent of the specific ionic/molecular species mediating the OER (*i.e.* OH^- *vs.* H_2O). We found that the degree of stability is always inversely proportional to the activity, and that stable surfaces are, in fact, not reactive.

Results and discussion

Relationship between stability of Ru oxide and reactivity for the OER

We begin by summarizing cyclic voltammograms (CVs) for a polycrystalline Ru electrode in both acidic and alkaline media. Inspection of Fig. 1a reveals that, independent of the pH of the solution, the adsorption of the oxygenated species on Ru occurs across the entire voltage window from the hydrogen evolution to the oxygen evolution potential regions. We note that the as-deposited polycrystalline Ru electrode surfaces are immediately converted to Ru oxide during the OER, and that this surface "hydrous oxide", formed by water electrooxidation at high anodic potentials, is distinct from a thermally-grown oxide in that it is amorphous, full of defects and contains hydroxyl species.[12] In order to understand how the electrode potential may affect the stability of the electrode surface before and during the OER, we measured the oxidation state of Ru using *in situ* X-ray absorption near edge structure (XANES) analysis. The inset in Fig. 1b shows typical XANES spectra recorded at 0.05 V, 1.25 V, and 1.45 V. The XANES spectra of Ru between 0.05 V and 0.8 V show no difference compared with standard Ru metal spectra, *e.g.*, Ru electrodes are in the $n = 0$ valence state. This indicates that reversible adsorption of hydroxyl species at this potential has no significant effect on the oxidation state of Ru. In acidic media, above 0.8 V the valence state systematically changes from $n = 0$ to $n = 3+$ between $0.8 < E < 1.0$ V, to mixtures of $n = 3+$ and $n = 4+$ between $1.0 < E$

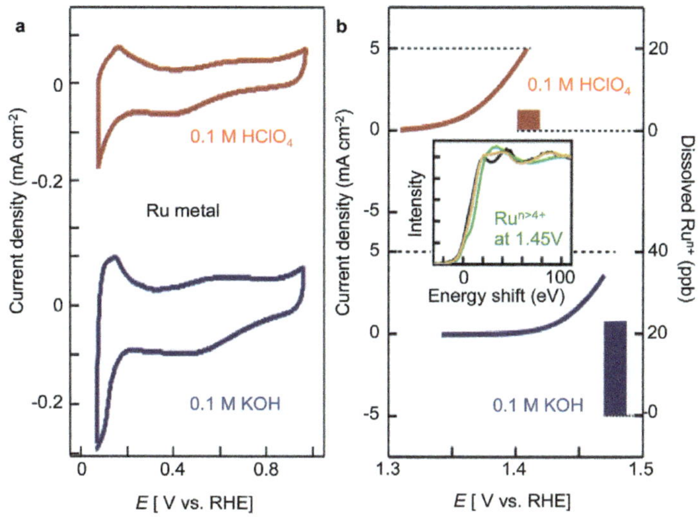

Fig. 1 (a) Cyclic voltammograms (CV) of Ru metal catalysts in acid (red line, 0.1 M $HClO_4$) and alkaline (blue line, 0.1 M KOH) electrolytes. (b) Activity of the oxygen evolution reaction (OER) (solid line) and Ru dissolution monitored by inductively-coupled plasma (ICP) measurements at a constant current density of 5 mA cm^{-2} (solid bars). The inset shows *in situ* XANES spectra at the potentials; 0.05 V (black), 1.25 V (yellow), and 1.45 V (green). The potentials are given *versus* the reversible hydrogen electrode (RHE).

< 1.3 V, and to $n > 4+$ for $E > 1.4$ V. Very similar changes in the valence state are also observed in the alkaline solution, with the only difference being that the change from $n = 4+$ to $n > 4+$ is shifted towards more positive potentials (1.47 V and above).

The relationships between the valence state of Ru, stability and reactivity were explored by monitoring the electrode potential-dependent Ru dissolution before and during the OER, by combining XANES with inductively-coupled plasma (ICP) measurements. As shown in Fig. 1b, the onset potential of the OER in acid is lower than that in alkaline solution, at the same disk current density of 1 mA cm^{-2}, implying that the OER is more active in acid than in alkaline solution. In addition, the dissolution of Ru in both electrolytes as measured by ICP is mirrored in the XANES data by a transition in the Ru valence state from $n = 4+$ to $n > 4+$, signaling that Ru dissolution is always triggered by the appearance of higher Ru oxidation states. We note that the ICP measurements were carried out at a constant current density of 5 mA cm^{-2}, as the difference in activity between the acid and alkaline solutions is large enough that it was very difficult to probe dissolution at the same overpotential. These observations highlight how potential-dependent variations in the oxidation state of the near-surface oxide can affect the stability of the surface atoms, as well as show that reactivity is intrinsically linked to the stability of these oxide materials.

Relationship between stability of Ru and $SrRuO_{3-x}$ polycrystalline electrodes and reactivity for the OER

Having established the functional links between Ru stability and activity for the OER, we investigated whether the same trends hold for Ru-based complex oxide materials such as $SrRuO_3$. To explore the role of Ru in these two different oxides, polycrystalline Ru and $SrRuO_{3-x}$ thin films (estimated thickness 10–30 nm) were sputter-deposited (AJA International, Inc.) at 200 °C onto conductive glassy carbon substrates. We verified that both samples showed characteristic Ru spectra in XANES, and that they exhibited high conductance using a probe station.[13] Polarization curves, shown in Fig. 2, show that the $SrRuO_{3-x}$ thin films are less electrocatalytically active for the OER than the monometallic Ru oxide. ICP measurements (Fig. 2, inset) were carried out for both samples at the same overpotential (rather than at constant current density as in Fig. 1b) in order to quantitatively compare the concentration of dissolved Ru during the OER. These data clearly show that the Ru concentration found in the electrolyte is much higher for the Ru oxide than for $SrRuO_{3-x}$, confirming that activity is controlled by the stability of the Ru active sites. Note that for the ICP measurements, the electrolyte was sampled at the end of the positive potentiodynamic OER sweep. Given that the same behavior is observed for experiments performed at a constant voltage (0.1 V from the reversible potential),[12] it is reasonable to suggest that the higher rates of Ru dissolution from monometallic oxides are associated with the lower stability of Ru oxide. In turn, dissolution is accompanied by the simultaneous formation of defect sites, which result in the higher reactivity of Ru surfaces relative to $SrRuO_{3-x}$. These results clearly show that there is a strong relationship between the activity and stability of the electrode surface, as well as that the density of the surface defects, present both intrinsically on the as-prepared surfaces and created during Ru dissolution, play a key role in determining the reactivity of the oxide materials. Nevertheless, the polycrystalline nature of the electrodes precludes a more precise understanding of the relationship between the nature and density of surface defects and the reactivity of the Ru-based oxides. In order to more precisely investigate this relationship and to establish fundamental links between activity and stability, it is necessary to synthesize oxide samples with highly controlled surface structures.

Relationship between stability of the $SrRuO_3$ single crystal films and reactivity for the OER

Toward this end, we synthesized single-crystalline $SrRuO_3$ (SRO) thin films for use as a model system to investigate the role of surface structure on electrocatalytic

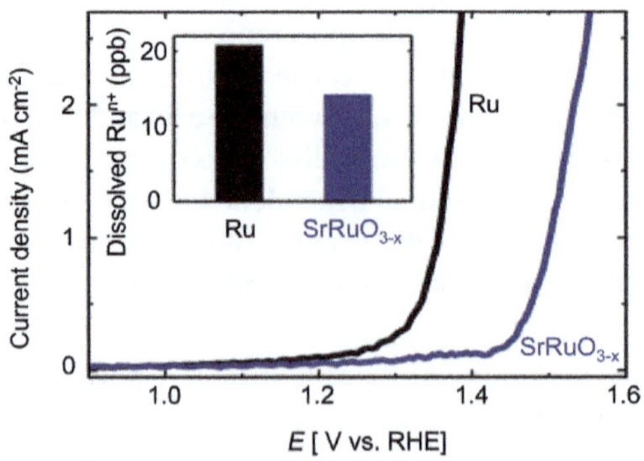

Fig. 2 Activity and stability of polycrystalline ruthenium-based catalysts for the OER, Ru (black) and $SrRuO_{3-x}$ (blue), on glassy carbon substrates. The inset shows the amount of dissolved Ru in the samples after OER processes, using ICP measurements within the same potential range.

activity and stability during cycling. Single crystal SRO is an ideal model system to study due to its high conductivity ($\sim$5000 Ω^{-1} m^{-1}) and its ability to readily grow high-quality thin films where it is possible to change the surface termination, as shown in Fig. 3a–c. For example, films grown along the (001) direction (Fig. 3a) are non-polar in the ionic limit, meaning that there is no net surface charge on the exposed surface layer. In contrast, films grown along the (110) and (111) directions (Fig. 3b and c) are polar, with alternating $(SrRuO)^{4+}/(O_2)^{4-}$ and $(Ru)^{4+}/(SrO_3)^{4-}$ layers, respectively. As a result, by simply varying the growth direction of the thin film it is possible to investigate the effects of the presence of surface energy differences, due to any residual polarity or under-coordinated polar surfaces, on the electrochemical activity and stability.[14]

The films were grown using radio-frequency (RF) magnetron sputter deposition (AJA International, Inc.), and conductive, 0.5 wt% Nb-doped $SrTiO_3$ (STO) substrates were used to improve the overall sample conductance. The SRO films exhibit a perovskite (ABO_3) crystal structure, with Ru ions surrounded by six oxygen ions (Fig. 3a). The lattice parameter of the SRO film is very close to that of the STO substrate ($\sim$0.64% with compressive strain), enabling the growth of high-quality, epitaxial thin films.[15,16] By controlling the orientation of the STO substrate, *e.g.* (001), (110), or (111), the growth direction and resulting surface structure of the SRO films were likewise able to be modified, and the CV measurements of the as-grown SRO (001), (110) and (111) thin films (Fig. 3d) show that the pseudocapacitance increases as (001) < (110) < (111).

Despite the choice of conductive, Nb-doped STO substrates, we found that the activity of the SRO films was hindered by the low conductance of the substrate. Note that conductivity of SRO is approximately 100 times higher than that of 0.5 wt% Nb:STO at room temperature. The current density–voltage characteristics of our SRO samples were measured through the bottom of the substrate using a conventional two-probe measurement system (Keithley 237 Source Measure Unit), and revealed nonlinear *I–V* behavior (Fig. 3e). Such nonlinear behavior is

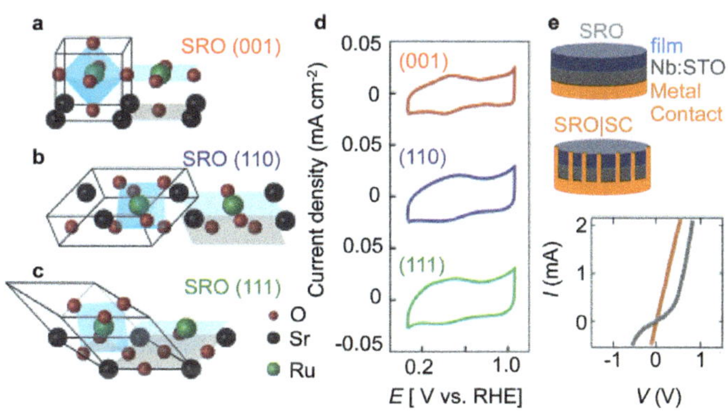

Fig. 3 (a–c) Schematics of the $SrRuO_3$ (SRO) films with different crystal structures grown on Nb-doped $SrTiO_3$ (STO) substrates (001), (110), and (111) respectively. (d) CVs of the SRO (001), (110), and (111) films. (e) Schematics and current–voltage (*I–V*) characteristics for the SRO/Nb:STO interface. Samples with and without a side contact (SC) between the SRO top surface and the backside of the substrate are denoted by SRO|SC (red line) and SRO (grey line), respectively.

indicative of high resistance as a result of a Schottky barrier at one of two locations: (1) the metal contact and Nb:STO substrate or (2) the SRO film and the Nb:STO substrate. The possibility of a Schottky barrier at the substrate–contact interface was eliminated by depositing a layer of Ti/Pt underneath the Ti metal contact, resulting in the formation of an Ohmic contact between Ti and STO. To avoid the formation of a contact between the SRO top surface and the backside of the substrate, we enclosed the perimeter of the samples with Kapton tape after mounting onto the sample chuck, prior to the start of the Ti/Pt room temperature deposition, and X-ray photoelectron spectroscopy (XPS) showed no indication of Ti or Pt contamination on the SRO surface after contact deposition. However, nonlinear *I–V* behavior was still observed even after ensuring Ohmic contact between the metal and substrate, suggesting that there is a Schottky barrier between the SRO and Nb:STO. Indeed, this is unsurprising given that there is a large difference between the work function of SRO (5.2 eV) and the electron affinity of the n-type semiconducting Nb:STO substrate (3.9 eV).[17] Unfortunately, it is very challenging to identify an interlayer material that can provide an Ohmic contact between SRO and STO while still enabling the growth of high-quality SRO thin films.

To circumvent this issue and verify the intrinsic activity of the SRO samples, a method was developed to directly contact the SRO thin film and enable analysis of the stability and activity of SRO. We prepared two sets of samples to compare the electrical behavior and verify the formation of an Ohmic contact: SRO with a metallic side connection (SRO|SC) and SRO without a side connection (SRO). Using these two samples, we observed that the presence of a metal side contact resulted in the formation of an Ohmic contact to the SRO thin film (Fig. 3e); however, the only difference in catalytic behavior observed between the two samples was that the kinetics of dissolution of the SRO thin film were slower on the SRO sample without a SC than for the SRO|SC sample. As a result, we made use of the slower kinetics of dissolution on the SRO samples without SC to enable more precise measurements of the kinetics of the OER, and investigate the relationship between the activity and stability of SRO. Detailed investigations of the stability of the SRO (001) films have been described elsewhere;[13] however, the results below provide new insight into the electrochemical behavior of this material and confirm the initial findings.

Polarization curves (Fig. 4a) were measured to examine the OER activity and stability of the SRO thin films on Nb:STO (001) in 0.1 M KOH electrolyte. The thin films were repeatedly cycled from 0 V to 1.95 V in order to trigger the OER and monitor changes in activity with cycling. We observed that the activity of the SRO film decreased gradually with the number of voltage cycles, with the OER current density decreasing by a factor of seven over 25 cycles. Atomic force microscopy (AFM) measurements (Fig. 4b) acquired before and after cycling indicate that the surface is initially terraced, with few intrinsic defects, but after cycling it becomes substantially roughened and exhibits many nano-sized islands that are randomly distributed on the surface. *In situ* X-ray scattering and XANES experiments confirmed that the SRO thin film was completely degraded after cycling, and revealed that the Ru oxidation state changes from $n = 4+$ to $n > 4+$ when holding at 1.45 V (ref. 13). Given these results, it is likely that the decrease in the OER current density after cycling originates from the dissolution of Ru ions after the formation of Ru active sites with $n > 4+$, which is consistent with our observations above on

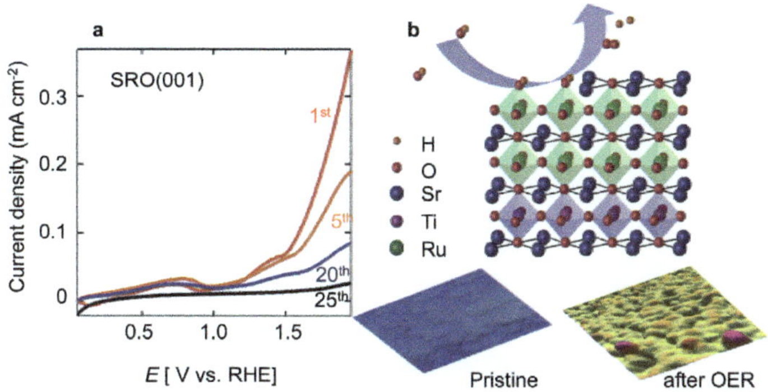

Fig. 4 (a) Successive changes in the current density on SRO (001) for the OER after various numbers of potential cycles. (b) Schematics of the model system and the OER process. AFM topography of the atomically flat SRO thin film surface ($1 \times 1\ \mu m^2$) with step terraces prior to cycling. After the OER, the film surfaces became rough, with RMS values increasing from 0.25 nm to 2.4 nm.

Ru metal (Fig. 1b). Note that the Ru active sites with an oxidation state of $n > 4+$ are easily dissolved and therefore do not remain at the surface indefinitely to serve as active sites. As our SRO films are very thin (10–30 nm), dissolution has a large impact on the observed activity due to the finite number of Ru atoms present in the film, relative to a bulk Ru sample. In fact, thicker polycrystalline SRO thin films (>100 nm) are more stable, with no difference in activity observed between the first and second potential sweep. As with the Ru metal, single crystal SRO surfaces are unstable and the OER on SRO is always accompanied by the dissolution of Ru cations.

The activity and stability for the OER in alkaline environments on Ru-based oxides – Ru, SRO polycrystalline films and SRO (001), (110) and (111) single crystalline films – are summarized in Fig. 5. This information provides insights into the material properties that influence the stability and activity of metal oxide surfaces during the OER – in particular, surface structure (polycrystalline *vs.* single crystal) and surface orientation. We observe that the activity of polycrystalline Ru metal is higher than that of polycrystalline SRO, whereas the stability of polycrystalline SRO is higher than that of Ru metal. A similar inverse relationship between activity and stability is observed for single crystalline SRO thin films, with the activity increasing in the order (001) < (110) < (111) and the stability increasing in the opposite order. Overall the activity of the crystalline SRO is lower than that of the polycrystalline samples, which is consistent with the expectation that single crystalline surfaces have a lower concentration of surface defects relative to polycrystalline samples.

Given these trends, we suggest that the density of the defects on the surface plays a significant role in determining the observed relationship between activity and stability. Interestingly, similar trends have recently been observed on other noble metal surfaces such as Au, Pt and Ir (ref. 12) – suggesting that the inverse relationship between activity and stability is a general phenomenon for electrocatalysts for the OER. In ref. 12, the role of the defects was probed by comparing activity *vs.* stability on the same materials with different surface structures. It was found that less defective single crystals of Ru are more stable, but less active, than

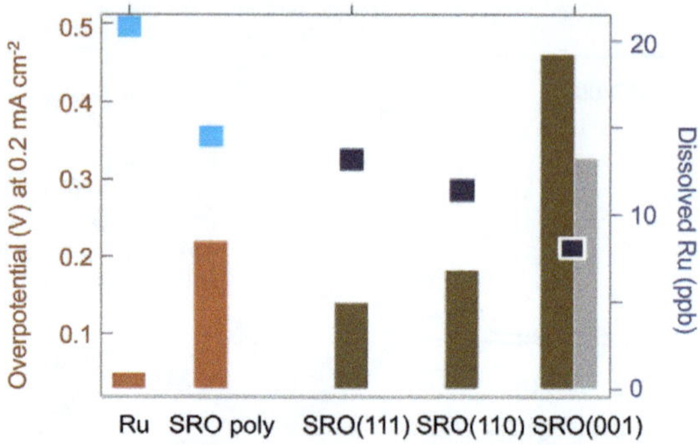

Fig. 5 Functional links between the stability and activity of Ru-based electrocatalysts (Ru, $SrRuO_{3-x}$ polycrystalline film and SRO single crystalline films from ref. 13) for the OER. The activity (bars) is measured by the OER overpotential at 0.2 mA cm^{-2}. [Note that data for SRO (001) at 0.2 mA cm^{-2} are an extrapolation from ref. 13 and the grey bar is the overpotential of SRO (001) at 0.1 mA cm^{-2}.] The stability (solid dots) is measured by the dissolved Ru species in the electrolyte (0.1 M KOH).

the polycrystalline Ru electrodes. Given that the same was observed for Ir in that work, we conclude that the inverse activity–stability relationship observed in this work is valid for many other oxide systems. Furthermore, *in situ* XANES measurements on all samples indicate that the onset of Ru dissolution into the electrolyte is accompanied by a change in the Ru oxidation state from $n = 4+$ to $n > 4+$, suggesting that the activity of Ru-based materials for the OER is dictated by the transition in Ru oxidation state. We propose that, in fact, the "active sites" for Ru-based electrocatalysts are locations on the electrode surface with Ru atoms in oxidation states greater than 4+.

In conclusion, the catalytic trends established thus far on high surface area materials should be revisited, and when possible, investigated on well-characterized single crystal surfaces. Targeted, detailed studies are necessary in order to unravel the relationship between stability and activity of oxide surfaces in electrochemical environments. Finally, although this work has focused on Ru-based materials, the fundamental and practical implications of the relationship between stability and activity likely extend well beyond this particular system, and will be extremely helpful in the design of new oxide materials that can serve as both active *and* stable electrocatalysts in aqueous environments and organic solutions.

Experimental

Thin film growth

The polycrystalline Ru and $SrRuO_{3-x}$ samples (5–200 nm) were grown at 200 °C, at a total pressure of 15 mTorr, using an RF-sputtering system on conducting glassy carbon substrates. Epitaxial $SrRuO_3$ films (10–30 nm) were also deposited by RF sputter deposition at 600 °C and 15 mTorr on 6 mm diameter, 0.5 wt% Nb-doped $SrTiO_3$ (001), (011), and (111) single-crystal substrates. Prior to deposition, the substrates were etched and annealed to produce TiO_2-terminated surfaces.

Electrochemical characterization

Electrochemical measurements were performed using an Autolab PGSTAT 302N potentiostat. Gold or platinum wires and Ag/AgCl were used as counter electrodes and reference electrodes, respectively. The electrolyte (0.1 M KOH or $HClO_4$) was prepared with Millipore deionized water. A typical three electrode, fluorinated ethylene propylene cell was used to avoid contamination. Potentials are reported with respect to that of the reversible hydrogen electrode (RHE).

In situ X-ray electrochemical study

X-ray measurements were performed at Sector 12BM of the Advanced Photon Source, Argonne National Laboratory. The *in situ* X-ray electrochemical cell was designed for use with 6 mm diameter, 0.5 mm thick substrates.

Acknowledgements

This work was supported by the U.S. Department of Energy, Basic Energy Sciences, Materials Sciences and Engineering Division. Use of the Advanced Photon Source and Center for Nanoscale Materials was supported by the U.S. Department of Energy, Office of Science, Office of Basic Energy Sciences, under Contract no. DE-AC02-06CH11357.

References

1 K. Kinoshita, *Electrochemical Oxygen Technology*, John Wiley & Sons, 1992.
2 W. Vielstich, H. Yokokawa, H. A. Gasteiger, *Handbook of Fuel Cells: Fundamentals Technology and Applications. Advances in Electrocatalysis, Materials, Diagnostics and Durability: Part 1*, John Wiley & Sons, 2009.
3 D. Strmcnik, *et al.*, *Nat. Chem.*, 2013, **5**, 300–306.
4 S. Trasatti, *Electrochim. Acta*, 1984, **29**, 1503–1512.
5 S. Trasatti and G. Buzzanca, *J. Electroanal. Chem. Interfacial Electrochem.*, 1971, **29**, A1–A5.
6 J. Greeley and N. M. Markovic, *Energy Environ. Sci.*, 2012, **5**, 9246–9256.
7 J. Rossmeisl, A. Logadottir and J. K. Nørskov, *Chem. Phys.*, 2005, **319**, 178–184.
8 J. O. Bockris and T. Otagawa, *J. Phys. Chem.*, 1983, **87**, 2960–2971.
9 Y. Matsumoto and E. Sato, *Mater. Chem. Phys.*, 1986, **14**, 397–426.
10 S. Ardizzone, G. Fregonara and S. Trasatti, *Electrochim. Acta*, 1990, **35**, 263–267.
11 R. Subbaraman, *et al.*, *Nat. Mater.*, 2012, **11**, 550–557.
12 N. Danilovic, *et al.*, *J. Phys. Chem. Lett.*, 2014, **5**, 2474–2478.
13 S. H. Chang, *et al.*, *Nat. Commun.*, 2014, **5**, 4191.
14 T. Sano, D. M. Saylor and G. S. Rohrer, *J. Am. Ceram. Soc.*, 2003, **86**, 1933–1939.
15 G. Koster, *et al.*, *Rev. Mod. Phys.*, 2012, **84**, 253–298.
16 S. H. Chang, Y. J. Chang, S. Y. Jang, D. W. Jeong, C. U. Jung, Y.-J. Kim, J.-S. Chung and T. W. Noh, *Phys. Rev. B: Condens. Matter Mater. Phys.*, 2011, **84**, 104101.
17 Y. Hikita, Y. Kozuka, T. Susaki, H. Takagi and H. Y. Hwang, *Appl. Phys. Lett.*, 2007, **90**, 143507.

Faraday Discussions

PAPER

Graphene-supported iron-based nanoparticles encapsulated in nitrogen-doped carbon as a synergistic catalyst for hydrogen evolution and oxygen reduction reactions

Jing Wang,[ab] Guoxiong Wang,*[a] Shu Miao,[c] Jiayuan Li[ab] and Xinhe Bao*[ac]

Received 3rd June 2014, Accepted 2nd July 2014
DOI: 10.1039/c4fd00123k

Electrolyzers and fuel cells have been extensively investigated as promising solutions for renewable energy storage and conversion. Hydrogen evolution reaction (HER) and oxygen reduction reaction (ORR) are important electrocatalytic processes in electrolyzers and fuel cells. Exploring efficient non-precious metal catalysts for HER and ORR in acidic medium remains a great challenge. Herein, we report that graphene-supported iron-based nanoparticles encapsulated in a nitrogen-doped carbon (Fe@N–C) hybrid material acts as an efficient HER and ORR catalyst. The hybrid material was synthesized by pyrolysis of graphene oxide and ammonia ferric citrate followed by acid-leaching. During the pyrolysis, nitrogen was doped into a graphene lattice, and the carbon nanoshell grown on graphene effectively suppressed the stacking of graphene sheets, exposing more active sites to reactants. The hybrid material showed higher electrocatalytic activities than graphene sheets or Fe@N–C alone, which is probably attributed to the synergetic role of nitrogen-doped graphene and Fe@N–C towards the electrocatalytic reactions.

Introduction

Renewable electricity from solar and wind energy has drawn much attention in recent years due to the increasing scarcity of fossil fuels and environmental deterioration.[1–3] A major challenge for renewable electricity is directly merging it into electricity grids because of its intermittent and localized nature. Therefore, it

[a]*State Key Laboratory of Catalysis, Dalian Institute of Chemical Physics, Chinese Academy of Sciences, 457 Zhongshan Road, Dalian, 116023, China. E-mail: wanggx@dicp.ac.cn; xhbao@dicp.ac.cn; Tel: +86-411-84686637*

[b]*University of Chinese Academy of Sciences, Beijing, 100039, China*

[c]*Dalian National Laboratory for Clean Energy, Dalian Institute of Chemical Physics, Chinese Academy of Sciences, Dalian, 116023, China*

is essential to develop efficient energy storage and conversion technologies towards the rapid development of renewable electricity.[4–6] Electrochemical technologies such as rechargeable batteries, supercapacitors, electrolyzers and fuel cells have been extensively investigated as promising solutions.[1,7,8] The combination of electrolyzers and fuel cells is a particularly interesting candidate. Hydrogen and oxygen are produced by water splitting in electrolyzers using renewable electricity, and then fed into the anode and cathode of proton exchange membrane fuel cell (PEMFC) for generating electricity as portable and transportation power sources.[4,9,10] Electrocatalysts are key materials in electrolyzers and PEMFCs for catalyzing hydrogen evolution/oxidation reactions and oxygen evolution/reduction reactions. Developing low-cost, efficient electrocatalysts for the above reactions is one of the most important challenges in the research and development of electrolyzers and PEMFCs.[11,12]

Carbon materials are usually the essential component in electrocatalysts, acting as a support to disperse the metal or metal oxide nanoparticles.[13,14] Graphene, a two-dimensional plane structure with a thickness of a single-layer carbon atom, has attracted great interest in the field of energy storage and conversion due to its high electronic conductivity, large surface area, and excellent chemical stability.[15–17] In PEMFC, electrochemical corrosion of carbon materials as catalyst supports leads to aggregation of Pt-based nanoparticles, and thus decreases the electrochemical surface area and eventually the long-term durability of PEMFC.[18,19] Traditional carbon supports widely used in PEMFC are carbon black (CB), carbon nanotube (CNT) and carbon nanofiber.[20,21] Due to the above unique properties, graphene has been investigated as a promising supporting material in carbon-supported Pt-based electrocatalysts.[22,23] Previous studies show that graphene-supported Pt-based electrocatalysts demonstrate enhanced catalytic activity.[24–26] However, the surface area of graphene sheets will be decreased during the electrochemical cycling due to the stacking of graphene sheets resulting from van der Waals forces and π–π interactions.[27,28] CB or CNT was often inserted between graphene sheets as nanospacers to suppress this stacking behavior.[29,30] The graphene–CB and graphene–CNT hybrid support materials provide a three-dimensional (3D) porous architecture for the deposition of Pt-based nanoparticles, expanding the electrochemical three-phase interface and thus enhancing the electrocatalytic activity and durability.

Although Pt-based electrocatalysts have demonstrated high electrocatalytic activities, the high cost and limited resources hinder their wide application in electrocatalytic reactions such as oxygen reduction reaction and hydrogen evolution reaction.[31–33] Meanwhile, the Pt-based electrocatalysts are vulnerable to poisonous species such as SO_2 and methanol, *etc.*[31–33] Many efforts have been devoted to develop non-precious metal catalysts for electrocatalytic reactions. Heteroatom-doped carbon materials have been considered as a new class of electrocatalysts. Due to the intrinsic chemical inertness of graphene, nitrogen is usually doped into graphene lattice to tailor its electronic structure and possess the electrocatalytic activity.[17,34,35] Various methods have been developed to synthesize nitrogen-doped graphene, such as thermal annealing of graphene or graphene oxide (GO) with NH_3, chemical vapor deposition, nitrogen plasma treatment of graphene and arc discharge of graphite with pyridine/NH_3.[36,37] However, these approaches have some limitations, such as high cost, inability to scale up, *etc.* In addition, the stacking behavior of nitrogen-doped graphene is even more serious than that of graphene-supported Pt-based nanoparticles, and the accessible active sites in nitrogen-doped graphene are exposed limitedly. It has been shown that some non-noble metal oxide nanoparticles, such as Co_3O_4 and MnO_x supported on graphene sheets, can suppress the stacking behavior,[29,38] however, the majority of the metal oxide/graphene hybrid materials can only work in alkaline media because of the acid-leaching of metal oxide.

Iron-based nanoparticles encapsulated in nitrogen-doped carbon (Fe@N–C) have been synthesized using ammonia ferric citrate as a single-source molecular precursor in our group.[39] The carbon phase with a hollow nanoshell structure is formed by pyrolysis and acid-leaching, which would be an ideal separator to suppress the stacking behavior of graphene sheets. Inspired by this, ammonia ferric citrate was mixed with GO uniformly and pyrolyzed in Ar atmosphere. The stacking of graphene sheets was effectively suppressed by growing hollow carbon nanoshell on the graphene surface, meanwhile, GO was doped with nitrogen and reduced to graphene by the decomposition species from ammonia ferric citrate during pyrolysis. The encapsulation structure prevents direct contact between the iron-based nanoparticles with oxygen and acidic medium. The graphene-supported Fe@N–C hybrid material demonstrates higher electrocatalytic activities towards hydrogen evolution and oxygen reduction reactions than Fe@N–C or graphene sheets alone.

Experimental

Materials synthesis

GO was prepared from natural graphite flakes using a modified Hummers method.[40] The hybrid material was prepared by freeze-drying an aqueous dispersion of ammonia ferric citrate and GO, followed by pyrolysis in Ar atmosphere and acid-leaching. In a typical experiment, 15 g ammonium ferric citrate ($C_6H_{11}FeNO_7$, J&K Chemical Ltd.) and 1.5 g GO were added to 300 mL of de-ionized water. After 10 min sonication, the stable suspension was freeze-dried. The mixture was placed in a quartz tube of a horizontal furnace and pyrolyzed in Ar atmosphere at a flow rate of 50 mL min^{-1} for 2 h at 600 °C. The product was leached in 0.5 M $HClO_4$ solution at 80 °C for 10 h to remove unstable Fe species, and washed with de-ionized water thoroughly. Finally, the sample was dried at 60 °C in an oven, which is denoted as Fe@N–C/RGO. For comparison, ammonia ferric citrate and GO were prepared under the same conditions, which are denoted as Fe@N–C and RGO, respectively. Dicyandiamide was added into the precursor mixture as an additional nitrogen source, and the mixture was pyrolyzed at 700 °C, followed by acid-leaching, which is denoted as H–Fe@N–C/RGO.

Physicochemical characterizations

The morphologies of the samples were investigated by an FEI Tecnai G^2 microscope at 120 kV, a JEM-2100 microscope at 200 kV and a QUANTA 200 FEG scanning electron microscope (SEM) at 20 kV. X-ray diffraction (XRD) was performed on a Rigaku D/Max-2500 diffractometer with a Cu Kα radiation source ($\lambda = 1.5418$ Å) at 40 kV and 200 mA at a scan rate of 5° min^{-1}. X-ray photoelectron spectroscopy (XPS) was carried out on a Thermo Scientific spectrometer with an Escalab 250 Xi X-ray as the excitation source. Thermogravimetry (TG) measurements were performed on a TGA-DSC analyzer (NETZSCH STA 449 F3 Jupiter®). Dry air provided by a pressured tank with a flow rate of 50 mL min^{-1} was used as the carrier gas. The experiments were carried out from room temperature to 900 °C with a heating rate of 10 °C min^{-1}. Nitrogen adsorption–desorption was carried out using a Quantachrome QUADRASORB SI system at 77 K, and specific surface areas of the samples were calculated by the Brunauer–Emmett–Teller (BET) equation.

Electrochemical measurements

Electrochemical measurements were carried out in a 150 mL three-electrode cell (AKCELL3, Pine Research Instrumentation) at 25 °C. A commercial glassy carbon (GC) electrode (AFE2M050 GC, 5 mm in diameter, 0.196 cm^2, Pine Research Instrumentation) covered by the sample with Nafion ionomer as a binder and Pt-wire electrode (Pine Research Instrumentation) were used as the working electrode and counter electrode. Ag/AgCl and Hg/HgO electrodes (Shanghai Yueci Electronic Technology Co. Ltd.) were used as the reference electrode in acidic and alkaline electrolyte solutions, respectively. The rotation rate and the potential of the working electrode were controlled by a MSR Electrode Rotator (Pine Research Instrumentation) and an Autolab potentiostat/galvanostat (PGSTAT 302N). The Ag/AgCl reference electrode was calibrated *vs.* reversible hydrogen electrode (RHE) in 0.5 M $HClO_4$ and 0.1 M $HClO_4$ solution, while the Hg/HgO reference electrode was calibrated *vs.* RHE in 0.1 M KOH solution. All potential values in this paper are referred to compared with the RHE. Five mg of the catalyst was dispersed in a mixture of 2 mL ethanol and 50 μL Nafion solution (5 wt%, Aldrich) with ultrasonic stirring to form a homogenous ink. The catalyst layer was prepared by dropping 25 μL of the ink onto a GC disk electrode by a micropipette and dried at room temperature. The hydrogen evolution reaction (HER) activities of the catalysts were evaluated by rotating disc electrode (RDE) measurement in Ar-saturated 0.5 M $HClO_4$ solution with a scan rate of 2 mV s^{-1}. The oxygen reduction reaction (ORR) activities of the catalysts were also evaluated by RDE measurements in O_2-saturated 0.1 M $HClO_4$ solution or 0.1 M KOH solution with a scan rate of 10 mV s^{-1}. Current–voltage data was recorded at a rotation rate of 2500 rpm.

Rotating ring disk electrode (RRDE) measurements of the samples in 0.1 M $HClO_4$ solution were conducted on a glassy carbon disk (Ø 5.7 mm) with a polycrystalline Pt biased at 1.06 V. The following equations were used for calculation of H_2O_2 (η) formation yield and the electron transfer number (n) per oxygen molecule:

$$\eta = \frac{200 \times I_r}{(N \times I_d) + I_r} \tag{1}$$

$$n = \frac{4 \times I_d}{I_d + I_r/N} \tag{2}$$

where I_r and I_d are ring and disk currents, respectively, and N is collection efficiency (0.38).

Results and discussion

Scheme 1 illustrates the typical procedures to synthesize the Fe@N–C/RGO hybrid material. Firstly, GO and ammonia ferric citrate are mixed into an aqueous solution. The presence of many oxygen-containing groups on the surface of the GO sheets makes GO uniformly disperse in the ammonia ferric citrate aqueous solution. Subsequently, the mixture is freeze-dried and then pyrolyzed in Ar atmosphere. Then ammonia ferric citrate is decomposed to form Fe@N–C, meanwhile, GO is reduced to nitrogen-doped graphene sheet. After acid-leaching, incompletely encapsulated iron-based nanoparticles are removed and hollow carbon nanoshells are left behind between graphene sheets. Some entirely encapsulated iron-based nanoparticles are supported on the graphene sheet.

SEM images in Fig. 1a and b display the obvious differences between RGO and Fe@N–C/RGO. Fig. 1a shows the very smooth surface of the typical graphene sheet. In sharp contrast, carbon nanoshells are uniformly decorated on the graphene sheet in Fe@N–C/RGO. The freeze-drying process can facilitate the uniform distribution of ammonia ferric citrate powder among GO sheets.[41] During the pyrolysis, carbon nanoshells grow on the surface of graphene sheets. This hybrid structure is considered to be much more stable than graphene/CNT and graphene/CB hybrid structures produced by mechanical mixing. Thanks to the unique encapsulation structure, the iron-based nanoparticles are prevented from contact with oxygen, the electrolyte and contaminants, and remains stable in a wide pH range. Therefore, Fe@N–C/RGO hybrid material can be used as an electrocatalyst under harsh reaction conditions.

Transmission electron micrograph (TEM) images in Fig. 1c and d confirm the obvious differences between RGO and Fe@N–C/RGO. Fig. 1c illustrates typical stacked graphene sheets, while hollow carbon nanoshells in Fe@N–C/RGO are supported on graphene sheets with a much smaller size compared with that in Fe@N–C.[39] The presence of GO separates the continuous distribution of ammonia ferric citrate, resulting in the formation of small carbon nanoshells on the graphene sheet during pyrolysis. Some nanoparticles are also observed to be entirely encapsulated within the carbon nanoshells. As shown in Fig. 1e, one nanoparticle

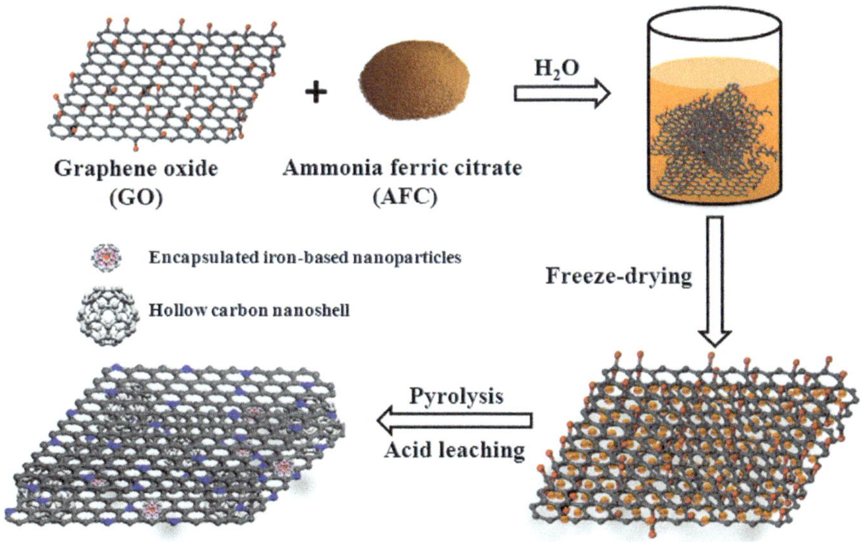

Scheme 1 The schematic preparation procedures of Fe@N–C/RGO.

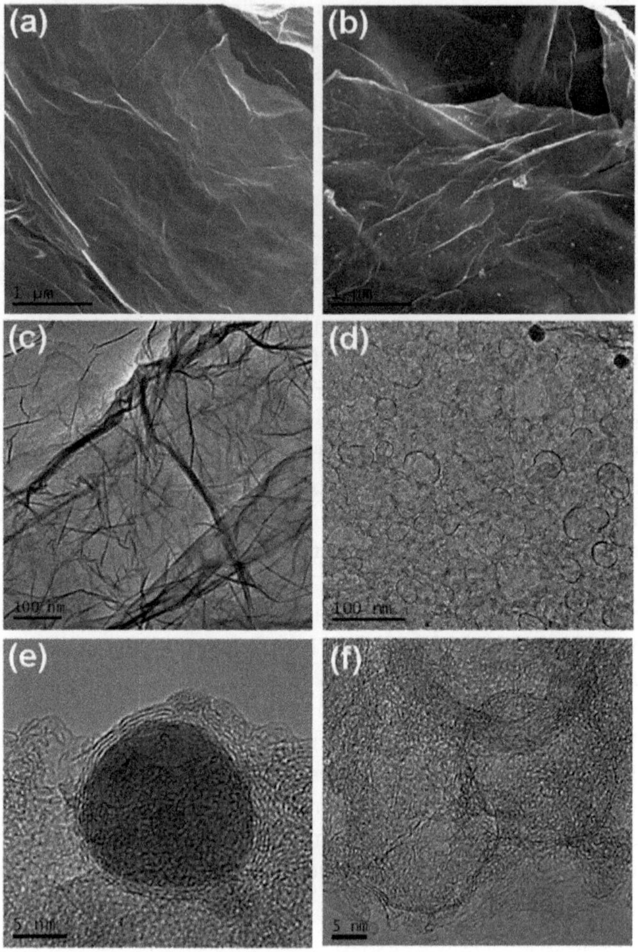

Fig. 1 SEM and TEM images of RGO (a), (c) and Fe@N–C/RGO (b), (d)–(f).

with a size of about 15 nm is completely encased in a carbon nanoshell with 3–4 graphene layers. Some hollow graphitic pores with a size of ~20 nm are also observed, which were generated by removing incompletely encapsulated iron-based nanoparticles during acid-leaching.

The present strategy for the synthesis of this unique Fe@N–C/graphene hybrid architecture has several advantages. Firstly, the preparation procedures are quite simple, consisting of freeze-drying, pyrolysis and acid-leaching using non-expensive equipment. Secondly, GO is reduced and doped with nitrogen uniformly by the decomposition of ammonia ferric citrate without the need of any other reductive or nitrogen-containing gas. Thirdly, the stacking of graphene sheets is suppressed effectively by the growth of carbon nanoshells on its surface, constructing an open 3D architecture, facilitating transport of reactants to active sites. Thus, this synthesis process exactly meets the requirements to scale up for mass production.

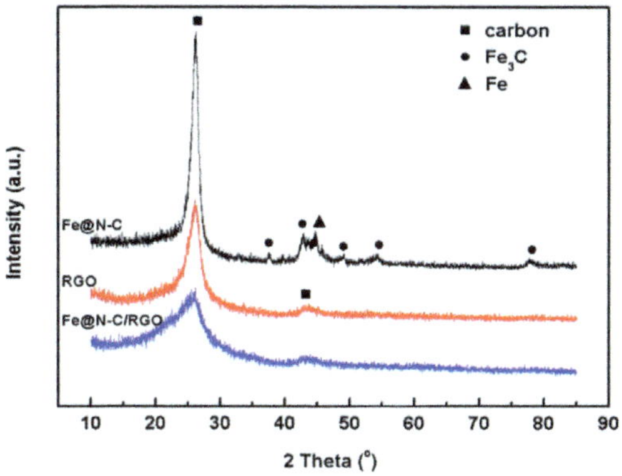

Fig. 2 XRD patterns of RGO, Fe@N–C and Fe@N–C/RGO.

Fig. 2 shows XRD patterns of Fe@N–C, RGO and Fe@N–C/RGO. The broad peaks at 26.1° and 43° are assigned to diffractions from the (002) and (10) planes of the hexagonal structure of graphite, indicative of a high graphitic structure in the three samples.[42] The thickness of the graphitic domains (L_c) in different samples is calculated according to the Scherrer equation using the full width at half maximum (FWHM) values of (002).[43] The calculated value of the thickness of graphitic domains (L_c) can be used to quantify the average thickness of the carbon nanoshells and graphene sheets. For Fe@N–C and RGO, the average thickness of the graphitic domains is 5.8 nm and 3.0 nm. It should be noted that the average value of L_c in Fe@N–C/RGO is only 1.5 nm, much smaller than those of Fe@N–C and RGO. This result indicates that pyrolysis of the mixture of GO and ammonia ferric citrate plays a synergistic role in suppressing the aggregation of graphene sheets and carbon nanoshells. The continuous distribution of GO and ammonia ferric citrate in the mixture is interrupted, and thus the thickness of graphitic domains is decreased. For Fe@N–C, the weak peaks at around 44.5° and 82.3° are assigned to Fe[44] and the other strong peaks are assigned to Fe_3C.[38] In contrast, diffraction peaks associated with Fe and Fe_3C are not observed for the hybrid material, probably due to overlapping with the diffraction peaks of graphene.

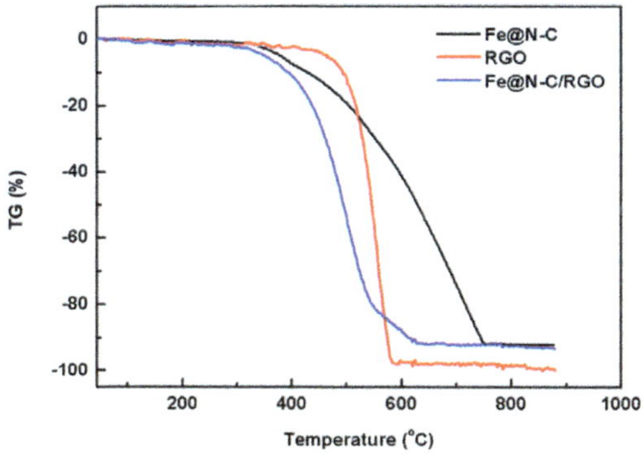

Fig. 3 TG curves of RGO, Fe@N–C and Fe@N–C/RGO in air atmosphere.

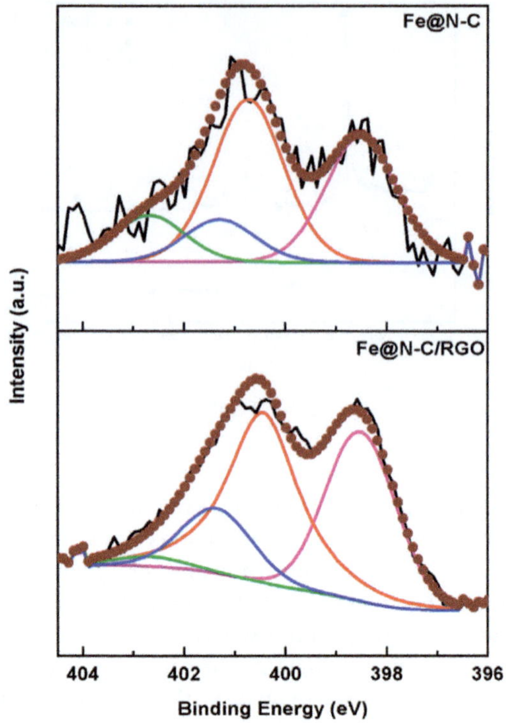

Fig. 4 The N 1s XPS spectra of Fe@N–C and Fe@N–C/RGO.

TG curves of different samples in air atmosphere are shown in Fig. 3. RGO has a 100% mass loss at the end of the TG experiment. In contrast, Fe@N–C and Fe@N–C/RGO have similar residual masses of about 8%. During the heating process, two types of oxidation reactions occurred.[39] One is the oxidation of iron-based species to Fe_2O_3, and the other is the oxidation of carbon to CO_2. CO_2 is released from the samples, and the residual material after TG experiments in air atmosphere is Fe_2O_3. The complete mass loss of RGO suggests that it contains negligible metal or metal oxide, and the similar mass loss of Fe@N–C and Fe@N–C/RGO confirms that there are nearly identical contents of encapsulated iron species in both samples. The mass loss of RGO mainly occurs in the temperature range of 410 °C and 590 °C, while that of Fe@N–C occurs in a temperature range of 300 °C and 750 °C. The mass loss of RGO is very fast,

Table 1 The total nitrogen content and percentage of different nitrogen species in the samples

Samples	Total nitrogen content (wt%)	Percentage of different nitrogen species (%)			
		Pyridinic	Pyrrolic	Quaternary	Oxidized
Fe@N–C	1.8	33.4	43.0	11.3	12.3
Fe@N–C/RGO	4.2	37.9	45.1	14.7	2.3

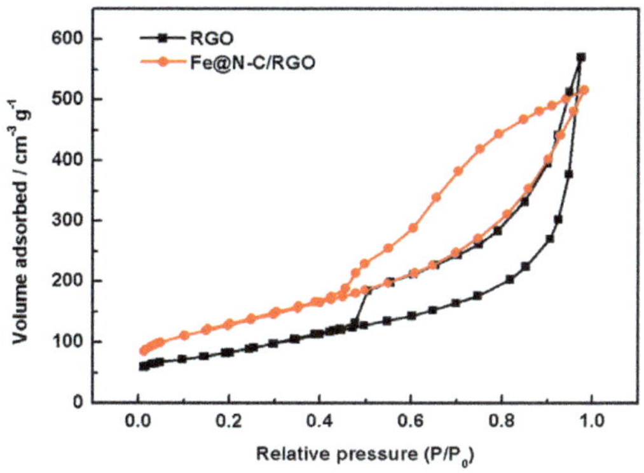

Fig. 5 Nitrogen adsorption–desorption isotherms of RGO and Fe@N–C/RGO.

indicating that RGO has a uniform composition distribution to activate oxygen in the air simultaneously and thus is oxidized to CO_2 in a narrow temperature window. Although Fe@N–C has a higher graphitization extent than RGO considering its high value of L_c, the onset oxidation temperature of Fe@N–C is much lower than that of RGO, suggesting the encapsulation structure of Fe@N–C can activate oxygen favorably. The mass loss at high oxidation temperature is probably ascribed to the oxidation of hollow carbon nanoshells without encapsulated iron-based nanoparticles. Interestingly, Fe@N–C/RGO hybrid material shows the lowest onset oxidation temperature and a narrow temperature range of carbon oxidation, suggesting improved oxygen-activating ability compared with RGO and Fe@N–C.

The total nitrogen content in Fe@N–C/RGO is found to be 4.2 wt% by XPS, much higher than that in Fe@N–C (1.8 wt%). During the pyrolysis, ammonia ferric citrate was decomposed and the nitrogen species was conveniently doped into the graphene due to an intimate contact between GO and ammonia ferric citrate in the mixture after freeze-drying. Therefore, the doped nitrogen content in Fe@N–C/RGO is greatly improved. The high resolution N 1s peaks in the XPS spectra are fitted into four components originating from pyridinic N (398.5 ± 0.2 eV), pyrrolic N (400.5 ± 0.2 eV), graphitic N (401.4 ± 0.2 eV), and oxidized N (402.6 ± 0.2 eV),[39] as shown in Fig. 4. The percentage of different nitrogen species is listed in Table 1, compared with Fe@N–C, the proportion of nitrogen species in Fe@N–C/RGO is different especially for the oxidized N. Both pyridinic-N and graphitic-N are reported to be responsible for the electrocatalytic activity.[45,46] The percentage of pyridinic-N and graphitic-N in Fe@N–C/RGO is 52.6% of the total nitrogen content, exceeding that of 44.7% in Fe@N–C, suggesting that a high content of nitrogen functional groups in Fe@N–C/RGO will contribute to the electrocatalytic activity.

The N_2 adsorption–desorption curves of RGO and Fe@N–C/RGO belong to the isotherms of type IV (based on IUPAC classification) with a hysteresis loop, indicative of the existence of mesopores, as shown in Fig. 5. The specific surface area and pore volume of RGO are 302 $m^2\ g^{-1}$ and 0.88 $cm^3\ g^{-1}$, while those of Fe@N–C/RGO are 453 $m^2\ g^{-1}$ and 0.80 $cm^3\ g^{-1}$, respectively. Compared with RGO, the specific surface area of Fe@N–C/RGO is greatly improved, indicative of a more

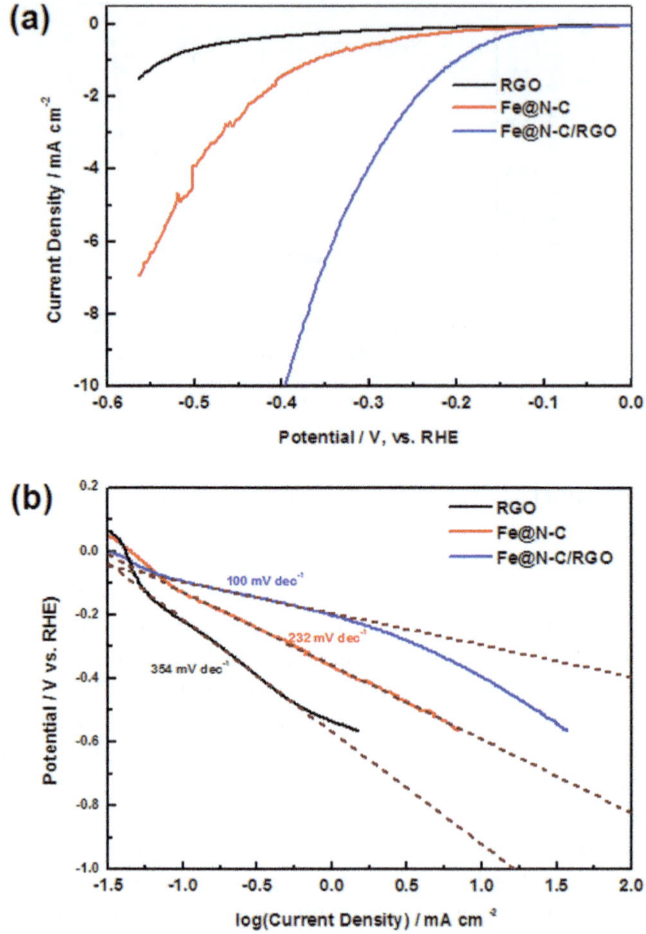

Fig. 6 Linear sweep voltammetries (a), and Tafel slopes (b) of RGO, Fe@N–C and Fe@N–C/RGO for hydrogen evolution reactions in Ar-saturated 0.1 M $HClO_4$ at 25 °C.

open structure. However, the pore volume of Fe@N–C/RGO is slightly decreased, probably because the majority of pore volume is contributed by the interspace of graphene sheets, and the carbon nanoshells supported on the graphene sheets decrease the interspace volume. The high specific surface area of Fe@N–C/RGO is expected to facilitate exposing more active sites to reactants with enhanced mass-transport properties during the electrocatalytic process.[39]

The electrocatalytic activities of RGO, Fe@N–C and Fe@N–C/RGO towards HER are evaluated in Ar-saturated 0.5 M $HClO_4$ solution. As shown in Fig. 6a, RGO has a negligible current density, indicating that pure carbon material is almost inert to HER. Fe@N–C shows an inferior activity for HER. In contrast, Fe@N–C/RGO demonstrates a greatly improved HER activity. The Tafel plots are shown in Fig. 6b. The slopes were calculated to be 100 mV dec^{-1} for Fe@N–C/RGO, 232 mV dec^{-1} for Fe@N–C and 356 mV dec^{-1} for RGO. In acidic medium, HER is composed of two out of three microscopic steps (Tafel/Volmer or Heyrovsky/Volmer):[47,48]

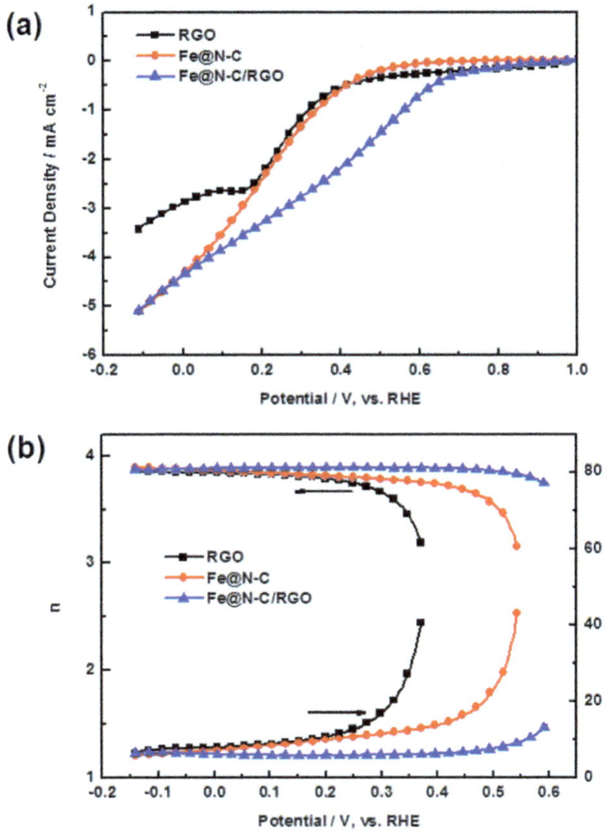

Fig. 7 Linear sweep voltammetries (a), electron transfer number and H_2O_2 yield (b) of RGO, Fe@N–C and Fe@N–C/RGO in O_2-saturated 0.1 M $HClO_4$ at 25 °C.

$$\text{Volmer step: } H^+ + e^- = H_{ad}$$

$$\text{Heyrovsky step: } H_{ad} + H^+ + e^- = H_2$$

$$\text{Tafel step: } 2H_{ad} = H_2$$

The Tafel curve of Fe@N–C/RGO clearly demonstrates that the Volmer step is the rate-determining step. The Volmer reaction involves the adsorption of a proton and the transfer of an electron to form the adsorbed H species. The comparatively lower onset potential and smaller Tafel slope of Fe@N–C/RGO than those of Fe@N–C and RGO alone denote that the presence of more nitrogen species favors proton adsorption kinetics. For HER in acidic medium, Pt/C has demonstrated the best activity.[49] Development of non-precious metal catalysts for HER has been one of the major challenges in renewable energy storage and conversion in recent years.[50–52] Metal sulfide (such as MoS_2, WS_2, *etc.*), metal nitride (such as Ni–Mo–N) and metal phosphides (such as Ni_2P) have demonstrated high HER activity in acidic medium.[50–52] However, the durability of these

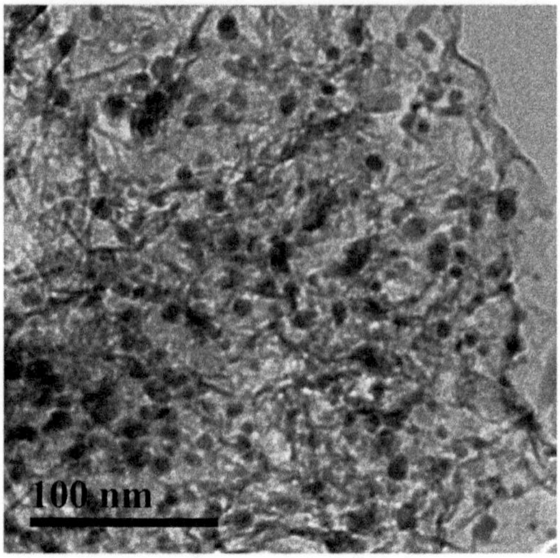

Fig. 8 TEM image of H–Fe@N–C/RGO.

catalysts is not satisfactory because the transition metal species are in direct contact with the acidic electrolyte solution and prone to dissolution. Carbon materials are very stable in acidic medium, but there are only a few reports on carbon materials for application in HER. Herein, Fe@N–C/RGO serves as a promising alternative for HER in acidic medium. Further functionalization of Fe@N–C/RGO hybrid material is essential to improve the HER activity.

The electrocatalytic activities of Fe@N–C, RGO and Fe@N–C/RGO towards ORR are also assessed in O_2-saturated 0.1 M $HClO_4$ solution. Fig. 7a shows linear sweep voltammetries of different samples. RGO and Fe@N–C have similar ORR activities. In contrast, Fe@N–C/RGO displays an obvious positive shift on the onset potential at 0.7 V and high cathodic current density. The ORR activity on Fe@N–C/RGO is much higher than that of RGO or Fe@N–C alone, confirming a synergetic effect of nitrogen-doped graphene and Fe@N–C for catalyzing ORR. The ORR pathway is further confirmed by RRDE, which is fit to monitor the number of transferred electron and the formation of peroxide species (HO_2^-) (Fig. 7b). The measured n is between 3.7 and 3.9 in the potential range of −0.15 V–0.6 V, and the H_2O_2 yields between 14% and 6% within the same potential window.

Dicyandiamide has been used as a suitable nitrogen source to increase the content of doped nitrogen.[53] We also incorporate dicyandiamide into the precursor mixture, which was then pyrolyzed and acid-leached, and the obtained material is denoted as H–Fe@N–C/RGO. Fig. 8 shows the TEM image of H–Fe@N–C/RGO. Compared with Fe@N–C, the density of encapsulated iron-based nanoparticles is obviously increased with the addition of dicyandiamide, and the graphene sheet can also be clearly observed, indicating that the encapsulated iron nanoparticles are supported on the graphene sheet. Several carbon encapsulated iron, iron–cobalt and iron carbide nanoparticles have been synthesized for catalyzing ORR in acidic medium.[54,55] It is proposed that the electronic interaction between the inner iron-based nanoparticle and outer carbon nanoshell promotes

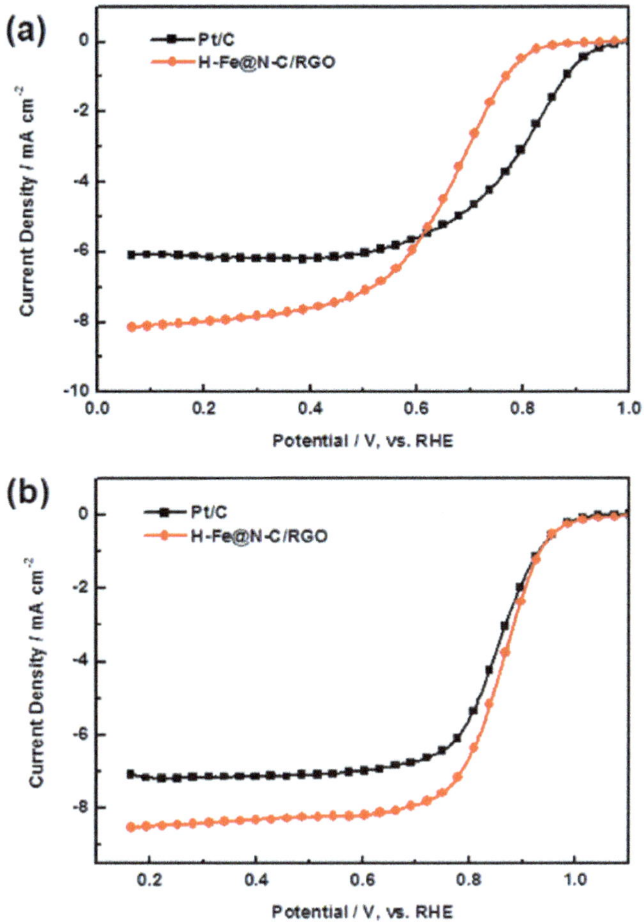

Fig. 9 Linear sweep voltammetries of H–Fe@N–C/RGO and Pt/C in O_2-saturated 0.1 M $HClO_4$ (a) and O_2-saturated 0.1 M KOH (b) solutions at 25 °C.

oxygen adsorption and association on the surface of the carbon nanoshell.[54,55] Therefore, high-density Fe@N–C would be expected to increase ORR activity. The ORR activities on H–Fe@N–C/RGO and Pt/C in acidic and alkaline mediums are shown in Fig. 9. The onset potential and half-wave potential ($E_{1/2}$) are 0.89 V and 0.67 V in 0.1 M $HClO_4$ solution, only slightly smaller than that of Pt/C catalyst. Interestingly, H–Fe@N–C/RGO shows even higher ORR activity in 0.1 M KOH solution. We have compared the ORR activities of the reported non-precious metal catalysts, as listed in Table 2. H–Fe@N–C/RGO shows high ORR activity both in acidic and alkaline mediums amongst the reported non-precious metal catalysts with encapsulation structure.[38,44,54–57] In the alkaline medium, H–Fe@N–C/RGO exhibits an ORR activity similar to the high-performance cobalt oxide/nitrogen-doped carbon hybrid materials[58,59] and a much higher ORR activity than metal-free nitrogen-doped graphene/CNT hybrid material.[60] Therefore, H–Fe@N–C/RGO has shown high ORR activities in a wide pH range and promising applications in fuel cells.

Table 2 The ORR activities on the non-precious catalysts in acidic and alkaline mediums. All the potential values from references were converted to *vs.* RHE for comparison

	Activity (*vs.* RHE)		
Sample	Acidic	Alkaline	References
H–Fe@N–C/RGO	$E_{onset} = 0.89$ V $E_{1/2} = 0.67$ V	$E_{onset} = 1.02$ V $E_{1/2} = 0.86$ V	This work
FeCo in pod-like CNT	$E_{onset} = 0.74$ V $E_{1/2} = 0.47$ V	Not measured	54
FeCo in thin pod-like CNTs	$E_{onset} = 0.78$ V $E_{1/2} = 0.59$ V	Not measured	56
Co in nitrogen-doped carbon nanoshells	$E_{onset} = 0.74$ V $E_{1/2} = 0.64$ V	$E_{onset} = 0.85$ V $E_{1/2} = 0.79$ V	57
Fe/Fe_3C in nitrogen-doped carbon	Not measured	$E_{onset} = 0.93$ V $E_{1/2} = 0.76$ V	44
Fe_3C in graphitic layers	$E_{onset} = 0.90$ V $E_{1/2} = 0.73$ V	$E_{onset} = 1.05$ V $E_{1/2} = 0.83$ V	55
Fe_3O_4 in 3D nitrogen-doped graphene aerogel	Not measured	$E_{onset} = 0.78$ V	38
Co_xO/CNT	Not measured	$E_{onset} = 0.93$ V $E_{1/2} = 0.85$ V	58
Co/N-doped nanoporous carbon	Not measured	$E_{onset} = 0.88$ V $E_{1/2} = 0.79$ V	59
Nitrogen-doped graphene/CNT	Not measured	$E_{onset} = 0.87$ V $E_{1/2} = 0.69$ V	60

Conclusions

A graphene-supported Fe@N–C hybrid material was obtained by pyrolysis of graphene oxide and ammonium ferric citrate in Ar atmosphere followed by acid-leaching. Hollow carbon nanoshells grown on graphene sheets by pyrolysis of ammonium ferric citrate suppressed the stacking of graphene sheet effectively, meanwhile, the decomposition species reduced the graphene oxide and doped nitrogen into the graphene lattice. The hybrid material showed higher electrocatalytic activities for HER and ORR in acidic medium than RGO or Fe@N–C alone, suggesting a synergistic role of nitrogen-doped graphene and Fe@N–C towards the electrocatalytic reactions. By incorporating dicyandiamide to the precursor mixture, the hybrid material exhibited high ORR activities both in acidic and alkaline mediums, which is amongst the best reported non-precious metal catalysts.

Acknowledgements

We gratefully acknowledge the financial support from the Ministry of Science and Technology of China (Grants 2012CB215500 and 2013CB933100), and the National Natural Science Foundation of China (Grants 21103178 and 21033009).

References

1 B. C. H. Steele and A. Heinzel, *Nature*, 2001, **414**, 345–352.

2 J. Zhang, K. Sasaki, E. Sutter and R. R. Adzic, *Science*, 2007, **315**, 220–222.
3 A. Hagfeldt, G. Boschloo, L. C. Sun, L. Kloo and H. Pettersson, *Chem. Rev.*, 2010, **110**, 6595–6663.
4 J. W. D. Ng, Y. Gorlin, T. Hatsukade and T. F. Jaramillo, *Adv. Energy Mater.*, 2013, **3**, 1545–1550.
5 H. Ibrahim, A. Ilinca and J. Perron, *Renewable Sustainable Energy Rev.*, 2008, **12**, 1221–1250.
6 S. M. Lukic, J. Cao, R. C. Bansal, F. Rodriguez and A. Emadi, *IEEE Trans. Ind. Electron. Control Instrum.*, 2008, **55**, 2258–2267.
7 I. Katsounaros, S. Cherevko, A. R. Zeradjanin and K. J. J. Mayrhofer, *Angew. Chem., Int. Ed.*, 2014, **53**, 102–121.
8 J. R. Miller and P. Simon, *Science*, 2008, **321**, 651–652.
9 A. Le Goff, V. Artero, B. Jousselme, P. D. Tran, N. Guillet, R. Metaye, A. Fihri, S. Palacin and M. Fontecave, *Science*, 2009, **326**, 1384–1387.
10 G. Y. Chen, S. R. Bare and T. E. Mallouk, *J. Electrochem. Soc.*, 2002, **149**, A1092–A1099.
11 M. K. Debe, *Nature*, 2012, **486**, 43–51.
12 J. Greeley, T. F. Jaramillo, J. Bonde, I. B. Chorkendorff and J. K. Norskov, *Nat. Mater.*, 2006, **5**, 909–913.
13 J. S. Guo, G. Q. Sun, Q. Wang, G. X. Wang, Z. H. Zhou, S. H. Tang, L. H. Jiang, B. Zhou and Q. Xin, *Carbon*, 2006, **44**, 152–157.
14 Y. H. Chang, C. T. Lin, T. Y. Chen, C. L. Hsu, Y. H. Lee, W. J. Zhang, K. H. Wei and L. J. Li, *Adv. Mater.*, 2013, **25**, 756–760.
15 Y. Zhao, C. G. Hu, Y. Hu, H. H. Cheng, G. Q. Shi and L. T. Qu, *Angew. Chem., Int. Ed.*, 2012, **51**, 11371–11375.
16 C. H. Zhang, L. Fu, N. Liu, M. H. Liu, Y. Y. Wang and Z. F. Liu, *Adv. Mater.*, 2011, **23**, 1020–1024.
17 Y. Wang, Y. Y. Shao, D. W. Matson, J. H. Li and Y. H. Lin, *ACS Nano*, 2010, **4**, 1790–1798.
18 R. Borup, J. Meyers, B. Pivovar, Y. S. Kim, R. Mukundan, N. Garland, D. Myers, M. Wilson, F. Garzon, D. Wood, P. Zelenay, K. More, K. Stroh, T. Zawodzinski, J. Boncella, J. E. McGrath, M. Inaba, K. Miyatake, M. Hori, K. Ota, Z. Ogumi, S. Miyata, A. Nishikata, Z. Siroma, Y. Uchimoto, K. Yasuda, K. I. Kimijima and N. Iwashita, *Chem. Rev.*, 2007, **107**, 3904–3951.
19 L. M. Roen, C. H. Paik and T. D. Jarvic, *Electrochem. Solid-State Lett.*, 2004, **7**, A19–A22.
20 M. Oezaslan, F. Hasche and P. Strasser, *J. Electrochem. Soc.*, 2012, **159**, B444–B454.
21 J. J. Wang, G. P. Yin, H. Liu, R. Y. Li, R. L. Flemming and X. L. Sun, *J. Power Sources*, 2009, **194**, 668–673.
22 C. C. Huang, C. Li and G. Q. Shi, *Energy Environ. Sci.*, 2012, **5**, 8848–8868.
23 R. I. Jafri, T. Arockiados, N. Rajalakshmi and S. Ramaprabhu, *J. Electrochem. Soc.*, 2010, **157**, B874–B879.
24 C. Xu, X. Wang and J. W. Zhu, *J. Phys. Chem. C*, 2008, **112**, 19841–19845.
25 E. Yoo, T. Okata, T. Akita, M. Kohyama, J. Nakamura and I. Honma, *Nano Lett.*, 2009, **9**, 2255–2259.
26 B. Seger and P. V. Kamat, *J. Phys. Chem. C*, 2009, **113**, 7990–7995.
27 Y. Q. Sun, Q. O. Wu and G. Q. Shi, *Energy Environ. Sci.*, 2011, **4**, 1113–1132.
28 B. Luo, S. M. Liu and L. J. Zhi, *Small*, 2012, **8**, 630–646.

29 H. Jiang, P. S. Lee and C. Z. Li, *Energy Environ. Sci.*, 2013, **6**, 41–53.
30 Z. J. Fan, J. Yan, L. J. Zhi, Q. Zhang, T. Wei, J. Feng, M. L. Zhang, W. Z. Qian and F. Wei, *Adv. Mater.*, 2010, **22**, 3723–3728.
31 N. M. Markovic, T. J. Schmidt, V. Stamenkovic and P. N. Ross, *Fuel Cells*, 2001, **1**, 105–116.
32 Y. Y. Shao, J. Liu, Y. Wang and Y. H. Lin, *J. Mater. Chem.*, 2009, **19**, 46–59.
33 H. A. Gasteiger, S. S. Kocha, B. Sompalli and F. T. Wagner, *Appl. Catal., B*, 2005, **56**, 9–35.
34 Z. Y. Lin, G. H. Waller, Y. Liu, M. L. Liu and C. P. Wong, *Carbon*, 2013, **53**, 130–136.
35 H. G. Wang, Z. Wu, F. L. Meng, D. L. Ma, X. L. Huang, L. M. Wang and X. B. Zhang, *ChemSusChem*, 2013, **6**, 56–60.
36 X. R. Wang, X. L. Li, L. Zhang, Y. Yoon, P. K. Weber, H. L. Wang, J. Guo and H. J. Dai, *Science*, 2009, **324**, 768–771.
37 X. L. Li, H. L. Wang, J. T. Robinson, H. Sanchez, G. Diankov and H. J. Dai, *J. Am. Chem. Soc.*, 2009, **131**, 15939–15944.
38 Z. S. Wu, S. B. Yang, Y. Sun, K. Parvez, X. L. Feng and K. Mullen, *J. Am. Chem. Soc.*, 2012, **134**, 9082–9085.
39 J. Wang, G. X. Wang, S. Miao, X. L. Jiang, J. Y. Li and X. H. Bao, *Carbon*, 2014, **75**, 381–389.
40 W. S. Hummers and R. E. Offeman, *J. Am. Chem. Soc.*, 1958, **80**, 1339.
41 Y. C. Xin, J. G. Liu, Y. Zhou, W. M. Liu, J. A. Gao, Y. Xie, Y. Yin and Z. G. Zou, *J. Power Sources*, 2011, **196**, 1012–1018.
42 M. Sevilla and A. B. Fuertes, *Carbon*, 2006, **44**, 468–474.
43 J. Ding, H. L. Wang, Z. Li, A. Kohandehghan, K. Cui, Z. W. Xu, B. Zahiri, X. H. Tan, E. M. Lotfabad, B. C. Olsen and D. Mitlin, *ACS Nano*, 2013, **7**, 11004–11015.
44 J. S. Lee, G. S. Park, S. T. Kim, M. L. Liu and J. Cho, *Angew. Chem., Int. Ed.*, 2013, **52**, 1026–1030.
45 D. H. Deng, X. L. Pan, L. A. Yu, Y. Cui, Y. P. Jiang, J. Qi, W. X. Li, Q. A. Fu, X. C. Ma, Q. K. Xue, G. Q. Sun and X. H. Bao, *Chem. Mater.*, 2011, **23**, 1188–1193.
46 C. V. Rao, C. R. Cabrera and Y. Ishikawa, *J. Phys. Chem. Lett.*, 2010, **1**, 2622–2627.
47 B. E. Conway and B. V. Tilak, *Electrochim. Acta*, 2002, **47**, 3571–3594.
48 N. Pentland, J. O. Bockris and E. Sheldon, *J. Electrochem. Soc.*, 1957, **104**, 182–194.
49 J. K. Norskov and C. H. Christensen, *Science*, 2006, **312**, 1322–1323.
50 G. J. Lu, P. Evans and G. Zangari, *J. Electrochem. Soc.*, 2003, **150**, A551–A557.
51 A. B. Laursen, S. Kegnaes, S. Dahl and I. Chorkendorff, *Energy Environ. Sci.*, 2012, **5**, 5577–5591.
52 W. F. Chen, K. Sasaki, C. Ma, A. I. Frenkel, N. Marinkovic, J. T. Muckerman, Y. M. Zhu and R. R. Adzic, *Angew. Chem., Int. Ed.*, 2012, **51**, 6131–6135.
53 A. Thomas, A. Fischer, F. Goettmann, M. Antonietti, J. O. Muller, R. Schlogl and J. M. Carlsson, *J. Mater. Chem.*, 2008, **18**, 4893–4908.
54 D. H. Deng, L. Yu, X. Q. Chen, G. X. Wang, L. Jin, X. L. Pan, J. Deng, G. Q. Sun and X. H. Bao, *Angew. Chem., Int. Ed.*, 2013, **52**, 371–375.
55 Y. Hu, J. O. Jensen, W. Zhang, L. N. Cleemann, W. Xing, N. J. Bjerrum and Q. F. Li, *Angew. Chem., Int. Ed.*, 2014, **53**, 1–6.

56 J. Deng, L. Yu, D. H. Deng, X. Q. Chen, F. Yang and X. H. Bao, *J. Mater. Chem. A*, 2013, **1**, 14868–14873.
57 Z. S. Wu, J. Z. Liu, K. Parvez, H. W. Liang, J. Shu, H. Sachdev, R. Graf, X. L. Feng and K. Mullen, *Adv. Mater.*, 2013, **26**, 1450–1455.
58 Y. Y. Liang, H. L. Wang, P. Diao, W. Chang, G. S. Hong, Y. G. Li, M. Gong, L. M. Xie, J. G. Zhou, J. Wang, T. Z. Regier, F. Wei and H. J. Dai, *J. Am. Chem. Soc.*, 2012, **134**, 15849–15857.
59 Z. Y. Liu, G. X. Zhang, Z. Y. Lu, X. Y. Jin, Z. Chang and X. M. Sun, *Nano Res.*, 2013, **6**, 293–301.
60 P. Chen, T. Y. Xiao, Y. H. Qian, S. S. Li and S. H. Yu, *Adv. Mater.*, 2013, **25**, 3192–3196.

Faraday Discussions

DISCUSSIONS

Electrochemical conversion and storage systems: general discussion

Andrew Mount, Shizhao Xiong, Xuyi Shan, Hyun-Wook Lee, Xiaoliang Yu, Yimin Chao, Galen Stucky, Gang Chen, Zhen Qi, Graham Hutchings, Yiren Zhong, Rudolf Holze, Wei Han, Lee Cronin, Shihe Yang, Hong Li, Xiang Hong, Erwin Reisner, Yong Yang, Weimin Xuan, Clare Grey, Ram Seshadri, Liqiang Mai, Jiafang Xie, Fuping Pan, Zhonghua Li, Joachim Maier, Zhongqun Tian, Yanxia Chen, Bingwei Mao, Heinz Frei, Changxu Lin, Fenglin Liao, Deyu Liu, Nanfeng Zheng, Rui Lin, Rose-Noelle Vannier, Dehui Deng, John M. Griffin, Nenad Markovic, Haimei Zheng and Ryoji Kanno

DOI: 10.1039/C5FD90001H

Lee Cronin opened the discussion of the paper by Joachim Maier: How could the understanding of point defect chemistry in materials apply to really large molecules, *i.e.* what would the relevant control parameters be for defined metal oxide 'molecules' with multiple redox states *etc.*?

Joachim Maier responded: Indeed, solids are large molecules (*e.g.* NaCl, a 3D Coulomb polymer). Large ensembles imply the existence of equilibrium point defects. Frozen-in structure elements such as substitutional dopants may be compared to kinetically stabilized substituents in organic molecules *etc.*

Hong Li asked: Could you give an example of a non-trivial size effect in the field of batteries?

Joachim Maier answered: Non-trivial size effects are effects that do not trivially scale with the spacing of interfaces. Examples are soggy sand electrolytes at high filler content and job-sharing storage in composites of sizes below the screening length.

Hong Li remarked: In all-solid lithium batteries, interfacial resistance between the electrolyte and cathode is quite often found. High interfacial resistance has been regarded to be related to the space charge layer at the interface. (Some other authors believe in the formation of a high resistance new interfacial phase). Some authors coat high dielectric constant material on the cathode to modify the space charge layer and decrease the interfacial resistance. In order to understand such a problem, it is necessary to understand the space charge layer effect at a complicated interface.

Is there any theory or method to calculate/estimate the space charge layer at interfaces in which each side is composed of more than two phases (A+B/C+D)? Do you believe that the high dielectric constant modification strategy could decrease the interfacial resistance?

Joachim Maier replied: If various curved phases meet at given contacts the local energies are different from those of planar interfaces, capillary pressures evolve and the geometry is complicated. Yet the general principles are similar to those given for planar contacts. The dielectric constant is important for the screening length, but is assumed to be constant in that phase.

Joachim Maier opened the discussion of the paper by Rose-Noelle Vannier: A good method to distinguish between k and D is the variation of the geometry of the oxides. Did you try that?

Rose-Noelle Vannier responded: Here k is orders of magnitude higher than D ($k = 1.9\times10^{-7}$ cm s^{-1} and $D = 2.5\times10^{-10}$ cm^2 s^{-1} for the $Sr_{0.10}$ doped sample at 649 °C, see Table 5 in our paper). The critical length under which the diffusion is limiting is therefore equal to 13.1 microns ($L_c=D/k$), which means that oxygen dissociation will be limiting only for a sample thickness lower than 13 μm. Because of the high kinetics towards oxygen transfer we have not been able to derive the k value using electrical conductivity relaxation (ECR). The sample we used was 17 mm in length with a thickness of about 1–2 mm. Because of the small critical length, we did not try to vary the sample thickness using ECR. However, we performed pulse isotope exchange (PIE) in collaboration with Henny Bouwmeester at University of Twente (NL), and we also used the isotope exchange depth profile (IEDP) technique developed by the group of John Kilner at Imperial College. Using IEDP, we managed to derive both k and D values by optimising the time of annealing of the dense sample under a labelled atmosphere. At least for the non-doped compound, the obtained k values were confirmed by PIE and the D values by ECR.[1]

1 Thoreton *et al.*, *J. Mater. Chem. A*, 2014, 2, 19717.

Ram Seshadri asked: You described some design principles, in your talk, for mixed oxide/electronic conductors, including the use of CoO_2 hexagonal slabs for the electronic conductivity, and rock-salt like slabs with vacancies for the oxide ion conductivity. Are there a finite set of such rules that could be outlined?

Rose-Noelle Vannier answered: One way to design a mixed ionic electronic conductor is to mix blocks displaying electronic conductivity and blocks displaying ionic conductivity. In this frame, the CoO_2 hexagonal slabs are of interest for electronic conductivity properties. Then, partial substitution may induce oxygen vacancies or interstitial oxygen in more ionic blocks. The nature of the outermost atoms at the surface is another key point that has to be taken into account in the design of such materials.

Clare Grey enquired: Can you explain the oxygen transport mechanism and the oxidation states of the different Co ions? What design principles do you use to improve the conductivity and how critical do you think the mismatch between the rock salt and CdI_2 layers is? Could you comment on the role of Sr^{2+} doping to vary the size mismatch between the two structural blocks? Can oxygen conductivity be improved by, for example, lanthanum doping to control the number of oxygen vacancies?

Rose-Noelle Vannier replied: The structure of this compound was first characterised by people from the CRISMAT laboratory in Caen (France). Using XANES and from structural considerations, they showed a partial disproportionation of Co^{3+} into Co^{4+} and Co^{2+}, Co^{4+} being located in the CoO_2 'hexagonal' layers and Co^{2+} in the rock salt layers.[1] With oxygen vacancies in the rock salt layers, oxygen diffusion is expected in these layers which are built upon CoO layers sandwiched in between two CaO layers, the CoO layers being highly disordered. We showed that the oxide diffusion was improved when calcium was partly substituted with strontium, which is likely due to an increase of the volume of the rock salt layers, in which the oxide ion diffusion may be facilitated due to the larger radius of strontium compared to calcium. Another parameter we should explore is the impact of the thickness of the rock salt layers on the oxide diffusion. About the mismatch between the rock salt layers and the hexagonal layers, the answer is not that easy. Does this mismatch induce constrains in the structure which could help the oxygen diffusion? The question is open. The characterisation of oxygen diffusion in the lanthanum doped and bismuth doped compound is currently in progress in our group. Since calcium substitution with lanthanum should lead to a decrease of the concentration of oxygen vacancies, we are expecting lower diffusion, as we showed a decrease of the oxygen diffusion coefficient of the parent compound $Ca_3Co_9O_{14}$ when the oxygen partial pressure was increased.[2]

1 Masset *et al.*, *Phys. Rev. B*, 2000, **62**, 166.
2 Thoreton *et al.*, *J. Mater. Chem. A*, 2014, **2**, 19717.

Lee Cronin asked: To what extent could you imagine designing/predicting the properties of the next generation of cathode materials using theory?

Rose-Noelle Vannier responded: It is ambitious indeed. However, thanks to low energy ion scattering, the information on the composition of the first atomic layer of a material can now be easily obtained. If one would have expected transition metals at the surface for their catalytic properties, alkaline earth or lanthanides are usually observed and are likely to play a key role in the kinetics of the oxygen reduction reaction (ORR) at the cathode of a solid oxide fuel cell. This is a first brick in the design of such materials. The second parameter to take into account is the oxide ion diffusion itself. Here, DFT calculations should be a powerful tool to guide the research. What is the impact of the number of perovskite slabs in the Ruddlesden–Popper structure? Where does oxygen diffusion take place in double perovskite? Does the misfit structure in $Ca_3Co_4O_9$ have an impact on the oxygen diffusion? Does the thickness of the rock salt block in such a structure have an impact? What about the number of oxygen vacancies and more generally the concentration of defects? By combining oxygen diffusion characterisation (isotope exchange, relaxation techniques, *etc.*), surface analysis and modelling, one should be able to derive key parameters which govern the

ORR and these should help in the design of new materials with better performance.

Fuping Pan said: The doping of strontium into $Ca_3Co_4O_9$ (CCO) can increase the diffusion coefficient of oxygen based on the results in your reports. How about the effect of strontium doping on the electrical conductivity of CCO? If strontium doping leads to a decrease in the electrical conductivity of CCO, this is unfavorable for the oxygen reduction reaction.

Rose-Noelle Vannier replied: As shown in Fig. 5 in our paper, a decrease of the electronic conductivity is indeed observed when strontium is added into the structure. However, it remains higher than 100 S cm^{-1}, which is the value which is commonly admitted in the field of SOFC cathodes. My opinion is that the key parameters are the kinetics of oxygen transfer at the surface, which is improved here when strontium is introduced in the structure, and the oxide ion diffusion itself, which has to be improved for these materials but is compensated for by the use of composites with ceria in application.

Clare Grey said: Following up on the strontium doping question, what effect does the volume of the rock salt layer have? Can you dope with lanthanum, for example, to improve the conductivity?

Rose-Noelle Vannier replied: The volume of the rock salt layers is increased when calcium is partly substituted with strontium. This leads to more space for oxide ion diffusion in these layers. Partial substitution is also possible with lanthanum. The characterisation of oxygen diffusion in lanthanum doped compounds is in progress.

Lee Cronin opened the discussion of the paper by John Griffin: Have you thought about studying systems with more than one cation, *i.e.* binary or even ternary systems? Such systems may give non-linear behaviour.

John Griffin answered: This would be a very interesting system to study because the different cations present could, in principle, be identified and quantified easily by NMR. It would be interesting to study if there is preferential adsorption of ions based on factors such as size, mobility or charge density.

Yong Yang asked: The work you have done is very nice, but I wonder if you could distinguish whether adsorbed ions are in fully or partially occupied pores from your recent spectra? In addition, I would like to know about whether we could obtain the same information from other nuclei such as Li, P or C.

John Griffin replied: Quantification of the adsorbed anions based on the adsorption spectra (Fig. 1 in the paper) shows that ions are densely packed inside the pores for a concentration of 1.5 M, whereas they are less densely packed for a lower concentration of 0.5 M. The difference in pore occupancy does not result in a significant difference in shift (the shift of the adsorbed species arises primarily due to the NICS effect associated with the delocalised electrons in the carbon surface).

Since the shift of the adsorbed species is nucleus independent, it is in principle possible to derive the same sort of information from NMR spectra of other nuclei in the system *e.g.*, ^{7}Li, ^{31}P, ^{13}C. However, NMR studies of these nuclei can be challenging from the point of view of sensitivity and also resolution, if strong relaxation processes are present (as can be the case for quadrupolar nuclei).

Yong Yang said: Could you comment on whether we could obtain some information about the adsorption of different ions in the inner or outer Helmholtz layers by using this technique?

John Griffin responded: The NICS effect that manifests itself in the shift of the adsorbed species should depend on the distance from the carbon surface. However, since the majority of the pores in this system are less than 2 nm in width, the concept of inner and outer Helmholtz layers is not really relevant.

Shizhao Xiong enquired: How do you calculate the radius of the solvated ions? How many layers did you put in the solvent shell?

John Griffin answered: The radii of the solvated ions used in this work are based on values given in ref. 1, which were calculated assuming a single layer solvation shell of 7 ACN molecules (NEt_4^+) and 9 ACN molecules (BF_4^-).

1 Kim *et al.*, *J. Electrochem. Soc.*, 2004, **151**, E199.

Liqiang Mai remarked: Based on your results, could you please give some suggestions to improve the voltages and the capacitances for supercapacitors?

John Griffin replied: The NMR results presented in this work show that in many cases co-ions remain inside the pores during charging. This will screen the charge of some of the counter-ions that are adsorbed during the charging process, meaning that they do not contribute to charge storage. If it could be possible to tailor the electrode/electrolyte design so that co-ions are more easily expelled, this could improve the capacitance.

Liqiang Mai asked: Could you comment on the difference between aqueous and non-aqueous electrolytes for supercapacitors?

John Griffin responded: Aqueous electrolytes are simpler, cheaper and typically less toxic than organic electrolytes. However, their major drawback is the lower voltage window, which limits the amount of energy that can be stored. Additionally, there can sometimes be problems associated with wetting of the often hydrophobic carbon surface. However, recent work has shown promising results for aqueous electrolytes in porous carbon supercapacitors.[1]

1 Fic *et al.*, *Energy Environ. Sci.*, 2012, **5**, 5842.

Andrew Mount said: It seems that in your experiments you are using a two-electrode system and controlling the voltage (potential difference) between these electrodes. Given that it is the potential that determines the charge state of and double layer structure at each electrode, is there anything that is fixing the potential of one electrode (and hence the other through the potential difference) in this system? Have you (or others) controlled the electrode potential with respect to a third (reference) electrode and what difference does this make to the observations?

John Griffin answered: The experimental setup used in this work comprises a two electrode system whereby only the total cell potential is controlled by the potentiostat. A three electrode system with a reference electrode would enable the measurement of the potential difference across each electrode independently. This would enable the determination of the charge stored on each electrode independently, which could be correlated with the change in ionic populations observed by NMR spectroscopy. However, incorporating a three electrode system into the experimental NMR setup would be quite challenging from a practical point of view.

In unpublished work using a three electrode setup on the same electrolyte/carbon system (outside of the NMR spectrometer), we have observed that the potential difference across each electrode is approximately half of the total cell potential, indicating that charge is distributed equally across the two electrodes.

Heinz Frei remarked: Given the need for additional operando spectroscopic techniques for elucidating the complex processes in supercapacitor electrodes, tender X-ray photoelectron spectroscopy at a synchrotron source (2–5 keV) might be useful. This electronic structure specific and environment sensitive technique can be used at practical pressures and probes a significantly deeper layer than standard XPS.

John Griffin responded: This certainly sounds interesting. However, if vacuum conditions are required for the experiment, this would probably lead to loss of solvent from the electrode surface.

Rudolf Holze communicated: Were all experiments done with the same carbon material? If yes, which criteria were applied in selecting this carbon, which property was particularly important?

John Griffin communicated in reply: Yes, all experiments were performed using YP50-F activated carbon. This carbon was chosen because it is commercially available and shows good electrochemical properties (high gravimetric capacitance and good power capabilities) in conventional supercapacitor devices. Further experiments on synthetic carbons will enable the investigation of *e.g.*, pore size effects and surface functionality.

Rudolf Holze communicated: Does the pore shape (cylindrical *vs.* funnel-type) make any difference?

John Griffin communicated in reply: In principle, the pore shape should make a difference to the capacitance on a microscopic level.[1] However, the pore structure in the activated carbon used in this work is thought to be highly disordered, and so it is difficult at this stage to relate the local pore shape to the capacitive properties.

1 Huang *et al.*, *Chem. Eur. J.*, 2008, **14**, 6614.

Rudolf Holze communicated: Does pore size distribution (monomodal *vs.* bimodal) have any effect?

John Griffin communicated in reply: The coexistence of larger and smaller pores in the same carbon structure may help the rate capabilities of the supercapacitor electrode, as the larger pores provide 'highways' for the ions to travel along, helping them move in and out of the carbon electrode more easily.

Rudolf Holze communicated: Do surface properties like 'acidity', concentration of functional groups and type of functional groups have any measurable or at least conceivable effect?

John Griffin communicated in reply: Surface functional groups can contribute to increased capacitance by taking part in redox reactions with electrolyte species. However, these reactions may not be reversible, or may result in breakdown of the electrolyte and so are not necessarily desirable in a supercapacitor.

We have not seen any noticeable effects relating to surface functional groups. However, these should in principle break up the ring currents that give rise to the shift of the adsorbed species, and so may be observable in NMR spectra of certain systems. More advanced NMR experiments such as correlations between electrolyte species and *e.g.* ^{1}H in OH groups on the surface may be an interesting way to study these groups.

Zhongqun Tian opened a general discussion of Joachim Maier's, Rose-Noelle Vannier's and John Griffin's papers: Your points on the classification and characteristics of (quasi-) stationary morphologies are really interesting. I have a question on your schematic representation of electronic and ionic excitation in band diagrams. I just wonder if these two excitations can be well described in the same way for solid state electrochemistry. In electronic excitation, the electron can be excited by a photon or electric field to a more energetic state and may lose or gain some energy when it comes to the ground state. But the process in ionic excitation could be more complex. The ions can be excited either for ionic transportation or without moving in the solid state. It would be useful if you could explain this diagram and physical picture more clearly, which may be important for solid-state chemistry and the controllable properties of solid phase materials.

Joachim Maier replied: The statistical correspondence is very far-reaching. I discussed this in detail in ref. 1.

1 J. Maier, *Z. Phys. Chem.*, 2005, **219**, 35.

Zhongqun Tian said: The term 'defect' has been widely used in solid state and materials chemistry for a long time and its meaning has evolved extensively, from imperfect/deleterious to beneficial types. For materials chemistry, the goal is to introduce localized disparate structure to the solid phase and surface which can enhance the properties and performance of materials. However, when the term 'defect' was translated into Chinese, it only represented the imperfect/deleterious meaning, so the meaning of the Chinese term is quite negative, I guess this situation also applies to other languages. This may lead to confusion among non-English students and new-comers, and also hinders us from unravelling and controlling the complexity that determines the functionality of various materials. As a consequence, the scientific connotation of 'defect' is difficult to clearly define. Its interesting to compare the cases of the solid state and liquid state. In the case of solution chemistry, when some impurities are found to be helpful for molecule reactions and assembly in the solution phase, those impurities can be quantified and changed from deleterious to a beneficial type, then further classified/renamed as a catalyst, additive or linker, according to their function in solution chemistry. So the scientific connotation of these species can be more clear. More importantly, the historical development of solid state and materials chemistry can be classified by several key stages, as shown in Fig. 1. At the earliest stage, study was focused on pure atomic, ionic or molecular crystals with perfect and ordered structure that contained no point, linear, or planar imperfections. Diffraction and spectroscopic techniques were used for structural characterization. Later, we focused on structurally well-defined surfaces of the above mentioned crystals with emphasis on molecule adsorption, reaction, nucleation and growth, and assembly at the surface. This was accomplished by preparing various facets of single crystals and developing different surface characterization techniques. These defect-free crystals and surfaces were easily studied and correlated with theoretical modelling systematically. Nowadays, we are at the third stage mainly focusing on more complex and useful materials. Some surface defects of metal and oxide crystals are well classified as vacancies, steps, kinks, B3 or B5 catalytic centers, *etc.* Some more complicated surface defects were investigated and then defined as adsorption/reaction active sites with multiple components and dynamic natures. More powerful *in situ* characterization techniques and theoretical methods are required for improving the sensitivity and energetic/spatial/temporal resolutions. With these developments, surface chemistry has moved forward to a more mature stage because most key subjects in this field are getting more well-defined and quantified in the energetic, spatial and temporal domains. Interestingly, historical development moves in an ascending spiral; the next big target will return to the solid bulk phase that has much more complexity and heterogeneity in structure and dynamics. It is highly desirable to rationally design structurally well-defined active sites located inside the bulk or at the solid/solid interface.

Prof. Maier made a very important point that solid state chemistry is playing a more and more important role in materials and solid/solid interfaces. I think it is necessary to develop new methods for creation, then *in situ* characterization of different types of active sites that can enhance the properties and functionality of new generation materials. This approach will be extended to soft matter (polymers and biological systems, *etc.*), which will be a very long way to go. Overall, for any new branch of science, its scientific maturity is based on well-defined terminology and the precise quantification of key subjects. I wonder if it is necessary to have two classes to define 'defect': a deleterious type and a beneficial type with a better scientific connotation. It seems to be needed to discuss a new word to describe the beneficial type, such as 'eu-defect' or 'eufect'.

Joachim Maier replied: This is a very good point. The notion of 'defects' sounds negative and prevents many people from appreciating their role. One could also term them 'ionic excitations', which then would emphasize the fact that to a certain degree and at finite temperatures they have to be realized due to entropy. (For higher-dimensional defects the equilibrium number though would be much less than unity.) In solids, frozen-in defects play a very decisive role as well. After all the term 'defects' is appropriate, and we might rather think in terms of protonic defects of the water structure when it comes to OH^- and H_3O^+, and of electronic defects when it comes to excess electrons and electron holes.

Galen Stucky asked: If you consider particles or surfaces with small dimensions so that the exterior surface energetics begin to dominate properties as opposed to the bulk energetics, *e.g.* nanoparticles, zeolite walls, mesoporous nanowires, how do you define the resulting control parameters for this transitional scaling dimension in your model?

Joachim Maier answered: These are typically constrained equilibria, *e.g.* point defect equilibria reacting *in situ* on the frozen (*ex situ*) morphological conditions.

Shihe Yang remarked: Prof. Maier put forth an interesting conceptual framework that concerns control parameters for arriving at electrochemically relevant materials. While the general idea is in place, details about how functionality can be achieved seems to be vague. Can we have a more tangible idea about this? The frozen defect states (with the so-called zero entropy production) Prof. Maier focused are actually metastable states. In this case, kinetics is a most important issue. Can the authors elaborate on it? In Table 2 of the paper, some parameters are not well defined and may even contain errors. Can the authors confirm this?

Joachim Maier replied: In the paper there is a detailed exact example mentioned of how to transfer from *in situ* parameters to *ex situ* parameters. The larger the deviation from equilibrium, the more qualitative the concept becomes. See the given references for more details.

Nanfeng Zheng enquired: Do you consider the influence of an external magnetic field on the ionic supercapacity? There are reports in the literature to enhance ionic supercapacities by applying an external magnetic field.

John Griffin answered: We have not considered the effect of the external magnetic field on the capacitance. We do not find a significant difference in the capacitance of devices cycled inside and outside of the magnet.

Changxu Lin asked: Does the chemical shift change over time? Is it reversible? Is it a reasonable subject to perform time-resolved or electro-stimuli-responsive experiments similar to a CV test in order to uncover the dynamic behaviors in your system?

John Griffin responded: The chemical shift changes as the capacitor is charged due to changes in the electronic structure of the carbon surface. We have demonstrated how the *in situ* NMR experiments can be performed during electrochemical cycling of the supercapacitor to study the charging dynamics.[1] In these experiments we observe reversible changes in the shift of the adsorbed species which illustrate the reversibility of the charge–discharge process in the supercapacitor device.

1 Wang *et al.*, *J. Am. Chem. Soc.*, 2013, **135**, 18968.

Shihe Yang commented: When the molecules are adsorbed on the inner surfaces of the pores, one would expect broadening of the NMR peaks. Can Prof. Griffin explain why this is not in this case observed? Can the authors also incorporate the electrical double-layer structure of the pores in their interpretations of the NMR data? A further question is whether one can find the optimal pore size by using NMR.

John Griffin replied: The ions are still very mobile inside the pores, and at room temperature they are in constant exchange between the in-pore and ex-pore environments. The mobility of the ions is on a timescale that is faster than the width of the NMR resonances (~kHz), and so we observe relatively narrow lines. However, we note that the in-pore resonance is broader than typical liquid resonances; this is most likely due to susceptibility broadening from the graphene-like carbon architecture.

Since the ions are highly dynamic inside the pores, and the pore structure is very disordered, it is difficult to define an electric double-layer structure in terms of local geometry. However, we can potentially characterise the structure of the double layer in terms of the amount of anions and cations using the intensities obtained from the NMR spectra (work is ongoing towards this end).

Correlations between the pore size and NMR observables have been discussed in recent experimental[1] and theoretical[2] work. In principle, by linking the pore size effects from NMR with electrochemical performance it is possible to say something about the optimal pore size; however, there may be other factors that also affect the NMR shifts, and so this may not provide a complete picture.

1 Borchardt *et al.*, *Phys. Chem. Chem. Phys.*, 2013, **15**, 15177.
2 Forse *et al.*, 2014, **118**, 7508.

Liqiang Mai said: Recently our group synthesized a nanowire templated semihollow bicontinuous graphene scroll architecture where a hollow exists between the graphene scroll and vanadium oxide nanowires.[1] What are the differences between this core–shell structure and pure nanowires?

1 M. Yan *et al.*, *J. Am. Chem. Soc.*, 2013, **135**, 18176–18182.

Joachim Maier replied: Clearly the transport pathways for electrons and ions are different. This alone guarantees very different rate performances.

Liqiang Mai asked: How do you characterize and distinguish the electrical transport and ionic conductivity of nanowires for electrodes?

Joachim Maier responded: There are two ways: either to use a percolating ensemble or to address a single nanowire. While measurement of electronic conductivity is straightforward, the measurement of ionic conductivity requires more sophisticated techniques.[1]

1 J. Maier, *Physical Chemistry of Ionic Materials: Ions and Electrons in Solids*, John Wiley and Sons, Ltd., Chichester, West Sussex/UK, 2004.

Clare Grey addressed Rose-Noelle Vannier and Joachim Maier: Joachim only touched very briefly on biologically relevant systems, under his category of 'dissipative morphology' in Table 1 of his paper. We also discussed the fact that extrinsic defects were critical to altering the properties of materials. Yet, in many practical devices, the 'outer fluxes' yield to structural changes that in the long term may result in degradation of the properties that our extrinsic doping/careful control of size/morphology has sought to control.

Can the panel comment on how they can build concepts of self-healing into their models? In other words, how can we design systems, which based on thermodynamics, will evolve to more stable structures, yet maintain desirable physical properties?

Joachim Maier answered: Self-healing or self-repairing is a special feature of dissipative structures. Think of organisms. Self-healing can, however, occur in near-equilibrium structures. The original heart pacemaker battery relied on the formation of a thin LiI electrolyte by contact of the two active masses Li and I_2. If interrupted, the electrolyte heals out. The same property is inherent to the SEI-passivation layer in Li-based batteries. For that reason I am not very optimistic as far as artificial SEI-layers are concerned.

Rose-Noelle Vannier responded: In the field of self-healing, my colleague Lionel Montagne at UCCS developed a new concept of self-healing glasses by adding boron, boron carbide, vanadium or vanadium carbide particles in the glass formulation which oxidise at high temperature and melt to make a new glass in the case of cracks.[1] We could imagine applying the same concept to electrode materials in a fuel cell: the addition of a second phase which could react with the degradation phase to give a new compound with the desired properties. This supposes the identification of the degradation products and the stability of all phases at high temperature. It is a strategy already used to trap chromium from the metallic Crofer current collector.

1 D. Coillot *et al.*, *Advanced Engineering Materials*, 2011, **13**, 426.

Xiaoliang Yu opened the discussion of the paper by Hyun-Wook Lee: The author used activated carbon as the counter electrode, so is the system in this work a battery or a hybrid capacitor?

Hyun-Wook Lee responded: We used a three electrode system, that is working, counter, and reference electrodes. So, although the activated carbon electrode was used as the counter electrode, the potential of the working electrode was controlled against the reference electrode. Therefore, this working electrode of NiHCFe works as a battery.

Xiaoliang Yu asked: The rate performance of the aqueous system in this work is superior and comparable to that of high-rate supercapacitors. The authors stated that the ion diffusion within the bulk is fast due to the large insertion site. But I wonder why the charge transfer is also so fast in this system?

Hyun-Wook Lee answered: The charge transfer is dependent on the hydration–dehydration (or solvation–desolvation) reactions. The hydration shells of ions in aqueous electrolytes are smaller than the channel size of nickel hexacyanoferrate, so the ions in aqueous electrolytes do not need to dehydrate fully. The hydrated ions can insert into the structure.

In the case of organic electrolytes, solvation and desolvation processes are necessary to insert ions into the structure. Hence, the charge transfer resistance is higher.

Joachim Maier remarked: Given the well-defined and limited space, one would expect a well-defined number of solvent molecules to enter. Is this number known?

Hyun-Wook Lee replied: In this study, we haven't studied solvent entering during the cation insertion process. We totally agree with the importance of the solvent during the electrochemical reaction, so are carefully performing solvent molecule studies.

Yiren Zhong enquired: Since some of the performances of the alkali ion insertion are quite different, will the lattice parameters of the PB analogues change during the repeated insertion and extraction process?

Hyun-Wook Lee answered: The expansion and shrinkage of the lattice parameters are consistent due to the state of the charge. We confirmed the lattice parameter changes by *ex situ* and *in situ* XRD.

Yiren Zhong commented: Basically, the electrochemical performance depends on the involvement of the ion access and electron transfer, so the active materials usually need to have good electron conductivity. Since the PB analogue has poor conductivity, how can it achieve good power ability? I mean the performance under different current densities.

Hyun-Wook Lee responded: PB analogues have better electrical and ionic conductivities than usual battery active materials. So, this material can achieve high rate capability. Generally, in a general battery electrode, conductive carbon (Super P) is mixed in the electrode in order to support electrical conductivity.

Clare Grey asked: Can you explain why you chose the NiHCFe prussian blue? Can you discuss the role of defects in the electrochemical performance?

Hyun-Wook Lee responded: Among the prussian blue analogues, NiHCFe has reasonable reaction potentials in Li, Na, K, Rb, and Cs aqueous electrolytes. Moreover, NiHCFe is stable in both organic and aqueous electrolytes.

Typical prussian blue analogues contain vacancies of $M(CN)_6$. This defect reduces the specific capacity in monovalent ion electrolytes. On the other hand, divalent ions also react with prussian blue analogues in vacancy sites. Hence, the vacancy can have a positive or negative role depending on the system.

Rudolf Holze remarked: Did you try X-ray diffraction after lithiation to identify structural changes related to ion insertion without/with solvent molecules?

Hyun-Wook Lee answered: Yes, we tried *ex situ* and *in situ* XRD after or during ion insertion. The lattice parameters changed less than 1% during the redox reaction.

Rudolf Holze said: Given the very high charge–discharge rate I wonder: is it a battery electrode or a capacitor electrode? How thick are the electrodes, and is the process mostly superficial – like in pseudocapacitance?

Hyun-Wook Lee replied: We have carefully thought about this, either battery or capacitor behavior. We have checked by EDS analysis at 10 C rate performance and the Na or K concentrations confirmed this issue. After full discharge, the electrode contains mobile cations, so we are sure the reaction was due to ion insertion into the structure at a very high C rate.

Xuyi Shan asked: Why are the trends in C rates different for the three ions in the two different electrolytes? The Li ion has the slowest kinetics among the three ions in aqueous electrolyte, but has the fastest kinetics in PC.

Hyun-Wook Lee responded: In aqueous electrolytes, the kinetics of Li, Na, and K ions are pretty similar. The reason that the Li ion looks to have slower kinetics compared to the other Na and K ions is that there are two reactions in the Li aqueous electrolyte. The reaction of H_3O^+ has a higher potential than that of Li, so although the concentration of H_3O^+ is much smaller than that of lithium, the system shows two redox reactions. Because of the complex system in Li aqueous conditions, the rate performance shows slower kinetics than Na and K. In organic electrolytes, the rate capabilities of the three cations depend on the charge transfer, so the trends follow the size of the ions due to the higher desolvation energy.

Ram Seshadri opened the discussion of the paper by Ryoji Kanno: Are there special considerations and procedures that must be employed in the creation of the solid Li-ion conducting electrolyte materials?

Ryoji Kanno replied: Yes. First, we should avoid using hazardous elements. Next, elements showing large polarizability will be preferred. In that sense, sulphides have an advantage over oxides. Then, crystal structure should be considered. A large framework with empty space is favored. Finally, for the synthesis process, it is very important to heat samples at higher temperatures to obtain higher crystallinity and to prevent a moisture attack to decrease impurities.

Joachim Maier said: Which materials, the Si, Ge or Sn based ones, sinter more easily?

Ryoji Kanno responded: There is little difference in sinterability. However, it seems that Ge-based electrolytes have higher melting points than the other two, so Ge-based ones could be sintered better.

Xiang Hong asked: Professor Kanno, so far as we know, a LiGePS-type solid electrolyte is not very stable *vs.* a lithium metal anode for application in solid-state Li metal batteries. When you replaced Ge by Sn or Si, did you observe an improvement of its stability?

Ryoji Kanno remarked: At present, we agree that it is difficult to apply LiGePS-type electrolytes to all solid-state batteries using an Li anode. However, improvement is observed in electrochemical stability to Li metal for Si-based solid electrolytes, and certain data suggest that LiGePS-type is stable to Li metal.[1] We would like to pursue the potential of the LiGePS-type Li-ion conductors in future work.

1 Hassoun *et al.*, *J. Power Sources*, 2013, **229**, 117.

Xiang Hong enquired: For the application of inorganic solid electrolytes in commercial batteries, how to process these electrolytes into thin films is always a big challenge. The conditions for making electrolyte films also have a big influence on their measured ionic conductivity and performance in cells. In your case, how did you solve this kind of problem to get reliable conductivities for your solid electrolytes with different compositions? For real application in commercial EV or CE batteries, how do you think you could facilitate the electrolyte-film fabricating process?

Ryoji Kanno said: Thank you for pointing that out. The important issue is how to protect solid electrolytes from moisture attack. Handling under an inert atmosphere will be the key, which is common to both the ionic conductivity measurement of electrolytes in thin films and application of solid electrolytes to EV or CE batteries. It is also very important to find the fabrication process for the sheet from the electrolyte. The combination of the solvents and the electrolytes is another important factor that makes the sheeting process easier. These are the current subjects to improve the sheet form for all-solid-state batteries.

Galen Stucky said: The capacitance is per gram of what? How does that compare with the similarly defined commercial capacitance?

Ryoji Kanno replied: The capacitance is calculated based on the weight of the cathode active material, $LiCoO_2$, as in the case of the performance of liquid-system batteries reported in the literature. Of course, the capacitance calculated based on the total weight or volume of the cell is important. Since solid electrolytes function as a separator, all solid-state batteries might have advantages over liquid-system batteries in terms of gravimetric or volumetric capacity, if the device fabrication process is developed to facilitate patterning and integration.

Graham Hutchings commented: Would there be any advantage for the applications you are investigating in making mixed metal compounds, *i.e.* adding a small amount of Si to the Ge compound or Sn to the Si compound? Would this induce any interesting effects?

Ryoji Kanno said: This is a very good point. For example, if we could obtain a similar compound as the Ge compound by mixing with Si and Sn, or by adjusting the ionic size by other combinations of elements, we could reduce the amount of Ge which might be expensive for practical applications. Since the topic is under investigation just now, we cannot present experimental results, but we will report some interesting effects of doping elements in the future.

Weimin Xuan asked: As you mentioned, the conductivity will be affected by different elements and lithium contents. Are there any other factors that could also cause conductivity change?

Ryoji Kanno answered: Thank you for your comments. Probably, we have to consider lithium distribution and transport characteristics, which are not discussed in the article due to the lack of sufficient experimental results for the Si and Sn systems. We now have diffraction data and also calculation data from references for the Ge system,[1] however, no data on other systems. Now we are collecting neutron diffraction data to clarify the lithium distribution in more detail. In order to discuss transport characteristics in detail, we think that we need to study the dynamics of ionic motion. Therefore, in the present study, we focused only on discussion based on the structural information and conductivity data.

1 O. Kwon *et al.*, *J. Mater. Chem. A*, 2015, **3**, 438–446.

Nanfeng Zheng said: Can the ionic conductivity be further increased if a small portion of sulfur is replaced by selenium?

Ryoji Kanno replied: For Si-related systems, we think a small portion of substitution by selenium might be favorable for the lattice volume expansion and ionic conduction because Si-based LGPS-type conductors suffer from small lattice parameters. For the other two systems, it is difficult to predict at the present stage.

Joachim Maier asked: How important are polarizability issues for ion conduction in these materials?

Ryoji Kanno responded: This is a very good point. Owing to the large polarizability of sulphur, sulphide lithium conductors exhibit a higher conductivity than oxides. However, the comparison should be made between materials with the same structure. Therefore, oxides or oxy-sulphides with LGPS-type structure should be studied to clarify this point in more detail.

Clare Grey enquired: Have you tried cycling your solid electrolytes *vs.* silicon, germanium and tin anodes? Could you comment on the electrode/electrolyte interfaces – would you expect differences if you, for example, cycled Si *vs.* the Si-containing electrolyte?

Ryoji Kanno said: This is a very interesting point for us. We are just now trying various types of combinations of electrolytes and anodes in all solid-state cells. Anodes we tested include the Si anode. In our results obtained so far, the performance of batteries largely depend on the combination, which is probably related to interphase formation.

Rudolf Holze remarked: How much contribution is there from grain boundary conductance? Some years ago a company (OHARA) produced something which looked to favour grain boundary conductance. Can you comment on the additional step in the process?

And two general remarks: comparison between specific conductance values (*i.e.* conductivities) of solids, solutions and ionic liquids are of very limited value only for practical applications. A solid electrolyte in almost all cases includes the function of the separator, resulting most likely in a much smaller electrode separation possibly compensating or even overcompensating the lower conductivity.

For thin layer batteries, printed batteries and batteries integrated into solid state electronics, most likely only solid electrolytes will be applicable.

Ryoji Kanno replied: This is very important. Unfortunately, we do not know the detail of the process that is employed by the company (OHARA). Our next step is to clarify the contribution of grain boundaries in LGPS-type Li conductors. For this purpose, ionic conductivity measurement at very low temperatures and investigation of ionic transportation characteristics using NMR are under consideration.

Zhongqun Tian opened a general discussion of the papers by Hyun-Wook Lee and Ryoji Kanno: Ionic liquids (ILs) for electrochemical applications, especially for batteries, must have high electrical conductivity. However, the electrical conductivity of pure ILs is relatively low at room temperature. I wonder if ILs could become the new generation of electrolyte widely used in batteries, as the temperature causes a dramatic increase in the values of the electrical conductivity for pure ILs. Could you two comment on it?

Hyun-Wook Lee answered: I don't know well about IL electrolytes so it is difficult to give a proper answer. In my opinion, the ionic conductivity of ILs is relatively lower than general organic carbonate electrolytes so the kinetics in IL electrolytes might be slower than that of organic electrolytes at room temperature. At high temperatures, prussian blue analogues might decompose water molecules in the structure and, at this moment, the kinetics might change from the current experimental results. The relation between temperature and the kinetics of the material will be tested in near future.

Zhongqun Tian remarked: Your work is really interesting. I would like to ask a general question. The theoretically predicted efficiency of (photo/electrical/chemical) energy conversion of different devices/systems is considerably higher than the practical ones at present. One of the reasons is that most studies have focused on the synthesis of various nanomaterials, especially nanoparticles. They are mixed with other conductive materials and components in a simple way when the electrode and device are constructed. In fact, many energy systems consist of three or four channels (for positive and negative charge carriers, liquid and/or gas) and multi-

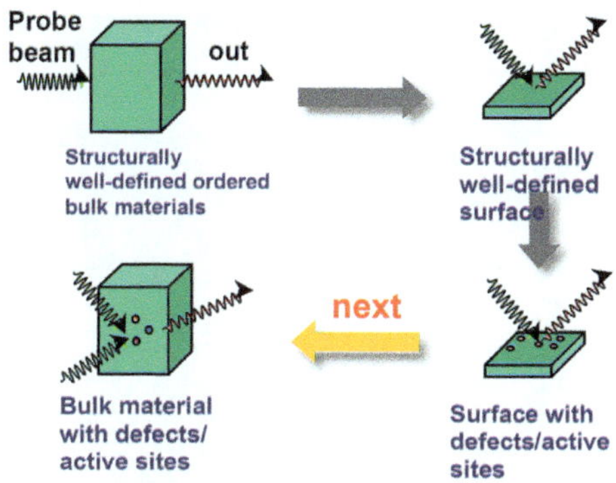

Fig. 1 The stages of the historical development of solid state and materials chemistry.

interfaces. Without rational assembly and construction of these channels and interfaces in mesoscale, their collaborative effect will be low or even poor. Mesoscale architecture does not only mean ordered structure, but new functionality and synergetic performance. To reach the full or maximum potential of device complexity and functionality, optimizing transport and response properties by design and control of hierarchical functional structure from atomic and nanoscale to mesoscale seems to be essentially important; to solve the problem of scalability in industrial production for such a delicate fabrication from nano- to mesoscale. One solution could be the use of old manufacturing lines for semiconductor device fabrication. For the IC industry, after developing about 15 nm chip processing technology, the facilities of 30-50 nm technology could be transferred partially for top-down manufacturing of energy devices. Could you comment on it?

Hyun-Wook Lee replied: In a battery electrode, high surface gives fast kinetics, such as high rate capability due to more reaction points between the electrolyte and electrode. In this case, the nanosized electrode materials give benefit for high performance. On the other hand, nanotechnologies also affect the cost, so we need to carefully consider the optimization between cost and performance.

Shihe Yang asked: When the lattice parameter shrinks, one would expect a conductivity increase if one just considers the decrease in the hopping distances.

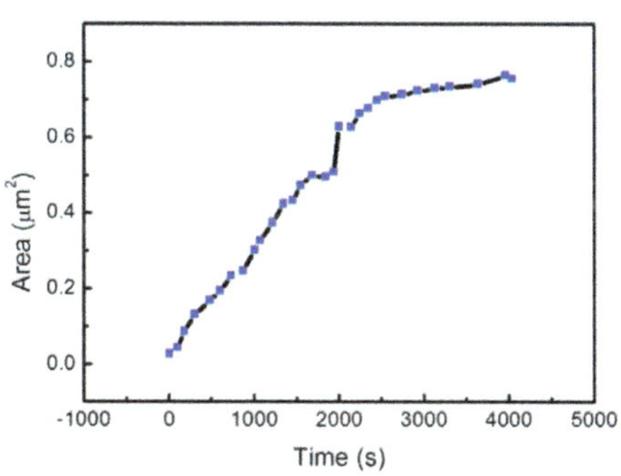

Fig. 2 The increase of the cross sectional area of the lithium dendrite.

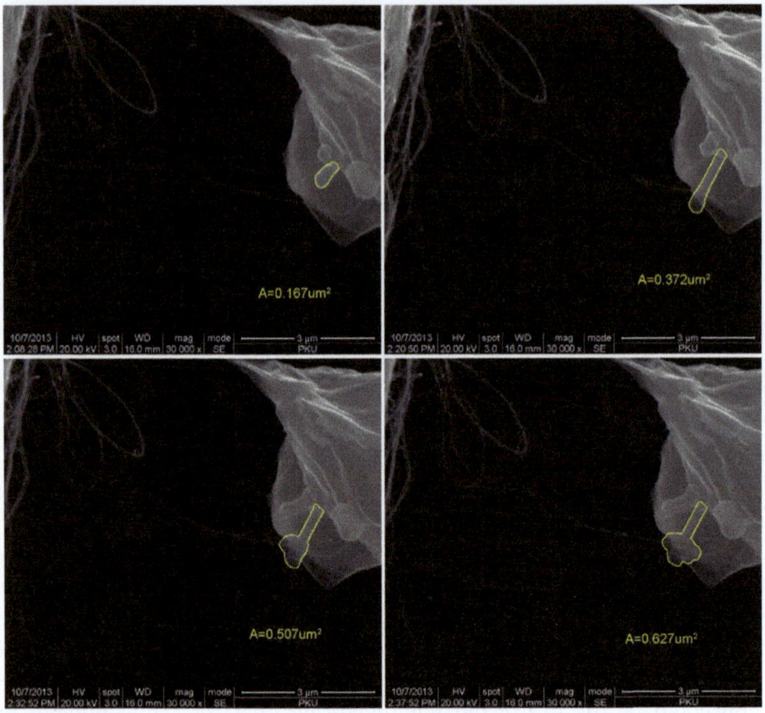

Fig. 3 Calculation of the area data indicated on the *y* axis of Fig. 2.

But this contradicts the experimental results. Clearly, other factors dominate. What are they? Can you rationalise the conductivity results in terms of enthalpy and entropy?

Ryoji Kanno replied: In addition to the hopping distances, we must consider the bottleneck size for the conduction pathway. As you say, lattice parameter shrinking will have a positive effect on hopping distance, but is unfavorable from

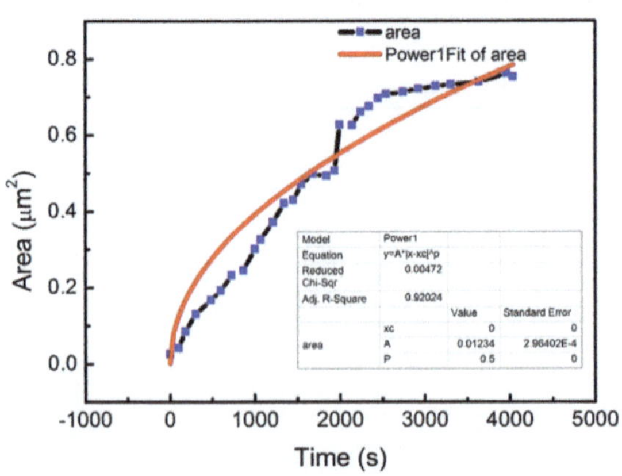

Fig. 4 Plot of area *vs.* time.

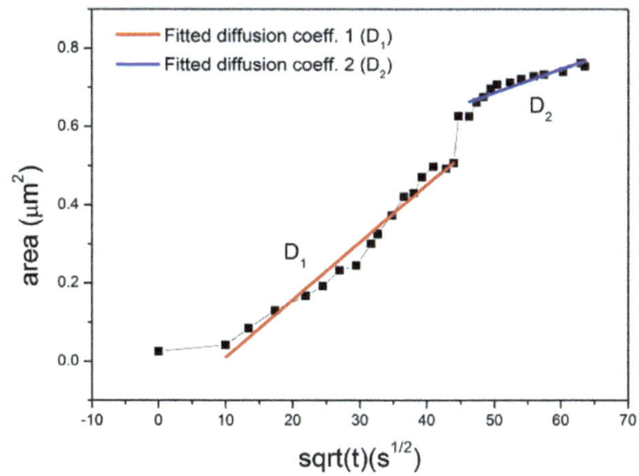

Fig. 5 Plot of area *vs.* $\sqrt{t}$.

the view point of increasing bottleneck size. Qualitative analysis in terms of enthalpy and entropy will require theoretical calculations, which we are carrying out in parallel with measurement of thermophysical properties.

Yong Yang enquired: Have you tested the composition and thickness of the interfacing layer between the solid electrolytes and electrode materials? In addition, could you comment on what is the role of the interphase layer in the solid electrolyte system, what is the driving force to form this interphase layer?

Ryoji Kanno said: This is a very good point. We wish to clarify the interaction mechanisms of the solid electrolyte/electrode interface. The composition and thickness of the interfacing layer will be studied using quantitative electrochemical measurement and *in situ* diffraction techniques. For the formation of the interphase layer in the solid electrolyte system, this might be caused by an impurity phase at the surface of the electrolyte, or the decomposition products of the electrolyte and/or electrode. The role of the interphase layer might be important; this may protect further decomposition of the electrolyte during the electrode reaction, and may cause closer contact between the electrode and the electrolyte at the interface.

Joachim Maier remarked: One should generally be careful in using diffusion equations derived for liquids for the solid state (often ionic–electronic coupling matters).

Hyun-Wook Lee replied: Thank you for your helpful comments. We will carefully consider the diffusion equation.

Wei Han opened the discussion of the paper by Haimei Zheng: What are the electron beam effects?

Haimei Zheng answered: The electron beam has strong effects on the electrochemical processes, including introducing solvated electrons, decomposing the electrolyte, local heating, *etc.* Thus, without applying an electric bias, reactions may occur in the electrolyte under the electron beam. However, the electron beam effects are dose dependent. In the current experiments, we use low dose imaging to limit the electron beam effects. For example, the observed Li–Au reactions can only be achieved when a cyclic voltammetry is applied. Without an applied electrochemical program, no electrochemical reaction occurs.

Liqiang Mai asked: Compared with gold, aluminum or copper are much closer to commercial application. Do you have any idea about changing gold to aluminum or copper to get closer to commercial application? Did you try to use a separator in your experiments?

Haimei Zheng responded: We have considered using aluminum and copper as the conductive electrodes in the electrochemical liquid cell, since aluminum and copper are the ideal current collectors for the cathode and anode of a battery cell. However, these electrodes are not compatible with the fabrication processes of electrochemical liquid cells, *i.e.*, Al and Cu are easily oxidized. As a result, the conductivity of the electrode deteriorates, which severely hinders the electron transfer in the electrochemical liquid cell. So, we did not use aluminum and copper as electrodes for actual experiments. We did not use any separator. In our current experiments, there is enough distance between the two Au electrodes to avoid the short circuit.

Gang Chen remarked: Since gas bubbles form around the Au electrode, and it is known that Au acts as an active catalyst only at the nanometer scale while the actual battery is of macroscopic size, how thick is your sample? Is your system able to reflect the actual physical and chemical processes taking place in a real system?

Haimei Zheng replied: The liquid cell has a liquid layer with a thickness of about 120 nm, which is estimated based on the spacer thickness of the liquid cell. The thickness of the Au electrodes is 100 nm. However, there could be Au nanoparticles, nanosized features or defects on the electrode surface, which can be active sites to catalyze the reaction of bubble formation. We think our electrochemical liquid cell does reflect actual physical and chemical processes. For instance, we did control experiments of bulk electrochemical reactions using two macroscopic Au electrodes (3 mm Au TEM grids) as working and counter electrodes, and gas bubble formation was also observed when cyclic voltammetry was applied.

Shihe Yang commented: I assume that the pressure in the bubbles is above environmental pressure. Can the authors make an estimation of it in terms of some well-known relations? Do you need surfactant molecules to stabilise the bubbles?

Haimei Zheng responded: We do not know the gas pressure inside the bubbles. Since many bubbles are of nanometer size and they are immersed in nanoscale thick liquid or in touch with the membrane wall, any theoretical estimation of the gas pressure inside a bubble can be inaccurate. We did not use surfactant molecules to stabilize the bubbles.

Bingwei Mao said: In your work, the evolution of the surface structure is attributed to Au–Li alloying. However, there have been a number of reports in the literature saying that noble metals such Au and Pt can be electrochemically reduced to generate Au and Pt anions, which can then form surface films with the cations present in solution. Therefore, formation and development of a surface film involving Au anions is also possible in your work. To check whether such a process is involved or not, it is necessary to investigate the surface morphological change after applying an anodic stripping. If only the surface alloy is involved, the alloy should be easily dealloyed after stripping at a reasonable anodic potential with characteristics holes on the surface. On the other hand, comparative investigations using copper or nickel electrodes seem also worth trying. Especially, the copper can form an alloy with lithium, but not easily be reduced to form a surface film. The characteristic behaviour of surface alloying–dealloying would help identify the surface film, if any, in the Au system.

Haimei Zheng answered: We agreed with the insightful comments. These could be valuable for future experiments.

Yiren Zhong remarked: You have studied the Au electrode regarding electrode dissolution and I noticed you have mentioned SEI formation on the surface. I wonder if you have studied the cobalt electrode, and have you observed what the SEI film is like? I mean the reversible formation and decomposition of the SEI film, or something else.

Haimei Zheng replied: We have not studied the cobalt electrode. Possible passivation film formation and decomposition on a cobalt cathode could be very interesting for future study.

Yong Yang opened the discussion of the paper by Hong Li: As your conclusion said, you conclude that the deposition of Li is a diffusion-controlled process; how could you deduce this conclusion from SEM measurement? Is the integrated area a two-dimensional image or not? In addition, could you comment on why the two-dimensional area still fits a diffusion-controlled law?

Hong Li answered: This is a good question. In order to answer your question, it is necessary to describe the data analysis process. The area data indicated on the y axis of Fig. 2 is calculated as in Fig. 3.

Consider the non-linear growth process of the lithium dendrite, *i.e.* the dendrite changes its growth direction and growth rate after contact with the CNT. We assumed that the diameter of the dendrite remains the same during growth, and therefore the area outlined by the yellow lines (shown in Fig. 3) is proportional to the volume of the dendrite. During the dendrite growth, the lithium diffusion can be treated mainly in 1D, so the volume of the dendrite can reflect the diffusion length x, *i.e.* area $\propto x$. In the 'area *vs.* time' plot (Fig. 4), we use the equation $x = \sqrt{(qDt)}$ to fit the whole curve. The fitting result does not match the experimental data very well. But if we plot the 'area *vs.* $\sqrt{t}$' curve instead (Fig. 5), we can use the linear fitting to fit to the equation $x = \sqrt{(qDt)}$ for two regions. It is obvious that there are two diffusion processes during the dendrite growth; the diffusion coefficients are D_1 and D_2 respectively.

It seems that the lithium diffusion processes are different before and after dendrite contact with the CNT. Further finite element simulation could be helpful for clarifying the growth of the lithium dendrites during *in situ* SEM measurement.

Joachim Maier remarked: Is it possible to successfully depress dendrite formation by additions to the electrolyte?

Hong Li responded: There are many efforts on searching additives in liquid electrolytes to depress the dendrite formation. Some are quite effective. However, it is difficult to suppress the dendrite formation at a high rate and after many cycles. The formation of the SEI influences the effectiveness of the additives. On the other hand, it is known that lithium dendrite formation could be suppressed effectively when a polymer or inorganic electrolyte is used.

Bingwei Mao remarked: You demonstrated that the morphology of the lithium anodes after the lithium deposition or dissolution process depends on the order of the process which corresponds to the working conditions of two kinds of batteries, and the morphology characteristics of the lithium anodes are recognized to be related to the inhomogeneity of the SEI film. In other words, the SEIs on the lithium anodes in different types of batteries are different in organic solvents. Would you like to give explanations on how the SEI can influence the lithium deposition and dissolution processes? Ionic liquids have wide electrochemical windows and low vapour pressure, and have been employed in the investigation of lithium secondary batteries. Especially, some ionic liquids can have cathodic electrochemical windows even lower than the potential for lithium bulk deposition. It is expected that the SEI formed on a lithium anode, if any, may be very different from those in organic solvents. Would you like to comment on the possible influence of ionic liquids on the lithium deposition and dissolution behaviour?

Hong Li replied: These are good questions and suggestions. Firstly, the surface of lithium is covered by Li_2CO_3 and Li_2O. It is still not clear whether the coverage of the surface film on the lithium is homogeneous with the same thickness. We would like to check this issue by Kelvin probe, non-contact conductivity measurement, SIMS and XPS. We would like to compare the composition, thickness and coverage of the surface film on lithium stored in different atmospheres and temperatures.

Secondly, the soaking of fresh lithium in different electrolytes will lead to spontaneous chemical reactions. According to previous investigations, Li_2CO_3 and alkali carbonate will grow gradually. This means that the native layer is not stable enough to prevent further reaction of lithium with the electrolyte.

Thirdly, a new SEI will form and grow on the metal lithium electrode during electrochemical reactions. Up to now, it is not clear how and whether the new SEI species will replace the native grown SEI layer completely during electrochemical reactions in different electrolytes and batteries. This is a difficult problem, needing careful and systematic investigations. According to our investigations, the formation of holes, cavities, dendrites and mossy like lithium on lithium electrodes in nonaqueous carbonate or ether based electrolytes is inhomogeneous. This will lead to poor coulumbic efficiency, cyclic performance and an increase of resistance and safety concerns. Ionic liquids have been used as additives in commercial carbonate electrolytes for Li-ion batteries to improve the safety and interfacial properties. Some ionic liquids could have wide electrochemical windows, which is essential to develop high voltage cathodes. Ionic liquids have also been used in Li-S and Li-air batteries and show promising performance. For developing rechargeable lithium batteries, as well as the electrochemical stability, other properties, including effective ionic conductivity, low viscosity, interfacial reactions with lithiated or delithiated anodes and cathodes, SEI properties, and corrosion reactions are equally important. Therefore, a balanced consideration is needed but very challenging. Currently, we have no clear answer on the influence of ionic liquids on lithium deposition and dissolution behaviour. We have selected a few types of ionic liquids for such investigations. We hope we can clarify these issues in the future.

Yiren Zhong commented: We have studied some Li-based batteries, such as Li-ion battery half cells and Li-O_2 batteries. Once we cycled the battery above a certain number of cycles, such as 1000 cycles in a Li-ion half cell, the performance started to decline. We have disassembled the battery and found thick coarse layers on the Li metal. We speculate it may be due to Li dentrite, so could you give me some advice on how to confine or prevent Li dentrites?

Hong Li answered: The thick coarse layers on the Li metal could very possibly be SEI covered/mixed on mossy like Li. In the cells using a metal lithium electrode and a non-aqueous electrolyte, dendrite formation is nearly unavoidable. Adding some additives into the electrolyte, such as $CsPF_6$ or $LiNO_3$ can suppress the formation and growth of dendrite lithium to a certain level, but cannot solve it completely. A possible strategy is to coat an inorganic solid electrolyte or polymer solid electrolyte on the surface of the lithium electrode. In some cases, a new solid electrolyte interphase could form between the lithium and coating layer. However, the formation of lithium dendrites can be significantly prevented. Such a strategy is called protected lithium, a key technology for developing next generation lithium batteries.

Liqiang Mai asked: What are the different influences when changing the ratio of oxygen and CO_2?

Hong Li responded: This is a good question. It seems that the color and the content of Li_2CO_3 are different when the ratio of O_2/CO_2 is changed. Up to now, we have not systematically investigated the influence of the molar ratio of O_2/CO_2 on the surface product on Li. We will investigate this issue in future.

Liqiang Mai commented: For Li-air battery investigation, did you try oxygen plus nitrogen, which is closer to air?

Hong Li replied: It is a good question. We have not tried testing a mix of O_2 and N_2. In the Li/O_2 cell, Li_2O_2 is the main reduction product in the air cathode. The electrochemical reaction of Li with N_2 will occur at 0.44 V *vs.* Li^+/Li. Therefore, N_2 will not react in the air cathode. However, naturally, N_2 will react with lithium chemically to form Li_3N. Therefore, in the Li-air electrode, the Li electrode should be protected or covered by stable SEI to avoid this side reaction.

Xuyi Shan enquired: Regarding the *in situ* SEM, how can you define the new forming structure as lithium, not lithium oxide?

Hong Li answered: In this experiment, no O_2 was purged. In our recent paper, 'New Insight in Understanding Oxygen Reduction and Evolution in Solid-State Lithium–Oxygen Batteries Using an *in situ* Environmental Scanning Electron Microscope',[1] a similar *in situ* SEM experiment was performed but the chamber was purged with O_2. The product formed after purging O_2 and an applied external voltage has a significantly different morphology compared to the result without purging O_2. We also did *in situ* TEM with and without O_2. Similar morphologies as with *in situ* SEM were obtained. SAED can identify metal lithium, Li_2O_2 and Li_2O phases. Accordingly, the dendrite structure obtained in the *in situ* SEM experiment can be identified as metal lithium instead of Li_2O_2 or Li_2O. Certainly, in the future, if *in situ* Auger microscopy could be performed for such an experiment, then metal Li formed during an *in situ* experiment can be identified directly.

1 H. Zheng *et al.*, *Nano Lett.*, 2014, **14**, 4245–4249.

Hyun-Wook Lee opened a general discussion of Haimei Zheng's and Hong Li's papers: Regarding bubble formation during the electrochemical reaction, I think that the bubbles might be due to lithium metal formation. When a bias was applied to the 2 Au electrodes, Li^+ ions move to one electrode following the applied bias. Then, when the electrode has an overpotential, lithium metal can deposit on the surface of Au (Li^+/Li^0). Lithium metal has a solid solution reaction with gold, so once lithium metal is deposited on the surface of Au, it is possible to dissolve Au into the electrolyte.

For this reason, I have following questions. Does only one or both electrodes react? Have you reversed the voltage? Have you checked any chemical confirmation, using for example EELS?

Haimei Zheng answered: The bubble formation is from electrolyte decomposition, not from 'lithium metal formation'. For example, we have also observed bubble formation in other electrolytes without Li. Only one Au electrode reacts with the electrolyte and gets lithiated. On the counter electrode side, the electrolyte gets oxidised on the surface of the Au electrode and forms a thin oxidation layer. We reversed the voltage from positive to negative on the working electrode. The Au working electrode still had the same reaction, and the counter Au electrode side did not get lithiated. In order to confirm the chemical species generated during the *in situ* lithiation process, we tried electron energy loss spectroscopy (EELS). But, the signal was not strong enough to identify the species, which is mainly because of the thick sample from the liquid layer plus membranes.

Graham Hutchings said: There has been a lot of discussion concerning Fig. 3 in your paper, namely concerning the origin of the features that you refer to as bubbles of CO_2. Do you have images of the stages labelled A and B in Fig. 3? For example, is the formation of the gas bubbles preceded by the formation of gold nanoparticles from the surface of the gold electrode? Such nanoparticles can act as active catalysts. However, it is also possible that it is defect site formation that creates the active centres.

Haimei Zheng replied: There are many intermediate states between stages A and B, since we captured the movie at 25 frames per second. We agreed that defects may form on the electrode surface, which can be the active sites for catalysis. We did not observe gold nanoparticles, nanosized features or defect site formation directly on the electrode, due to the limited spatial resolution under the current experimental conditions.

Zhongqun Tian addressed Haimei Zheng and Hong Li: You two have shown the power and bright future of *in situ* TEM to study energy systems and devices at nano and subnano scales. The combination with other electron-, ion- and/or photon-beam techniques could be a new direction. Is there a sufficient space for combination of other *in situ* techniques?

Hong Li answered: Yes, this is a good suggestion. Currently, it is possible to investigate local structure variation and the electronic state at the atomic level by ABF–STEM–EELS. For ABF–STEM–EELS, special sample holders are designed for investigating temperature effects, *in situ* charging and discharging and IV tests. Some TEM instruments can integrate a SPM nanoprobe to test local transport properties. In an SEM equipped with an environmental chamber, nanoprobe, optical fiber and gas purge, a temperature varied platform could be integrated, which is suitable for investigating morphology change, especially for air batteries. Auger spectroscopy is suitable for mapping lithium. A time-resolved TEM technique is also developed. It is believed that an integrated TEM microscope could collect structure, chemical and electronic information with high spatial and time resolution in the near future for broad research fields.

Haimei Zheng replied: Yes, the combination of *in situ* TEM and *in situ* X-ray imaging and spectroscopy can be very powerful. Using *in situ* TEM, we can directly visualize the dynamic reaction with structure, morphology and electronic states at the nanometer or atomic scale. The *in situ* X-ray technique provides ensemble electronic structure changes or imaging at a larger length scale (50 nm and above). The two techniques combined allows the study of the structure, phases, morphology and electronic states of a reaction in real time at different length scales.

Zhongqun Tian commented: At present the main concern with *in situ* TEM is the electron beam effect that may cause damage of the sample and distort the result. There is a way to avoid/investigate the electron beam effect with a systematic measurement by snapping the TEM images with different time periods and by changing the beam intensity. As always, each effect has its weaknesses and strengths, the question is how to maximize the positive effect? For those systems with the electron beam effect, can you use the beam as a tool to initiate selective surface chemistry and electrochemical reactions at the nanoscale?

Haimei Zheng responded: We agreed that the electron beam can cause damage and distort the experimental results. Controlling electron beam effects is necessary in most cases. The electron beam can be utilized to initiate reactions, as we have demonstrated in many studies of colloidal nanocrystal growth. We believe the electron beam can also be used as a tool in the study of other chemical reactions.

Galen Stucky remarked: Regarding the CO_2 bubble formation chemistry, do you see LiF in your electron microscopy images? Is there moisture in your system? Where is the residual LiF relative to the particles and bubbles?

Haimei Zheng answered: Yes, we did observe the growth of LiF crystals in our electron microscopy image, which has been confirmed by selected area diffraction patterns. There could be some moisture in our electrochemical liquid cell during sample preparation. But as the sample preparation was handled inside the glovebox, the exposure to moisture may not be significant. Based on our experimental results, there is no preferential nucleation point of LiF relative to the particles and bubbles.

Zhen Qi opened the discussion of the paper by Xinhe Bao:† How about the corrosion resistance of your carbon materials in application? Is it very stable or can it be corroded, as normal carbon supported nanoparticles always suffer from carbon corrosion during application?.

Dehui Deng responded: These carbon encapsulated metals show a good corrosion resistance during reactions. For example, the pod-like carbon nanotubes show a long term durability with more than 200 h in PEMFCs, indicating the carbon is still intact after the stability test. For traditional carbon supported nanoparticles, the metal particles can directly contact the oxygen, which therefore can catalyze the oxidation of carbon, leading to carbon corrosion happening more easily.

Rudolf Holze asked: Regarding the corrosion resistance of the carbon material, are you sure the iron is completely encapsulated?

Dehui Deng replied: Firstly, the outside iron has been removed in acid at 90 °C before the reaction test. We found that all iron left in the pod-Fe presents a metallic state according to our XPS, XRD and XAFS characterizations, which means that all iron has been well encapsulated inside the carbon nanotubes. If there are iron particles on the outside of the carbon nanotubes, they will be very easily oxidized in air. So we think the iron is completely encapsulated inside of the carbon nanotubes, and therefore these materials show a good corrosion resistance during PEMFCs.

† Xinhe Bao's paper was presented by Dehui Deng, Dalian Institute of Chemical Physics, China.

Heinz Frei asked: Can protons go through the N-doped carbon cage (shell) walls? Is the pH on the inside of the cage the same as on the outside?

Dehui Deng responded: The protons cannot go through the carbon walls. It is demonstrated by others that the proton cannot go through a perfect graphene layer, even it is a single layer. We think there is no solution inside the cage since any reaction medium or air cannot go through the cage. Therefore, the pH on the inside of the cage should be different from that on the outside.

Rudolf Holze asked: Do you have an idea of how to measure the number of protons (*i.e.* pH) inside the nanotubes?

Heinz Frei replied: From the O_2 yield measured in solution electrochemically and in the head space by mass spectroscopy, taking into account the known inner surface area of the nanotubes, we can determine the number of protons generated per unit time (4 protons per O_2 molecule). In addition, we have measured the proton flux through silica layers for thicknesses ranging from 3 nm to 10 nm electrochemically in pH 4 solution as described in Fig. 6 of our paper. According to the measured proton flux of 2.5 protons s^{-1} nm^{-2} (4 nm thick layer), proton removal from the catalyst zone should be adequate.

Erwin Reisner remarked: Do you have any experimental evidence such as Raman spectroscopy with ^{18}O-labelled O_2 for the oxygen-intermediates discussed during your presentation? Also, do you have any Moessbauer data for the encapsulated iron nanoparticles pre-, post- and during catalysis?

Rudolf Holze also commented: The state of oxidation, further elemental and structural/morphological properties of iron can be probed *in situ* using Mossbauer spectroscopy. Did you try this?

Dehui Deng answered: It is a very good suggestion. We have not used Mossbauer now and we will try this tool in the future.

Rui Lin enquired: It looks like the graphene-supported iron-based catalyst has the potential to be used in fuel cell applications, but how far is your catalyst from a real application in a fuel cell? What would be the big challenge of this kind of non-precious catalyst compared to Pt based catalysts in fuel cells?

Dehui Deng replied: As you said, these graphene or carbon nanotube encapsulated non-precious metals can indeed promote the performance in fuel cells. However, I think there is still a long way before real application in a fuel cell. The biggest challenge compared with Pt based catalysts is how to further increase the ORR activity with the premise of keeping the stability of the fuel cell. The highest power density of current catalysts is about one third of that of Pt based catalysts. So I think these is still much room to increase the catalytic activity by optimizing the catalysts.

Yimin Chao asked: This is a great idea to encapsulate catalysts with graphene/carbon nanotubes. In you paper you have describe the procedure to remove excess catalyst by washing with acids. Your experimental results have proved the existence of catalysts in the product after the encapsulation. However, there is no evidence that all the catalysts are fully encapsulated within the nanotubes. How can you make sure there is no such material left outside the carbon nanotubes? Is there any solid data to support your statement?

Dehui Deng answered: In the originally synthesized catalyst, there are two types of metals, that is, one located on the outside of the carbon nanotubes and the other encapsulated inside the carbon nanotubes. We have used strong acids to remove the outside metal particles. So all the remaining metal particles should be well encapsulated inside the carbon nanotubes. TEM images have demonstrated that all the observed metal particles are located inside the carbon nanotubes. Furthermore, we also find that the metal particles present a metallic state by XRD, XPS and XAFS, indicating the metal particles have been well protected by the graphene layer. If there is no graphene layer on the outside of the metal particles, it is known that these exposed metal nanoparticles are very easily oxidized in air.

Fenglin Liao said: Did you try some other method to increase the electron density of your graphene and confirm that the catalytic performance really increases with increasing electron density, like the doping or hetero-junction methods?

Dehui Deng replied: We have actually done this related study in our previous publications in ref. 1 and 2. According to the DFT calculations, we found the hetero-atoms such as nitrogen can further increase the electron density of graphene around the Femi level. Accordingly, we found N-doped pod-like carbon nanotube encapsulated non-precious metals can further promote the catalytic performance.

1 D. Deng *et al.*, *Angew. Chem. Int. Ed.*, 2013, **52**, 371.
2 J. Deng *et al.*, *Energy Environ. Sci.*, 2014, **7**, 1919–1923.

Rudolf Holze remarked: The concluding cartoon you showed implies that the catalytically active sites are those on the surface of the graphene, where the iron particles are located underneath. Is there any proof of this? And does this imply an electronic effect without direct iron–dioxygen interaction?

Dehui Deng answered: This is a very good question. We have demonstrated by *in situ* XAFS that the iron valence state is unchanged during the reaction. That means the iron cannot contact the dioxygen during the reaction. So the only possibility for the reaction is on the outside of the carbon. We have used PEEM and soft X-ray imaging and spectroscopy tools to demonstrate that the iron electron can penetrate the graphene layer and further affect the reactions.

Deyu Liu commented: In Fig.1(e) of the paper, I noticed that the iron nanoparticle is wrapped with 4 or 5 layers of graphene. I am wondering, is there any specific relationship between the RGO thickness and the catalytic performance? Or is there an optimal number of carbon layers on the surface?

Dehui Deng responded: According to our study in ref. 1 and also the work being carried out, we find there is a specific relationship between the graphene layer thickness and the catalytic performance, *i.e.* reducing the graphene layer thickness will promote electron transfer and further reduce the work function at the graphene surface, thereby leading to a higher catalytic performance. We have demonstrated this trend in both ORR and HER. The ideal graphene layer is suggested to be 1–3 layers according to our study, which can promote both the catalytic activity and stability.

1 J. Deng *et al.*, *J. Mater. Chem. A*, 2013, **1**, 14868.

Andrew Mount communicated: In Fig. 6 and the associated discussion in the paper, it is stated that Fe@N–C/RGO has a smaller Tafel slope and a lower onset potential (which I take to mean a less negative onset reduction potential for hydrogen evolution). It indeed has a lower Tafel slope, but the data in Fig. 6 actually appear to show crossover, with Fe@N–C/RGO having the most negative onset potential from the data at low currents. This seems to suggest favoured proton reduction kinetics at the expense of catalytic thermodynamics.

The authors state that the Tafel curve in Fig. 6 of the paper 'clearly demonstrates that the Volmer step is the rate-determining step' for Fe@N–C/RGO. Is this inferred from the transfer coefficient? Radically different slopes are observed for the three species in Fig. 6.

Heinz Frei opened the discussion of the paper by Nenad Markovic: I think the local structural motif(s) on the oxide catalyst surface determines the catalytic activity. For example, in the case of Co oxides for water oxidation, there is strong evidence emerging from spectroscopic and electrokinetic studies that the most active structural motif involves adjacent oxo-bridged Co–OH centers. The introduction of defects may increase the probability of forming motifs that are active sites in many cases, but this does not exclude cases of stable surfaces with motifs that constitute very active catalytic sites.

Nenad Markovic answered: We do agree that local structures are indeed very important for determining the activity of surface atoms; however, these motifs are also very important for defining the stability of these atoms, which is reflected in the observed activity of the system. Without direct measurements of the stability of the material (*e.g.*, ICP measurements of dissolved active species), it is very difficult, if not impossible, to claim that the proposed active centers are stable. We recently published results for the OER on Co–OH–O clusters[1] that show clearly during the OER that there is a strong sintering of Co–OH–O, and we believe that the same is applicable for other Co and 3d oxide compounds. Thus far we have studied the OER on over 15 different oxides, and for all of these systems we find that there are strong functional links between activity and stability, namely that more active materials are always less stable under the same experimental conditions.

1 R. Subbaraman *et al.*, *Nat. Mater.*, 2012, **11**, 550.

Erwin Reisner commented: You conclude that the best materials for the OER should balance stability and activity. But would a highly active and subsequently unstable material with an efficient self-repair mechanism of damage be a better solution?

Nenad Markovic replied: Certainly, a self-repair (healing) process during the reaction would be an ideal case, and there are examples of systems that come close to this type of behavior (most notably Ru–Ir); however, in the case of the OER, truly 'self-healing' processes are not possible due to the inevitable dissolution of active cation species at the large potentials necessary to carry out the reaction. In our paper we present an example of a material where only one of the active cation species (Ru) participates in the OER. For this type of material, it is very difficult for self-healing processes to take place, as Ru cations are constantly dissolved during the reaction. Furthermore, there is no re-deposition of Ru and Sr at positive bias voltages above the onset potential for the OER, which would protect active Ru sites.[1] In the case of Ru–Ir oxides, where both cation species are active for the OER, we do observe that these surfaces exhibit a quasi 'self-healing' behavior.[2] Dissolution of Ru from the alloy system resulted in an increased surface area, which was followed by a rearrangement of the more noble (and thus more stable) Ir atoms to increase their coordinate number and re-stabilize the active surface. Nevertheless, even in this case the more stable Ir atoms are also slowly dissolved during the OER, meaning that these surfaces, and any other active catalyst surfaces for the OER, are not and cannot be truly self-healing.

1 S. H. Chang *et al.*, *Nat. Commun.*, 2014, **5**, 4191.
2 N. Danilovic *et al.*, *J. Phys. Chem. Lett.*, 2014, **5**, 2474.

Yanxia Chen remarked: In acid media, if the material is more unstable (*e.g.* has more stepped sties, like Pt(332), Pt(221) *etc.*), the ORR activity of it increases. But in alkaline media, we can observe that the electrode that has more ORR activity is always more stable (*e.g.* Pt(111) and Au(100)), so why is that? What's the physics behind this phenomenon?

Why is almost every electrode's activity of ORR better in alkaline media than acid media?

Nenad Markovic answered: Unfortunately, there are no systematic results focusing on activity-stability correlations for metal single crystal surfaces during the ORR in acidic and alkaline environments. Therefore, it is again very difficult to suggest any guiding rules that may control correlations between the surface structure, stability and activity. The observed differences in activity at different pH are more related to anion effects on the availability of active sites than the stability of surface atoms. The unique activity of Au (100) is certainly still a big puzzle that we would like to investigate in the near future; however, the attempt to uncover some of the 'physics' behind this phenomenon can be found in ref. 1.

The activity of the ORR is strongly coupled to the pH of the solution *via* surface coverage by spectator species, and thus controls the availability of the active metal sites. See ref. 1 for more discussion of this topic.

1. N. Markovic and P. Ross, *Surf. Sci. Rep.*, 2002, **45**, 117.

Zhonghua Li commented: What is key, the defects on different facets, or the nature of the different facets?

Nenad Markovic responded: It is difficult to assign which is more important, as both of these factors are intrinsically tied to one another. Each crystallographic orientation possesses its own unique set of defect structures that will determine functional links between activity and stability. As a general rule, (111)-type single crystal surfaces for Pt-like metals have inherently fewer defects than (100) or (110); however, this might not be true for nanoparticles because it is very difficult to say with great certainty what the defect populations are, even on highly faceted particles. In contrast to metal surfaces, our results show that for perovskite-type complex oxides, the (111) surface contains more defects than (100), suggesting that not only the nature of the facets, but also the nature of the materials, may control the density of surface defects.

Jiafang Xie communicated: In your experiment about the stability of Pt nanoparticles with different sizes, you concluded that 7 nm was the most stable size for Pt. I wonder if is there a possibility that each metal nano-catalyst has a certain size of the most stable, which we can search through experiments?

Nenad Markovic communicated in reply: Although there is no general rule for an 'optimal' size of nanoparticle catalysts for the ORR, our experiments and results[1] suggest that 7 nm for Pt nanoparticles is an optimal size to balance activity with stability.

1 D. Li *et al.*, *Energy Environ. Sci.*, 2014, 7, 4061.

Jiafang Xie communicated: Smaller nanoparticles with higher surface area and density of defects usually show instability and always reconstruct to larger NPs during electrolysis, and the density of surface defects of these larger NPs lowers; in the end, a certain size of NPs with the least density of defects is obtained. Many studies about metal nanoparticles electrocatalyzing the ORR or OER only present the initial morphology of the materials, and discuss their high activity based on these initial states. I wonder if this is correct or not, but I would like to know your comments on it.

Nenad Markovic communicated in reply: Unfortunately, many researchers have investigated and just focused on the initial shape and size of their nanoparticles without considering the possibility that the structure might change during the OER and ORR. It is very important to know the history of the experiments (*e.g.* number of cycles, potential window, *etc.*) as well as the morphology of the particles before and after the experiment. An interested reader can find more information about this in our recent paper.[1]

1 D. Li *et al.*, *Energy Environ. Sci.*, 2014, 7, 4061.

Jiafang Xie communicated: I'm curious about the inherent driven force for dissolution and redeposition of surface atoms during electrolysis, especially in aqueous solution, for this would help us better understand the mechanism and optimize the design of electrocatalysts. But I can only imagine that the surface atoms linked to fewer neighboring atoms on the surface and in the bulk phase are likely to have more strong bonds with molecules from the electrolyte, so these more active atoms are much easier to be pulled away from the surface, driven by bonds with molecules in the electrolyte. Do these atoms obey a certain rule or do they just leave the substrate randomly?

Nenad Markovic communicated in reply: Dissolution depends on many factors, including thermodynamics, kinetics, the applied surface potential, and the oxidation state of active components (in our case Ru^{4+} *vs.* $Ru^{n>4+}$). It is absolutely correct that the nature of adsorbates also plays a strong role in determining the dissolution of atomic species from the electrode surface. Although it is very difficult to develop a rigorous 'rule' for which species dissolve first, it does appear that atoms with low coordination numbers are inherently unstable – a phenomenon that is demonstrated in our system for Ru and SRO polycrystalline films *vs.* SRO single crystal-films.

Jiafang Xie communicated: In the end of the paper, you emphasized that the work would be extremely helpful in the design of new oxide materials that could serve as both active and stable electrocatalysts in aqueous environments. My previous work is about CuO nanomaterials with flower structures prepared by simple electrochemical anodic oxidation of polycrystalline copper foil showing enhanced activity and stability in electrocatalyzing CO_2 reduction over polycrystalline copper foil.[1] Though I'm still digging into the deeper mechanism of this enhancement, I gain little until now. I wish for your comments.

1 J.-F. Xie *et al.*, *Electrochim. Acta*, 2014, **139**, 137.

Nenad Markovic communicated in reply: Every system has its own unique properties and requires the development of a set of descriptors that define the behavior of the specific material. However, for any material it is necessary to develop very well-defined systems that make it possible to reliably measure these descriptors. To investigate the enhanced activity and to elucidate the mechanism, we would like to suggest controlled experiments with well-defined systems. Systematic studies focused on stability and defect density can provide further insight on the detailed mechanism of the enhanced activity observed for your particular system. Here, we would like to suggest our recent approach to determining stability-activity relationships between highly defective amorphous electrochemical oxides and crystalline thermal chemical oxides as a possible roadmap.[1]

1 N. Danilovic *et al.*, *J. Phys. Chem. Lett.*, 2014, **5**, 2474.

Faraday Discussions

PAPER

A facile two-step method for fabrication of plate-like WO_3 photoanode under mild conditions†

Nan Wang,[ab] Jian Zhu,[ab] Xiaojia Zheng,[ab] Fengqiang Xiong,[ab] Baokun Huang,[a] Jingying Shi*[a] and Can Li*[a]

Received 28th June 2014, Accepted 14th July 2014
DOI: 10.1039/c4fd00139g

Fabrication of photoelectrodes on a large-scale, with low-cost and high efficiency is a challenge for their practical application in photoelectrochemical (PEC) water splitting. In this work, a typical plate-like WO_3 photoanode was fabricated with chemical etching of the as-prepared mixed tungsten–metal oxides (W–M–O, M = Cu, Zn or Al) by a reactive magnetron co-sputtering technique, which results in a greatly enhanced PEC performance for water oxidation in comparison with that obtained from a conventional magnetron sputtering method. The current approach is applicable for the fabrication of some other semiconductor photoelectrodes and is promising for the scaling up of applications for highly efficient solar energy conversion systems.

1 Introduction

Photoelectrochemical (PEC) water splitting has been regarded as a promising way to produce hydrogen from water and solar energy.[1–3] The photoelectrode, a conductive, substrate-supported semiconductor thin-film material, in general, is the crucial part of a photoelectrolysis cell. Depending on the techniques and preparation conditions, photoelectrodes may present different structural, optical and electrical behaviors, and consequently different PEC performances.

The reactive magnetron sputtering technique has been used widely in industry for preparing large-area and uniform films well connected to the substrates. The structure of the deposited film can be easily controlled by adjusting the process parameters, and it is also quite easy to repeat the production process. Hence, this technique shows huge potential for practical, large-scale manufacturing of thin-film materials. Some metal oxide or oxynitride photoelectrodes such as TiO_2,[4]

[a]*State Key Laboratory of Catalysis, Dalian Institute of Chemical Physics, Dalian National Laboratory for Clean Energy, Chinese Academy of Sciences, 457 Zhongshan Road, Dalian 116023, China. E-mail: canli@dicp.ac.cn; jingyingshi@dicp.ac.cn; Fax: +86 411 84694447; Tel: +86 411 84379070*

[b]*University of Chinese Academy of Sciences, Beijing, 100049, China*

† Electronic supplementary information (ESI) available: Raman spectra, SEM images, linear sweep voltammetric scans curves and tables. See DOI: 10.1039/c4fd00139g

WO_3,[5] $BiVO_4$,[6] and TaON[7] have been prepared by this method for PEC water splitting, and showed good stability and anti-corrosion properties in electrolyte. However, these films generally present a mirror-like surface as well as a compact structure that result in high reflectivity and a small contact surface area with the electrolyte, which make their PEC water-splitting performances greatly limited. To overcome this shortcoming, we intentionally introduce other metal oxides through magnetron co-sputtering and subsequently remove the impurity phase by chemical etching to fabricate porous photoelectrodes with higher PEC activity.

WO_3 is a well-known semiconductor material for photoanodes as it can absorb within the visible light region and shows a remarkable stability in acidic solutions.[8,9] Porous WO_3 photoanodes can be fabricated generally through liquid-phase routes. For example, Amal *et al.*[10] prepared flower-like nanostructured tungsten oxide hydrate thin films by electrochemical anodization of tungsten foil at high potentials. Widenkvist *et al.*[11] obtained plate-like $WO_3 \cdot H_2O$ films on a tungsten substrate in acid medium at temperatures higher than 50 °C. The groups of Yang[12] and Demir[13] synthesized tungsten trioxide hydrate ($WO_3 \cdot 1/3H_2O$) films *via* conventional hydrothermal technology at higher than 120 °C. Recently our group successfully fabricated the hexagonal nanoflower $WO_3 \cdot 1/3H_2O$ array on fluorine-doped tin oxide (FTO) substrates by a microwave-assisted hydrothermal method.[14] All of the above as-prepared WO_3 hydrate ($WO_3 \cdot nH_2O$) thin films can be converted into the WO_3 phase *via* calcination at about 450 °C without changing the morphology. In other words, nanocrystalline WO_3 films prepared by these liquid-phase synthesis methods possess porous network structures, and often exhibit a large surface area as well as enhanced light absorbance. However, these wet chemical synthesis routes usually need high reaction temperatures or high anodization voltages which are difficult to operate in a large scale.

In this work, some impurity metals (M = Cu, Zn or Al) were sputtered together with tungsten to produce mixed tungsten–metal oxides (W–M–O) thin films on FTO glass substrates. After soaking in an acidic solution of HCl, H_2SO_4 or HNO_3 for etching at room temperature, these extra metal oxides disappeared due to dissolution in the acidic media while a porous plate-like WO_3 phase, distinct from conventional compact WO_3 thin films prepared by a magnetron sputtering, was left behind with twice photocurrent enhancement for PEC water oxidation. This approach can be applied generally and useful for the fabrication of some other semiconductor photoelectrodes.

2 Experimental

Sample preparation

Samples of W–M–O films (M = Cu, Al or Zn) were deposited on commercial FTO (NSG, sheet resistance of 20 $\Omega \ \square^{-1}$) glass substrates by reactive magnetron co-sputtering system (PVD75, Kurt J. Lesker Company). The FTO glass substrates were cleaned in acetone–isopropanol–ethanol–DI water in sequence. The co-sputtering system consists of two DC guns, loaded with a 99.95% pure W target and 99.99% pure Cu (or 99.99% pure Zn, 99.99% pure Al) target, respectively. The target–substrate distance was fixed at 170 mm. Both targets have a diameter of 3 inches and a thickness of 3 mm. After pressure had been pumped down to 7×10^{-7} torr, the O_2 and Ar gases (99.99% purities) were introduced into the chamber through mass flow controllers and the gas flow was set for an $[O_2]/([Ar]+[O_2])$ ratio of 38%. The oxygen partial pressure and the total sputtering pressure were 2 and 10 mtorr, respectively. The sputtering power of the tungsten target was fixed at about 150 W while the M content in the films was controlled by varying the sputtering power applied to the M target. The sputtering power was 60, 60 and 70 W for Zn, Al and Cu targets, respectively. The as-sputtered samples were immersed in an aqueous solution of H_2SO_4, HNO_3 or HCl for various periods of time. Subsequently, the as-etched photoelectrodes were washed with deionized (DI) water and then calcined in air at 500 °C for 2 h.

Characterization

The crystal structure was measured by X-ray powder diffraction (XRD, Rigaku D/Max-2500/PC powder diffractometer) using Cu Kα radiation ($\lambda = 0.15406$ nm) with an operating voltage of 40 kV and an operating current of 200 mA. The scan rate is 5° min^{-1} in the range of 10–60°. The surface morphology and dopant concentration were measured by scanning electron microscopy (SEM, Quanta 200 FEG). The surface profile was used to characterize the film thickness and the roughness (Dektak 150 Surface Profiler, Veeco). Optical properties were measured using an Agilent Cary 5000 UV-Vis spectrophotometer. Raman spectra were collected using a Renishaw inVia Raman spectrometer excited by a solid laser (532 nm), and the spectral resolution was 1.5–2 cm^{-1}. In order to investigate the chemical bonding and dopant concentration, X-ray photoelectron spectroscopy (XPS) were made using a KROTOS AMICAS spectrometer (Shimadzu, Japan) with a magnesium Kα X-ray source. The carbon C 1s line position at 284.8 eV was used as a reference to correct the charging effect. The as-sputtered W–Zn–O and W–Cu–O films and the samples after etching in 1 M H_2SO_4 for 12 h were washed in 1 M KOH solution (10 mL). The size of each electrode was fixed at 15 mm × 20 mm and the thickness was about 1 μm. W, Cu and Zn concentrations of leached solutions were analyzed by inductively coupled plasma atomic emission spectrometry (ICP-AES) on a Shimadzu ICPS-8100 spectrometer.

Photoelectrochemical measurement

The photoelectrochemical properties were measured using a CHI electrochemical analyser (CHI 440B, CH Instruments, Inc.). Chopped I–V curves of the as-prepared thin films were obtained under simulated AM 1.5 solar illumination (100 mW cm^{-2}) using a three-electrode setup which had the as-prepared samples as the working electrode, platinum foil as the counter electrode, and a saturated calomel electrode (SCE) as the reference electrode in 0.5 M Na_2SO_4 (pH 3.0) with scan rate of 20 mV s^{-1}. Electrochemical impedance spectra (EIS) were measured in the frequency range of 100 kHz to 100 mHz and with an AC amplitude of 10 mV using an electrochemical workstation (IM6 Zahner, Germany). The relative electrochemical surface area (ECSA) was determined from capacitance measurements carried out in the dark by cyclic voltammetry in the region between 0.15 and 0.35 V *vs.* SCE in 0.5 M Na_2SO_4 (pH 3.0), and the scan rates were varied between 25 and 300 mV s^{-1}.

3 Results and discussion

Fig. 1a exhibits the XRD patterns for the as-sputtered samples etched in 1 M H_2SO_4 aqueous solution for different time intervals. Besides the characteristic diffraction peaks represent for the FTO substrate, a broad peak in the range of 15–35° clearly indicates an amorphous phase for the as-sputtered film. Hardly any new phase can be detected when the sample was etched for only 1 h, whereas a new diffraction peak located at about 12.7° appeared after soaking for 2 h. Extending the etching time to 3–12 h, the diffraction peaks corresponding to $WO_3 \cdot 2H_2O$ (JCPDS card no. 18-1420, monoclinic, *P*2/*m*(10)) can be clearly observed. The increased diffraction intensity and the narrowed full width at half-maximum (FWHM) of the characteristic peaks with etching time suggest an improved crystallinity and a larger grain size. After being calcined at 500 °C for 2 h, the XRD patterns for the corresponding samples are shown in Fig. 1b. Clearly, the mixed crystallized phases of WO_3 and $CuWO_4$ (JCPDS card no. 43-1035 and no. 72-0616) are present when the etching time is shorter than 5 h, whereas a pure monoclinic WO_3 phase is be achieved after the as-sputtered sample was further corroded for a period of 6 h or more. These results may suggest that the residual copper can react with tungsten oxide to produce $CuWO_4$ in the calcination process and that the impurity copper can be completely removed by etching.

The phase transformation of the surface region was further investigated by Raman spectroscopy (in Fig. 2). Compared with the Raman spectrum of the blank FTO substrate (Fig. S1†), the bands at 552 and 1097 cm^{-1} originate from the FTO substrate so those at 798 and 956 cm^{-1} should be assigned to W–Cu–O compound for the as-sputtered sample, which is also different from the Raman bands for the as-sputtered tungsten oxide (Fig. S1†). This result obviously indicates the successful introduction of the copper impurity by co-sputtering. The signals from

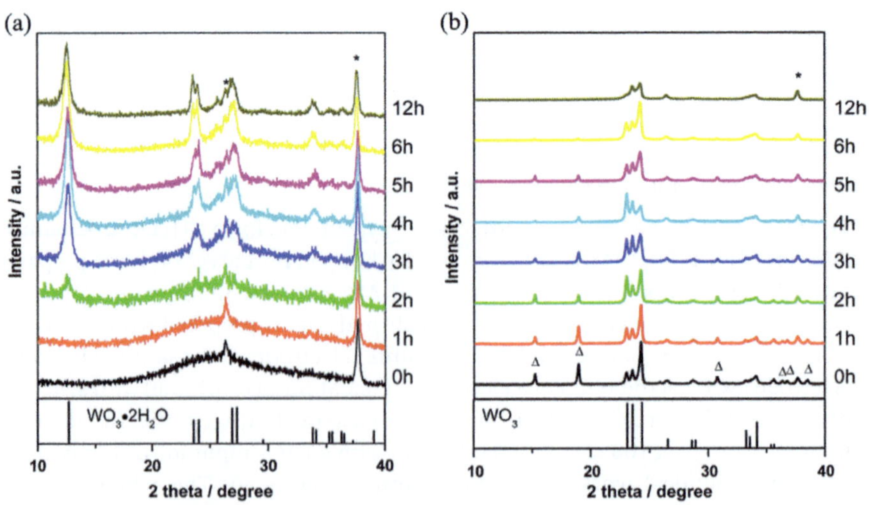

Fig. 1 (a) XRD patterns for the as-sputtered samples etched in 1 M H_2SO_4 aqueous solution for various periods. (b) XRD patterns of etched W–Cu–O samples after calcination in air at 500 °C for 2 h. The peaks marked with △ are assigned to $CuWO_4$ and the peaks marked with * to the FTO substrate.

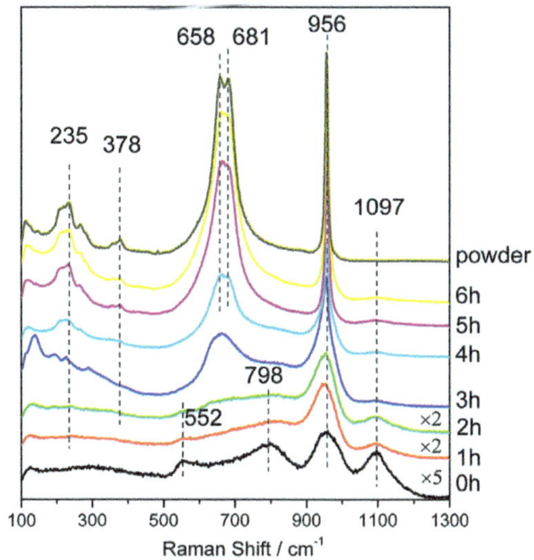

Fig. 2 Raman spectra of samples prepared at different etching times and the precipitation in the solutions.

the substrate as well as the intensity of the band at 798 cm^{-1} become weak when the etching time is within 2 h. However, the Raman scattering at 956 cm^{-1}, which originates from the symmetric stretching mode of the terminal W=O bond, stays at the same intensity. Extending the etching time to 3 h, the band at about 645 cm^{-1} appears, indicating that the oxide is gradually converted into $WO_3 \cdot H_2O$.[15] This broad band is blue-shifted to 670 cm^{-1} which is assignable to the stretching mode of the W–O–W bonds, gaining higher intensity with increasing etching time to 4 h. As the time is prolonged to 5 h, this band splits into two bands at 658 and 681 cm^{-1}, respectively, which may indicate the $WO_3 \cdot H_2O$ phase had evolved into $WO_3 \cdot 2H_2O$. Moreover, the signal from the W=O bond at 956 cm^{-1} is greatly enhanced. On the other hand, the lower frequency bands at 378 cm^{-1} (W–OH_2 vibration) and 235 cm^{-1} (W–O–W bending vibration) appear. These bands are significantly influenced by the degree of hydration and can be used as a spectral marker to determine the hydration level of WO_3.[15,16] The yellowish precipitate present in the etching solution was also collected for Raman measurement and it shows a doublet of narrow bands at 658 and 681 cm^{-1} corresponding to $WO_3 \cdot 2H_2O$.

The surface morphologies observed by SEM at different chemical etching time intervals are illustrated in Fig. 3. The as-sputtered sample shows a compact morphology (Fig. 3a) which is similar to film materials deposited using a single magnetron sputtering technique.[5,17] After one hour's etching, small pinholes and cracks appear at random, together with some platelets present on the surface as shown in Fig. 3b. When the etching time was increased to 2 h, the number of cracks and the width of gaps developed. The platelets aggregated to form spheroidal structures (Fig. 3c). Upon further extension of the etching time, these spheroidal structures became larger until they covered the whole surface with a diameter of about 1 μm and the thickness of each platelet is about 80 nm, while pores and cracks vanish gradually when the etching time is extended to 5 h

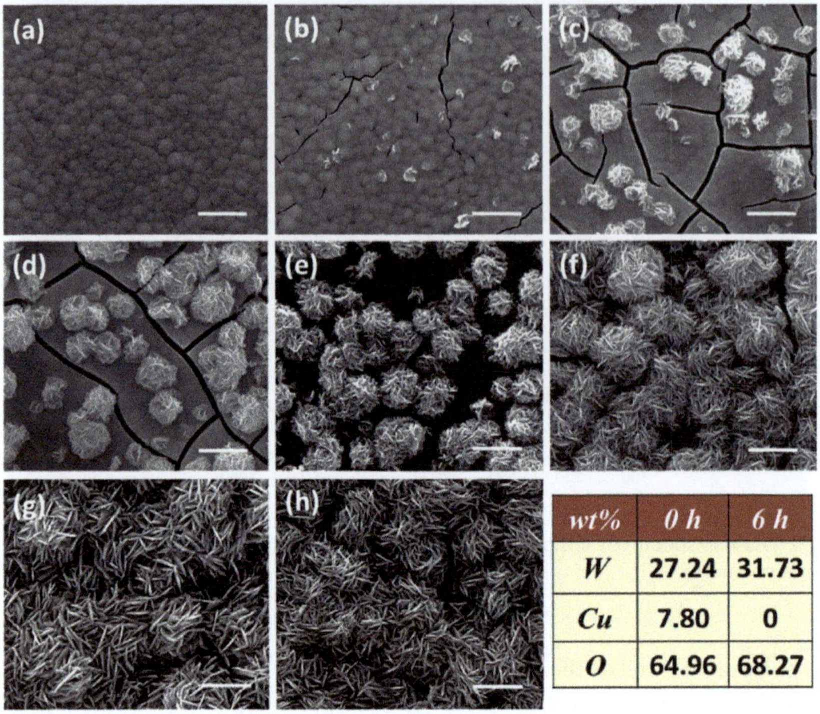

wt%	0 h	6 h
W	27.24	31.73
Cu	7.80	0
O	64.96	68.27

Fig. 3 SEM images of amorphous W–Cu–O films with different chemical etching durations to investigate the growth of plate-like structures: (a) 0 h, (b) 1 h, (c) 2 h, (d) 3 h, (e) 4 h, (f) 5 h, (g) 6 h, (h) 48 h and EDX elemental analysis. Scale bar: 1 μm.

(Fig. 3d–f). The plate-like morphology and particle size remain almost unchanged even after a longer etching time of up to 48 h (Fig. 3g and 3h). The element atom ratios detected by EDX analysis are listed in the inserted table, and it can be seen that the W element remains present while the content of Cu decreases to the detection limit after etching for 6 h. The cross-sectional SEM images for samples in Fig. 3 were also detected and the are results shown in Fig. 4. It is observed that the platelets appear on the surface and then permeate to the bulk of film gradually without significant change of the film thickness, which implies that the dissolution and crystallization proceed simultaneously.

For comparison, the as-sputtered tungsten oxide film was also immersed into acid solution for several hours. The compact film seems to be marginally dissolved after treatment for 6 h, as seen *via* SEM observation in Fig. S2,† showing a much slower reaction rate than the amorphous W–M–O films etched in acid solution presented in Fig. 3, Fig. S3 and S4,† which means the presence of the impurity metal can accelerate the corrosion process.

The chemical surface states of the as-sputtered W–Cu–O films before and after etching were characterized by X-ray photoelectron spectroscopy (XPS) as shown in Fig. 5. Before etching, the two main peaks of Cu 2p located at 932.0 and 951.9 eV correspond to Cu $2p_{3/2}$ and Cu $2p_{1/2}$, and the satellite peaks on the high-energy side are clearly observed. After etching, there are no Cu signals detected (in Fig. 5a). Spectra of the W 4f peaks presented in Fig. 5b are dominated by a

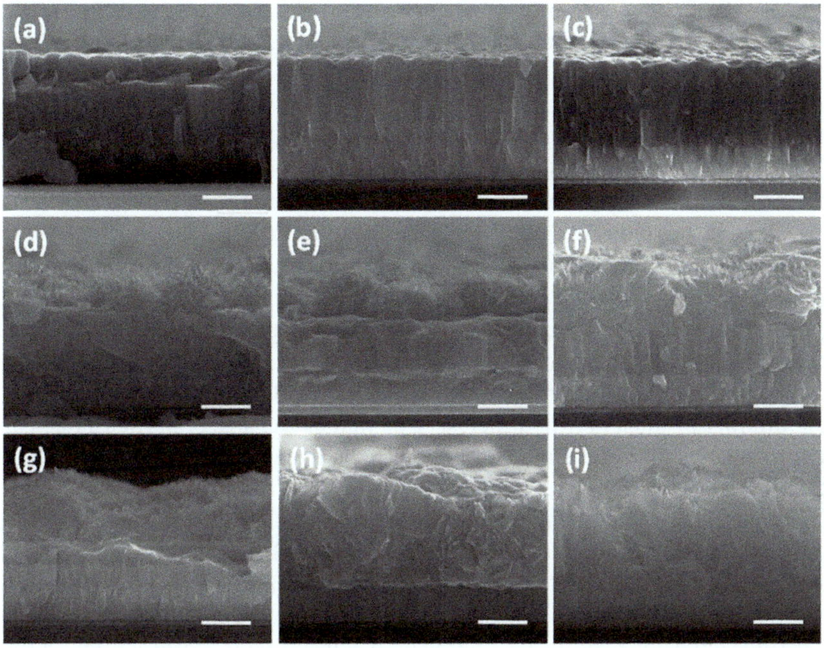

Fig. 4 Cross-section SEM images of amorphous W–Cu–O films after different chemical etching durations: (a) 0 h, (b) 1 h, (c) 2 h, (d) 3 h, (e) 4 h, (f) 5 h, (g) 6 h, (h) 12 h and (i) 48 h. Scale bar: 500 nm.

spin–orbit doublet at binding energies of 37.85 eV (W $4f_{5/2}$) and 35.70 eV (W $4f_{7/2}$), which are associated with the W^{6+} oxidation state. The O 1s spectrum of the as-prepared sample shows a peak located at 530.7 eV in Fig. 5c, which corresponds to W=O bonding modes. And a small peak appears in the region at 533.1 eV after etching, which is also the signal from H_2O molecules bound with WO_3.[15]

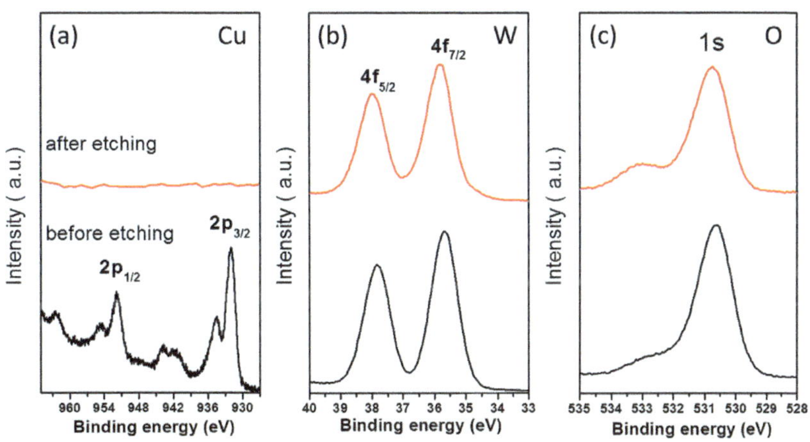

Fig. 5 XPS spectra of the as-sputtered film before and after etching: (a) Cu 2p, (b) W 4f and (c) O 1s.

Instead of copper, aluminum and zinc were also used as the impurity metal sources for reactive magnetron co-sputtering to deposit W–Al–O and W–Zn–O mixed oxide films onto FTO substrates. During the chemical etching process in acidic solution, similar morphology evolutions were observed, from the compact surface, to pinholes and cracks, to a final plate-like aggregated crystal structure (as shown in Fig. S2 and S3†). Further SEM studies investigated the influence of the acid concentration on the shape and size variation, as shown in Fig. 6. It reveals that the quantity of acid plays an important role in determining the final morphology of the electrode. In 0.5 M HCl aqueous solution, some large plate structures are obtained (Fig. 6a). Increasing the concentration of HCl to 1.0 M, there are middle-sized agglomerated plates on the surface (Fig. 6b). It is also observed that aggregation of plates with a smaller size are obtained when the concentration of HCl is further increased (Fig. 6c and d), which demonstrates that the number of nuclei formed increases as well as the size of particles decreases. The difference of shape can be explained in the form of the different growth rates of different crystal faces.[18] The SEM images of the obtained photoanodes prepared using different impurity metals under the same acid solution concentration in the etching process are compared in Fig. S5,† which reveals almost the same size and thickness with similar plate-like morphologies. These results indicate that this method is a general approach toward the fabrication of plate-like tungsten oxide hydrate. Compared with conventional hydrothermal synthesis or anodizing approaches, the etching method is effective and operative at ambient temperature and atmospheric pressure.

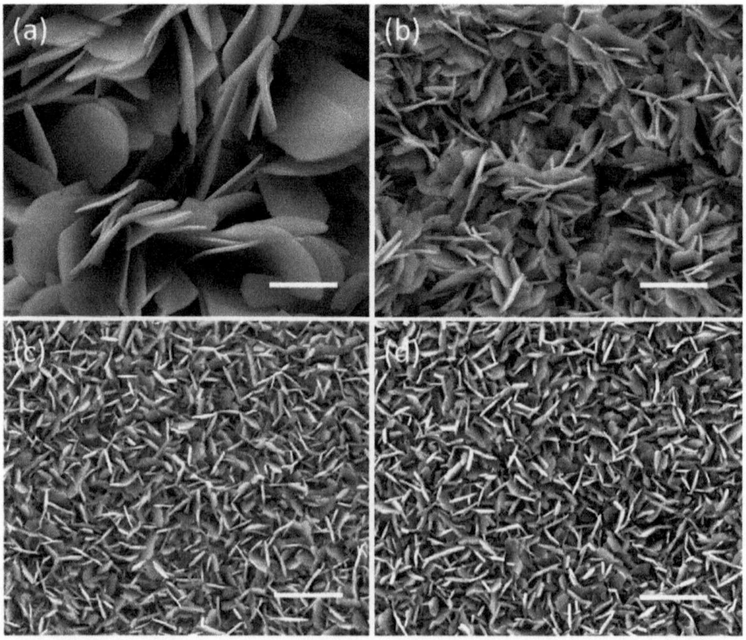

Fig. 6 SEM images of W–Cu–O film etched in (a) 0.5 M HCl, (b) 1 M HCl, (c) 2 M HCl and (d) 4 M HCl at room temperature; scale bars: 1 μm.

In order to verify the purity of product, the compositions of the as-sputtered sample and etched sample (12 h) were further detected *via* ICP-AES analysis. The Cu/W molar ratio of the W–Cu–O film decreased from 32.65 to 0.3%, and similarly the Zn/W molar ratio of the W–Zn–O film decreased from 17.72 to 0.03% (as listed in Table S1†), which confirms the removal of the impurity metal sources as indicated above.

Based on the above experimental results, we propose that the growth of $WO_3 \cdot 2H_2O$ crystal follows the dissolution–recrystallization mechanism during chemical etching at room temperature, which can be described by the following reactions:

$$W\text{–}M\text{–}O + (n + 2)H^+ \rightarrow WO_2^{2+} + M^{n+} + H_2O \quad (M = Cu, Al, Zn; n = 2 \text{ or } 3) \quad (1)$$

$$WO_2^{2+} + 2H_2O \rightarrow WO_3 \cdot H_2O + 2H^+ \quad (2)$$

$$WO_3 \cdot H_2O + H_2O \rightarrow WO_3 \cdot 2H_2O \quad (3)$$

When soaking in the strongly acidic aqueous solution, the as-the sputtered amorphous W–M–O film was gradually dissolved and tungsten element exists in the form of WO_2^{2+} (as has been reported in refs 19 and 20) and M^{n+} cations through corrosive pitting. The released WO_2^{2+} species quickly bond with the surrounding water molecules to form the insoluble $WO_3 \cdot H_2O$ nucleus while M^{n+} cations spread into the solution resulting in the leaching of impurity metal oxide. As the etching time goes on, WO_2^{2+} and M^{n+} cations are continuously produced so that the grain size of the $WO_3 \cdot H_2O$ crystals becomes increasingly larger and the M^{n+} cations simultaneously dissolve into the solution, leading to a final pure phase of tungsten oxide hydrate deposited on the surface. Certainly, it is reasonable for the $WO_3 \cdot H_2O$ to bond further with another H_2O molecule to convert into $WO_3 \cdot 2H_2O$ completely, as indicated by the above Raman analysis. A scheme for the formation mechanism of plate-like tungsten oxide hydrate ($WO_3 \cdot 2H_2O$) is shown in Fig. 7. In addition, it is noticed that the crystal size of the obtained $WO_3 \cdot 2H_2O$ increases with more dilute acidic solution as the etching medium, as observed above in Fig. 6. This is because the amount of WO_2^{2+} ions initially produced is smaller in a dilute acidic medium than that obtained in a more concentrated one, implying a smaller quantity of nucleation centers and a larger crystal size as a result. This phenomenon further confirms the above dissolution–recrystallization mechanism.

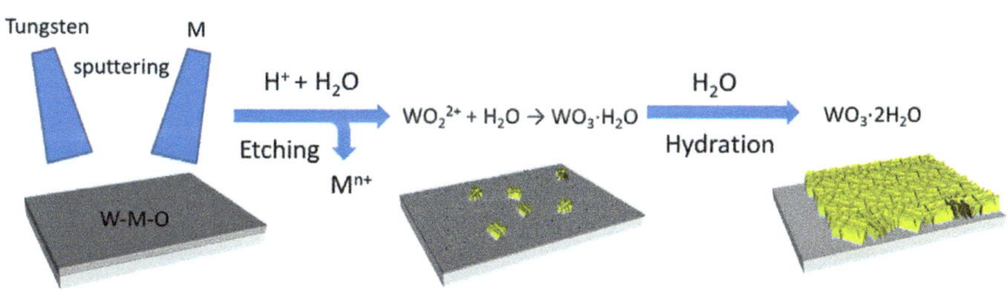

Fig. 7 Scheme for the formation mechanism of plate-like tungsten oxide hydrate ($WO_3 \cdot 2H_2O$).

Fig. 8 compares the optical properties of the plate-like WO_3 film prepared by the current two-step method (Cu as the impurity source and etched in 1 M H_2SO_4 for 12 h) and the compact WO_3 deposited *via* the conventional magnetron sputtering technology, both thicknesses of which are controlled to be ~1 μm. It is seen from Fig. 8a that the optical transmittance for the plate-like films is under 20% in the visible-light range, which is much lower than that of the compact one (~60%). The surface roughness estimated from the reflectance property (Fig. 8b) by the profiler is 6.9 and 68.6 nm for compact and plate-like WO_3 films, respectively (shown in Fig. S6†), indicating an effective surface roughening by the etching treatment. Despite the higher light reflectance as shown in Fig. 8b, the porous WO_3 still has a much higher light absorption ability than the compact one (shown in Fig. 8c, evaluated according to the formula: $A = 100 - R - T$), which can be attributed to increased absorption paths among the plate-like structures.

The morphological change not only affects the roughness but also the surface area, which could play a major role in the reaction activity. The relative electrochemical surface area (ECSA) values of the electrodes are determined from the capacitive region of the cyclic voltammograms, as shown in Fig. 9. The measurements were carried out in the region between 0.15 and 0.35 V *vs.* SCE in 0.5 M Na_2SO_4 (pH 3.0) in the dark. Fig. 9a shows the capacitive current for the plate-like WO_3 film at six different scan rates from 25 to 300 mV s^{-1}, whereas the *C–V* curves of compact WO_3 and plate-like WO_3 photoanodes (scan rate = 50 mV s^{-1}) are compared in Fig. 9b. The linear relationships between the capacitive current and scan rate for the two electrodes are plotted in Fig. 9c. Assuming that the intrinsic specific surface capacitance of all the films is approximately the same, the relative surface area of the films can be determined.[21] The relative electrochemically active surface area of plate-like WO_3 is twice as large as the sputtered compact WO_3 film.

The photoelectrochemical behavior of the plate-like and compact WO_3 photoanodes was measured by obtaining the chopped light linear sweep voltammograms in 0.5 M Na_2SO_4 solution (pH 3.0), and these are compared in

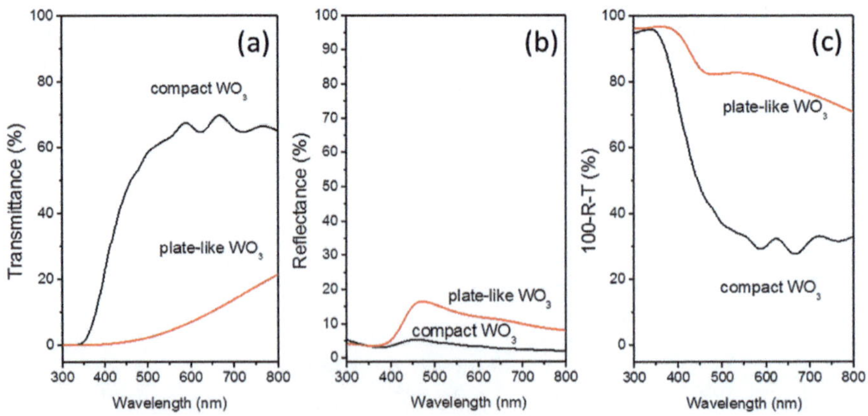

Fig. 8 UV-Vis spectra of compact WO_3 (black) and plate-like (red) WO_3 films grown on the FTO substrate. (a) Transmittance, (b) reflectance and (c) absorption ($100 - R - T$) spectra of the prepared photoelectrodes.

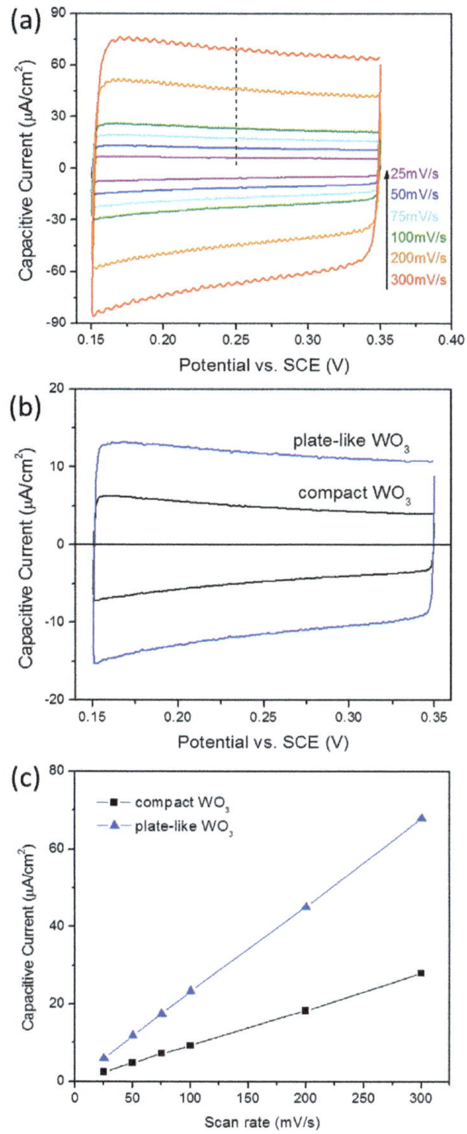

Fig. 9 (a) Cyclic voltammograms of plate-like WO_3 electrode tested at different scan rates. (b) *C–V* curves of compact WO_3 and plate-like WO_3 electrodes (scan rate = 50 mV s^{-1}). (c) The linear relationship between the capacitive current and scan rate.

Fig. 10a. Clearly, the plate-like WO_3 electrode shows a higher saturation photocurrent density than the compact one although nearly the same onset potential of about 0.2 V *vs.* SCE is found for the two photoanodes. At the bias of 1.2 V *vs.* SCE, the photocurrent density generated over the porous sample is 0.95 mA cm^{-2}, which is almost twice of that generated over the compact one (0.58 mA cm^{-2}). In addition, the PEC performance of the samples prepared at different etching intervals following calcination in air at 500 °C was also measured, as shown in Fig. S7.† The photocurrent of the electrodes decreases with the shortening of the

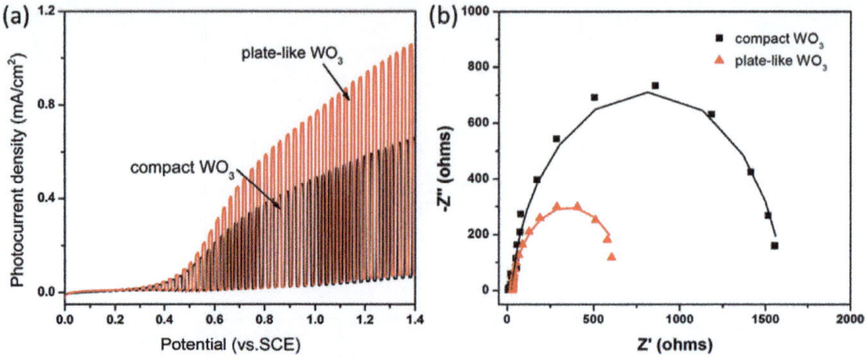

Fig. 10 (a) Linear sweep voltammetric scans of compact and plate-like WO_3 photoanodes in 0.5 M Na_2SO_4 (pH 3.0) under chopped AM 1.5G illumination; scanning rate: 20 mV s^{-1}. (b) Electrochemical impedance spectra and Nyquist plots of the compact and plate-like WO_3 photoanodes measured at 1.0 V *vs.* SCE in 0.5 M Na_2SO_4 solution (pH 3.0) under AM 1.5G illumination.

etching time and the as-sputtered W–Cu–O electrode without etching has the lowest photocurrent density, which reveals that the existence of Cu could be harmful to the PEC water oxidation reaction. After complete removal of the impurity metal, the saturation photocurrent density for the sample is 5-fold higher than that of un-etched one.

Electrochemical impedance spectroscopy was used to gain more insight into the reasons for the high photocurrent of the plate-like photoanode. The EIS spectra of compact WO_3 and plate-like WO_3 photoelectrodes measured under irradiation by AM 1.5G light (100 mW cm^{-2}), with a bias voltage of 1.0 V *vs.* SCE, are shown in Fig. 10b. The semicircle for compact WO_3 is larger than the plate-like WO_3 sample, indicating that the latter has a lower charge-transfer resistance. The fitted values of the solution resistance (R_s), charge-transfer resistance across the interface (R_p) and constant phase elements (CPE-T and CPE-P) were calculated and are listed in Table S2.† The fitted values of R_p are 1576 and 655.5 Ω for the compact WO_3 and plate-like WO_3 samples, respectively, implying that the plate-like structure is beneficial for this charge-transfer process.

Based on the above analysis, the improvement in the PEC water oxidation activity for the porous plate-like WO_3 photoanode can be mainly attributed to the increase of exposed surface area, light absorption and lower electrode/electrolyte interface charge-transfer resistance.

4 Conclusions

In this work, a porous plate-like tungsten oxide photoelectrode was successfully prepared on an FTO substrate by a facile two-step route including magnetron co-sputtering and chemical etching methods. The process of growth includes the dissolution of the as-sputtered amorphous W–M–O films in acidic medium, species precipitation and the recrystallization process. Compared to the compact WO_3 film counterparts, the photocurrent of PEC water oxidation with the plate-like WO_3 photoanode is improved significantly because the porous structure provides more exposed surface area, increased light absorption and a lower electrode/electrolyte interface charge-transfer resistance.

Acknowledgements

This work was financially supported by the National Natural Science Foundation of China (no. 21090340) and 973 National Basic Research Program of the Ministry of Science and Technology (no. 2014CB239400).

References

1 M. G. Walter, E. L. Warren, J. R. McKone, S. W. Boettcher, Q. X. Mi, E. A. Santori and N. S. Lewis, *Chem. Rev.*, 2010, **110**, 6446–6473.
2 B. D. Alexander, P. J. Kulesza, I. Rutkowska, R. Solarska and J. Augustynski, *J. Mater. Chem.*, 2008, **18**, 2298.
3 H. M. Chen, C. K. Chen, R. S. Liu, L. Zhang, J. Zhang and D. P. Wilkinson, *Chem. Soc. Rev.*, 2012, **41**, 5654–5671.
4 H. Kikuchi, M. Kitano, M. Takeuchi, M. Matsuoka, M. Anpo and P. V. Kamat, *J. Phys. Chem. B*, 2006, **110**, 5537–5541.
5 B. Marsen, E. L. Miller, D. Paluselli and R. E. Rocheleau, *Int. J. Hydrogen Energy*, 2007, **32**, 3110–3115.
6 L. Chen, E. Alarcón-Lladó, M. Hettick, I. D. Sharp, Y. Lin, A. Javey and J. W. Ager, *J. Phys. Chem. C*, 2013, **117**, 21635–21642.
7 D. Yokoyama, H. Hashiguchi, K. Maeda, T. Minegishi, T. Takata, R. Abe, J. Kubota and K. Domen, *Thin Solid Films*, 2011, **519**, 2087–2092.
8 X. Liu, F. Wang and Q. Wang, *Phys. Chem. Chem. Phys.*, 2012, **14**, 7894–7911.
9 C. Santato, M. Odziemkowski, M. Ulmann and J. Augustynski, *J. Am. Chem. Soc.*, 2001, **123**, 10639–10649.
10 C. Ng, C. Ye, Y. H. Ng and R. Amal, *Cryst. Growth Des.*, 2010, **10**, 3794–3801.
11 E. Widenkvist, R. A. Quinlan, B. C. Holloway, H. Grennberg and U. Jansson, *Cryst. Growth Des.*, 2008, **8**, 3750–3753.
12 W. Li, C. Liu, Y. Yang, J. Li, Q. Chen and F. Liu, *Mater. Lett.*, 2012, **84**, 41–43.
13 Z. Jiao, J. Wang, L. Ke, X. W. Sun and H. V. Demir, *ACS Appl. Mater. Interfaces*, 2011, **3**, 229–236.
14 N. Wang, D. Wang, M. Li, J. Shi and C. Li, *Nanoscale*, 2014, **6**, 2061–2066.
15 M. F. Daniel, B. Desbat, J. C. Lassegues, B. Gerand and M. Figlarz, *J. Solid State Chem.*, 1987, **67**, 235–247.
16 R. S. Lillard, G. S. Kanner and D. P. Butt, *J. Electrochem. Soc.*, 1998, **145**, 2718–2725.
17 V. S. Vidyarthi, M. Hofmann, A. Savan, K. Sliozberg, D. König, R. Beranek, W. Schuhmann and A. Ludwig, *Int. J. Hydrogen Energy*, 2011, **36**, 4724–4731.
18 B. L. Cushing, V. L. Kolesnichenko and C. J. O'Connor, *Chem. Rev.*, 2004, **104**, 3893–3946.
19 M. S. Elbasiouny, S. A. Hassan and M. M. Hefny, *Corros. Sci.*, 1980, **20**, 909–917.
20 J. Johnson and C. Wu, *J. Electrochem. Soc.*, 1971, **118**, 1909–1912.
21 B. A. Pinaud, P. C. K. Vesborg and T. F. Jaramillo, *J. Phys. Chem. C*, 2012, **116**, 15918–15924.

Faraday Discussions

PAPER

Comparison of photoelectrochemical water oxidation activity of a synthetic photocatalyst system with photosystem II

Yi-Hsuan Lai, Masaru Kato, Dirk Mersch and Erwin Reisner*

Received 4th April 2014, Accepted 12th June 2014
DOI: 10.1039/c4fd00059e

This discussion describes a direct comparison of photoelectrochemical (PEC) water oxidation activity between a photosystem II (PSII)-functionalised photoanode and a synthetic nanocomposite photoanode. The semi-biological photoanode is composed of PSII from the thermophilic cyanobacterium *Thermosynechococcus elongatus* on a mesoporous indium tin oxide electrode (*meso*ITO|PSII). PSII embeds all of the required functionalities for light absorption, charge separation and water oxidation and ITO serves solely as the electron collector. The synthetic photoanode consists of a TiO_2 and NiO_x coated nanosheet-structured WO_3 electrode (*nano*WO_3|TiO_2|NiO_x). The composite structure of the synthetic electrode allows mimicry of the functional key features in PSII: visible light is absorbed by WO_3, TiO_2 serves as a protection and charge separation layer and NiO_x serves as the water oxidation electrocatalyst. *Meso*ITO|PSII uses low energy red light, whereas *nano*WO_3|TiO_2|NiO_x requires high energy photons of blue-end visible and UV regions to oxidise water. The electrodes have a comparable onset potential at approximately 0.6 V *vs.* reversible hydrogen electrode (RHE). *Meso*ITO|PSII reaches its saturation photocurrent at 0.84 V *vs.* RHE, whereas *nano*WO_3|TiO_2|NiO_x requires more than 1.34 V *vs.* RHE. This suggests that *meso*ITO|PSII suffers from fewer limitations from charge recombination and slow water oxidation catalysis than the synthetic electrode. *Meso*ITO|PSII displays a higher '*per active*' site activity, but is less photostable and displays a much lower photocurrent per geometrical surface area and incident photon to current conversion efficiency (IPCE) than *nano*WO_3|TiO_2|NiO_x.

1. Introduction

A possible route for the conversion of energy in sunlight into a storable hydrogen-based fuel is solar water splitting.[1] Photosynthesis uses sunlight, water and CO_2 to generate readily useable carbohydrates. At the heart of this process stands the protein complex PSII, which catalyses efficiently the most thermodynamically

Christian Doppler Laboratory for Sustainable SynGas Chemistry, Department of Chemistry, University of Cambridge, Lensfield Road, Cambridge CB2 1EW, UK. E-mail: reisner@ch.cam.ac.uk

demanding process in biology: solar light-driven water oxidation. As such, PSII not only sets the benchmark for photo-water oxidation catalysis, but also provides an invaluable inspiration for the design of artificial photosynthetic systems.[2] Essential to the function of PSII is its fine-tuned ability to combine light absorption, multi-charge separation and efficient water oxidation catalysis. Electron transfer in PSII occurs from the excited primary electron donor (P680*), which can be generated upon red light absorption at 680 nm. The electrons are further transferred to the electron acceptor plastoquinones, Q_A and Q_B, *via* a pheophytin (Pheo). The generated holes are captured by a tyrosine (Tyr_Z) and subsequently transferred to the $CaMn_4$ oxygen evolving catalyst (OEC) of PSII to oxidise water (Scheme 1a).[2,3]

Future solar fuel technologies are likely to rely on PEC cells and many electrodes for photo-water oxidation have recently been reported.[1b,5] Despite the immense interest in mimicking PSII,[6] a direct comparison of a PSII-immobilised electrode with a purely synthetic, PSII-inspired water oxidation photoanode has not yet been presented.

In this discussion, we describe a side-by-side comparison between a PSII-electrode and a purely synthetic, PSII-inspired water oxidation photoelectrode recently developed in our laboratory. The semi-biological photoanode is composed of PSII isolated from the thermophilic cyanobacterium *Thermosynechococcus elongatus* on a mesoporous indium tin oxide electrode (*meso*ITO|PSII, Fig. 1a).[7] PSII embeds the required functionalities for photo-water oxidation (Scheme 1a) and direct electron transfer is observable from the plastoquinones in PSII to *meso*ITO, allowing us to use electrochemical methods to study the photocharacteristics of PSII.[7]

The PSII-inspired synthetic photoanode consists of a TiO_2 and NiO_x composite on nanosheet-structured WO_3 electrode (*nano*$WO_3|TiO_2|NiO_x$, Fig. 1b). The composite structure of the synthetic electrode allows for partial mimicry of functional key features in PSII: WO_3 serves as the visible light harvester, TiO_2 acts as a protection and charge separation layer and NiO_x behaves as a water oxidation electrocatalyst in borate solution (Scheme 1b).[4] NiO_x was reported to have a similar cubane-like structure to the $CaMn_4$ OEC of PSII.[8] In addition, this

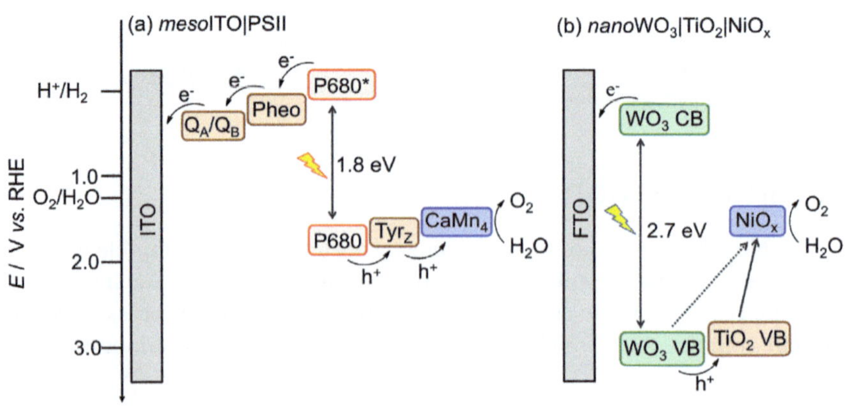

Scheme 1 Energy diagrams for solar light driven water oxidation with (a) *meso*ITO|PSII and (b) the bio-inspired *nano*$WO_3|TiO_2|NiO_x$.[4]

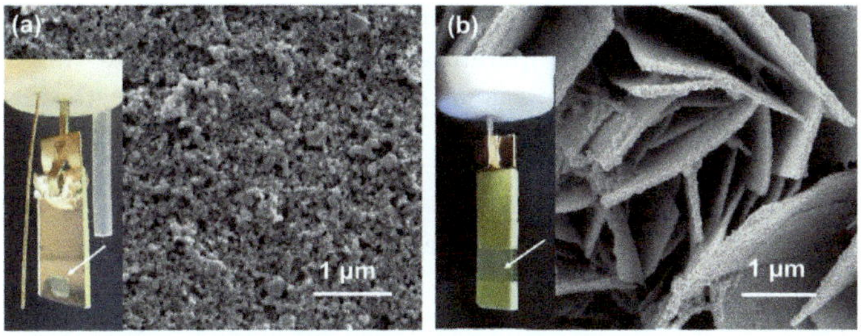

Fig. 1 SEM images of (a) *meso*ITO and (b) *nano*$WO_3|TiO_2|NiO_x$. The insets in (a) and (b) show photograph images of the *meso*ITO|PSII and *nano*$WO_3|TiO_2|NiO_x$ electrodes, respectively (arrows indicate the exposed working areas).

synthetic electrode is composed of inexpensive materials and is functional under benign conditions (room temperature and pH 8 to 9 aqueous solution).[4] Functional features of the enzymatic and synthetic systems allow us to compare their performance in PEC O_2-evolution activity, efficiency and stability.

2. Experimental

Preparation of *meso*ITO|PSII electrode

*Meso*ITO|PSII was prepared according to literature procedures (Scheme 2a).[7a,9] *Meso*ITO was loaded onto an ITO-coated glass slide (VisionTek Systems Ltd) by spreading ITO nanoparticles (Aldrich; <50 nm particle size; 27 m^2 g^{-1} surface area; 90% In_2O_3 and 10% SnO_2) with a geometric surface area of 0.25 cm^2 (using Scotch® tape (3M) as spacers to control the active surface area and the film thickness), followed by annealing at 450 °C for 30 min.[9] PSII dimers were isolated from a thermophilic cyanobacterium, *Thermosynechococcus elongatus*, and purified following a published protocol.[10] To assemble *meso*ITO|PSII, 2 μL of a

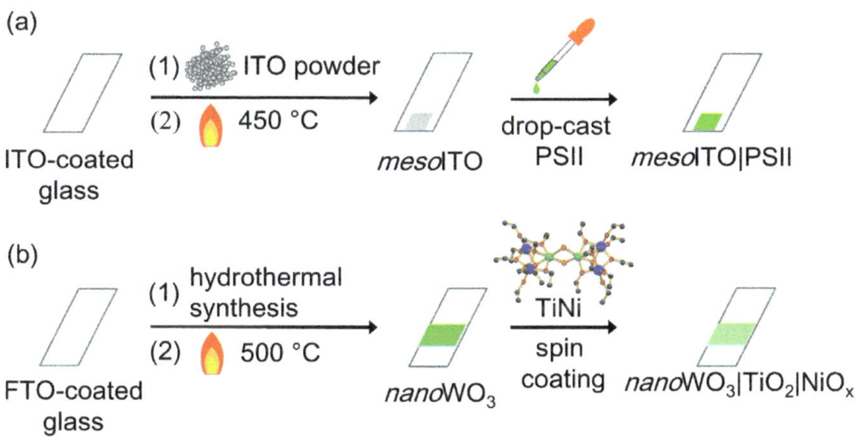

Scheme 2 Schematic representation of the preparation of (a) *meso*ITO|PSII and (b) *nano*$WO_3|TiO_2|NiO_x$ electrodes. The atoms in TiNi are colour-coded as follows (H atoms omitted for clarity): Ni (green), Ti (blue), Cl (orange), O (red), C (grey).

0.54 (mg Chlorophyll *a*) mL^{-1} PSII solution was drop-cast onto the *meso*ITO surface and kept at room temperature in the dark. After 30 min, the PSII-modified *meso*ITO electrode was used as a working electrode for PEC measurements.[7a,9]

Preparation of *nano*$WO_3|TiO_2|NiO_x$ electrode

Nano$WO_3|TiO_2|NiO_x$ was prepared as described previously (Scheme 2b).[4] To prepare a WO_3 seed layer on a fluoride-doped tin oxide (FTO)-coated glass electrode, a precursor solution containing H_2WO_4 (0.625 g, 99%; Sigma-Aldrich) and polyvinyl alcohol (0.5 g, 98–99%, medium molecular weight; Alfa Aesar) in H_2O_2 (10 mL, 30%; Sigma-Aldrich) was spin-coated on the FTO-coated glass (Pilkington; TEC Glass™ 7; sheet resistance 7 ohm sq^{-1}), followed by annealing in air at 500 °C for 2 h. To grow the WO_3 nanosheets on the WO_3 seed layer-modified FTO-coated glass, a second precursor solution was prepared: a solution of H_2WO_4 (3 mL of 0.25 M) in H_2O_2 was added to an aqueous HCl solution (3 mL, 1 M) containing oxalic acid (0.2 g, 99%; Sigma-Aldrich), and then acetonitrile (10 mL, HPLC grade; Fisher Chemicals) was added to this acidic solution. The second precursor solution was put in a 23 mL Teflon-lined stainless steel autoclave (model 4749, Parr). The WO_3 seed layer-modified FTO-substrate was immersed vertically into the second precursor solution in the autoclave, whereupon the autoclave was sealed and heated at 180 °C for 2.5 h. After growth of the WO_3 nanosheets, the resultant electrodes were rinsed with ethanol and then annealed in air for 1 h at 500 °C.

$[Ti_2(OEt)_9(NiCl)]_2$ (TiNi) was synthesised by heating titanium ethoxide (7.00 mL, 33.4 mmol; 99.99%; Sigma-Aldrich) and nickel(II) chloride (271 mg, 2.09 mmol; >97%; Sigma-Aldrich) in anhydrous ethanol (7.00 mL, 120 mmol) in a Teflon-lined autoclave at 150 °C for 24 h.[11] After removal of ethanol, dark yellow crystals of the product were grown from dry toluene at −14 °C. *Nano*$WO_3|TiO_2|NiO_x$ was assembled by spin coating a fresh solution of TiNi (5 mM in toluene) on *nano*WO_3. *Nano*$WO_3|TiO_2|NiO_x$ was dried for at least 30 min in air at room temperature and then washed with water prior to use as a working electrode.

Physical characterisation

The surface morphology of *meso*ITO|PSII and *nano*$WO_3|TiO_2|NiO_x$ was characterised with a Philips XL30-SFGE scanning electron microscope (SEM). UV-vis absorption spectra of the electrodes were recorded on a Varian Cary 50 UV-vis spectrophotometer equipped with an energy diffuse reflectance accessory (Barrelino™). The amount of PSII adsorbed on *meso*ITO was estimated by scraping off the PSII-modified ITO nanoparticles from the ITO substrate, suspending them in MeOH, centrifuging the suspension and analysing the supernatant by UV-vis spectrophotometry.[7a] The concentration of Chlorophyll *a* (Chl *a*) in the supernatant was determined using an extinction coefficient equal to 79.95 (Chl *a* mg)$^{-1}$ mL cm^{-1} at 665 nm. The amount of PSII on *meso*ITO was calculated to be approximately 21.2 pmol cm^{-2} assuming that the PSII dimer contains 35 Chl *a* molecules.[3a] This PSII loading allowed us to calculate a turnover frequency (TOF) of 0.03 (mol O_2) (mol PSII)$^{-1}$ s^{-1} for *meso*ITO|PSII based on photocurrents obtained at 1.23 V *vs.* RHE with four electrons per O_2 molecule.

PEC measurements

PEC measurements were recorded with an Ivium CompactStat potentiostat using a conventional three-electrode system. *Meso*ITO|PSII (exposed projected geometrical area of 0.25 cm^2) or *nano*$WO_3|TiO_2|NiO_x$ (exposed projected geometrical area of 0.5 cm^2) were used as the working electrodes connected to a Pt foil counter electrode and a Ag/AgCl/KCl(sat) reference electrode. All the potentials were converted to the reversible hydrogen electrode (RHE) by using the following equation: $E(\text{V } vs. \text{ RHE}) = E(\text{V } vs. \text{ Ag/AgCl}) + 0.197 + 0.059 \times \text{pH}$.[12] An aqueous buffered solution (pH 6.5) containing 50 mM KCl, 15 mM $CaCl_2$, 15 mM $MgCl_2$, and 40 mM 2-(*N*-morpholino)ethanesulfonic acid (MES) was used for the *meso*ITO|PSII system, whereas the photocurrent responses of *nano*$WO_3|TiO_2|NiO_x$ were measured in a potassium borate buffer solution (B_i, pH 9.2). A solar light simulator (Newport Oriel, Xenon 150 W) was used as a light source unless otherwise noted. The light intensity was calibrated to 100 mW cm^{-2} (1 sun). An air mass 1.5 global (AM 1.5G) filter and an IR water filter were used. A 590 nm cut-off filter (UQG Optics) was used for all PEC experiments with *meso*ITO|PSII to avoid photoexcitation of ITO. Representative transient photocurrent responses at different potentials of *meso*ITO|PSII and *nano*$WO_3|TiO_2|NiO_x$ with two cycles of a dark period followed by irradiation for 20 s each are shown in Fig. 2a and b, respectively.

Photon to current conversion efficiency measurements

A 300 W Xenon lamp coupled to an MSH300 monochromator (LOT Quantum design) was used for IPCE measurements. The light intensity for the IPCE measurements was measured as a function of wavelength with a photodetector (SEL033/F/QNDS1/W) and power meter (ILT1400). The recorded photocurrents and light intensities at various wavelengths were used to calculate IPCE (η) according to the following equation:

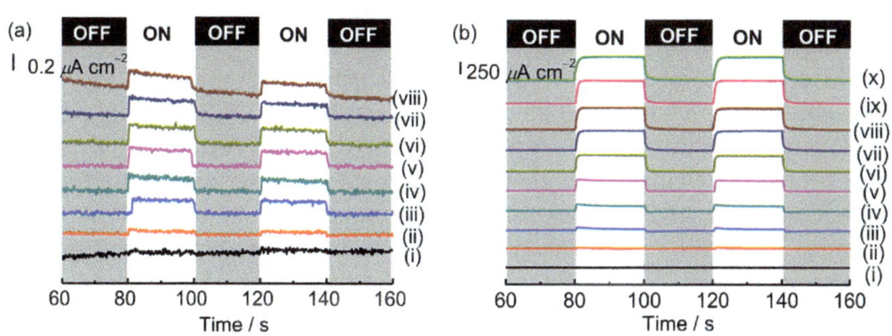

Fig. 2 Representative transient photocurrent response of (a) *meso*ITO|PSII and (b) *nano*$WO_3|TiO_2|NiO_x$ under standardised solar light irradiation (AM 1.5G; ON) and in the dark (OFF) at different applied potentials: (i) 0.54, (ii) 0.64, (iii) 0.74, (iv) 0.84, (v) 0.94, (vi) 1.04 V, (vii), 1.14, (viii), 1.24, (ix) 1.34 and (x) 1.44 V *vs.* RHE. Measurements with *meso*ITO|PSII were recorded in an aqueous pH 6.5 electrolyte solution, whereas *nano*$WO_3|TiO_2|NiO_x$ was employed in an aqueous B_i buffer (0.1 M, pH 9.2).

$$\eta(\%) = \frac{I_{el}}{I_{ph}} \times 100 = \frac{\dfrac{J \times 10^{-2}}{F}}{\dfrac{W \times \lambda \times 10^{-9}}{N_A \times h \times c}} \times 100 \approx \frac{1240 \times J}{\lambda \times W} \tag{1}$$

where I_{el} is the electron flux at the external circuit (mol m^{-2} s^{-1}), I_{ph} is the incident photon flux (mol m^{-2} s^{-1}), J is the measured photocurrent density (μA cm^{-2}), F is the Faraday constant (96 484 A s mol^{-1}), λ is the wavelength of light (nm), W is the incident power of the monochromated light (W m^{-2}), N_A is Avogadro's number (6.022×10^{23} mol^{-1}), h is Planck's constant (6.626×10^{-34} J s) and c is the speed of light (2.998×10^{8} m s^{-1}).

Absorbed photon to current conversion efficiency (APCE) was obtained by dividing the IPCE by the light harvesting efficiency (LHE) at each wavelength:[13]

$$\text{APCE (\%)} = \text{IPCE (\%)/LHE} \tag{2}$$

$$\text{LHE} = 1 - 10^{-A(\lambda)} \tag{3}$$

where $A(\lambda)$ is the absorbance at wavelength λ.

3. Results and discussion

Surface morphologies of *meso*ITO|PSII and *nano*$WO_3|TiO_2|NiO_x$

*Meso*ITO has a pore diameter of up to 100 nm (Fig. 1a) and therefore provides a high surface area to integrate PSII (approximately $20.5 \times 10.5 \times 11.0$ nm^3 for the PSII dimer).[3a] *Meso*ITO|PSII was assembled by drop-casting the PSII solution on the *meso*ITO electrode surface (geometric active surface area: 0.25 cm^2, inset of Fig. 1a).

Nano$WO_3|TiO_2|NiO_x$ was studied as a PSII-inspired synthetic photoanode. This synthetic photoanode consists of green WO_3 nanosheets coated with a uniform composite film containing TiO_2 and NiO_x nanoparticles (Fig. 1b). WO_3 is an n-type semiconductor with a suitable band gap (approximately 2.7 eV) and is capable of photo-oxidising water with blue-end visible light (valence band potential at approximately 3 V *vs.* RHE, Scheme 1b). The nanosheet structure of WO_3 gives a high surface area for light absorption, electrocatalyst loading and a decreased hole diffusion length in the photoanode.

A uniform decoration with Ti and Ni containing nanoparticles on WO_3 can be obtained by spin-coating the molecular TiNi precursor solution onto *nano*WO_3 and subsequently hydrolysing the metal-oxide precursor.[4] TiNi thereby serves as a single source precursor for the formation of TiO_2 and NiO_x on *nano*WO_3. The TiO_2 layer protects WO_3 from the alkaline electrolyte solution and supports charge separation by receiving holes from the photoexcited *nano*WO_3. The NiO_x is assembled *in situ* by photo-generated holes in borate buffer (B_i) and promotes water oxidation catalysis. We note that at least some NiO_x is likely to be in close contact with WO_3 and direct hole transfer from WO_3 to NiO_x is therefore also possible (Scheme 1b).

PEC characteristics

Fig. 3a shows a schematic *J–V* curve of an efficient (red dotted line) and inefficient photoanode (black solid line). First, we will discuss (i) the onset potential, (ii) the

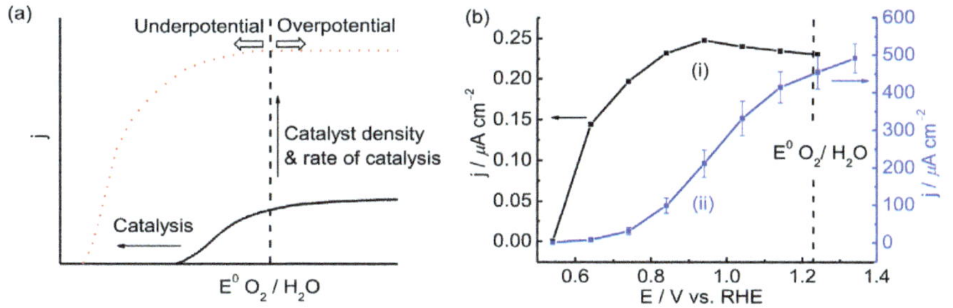

Fig. 3 (a) Performance of an idealised photoanode compared with a photoanode deviated from its ideal case. The solid arrows indicate the parameters that need to be addressed to improve the photoanode. (b) Photocurrent responses at various potentials of a (i) *meso*ITO|PSII electrode and (ii) *nano*WO_3|TiO_2|NiO_x under standardised solar light irradiation (AM 1.5G, 100 mW cm^{-2}).[2,4] A 590 nm cut-off filter was used for *meso*ITO|PSII. Measurements with electrode (i) were recorded in an aqueous pH 6.5 MES buffer solution and electrode (ii) were recorded in an aqueous B_i buffer (0.1 M, pH 9.2). The standard deviation for measurements with the *meso*ITO|PSII electrode (i) was approximately 20%. E(V *vs.* NHE) = E(V *vs.* RHE) − 0.059 × pH.[12]

potential required for reaching plateau current ('saturation photocurrent') and (iii) the plateau current as parameters to characterise the photoelectrodes and compare it to an ideal photoanode. An optimal semiconductor photoanode should have an onset potential close to the conduction band position of WO_3 during illumination, whereas we would expect an ideal PSII-based electrode to display photocurrents at slightly more positive potentials than the Q_B redox potential. At the same time the saturation photocurrent should readily be reached well within the "underpotential" region (at more negative potential than the thermodynamic water oxidation).

In practice however, the onset and saturation potential are shifted to considerably more anodic potentials due to slow catalysis and charge recombination events at the photoanode. The PEC responses of *meso*ITO|PSII and *nano*WO_3|TiO_2|NiO_x were recorded at different potentials under standardised solar light irradiation. A 590 nm long-pass filter was used for *meso*ITO|PSII to prevent photoexcitation of ITO. The mean values of the net photocurrent density (J) at specific potentials were obtained by subtracting peak values of the photocurrent density from the background.

The redox potential of plastoquinone Q_A and Q_B is approximately 0.3 V *vs.* RHE (∼−0.1 V *vs.* NHE in pH neutral environment).[14] *Meso*ITO|PSII was studied in a pH 6.5 MES buffer solution (in the absence of soluble redox mediator) and has an onset potential of 0.60 V *vs.* RHE during irradiation, which is approximately 0.3 V more anodic than its theoretical onset potential. We note that direct electron transfer is required from the plastoquinones to ITO for this semi-biological electrode. Coupling of the plastoquinones to the ITO surface is likely to be non-ideal, which would explain the anodic shift of the onset potential. The saturation photocurrent of approximately 0.25 μA cm^{-2} was reached close to the onset potential at 0.84 V *vs.* RHE (trace i, Fig. 3b). This small difference between the onset potential and the saturation photocurrent potential may be explained by the efficient charge separation and facile kinetics for water oxidation in PSII (Scheme 1a).

Table 1 Comparison of PEC water oxidation activity of *meso*ITO|PSII and *nano*WO_3|TiO_2|NiO_x

System[a]	Onset potential, E_O (V *vs.* RHE)	ΔE_{O-P}[b] (V)	J_{max}[c] (μA cm^{-2})	TOF[d] (s^{-1})	Stability, $t_{1/2}$[e] (min)	$IPCE_{max}$[f] (%)	$APCE_{max}$[f] (%)
*meso*ITO|PSII	0.60	0.2	0.3	0.03	~4	0.125	0.70
nanoWO_3|TiO_2|NiO_x	0.64	0.7	500	8×10^{-4}	~240	50	80

[a] *Meso*ITO|PSII was employed as working electrode in a pH 6.5 MES solution. *Nano*WO_3|TiO_2|NiO_x was studied in an aqueous pH 9.2 B_i buffer (0.1 M). Both electrodes were studied at room temperature with a Pt counter and a Ag/AgCl/KCl(sat) reference electrode. [b] Potential difference between onset potential (E_O) and the potential required to reach the plateau photocurrent (E_P). [c] Saturation photocurrent. [d] TOF [in mol O_2 (mol catalyst)$^{-1}$ s^{-1}] calculated based on photocurrents obtained at 1.23 V *vs.* RHE. For PSII, 100% Faradaic efficiency was assumed. For NiO_x, the TOF was calculated based on evolved O_2 gas.[4] [e] Calculated at an applied potential of 0.94 V *vs.* RHE under solar light irradiation. [f] Recorded at an applied potential of 0.94 V *vs.* RHE (680 nm for *meso*ITO|PSII and 375 nm for *nano*WO_3|TiO_2|NiO_x).

The conduction band of WO_3 is situated at approximately 0.3 V *vs.* RHE,[15] which is similar to the redox potential of Q_A and Q_B in PSII. Without modification of the surface with TiO_2 and NiO_x, the onset potential of *nano*WO_3 is 0.74 V *vs.* RHE, which is notably larger than the onset potential of *meso*ITO|PSII due to poor catalysis.[4] Interfacing TiO_2 and NiO_x on *nano*WO_3 reduces the onset potential to 0.64 V *vs.* RHE, a potential which is comparable with *meso*ITO|PSII. A significant difference in the bias potential required to reach the plateau photocurrent is observed between the two electrode systems. A saturation photocurrent of 500 μA cm^{-2} is reached at 1.34 V *vs.* RHE with *nano*WO_3|TiO_2|NiO_x compared to 0.84 V *vs.* RHE for *meso*ITO|PSII (Fig. 3b). TiO_2 and NiO_x offer only a partial offset for the poor charge recombination kinetics and water oxidation catalytic activity of *nano*WO_3. The plateau photocurrent density is dominated by the amount of the photogenerated holes reaching the surface and oxidising water. Enhancing the catalyst density on the electrode (the amount of photocatalyst based on the geometrical surface area), using a more efficient catalyst or reducing charge recombination within the photoanode increases the photocurrent.[1b,5,13,16] *Meso*ITO|PSII therefore displays more efficient catalysis and suffers from less charge recombination below 1.23 V *vs.* RHE (the underpotential region relative to $E^\circ(H_2/H_2O)$). PSII has an evolutionarily well-developed machinery for the photo-oxidation of water and immobilised PSII retains these features and behaves more like an ideal photoanode compared to the synthetic electrode. The amount of immobilised PSII on *meso*ITO is only 21.2 pmol cm^{-2} due to the large geometrical footprint of the enzyme photocatalyst, whereas 0.78 μmol cm^{-2} of NiO_x is deposited on *nano*WO_3 (Table 1). Taking into account the photocurrent density, a much higher single-site catalytic activity was observed for the OEC in PSII with a turnover frequency (TOF) of approximately 0.03 s^{-1} at 1.23 V compared to 8×10^{-4} s^{-1} at 1.23 V *vs.* RHE for NiO_x in the synthetic system (Table 1).[4] On the other hand, the low PSII loading dramatically limits the photocurrent density with *meso*ITO|PSII and a three-order of magnitude higher photocurrent density per geometric surface area is observed with *nano*WO_3|TiO_2|NiO_x.

The electronic absorption spectra in Fig. 4 reveal another key difference between the electrodes. *Meso*ITO|PSII generates O_2 during red light irradiation,[10b,17] whereas WO_3 is only capable of absorbing photons in the blue and UV regions of the solar spectrum. The synthetic electrode therefore requires substantially more energetic photons to oxidise water. The IPCE spectra show a good agreement with the respective absorption spectra for both electrode systems. IPCE is the measure of the ratio of the photocurrent *versus* the rate of incident photons as a function of wavelength (eqn (1)). IPCE thereby confirms that the photocurrent of *meso*ITO|PSII is the result of red light excitation (Fig. 4a), whereas the photocurrent of *nano*$WO_3|TiO_2|NiO_x$ is generated by blue and UV light (Fig. 4b).

An optimum IPCE of PSII on *meso*ITO of 0.125% was obtained at 680 nm, corresponding to the estimated excited-state energy of P680 (1.825 eV, E_{680}). For any photon absorbed by PSII, the energy driving the photo-water oxidation is equal to E_{680}. The APCE can be readily obtained from the IPCE and the LHE by using eqn (2). LHE quantifies the absorbance of monochromatic light by the electrodes as a function of absorption coefficient (eqn (3)). The LHE of PSII on *meso*ITO at 680 nm is approximately 17%, and the corresponding APCE is 0.70% (Fig. 4 and Table 1). The low quantum efficiency of *meso*ITO|PSII presumably stems from the low loading of PSII and the random orientation of PSII on the *meso*ITO electrode, hindering the injection of electrons from PSII to the *meso*ITO.[7] However, the APCE of *meso*ITO|PSII is comparable with the APCE of PSII on a Au substrate.[18] In contrast, *nano*$WO_3|TiO_2|NiO_x$ can only use high-energy photons provided by solar light to promote electron–hole pair separation. *Nano*$WO_3|TiO_2|NiO_x$ has a maximum IPCE and APCE of 50% and 80% at 375 nm at an applied potential of 0.94 V *vs.* RHE, respectively. The IPCE drops to zero at wavelengths longer than 465 nm, which is consistent with the band gap of WO_3 (2.7 eV, E_{WO3}). Thus, more light energy input ($E_{WO3} - E_{680} = 0.9$ eV) is required to photo-oxidise water and to promote electrons at approximately 0.64 V *vs.* RHE with the synthetic hybrid electrode. The *nano*$WO_3|TiO_2|NiO_x$ electrode requires an additional energy input to compensate for less efficient water oxidation by NiO_x compared to $CaMn_4$ OEC in PSII.

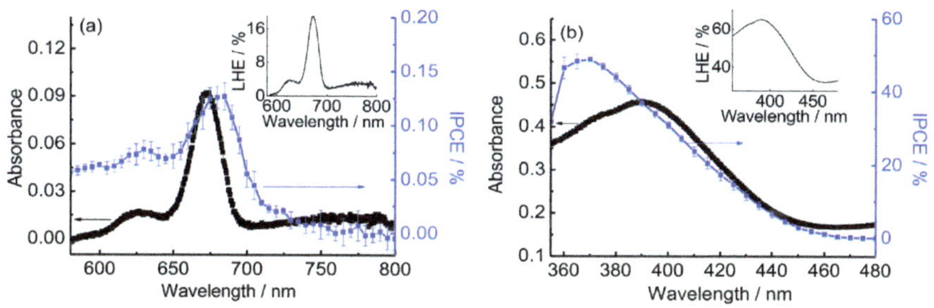

Fig. 4 Diffuse reflectance UV-vis absorption and IPCE spectra of (a) *meso*ITO|PSII and (b) *nano*$WO_3|TiO_2|NiO_x$. IPCE spectra were recorded at an applied potential of 0.94 V *vs.* RHE. Measurements were recorded in an aqueous pH 6.5 MES buffer with *meso*ITO|PSII and in an aqueous pH 9.2 B_i buffer with *nano*$WO_3|TiO_2|NiO_x$. Insets show the corresponding LHE.

Photostability

From a practical standpoint, the long-term photostability of the electrodes in aqueous solution is an important criterion. *Meso*ITO|PSII has a half-life time of approximately 4 min and complete loss of activity is observed within 30 min at an applied potential of 0.94 V *vs.* RHE (Fig. 5a and Table 1). The short lifetime of *meso*ITO|PSII can be attributed to the intrinsic photo-instability of PSII in the presence of intense light irradiation (with no repair machinery being available in isolated PSII) as well as the fragility of the PSII-ITO interface (film loss of PSII). Under strong light irradiation, over-accumulation of excited states leads to excessive charge generation in PSII and results in the generation of Chl *a* triplet states. These react with triplet oxygen to produce singlet oxygen, which is understood to be one of the reasons for the enhanced damage of PSII.[19]

The synthetic *nano*WO_3|TiO_2|NiO_x displays much greater photostability. A half-life time of approximately four hours was observed during solar light irradiation, and a complete loss of activity was only observed after 24 h (Fig. 5b and Table 1). Bare WO_3 is known to be soluble in alkaline solution and easily damaged by side reactions during photo-O_2 evolution.[4,20] However, the TiO_2 and NiO_x coatings stabilise WO_3 by acting as a protective layer and an electrocatalyst, respectively.

Comparison of the semi-biological and synthetic electrode

*Meso*ITO|PSII shows more efficient 'single-site' catalysis and superior resistance to charge recombination than *nano*WO_3|TiO_2|NiO_x. This is evident from the higher TOF of PSII and a 0.5 V more cathodic potential to reach the photocurrent plateau for *meso*ITO|PSII (Table 1). In this respect, the PEC response of *meso*ITO|PSII shown in Fig. 3 is closer to an "ideal" photoanode. Moreover, PSII is capable of photo-oxidising water with low energy "red photons", whereas WO_3 requires higher energy photons to compensate for significantly higher charge recombination and less efficient NiO_x OEC. Interfacing *nano*WO_3|TiO_2|NiO_x with another

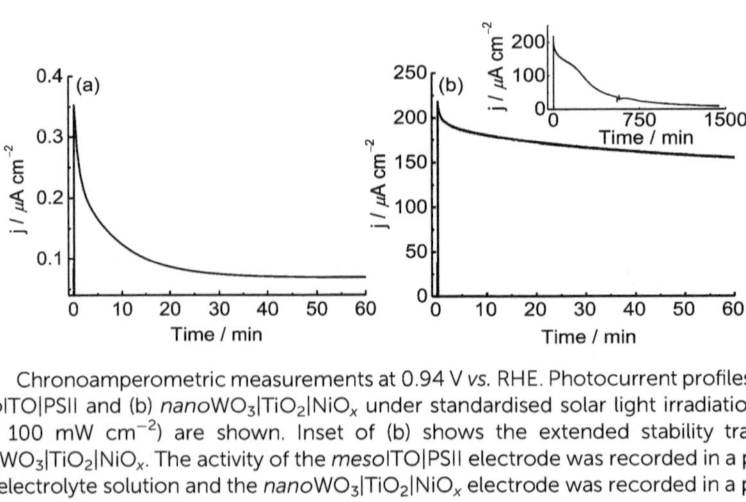

Fig. 5 Chronoamperometric measurements at 0.94 V *vs.* RHE. Photocurrent profiles of (a) *meso*ITO|PSII and (b) *nano*WO_3|TiO_2|NiO_x under standardised solar light irradiation (AM 1.5G, 100 mW cm^{-2}) are shown. Inset of (b) shows the extended stability trace of *nano*WO_3|TiO_2|NiO_x. The activity of the *meso*ITO|PSII electrode was recorded in a pH 6.5 MES electrolyte solution and the *nano*WO_3|TiO_2|NiO_x electrode was recorded in a pH 9.2 B_i buffer solution (0.1 M).

more efficient co-catalyst might accelerate the water oxidation kinetics and alleviate the surface recombination in the low bias region.

Conversely, *nano*$WO_3|TiO_2|NiO_x$ has a clear advantage in terms of photocurrent density per geometrical surface area due to its higher catalyst density compared to *meso*ITO|PSII, which is important for applications. Although *nano*$WO_3|TiO_2|NiO_x$ displays a higher photocurrent density than *meso*ITO|PSII, the photocurrent is still much lower than its theoretical photocurrent (approximately 5 mA cm^{-2}).[21] To boost further the photocurrent plateau, charge recombination within the *nano*$WO_3|TiO_2|NiO_x$ must be minimised. Combination with another semiconductor to form a p–n junction,[22] increasing intrinsic conductivity by elemental doping,[23] and suppression of electron back injection with an oxide underlayer[24] are possible strategies to enhance further the photocurrent of this synthetic electrode.

Inorganic semiconductors offer a more robust approach to provide the oxidising potentials necessary to photo-oxidise water than semi-biological systems. In this study, WO_3 was interfaced with the TiO_2 and NiO_x composite by a simple and scalable deposition of a molecular precursor to improve its photo-activity and stability. *Nano*$WO_3|TiO_2|NiO_x$ is considerably more stable than *meso*ITO|PSII at photo-oxidative conditions (Fig. 5 and Table 1).

4. Conclusions

An enzymatic photoanode consisting of PSII immobilised on *meso*ITO and a synthetic photoanode consisting of a TiO_2 and NiO_x composite on a WO_3 semiconductor operational under near pH-neutral conditions have been discussed. *Nano*$WO_3|TiO_2|NiO_x$ mimics the key functionalities of PSII, namely, light absorption, charge separation and water oxidation catalysis. As expected, the highly evolved PSII displays a higher turnover frequency than NiO_x on *nano*$WO_3|TiO_2|NiO_x$. Bulk and surface charge recombination is currently a major hurdle in achieving high photon-to-current conversion efficiencies. The resistance to charge recombination of PSII is evident from the lower applied potential required to reach its saturation photocurrent in the *meso*ITO|PSII system compared with *nano*$WO_3|TiO_2|NiO_x$. Thus the enzymatic system exhibits less surface charge recombination events at a low overpotential region compared with the synthetic system. In this respect, *meso*ITO|PSII behaves more like an "ideal" photocatalyst. The IPCE spectra confirm that PSII has the ability to promote electron transfer with low energy photons (680 nm), whereas WO_3 is only capable of utilising high-energy photons (<465 nm) to achieve water oxidation at a similar onset potential.

Nevertheless, the *nano*$WO_3|TiO_2|NiO_x$ electrode displays higher IPCEs and a more than three orders of magnitude higher photocurrent density than *meso*ITO|PSII due to the high NiO_x catalyst density on the synthetic electrode compared to the low PSII loading. Another distinct advantage of the synthetic over the enzymatic photoanode is the greater photo-stability under solar light irradiation due to the lack of repair machinery in the isolated PSII. In summary, our synthetic system will show promise for applications in water splitting in the long-term if important lessons from PSII can be learnt: longer wavelength absorption will boost the overall solar energy conversion efficiency and saturation currents at a more negative potential decrease heat losses in an operating device.

Acknowledgements

Financial support from EPSRC (EP/H00338X/2 to E. R. and nanoDTC to D. M.), the Christian Doppler Research Association (Austrian Federal Ministry of Science, Research and Economy and National Foundation for Research, Technology and Development), the OMV Group (to E. R.) and Cambridge Trust (to Y.-H. L.) are gratefully acknowledged. We thank Mr Timothy C. King and Prof. Dominic S. Wright for a kind gift of $[Ti_2(OEt)_9(NiCl)]_2$. Dr Tanai Cardona and Prof. A. William Rutherford provided us with a sample of photosystem II. We also thank Dr Hyun S. Park, Miss Claire Wombwell and Prof. Rutherford for valuable comments on the manuscript.

References

1 (*a*) Y. Tachibana, L. Vayssieres and J. R. Durrant, *Nat. Photonics*, 2012, **6**, 511; (*b*) S. Y. Reece, J. A. Hamel, K. Sung, T. D. Jarvi, A. J. Esswein, J. J. H. Pijpers and D. G. Nocera, *Science*, 2011, **334**, 645; (*c*) C.-Y. Lin, Y.-H. Lai, D. Mersch and E. Reisner, *Chem. Sci.*, 2012, **3**, 3482; (*d*) L. Liao, Q. Zhang, Z. Su, Z. Zhao, Y. Wang, Y. Li, X. Lu, D. Wei, G. Feng, Q. Yu, X. Cai, J. Zhao, Z. Ren, H. Fang, F. Robles-Hernandez, S. Baldelli and J. Bao, *Nat. Nanotechnol.*, 2014, **9**, 69; (*e*) J. R. McKone, N. S. Lewis and H. B. Gray, *Chem. Mater.*, 2014, **26**, 407; (*f*) W. D. Chemelewski, H.-C. Lee, J.-F. Lin, A. J. Bard and C. B. Mullins, *J. Am. Chem. Soc.*, 2014, **136**, 2843; (*g*) P. M. Rao, L. Cai, C. Liu, I. S. Cho, C. H. Lee, J. M. Weisse, P. Yang and X. Zheng, *Nano Lett.*, 2014, **14**, 1099; (*h*) B. M. Klepser and B. M. Bartlett, *J. Am. Chem. Soc.*, 2014, **136**, 1694; (*i*) C.-Y. Lin, D. Mersch, D. A. Jefferson and E. Reisner, *Chem. Sci.*, 2014, **5**, 4906.

2 M. Kato, J. Z. Zhang, N. Paul and E. Reisner, *Chem. Soc. Rev.*, 2014, **43**, 6485.

3 (*a*) K. N. Ferreira, T. M. Iverson, K. Maghlaoui, J. Barber and S. Iwata, *Science*, 2004, **303**, 1831; (*b*) H. Dau and I. Zaharieva, *Acc. Chem. Res.*, 2009, **42**, 1861; (*c*) Y. Umena, K. Kawakami, J.-R. Shen and N. Kamiya, *Nature*, 2011, **473**, 55; (*d*) T. Cardona, A. Sedoud, N. Cox and A. W. Rutherford, *Biochim. Biophys. Acta*, 2012, **1817**, 26.

4 (*a*) Y.-H. Lai, T. C. King, D. S. Wright and E. Reisner, *Chem.–Eur. J.*, 2013, **19**, 12943; (*b*) Y.-H. Lai, C.-Y. Lin, Y. Lv, T. C. King, A Steiner, N. M. Muresan, L. Gan, D. S. Wright and E. Reisner, *Chem. Commun.*, 2013, **49**, 4331.

5 (*a*) S. D. Tilley, M. Cornuz, K. Sivula and M. Grätzel, *Angew. Chem., Int. Ed.*, 2010, **49**, 6405; (*b*) W. J. Youngblood, S.-H. A. Lee, Y. Kobayashi, E. A. Hernandez-Pagan, P. G. Hoertz, T. A. Moore, A. L. Moore, D. Gust and T. E. Mallouk, *J. Am. Chem. Soc.*, 2009, **131**, 926.

6 (*a*) Y. Jiang, F. Li, B. Zhang, X. Li, X. Wang, F. Huang and L. Sun, *Angew. Chem., Int. Ed.*, 2013, **52**, 3398; (*b*) J. D. Megiatto Jr, A. Antoniuk-Pablant, B. D. Sherman, G. Kodis, M. Gervaldo, T. A. Moore, A. L. Moore and D. Gust, *Proc. Natl. Acad. Sci. U. S. A.*, 2012, **109**, 15578; (*c*) M. M. Najafpour, T. Ehrenberg, M. Wiechen and P. Kurz, *Angew. Chem., Int. Ed.*, 2010, **49**, 2233; (*d*) M. Wiechen, I. Zaharieva, H. Dau and P. Kurz, *Chem. Sci.*, 2012, **3**, 2330.

7 (*a*) M. Kato, T. Cardona, A. W. Rutherford and E. Reisner, *J. Am. Chem. Soc.*, 2012, **134**, 8332; (*b*) M. Kato, T. Cardona, A. W. Rutherford and E. Reisner, *J. Am. Chem. Soc.*, 2013, **135**, 10610.

8 M. Risch, K. Klingan, J. Heidkamp, D. Ehrenberg, P. Chernev, I. Zaharieva and H. Dau, *Chem. Commun.*, 2011, **47**, 11912.

9 P. G. Hoertz, Z. Chen, C. A. Kent and T. J. Meyer, *Inorg. Chem.*, 2010, **49**, 8179.

10 (*a*) A. Boussac, F. Rappaport, P. Carrier, J. M. Verbavatz, R. Gobin, D. Kirilovsky, A. W. Rutherford and M. Sugiura, *J. Biol. Chem.*, 2004, **279**, 22809; (*b*) M. Sugiura and Y. Inoue, *Plant Cell Physiol.*, 1999, **40**, 1219.

11 S. Eslava, M. McPartlin, R. I. Thomson, J. M. Rawson and D. S. Wright, *Inorg. Chem.*, 2010, **49**, 11532.

12 (*a*) A. J. Bard and L. R. Faulkner, *Electrochemical Methods: Fundamentals and Applications*, John Wiley & Sons, Inc., NewYork, 2nd edn, 2001; (*b*) A. Kay, I. Cesar and M. Grätzel, *J. Am. Chem. Soc.*, 2006, **128**, 15714.

13 T. W. Kim and K.-S. Choi, *Science*, 2014, **343**, 990.

14 T. Shibamoto, Y. Kato, M. Sugiura and T. Watanabe, *Biochemistry*, 2009, **48**, 10682.

15 R. M. Navarro, M. Consuelo Alvarez-Galvan, J. A. V. de la Mano, S. M. Al-Zahrani and J. L. G. Fierro, *Energy Environ. Sci.*, 2010, **3**, 1865.

16 L. Zhang, E. Reisner and J. J. Baumberg, *Energy Environ. Sci.*, 2014, **7**, 1402.

17 G. Raszewski, W. Saenger and T. Renger, *Biophys. J.*, 2005, **88**, 986.

18 O. Yehezkeli, R. Tel-Vered, J. Wasserman, A. Trifonov, D. Michaeli, R. Nechushtai and I. Willner, *Nat. Commun.*, 2012, **3**, 742.

19 A. W. Rutherford, A. Osyczka and F. Rappaport, *FEBS Lett.*, 2012, **586**, 603.

20 J. A. Seabold and K.-S. Choi, *Chem. Mater.*, 2011, **23**, 1105.

21 Z. Chen, T. F. Jaramillo, T. G. Deutsch, A. Kleiman-Shwarsctein, A. J. Forman, N. Gaillard, R. Garland, K. Takanabe, C. Heske, M. Sunkara, E. W. McFarland, K. Domen, E. L. Miller, J. A. Turner and H. N. Dinh, *J. Mater. Res.*, 2010, **25**, 3.

22 A. Paracchino, V. Laporte, K. Sivula, M. Grätzel and E. Thimsen, *Nat. Mater.*, 2011, **10**, 456.

23 A. J. E. Rettie, H. C. Lee, L. G. Marshall, J.-F. Lin, C. Capan, J. Lindemuth, J. S. McCloy, J. Zhou, A. J. Bard and C. B. Mullins, *J. Am. Chem. Soc.*, 2013, **135**, 11389.

24 T. Hisatomi, H. Dotan, M. Stefik, K. Sivula, A. Rothschild, M. Grätzel and N. Mathews, *Adv. Mater.*, 2012, **24**, 2699.

Faraday Discussions

ROYAL SOCIETY OF CHEMISTRY

PAPER

Photoelectrochemical properties of $SrNbO_2N$ photoanodes for water oxidation fabricated by the particle transfer method

Haruki Urabe, Takashi Hisatomi, Tsutomu Minegishi, Jun Kubota and Kazunari Domen*

Received 3rd June 2014, Accepted 21st July 2014
DOI: 10.1039/c4fd00122b

$SrNbO_2N$ photoanodes, which have a band gap energy of 1.8 eV, were tested for photoelectrochemical water oxidation in water splitting. The photoanodes were fabricated either by a particle transfer method or an electrophoretic deposition method. The effects of the precursors, fabrication method, and CoO_x catalyst loading were studied in order to identify the shortcomings of the photoanodes and improve their photoelectrochemical properties for water oxidation. $SrNbO_2N$ photoanodes fabricated by particle transfer generated a photocurrent that was one order of magnitude higher than that of photoanodes prepared *via* electrophoretic deposition. The stoichiometric oxide precursor ($Sr_2Nb_2O_7$) was found to be preferable to the Sr-rich oxide precursor ($Sr_5Nb_4O_{15}$). CoO_x increased the photoanodic current on $SrNbO_2N$ photoelectrodes. Nevertheless, the incident photon-to-current efficiency was still limited to 10% at most. Potential problems with $SrNbO_2N$ photoanodes were discussed.

1 Introduction

Hydrogen has attracted much interest as a clean energy medium because it generates only water as a by-product in the oxidation reaction. Production of hydrogen from water using renewable solar energy would enable a clean and renewable hydrogen cycle and contribute to the resolution of concerns over global warming and the depletion of fossil fuels, because the amount of solar energy reaching the Earth's surface far exceeds our global energy consumption. Photoelectrochemical (PEC) water splitting using semiconducting electrodes has been studied as one of the most promising means of solar hydrogen production.[1–3] A solar-to-hydrogen conversion efficiency (STH) of 10% is regarded as a benchmark, and a higher STH is demanded for cost-competitive solar hydrogen production on a large scale.[3,4] It is clearly necessary to harvest sunlight at wavelengths up to

Department of Chemical System Engineering, School of Engineering, The University of Tokyo, 7-3-1 Hongo, Bunkyo-ku, Tokyo 113-8656, Japan. E-mail: domen@chemsys.t.u-tokyo.ac.jp; Fax: +81 3 5841 8838; Tel: +81 3 5841 1148

600 nm or much longer, and to harvest it more effectively, in order to attain the target STH of 10% at reasonable quantum efficiencies.[4] This necessitates the development of active narrow-band-gap semiconductors.

Many oxide photoanodes are stable during PEC water oxidation. However, the top edge of the valence band of oxide semiconductors is mainly composed of O 2p orbitals and is located at around +3.0 V *vs.* NHE at pH 0.[5] This is much more positive than the reversible oxygen evolution potential (+1.23 V *vs.* NHE at pH 0), causing a large overpotential loss on oxide photoanodes in PEC water oxidation. In addition, the potential of the bottom of the conduction band, or more strictly, the Fermi level, for such oxides tends to be too positive relative to the reversible hydrogen evolution potential (0 V *vs.* NHE at pH 0) when the band gap is narrow enough (<2 eV) to harvest sunlight. Consequently, an applied voltage from an external power source or photovoltaic is necessary. N 2p orbitals are located at more negative potentials than O 2p orbitals and form the top of the valence band of (oxy)nitride semiconductors. The band gap of certain (oxy)nitrides is small enough to absorb visible light while straddling the reversible hydrogen and oxygen evolution potentials.[6] In fact, some particulate (oxy)nitride semiconductors can generate hydrogen and oxygen from water at the stoichiometric ratio.[7–9] This suggests that PEC water splitting could be achieved using (oxy) nitride photoelectrodes with minimal or no external power input.

$SrNbO_2N$ is a perovskite-type n-type (oxy)nitride semiconductor with a band gap energy of 1.8 eV that is applicable to sacrificial photocatalytic oxygen evolution.[10] It becomes applicable to PEC water oxidation up to 700 nm upon being immobilized onto a conductive substrate such as fluorine-doped tin oxide (FTO) coated glass by electrophoretic deposition.[11] However, the incident photon-to-current efficiency (IPEC) for PEC water oxidation in a neutral aqueous Na_2SO_4 solution was less than 1% at 1.2 V *vs.* RHE under illumination of 400 nm photons. This is much lower than other photoanodes active under visible light, such as TaON,[12,13] Ta_3N_5,[13–15] $BaTaO_2N$,[16] $BiVO_4$,[17] and Fe_2O_3.[18] The low IPCE of $SrNbO_2N$ photoanodes can in part be attributed to (i) the insufficient transport of photoexcited carriers between $SrNbO_2N$ particles and the conductive substrate and (ii) the slow kinetics of water oxidation on the (oxy)nitride surface.[11] A necking treatment using Nb-containing compounds under a heated ammonia flow has been carried out to improve the electronic and physical contact.[11] However, the treatment temperature was limited to 753 K because of the instability of FTO, and thus the effect was rather small. The problem of slow water oxidation could be rectified by the use of oxygen evolution catalysts that could enhance the separation and collection of holes generated in $SrNbO_2N$. Adsorption of colloidal iridium oxide onto $SrNbO_2N$ photoanodes actually increased the photocurrent to some extent, which suggested that water oxidation catalysts that were more efficient in charge collection and more active in water oxidation could further improve the performance of the photoanode.[11]

In the present study, the particle transfer method was employed to fabricate $SrNbO_2N$ photoanodes. Using this method, $SrNbO_2N$ particles can be rigidly immobilized on a metal film deposited by sputtering.[19] In addition, $SrNbO_2N$ particles that are physisorbed onto a photoelectrode can be removed by ultrasonication. Thus, it is expected that the particle transfer method would enable better electrical contact between $SrNbO_2N$ particles and a conductive layer, as compared to electrophoretic deposition onto FTO-coated glass. Using $SrNbO_2N$

photoanodes fabricated by the particle transfer method, the effects of precursors and modification with the catalyst CoO_x, which has been demonstrated to function well as an oxygen evolution catalyst on other (oxy)nitride photocatalysts and photoelectrodes,[12,16,20,21] were studied.

2 Experimental

2.1 Preparation of $SrNbO_2N$ photocatalyst powder

$Sr_2Nb_2O_7$ powder was first prepared as an oxide precursor using a polymerized complex method.[10] In a typical synthesis, $NbCl_5$ (High Purity Chemicals, 99.9%) and $SrCO_3$ (Kanto Chemical, 99.9%) were dissolved at the stoichiometric ratio (Sr/Nb = 1.0) in methanol. Citric acid (Wako Pure Chemicals, 98.0%) and ethylene glycol (Kanto Chemical, 99.5%) were added to the solution, which was kept at 473 K overnight to promote polymerization. After the mixture became a yellow resin, it was pyrolyzed at 623 K and calcined in air at 973 K for 2 h. The $Sr_2Nb_2O_7$ obtained was nitrided under an NH_3 flow (250 mL min^{-1}) at 1173 K for 15 h. $Sr_5Nb_4O_{15}$ was also prepared as a precursor for $SrNbO_2N$ by changing the Sr/Nb ratio to 1.25 in the oxide synthesis. In this case, the samples were rinsed with distilled water after nitridation to remove excess Sr species.

The $SrNbO_2N$ samples were modified with CoO_x (up to 10 wt% as Co) by impregnation with an aqueous $Co(NO_3)_2$ solution followed by a nitridation treatment at 773 K for 0.5 h under an NH_3 flow (100 mL min^{-1}). $SrNbO_2N$ loaded with CoO_x (*y* wt% as Co) is denoted by CoO_x(*y* wt%)/$SrNbO_2N$.

2.2 Fabrication of $SrNbO_2N$ photoelectrodes by particle transfer

Photoelectrodes of CoO_x/$SrNbO_2N$ particles were fabricated by particle transfer.[19] CoO_x/$SrNbO_2N$ powder (30 mg) suspended in 2-propanol (1 mL) was dropped onto a glass plate (30 × 30 mm) to pile up the photocatalyst powder. After the solvent was evaporated, thin layers of Nb and Ti were sputtered on the surface as contact and conductor layers, respectively, by radio-frequency (RF) magnetron sputtering at an Ar pressure of 1.0×10^{-1} Pa. The contact and conductor layers were deposited at 473 K at an RF power of 100 W for 5 min and 200 W for 3 h, respectively. In some cases, Ti was also used as a contact layer. The sputtered film was transferred onto another glass plate using an adhesive and ultrasonicated sufficiently to remove photocatalyst particles loosely attached to the metal layers. Finally, an electrical contact was established between the metal layers and a conductive wire using In solder. CoO_x/$SrNbO_2N$ photoelectrodes prepared by the particle transfer method are denoted by CoO_x/$SrNbO_2N$/Nb/Ti-PT.

2.3 Fabrication of $SrNbO_2N$ photoelectrodes by electrophoretic deposition

Photoelectrodes of CoO_x/$SrNbO_2N$ particles were prepared *via* electrophoretic deposition onto FTO, necking treatment with $NbCl_5$ methanol solution, followed by a nitridation treatment according to our previous work.[11] CoO_x(2.5 wt%)/$SrNbO_2N$ powder (36 mg) was suspended in acetone (45 mL) containing iodine (9 mg). Two parallel FTO electrodes were immersed in this solution with a 15 mm spacing between them, and a 100 V bias was applied between the two electrodes for 20 s. The $SrNbO_2N$-coated area was typically 1 × 3 cm. The $NbCl_5$ methanol solution (10 mM, 25 μL) was dropped onto the as-prepared electrodes and then dried. After this procedure was repeated six times, the electrode was heated under an NH_3 flow (100 mL min^{-1}) at 753 K for 0.5 h. $CoO_x/SrNbO_2N$ photoanodes prepared by the electrophoretic deposition method are denoted by CoO_x/$SrNbO_2N$/FTO-EPD.

2.4 Characterization

The crystal phases of the samples were analysed by powder X-ray diffraction (XRD, RINT-Ultima III, Rigaku). Diffuse reflectance spectroscopy (DRS) was performed using a UV-Vis-NIR spectrometer (V-670, Jasco) and the spectra were converted from reflectance to absorbance by the Kubelka–Munk method. The morphology of the samples was characterized by scanning electron microscopy (SEM, S-4700, Hitachi). The BET surface area was determined by nitrogen adsorption at liquid-nitrogen temperature in a constant-volume adsorption apparatus (BELSORP-mini, BEL JAPAN).

2.5 Photoelectrochemical measurements

PEC properties of the $SrNbO_2N$ photoanodes were measured in the three-electrode configuration under illumination from a 300 W Xe lamp equipped with a cut-off filter ($\lambda > 420$ nm) or from a solar simulator (SAN-EI ELECTRIC CO., LTD., XES-40S2-CE). A Pt wire and a Ag/AgCl electrode were used as counter and reference electrodes, respectively. The electrolyte solutions used were Ar-purged 0.2 M aqueous phosphate solutions whose pH values were adjusted to 13 using an aqueous NaOH solution (5 M). The potential of the $SrNbO_2N$ photoanodes was swept at 20 mV s^{-1} from negative to positive potential unless otherwise noted. In some cases, hydrogen peroxide (H_2O_2) was added to the electrolyte solution (*ca.* 0.1 M) as an electron donor in order to estimate the maximum photocurrent that could be obtained when the injection of holes from the photoanode surface to the electrolyte solution was ideally facilitated.[22]

3 Results and discussion

First, the effect of the precursors on the physical properties and PEC activity of $SrNbO_2N$ is discussed. Fig. 1B shows the diffuse reflectance spectra of $SrNbO_2N$ samples nitrided from $Sr_2Nb_2O_7$ and $Sr_5Nb_4O_{15}$. $SrNbO_2N$ prepared from $Sr_5Nb_4O_{15}$ exhibited weaker background absorption than that prepared from $Sr_2Nb_2O_7$ at wavelengths longer than the absorption edge (*ca.* 670 nm), suggesting that the reduction of Nb^{5+} species was suppressed owing to excess Sr species in the precursor. A similar effect was previously observed during the synthesis of $BaNbO_2N$.[21] However, the sample prepared from $Sr_5Nb_4O_{15}$ had weaker and broader X-ray diffraction (XRD) peaks (see Fig. 1A), which was indicative of the lower crystallinity of that sample. SEM observations confirmed that $SrNbO_2N$ prepared from $Sr_5Nb_4O_{15}$ had smaller particle sizes than that prepared from $Sr_2Nb_2O_7$ (see Fig. 1C), even though the nitridation conditions used were identical. In accordance with the particle morphologies, $SrNbO_2N$ prepared from $Sr_5Nb_4O_{15}$ had a larger BET surface area (27 m^2 g^{-1}) than that from $Sr_2Nb_2O_7$ (11 m^2 g^{-1}). During the phase transformation from $Sr_5Nb_4O_{15}$ to $SrNbO_2N$, an excess of Sr species was generated among $SrNbO_2N$ domains, which seemingly prevented their growth. Such Sr species dissolved in water after the nitridation. As

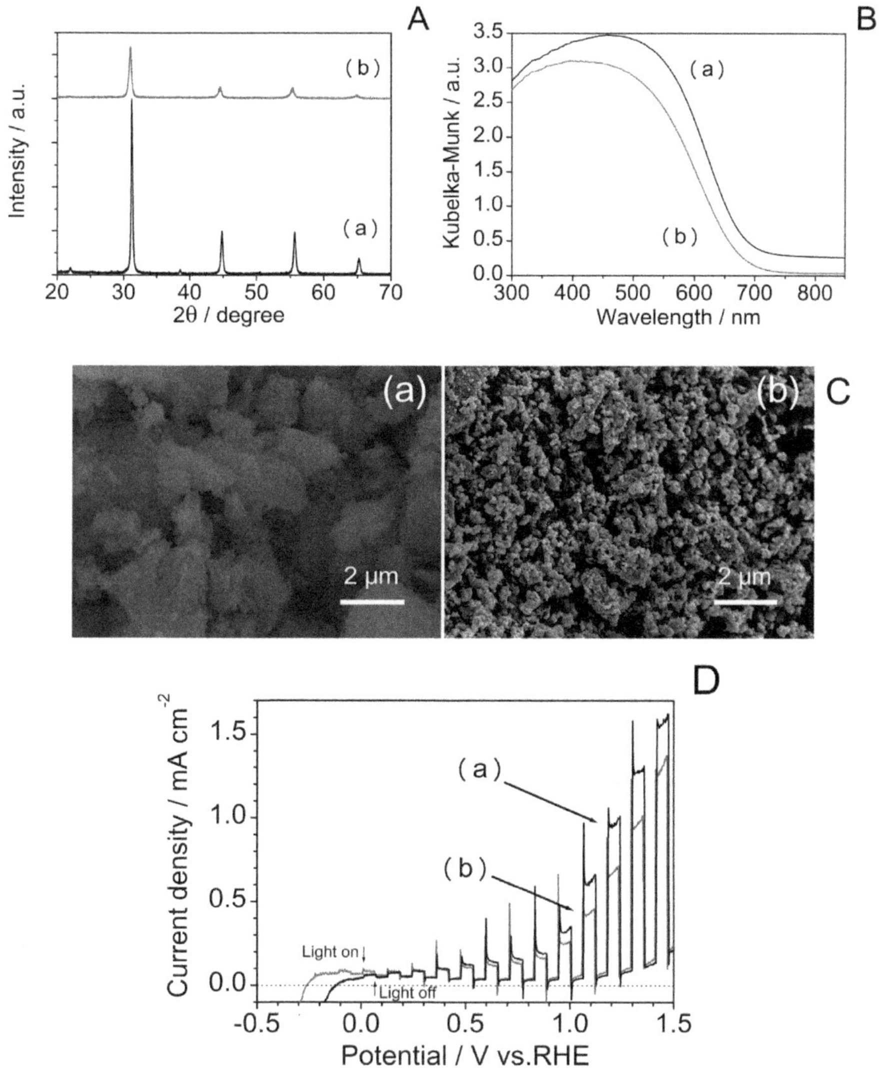

Fig. 1 (A) XRD patterns, (B) diffuse reflectance spectra and (C) SEM images of $SrNbO_2N$ powders, and (D) i–E curves of CoO_x(2 wt%)/$SrNbO_2N$/Ti-PT photoanodes prepared from (a) $Sr_2Nb_2O_7$ and (b) $Sr_5Nb_4O_{15}$ in 0.2 M aqueous phosphate solutions (pH = 13) under chopped Xe lamp illumination (λ > 420 nm).

a result, fragmented $SrNbO_2N$ particles were obtained from $Sr_5Nb_4O_{15}$. Fig. 1D shows the current–potential (i–E) curves of the $SrNbO_2N$ electrodes. The $SrNbO_2N$ electrode prepared from $Sr_5Nb_4O_{15}$ showed a lower photocurrent than that prepared from $Sr_2Nb_2O_7$. Lower crystallinity generally results in a lower photocurrent, owing to the larger density of trap states and lower mobility of carriers. Besides, fragmented particles would not be suitable for photoelectrodes prepared by the particle transfer method because it is difficult to decrease the number of grain boundaries for efficient carrier transport and, at the same time, maintain a sufficient optical thickness for light absorption. Accordingly, in the following discussion, $Sr_2Nb_2O_7$ was used as a precursor for $SrNbO_2N$.

The influence of the fabrication method of the $SrNbO_2N$ photoanodes was investigated. Fig. 2 shows *i–E* curves of the CoO_x(2.5 wt%)/$SrNbO_2N$ photoanodes fabricated by the particle transfer method and the electrophoretic deposition method. CoO_x/$SrNbO_2N$/Ti-PT and CoO_x/$SrNbO_2N$/FTO-EPD generated photoanodic currents of, respectively, 350 μA cm^{-2} and 41 μA cm^{-2} at a potential of +1.23 V *vs.* RHE under simulated sunlight illumination. Cross-sectional SEM images of these photoanodes are presented in Fig. 3. $SrNbO_2N$ particle layers several micrometres and a few hundred nanometers in thickness were observed for CoO_x/$SrNbO_2N$/FTO-EPD and CoO_x/$SrNbO_2N$/Ti-PT, respectively. Apparently, the number of $SrNbO_2N$ particles in the particle layer was much larger for CoO_x/$SrNbO_2N$/Ti-EPD. However, electrons generated in the $SrNbO_2N$ layer needed to travel over a number of grain boundaries to reach the back contact. In addition, the $SrNbO_2N$ particle layers were not firmly attached to the FTO substrate. The photocurrent was small, likely because of the large resistance at each grain boundary and at the interface of the $SrNbO_2N$ particles and FTO substrate. In contrast, CoO_x/$SrNbO_2N$/Ti-PT had only a few grain boundaries. $SrNbO_2N$ particles that persisted through ultrasonication were rigidly embedded onto the sputtered metal layer. These factors would lower the series resistance and contribute to more efficient migration of photoexcited electrons, leading to a higher photocurrent. Note that the kind of back contact metal had a significant impact on the photocurrent of $SrNbO_2N$ photoanodes fabricated by the particle transfer method, similar to our first report on the particle transfer of a $LaTiO_2N$ photocatalyst.[19] As seen in Fig. 2c, the photocurrent increased to 570 μA cm^{-2} when Nb was used as a contact layer instead of Ti. The metal species used for the contact layers might alter the height of the energy barrier for electron transfer from the $SrNbO_2N$ particles to the metal back contact at the interface because of the different work functions and affinities to anions, as well as the possibility of interfacial doping during sputtering, although the details are still under investigation.

Fig. 4A shows the dependence of the photocurrent of $SrNbO_2N$/Nb/Ti (PT) at +1.23 V *vs.* RHE on the amount of CoO_x loading. Selected *i–E* curves are shown in

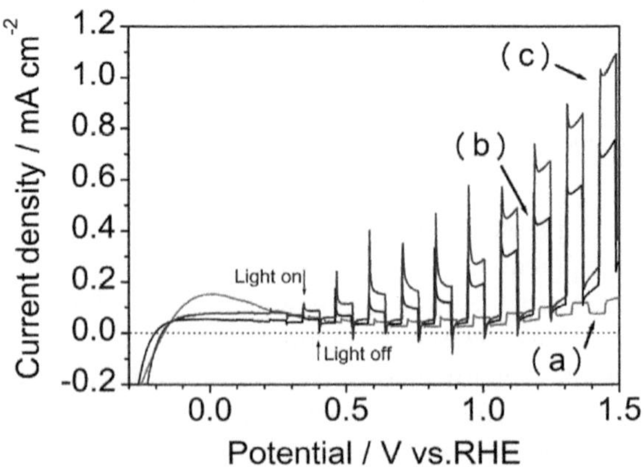

Fig. 2 *i–E* curves of (a) CoO_x(2.5 wt%)/$SrNbO_2N$/FTO-EPD, (b) CoO_x(2.5 wt%)/$SrNbO_2N$/Ti-PT and (c) CoO_x(2.5 wt%)/$SrNbO_2N$/Nb/Ti-PT in 0.2 M aqueous phosphate solutions (pH = 13) under chopped simulated sunlight illumination.

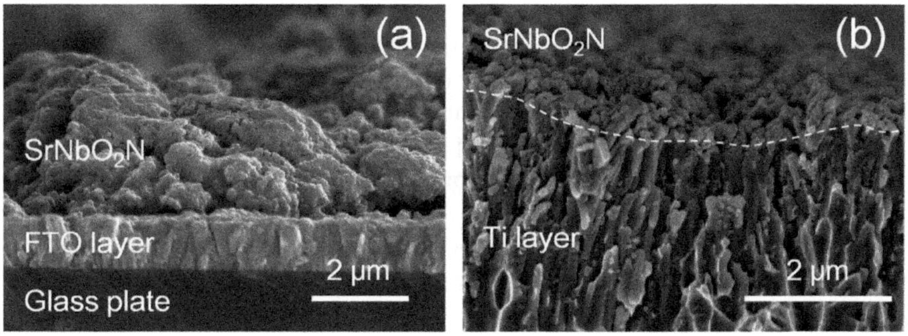

Fig. 3 Cross-sectional SEM images of $SrNbO_2N$ photoanodes prepared *via* (a) electrophoretic deposition and (b) particle transfer.

Fig. 4B. The photocurrent improved with increasing loading amount up to 2–3 wt%, particularly at potentials more positive than +0.9 V *vs.* RHE. CoO_x worked as a catalyst to enhance charge separation and facilitate water oxidation by $SrNbO_2N$, similar to the cases of TaON,[12] $BaTaO_2N$,[16] and $BaNbO_2N$.[21] The initial IPCE of CoO_x(2.5 wt%)/$SrNbO_2N$/Nb/Ti-PT at +1.2 V *vs.* RHE was estimated to be 10% under illumination of 400 nm photons, which was an order of magnitude higher than that reported in our previous work.[11] The increase of the CoO_x loading amounts to 2.5 wt% did not shift the potential of the steep onset of the

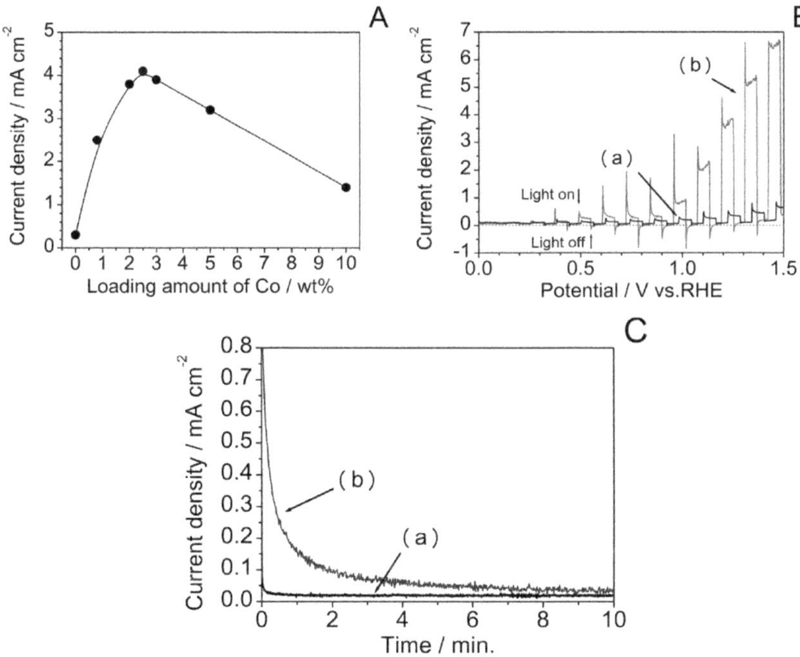

Fig. 4 (A) Photocurrent of CoO_x/$SrNbO_2N$/Nb/Ti-PT modified with varying amounts of CoO_x at +1.23 V *vs.* RHE in 0.2 M aqueous phosphate solutions (pH = 13). (B) *i*–*E* curves and (C) *i*–*t* curves of (a) $SrNbO_2N$/Nb/Ti-PT and (b) CoO_x(2.5 wt%)/$SrNbO_2N$/Nb/Ti-PT. The measurements were performed under Xe lamp illumination (λ > 420 nm).

photoanodic current at approximately +0.9 V *vs.* RHE, but rather simply increased the current density, likely owing to the increased density of CoO_x particles. Excessive loading of the CoO_x catalyst lowered the photocurrent, presumably because of shading of the incident light. One of the characteristic features of the $CoO_x/SrNbO_2N$ photoanodes was the photoanodic current at relatively negative potentials (0.3–0.9 V *vs.* RHE). This photoanodic current was not observed during potential scans from positive to negative. Analogous results were observed when Co-containing catalysts were used in earlier studies.[23] It was reported that the photoanodic current peak of $CoPi/Fe_2O_3$ photoanodes at +0.8–1.1 V *vs.* RHE became larger as the scan rate of the potential from negative to positive increased. The photoanodic current was in part attributed to oxidation of the redox-active cobalt catalyst. In the present study, the photoanodic current at potentials more negative than the steep onset potential of the photocurrent were mostly attributed to CoO_x oxidation.

Stabilization of the photoanodes could be another positive effect of loading oxidation catalysts, as in the cases of TaON,[12] $BaTaO_2N$,[16] and $BiVO_4$.[24] Time courses of the photocurrent (*i*–*t* curves) of $SrNbO_2N$/Nb/Ti-PT and CoO_x(2.5 wt%)/$SrNbO_2N$/Nb/Ti-PT at +1.0 V *vs.* RHE are presented in Fig. 4C. The photoanodic current decayed rapidly, regardless of the catalyst loading. However, the decay was relatively slower and a perceivable photoanodic current (50 μA cm^{-2}) was observed after 5 min of light illumination with the catalyst. Although the poor stability of $CoO_x/SrNbO_2N$/Nb/Ti-PT is a critical problem, the present results imply that the stability problem could be moderated by the use of appropriate catalysts. The methods of modifying $SrNbO_2N$ particles necessitate further refinement in future studies.

i–*E* curves of CoO_x(2.5 wt%)/$SrNbO_2N$/Nb/Ti-PT were recorded in the presence and absence of hydrogen peroxide, as shown in Fig. 5, to discuss factors that limit PEC performance. Hydrogen peroxide is believed to capture photoexcited holes that reach the photoanode surface successfully without significant recombination.[22] The addition of hydrogen peroxide caused generation of a cathodic dark current at potentials more negative than +0.5 V *vs.* RHE and an anodic dark current at potentials more positive than +1.1 V *vs.* RHE. $SrNbO_2N$ is regarded as

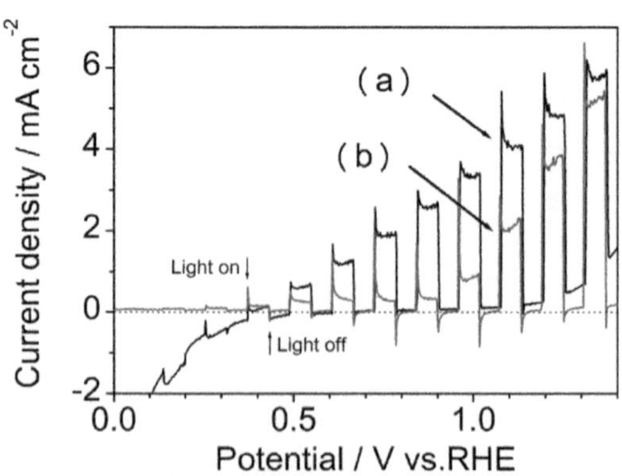

Fig. 5 Current–potential curves of CoO_x(2.5 wt%)/$SrNbO_2N$/Nb/Ti-PT in 0.2 M aqueous phosphate solutions (pH = 13) (a) in the presence and (b) absence of H_2O_2 (*ca.* 0.1 M). The measurements were performed under chopped Xe lamp illumination (λ > 420 nm).

an n-type semiconductor with a conduction band edge more negative than 0 V *vs.* RHE.[11] Therefore, these dark currents were attributable to electrochemical reactions on the metal layer exposed to the electrolyte solution below the $SrNbO_2N$ particle layers and would not reflect the semiconducting properties of $SrNbO_2N$, such as the flat band potential. It was found that the photocurrent at potentials more negative than +1.23 V *vs.* RHE was improved significantly by adding H_2O_2. Therefore, it is considered that there still remains room for enhancement of the surface redox reactions in the negative potential region despite the CoO_x loading. This agrees with the result that redox-active Co species could be transiently oxidized at potentials more negative than +0.9 V *vs.* RHE, whereas water could not. To achieve a high solar energy conversion efficiency, it is crucial to improve the reaction kinetics at the negative potentials where the voltage gain is sufficiently large.[25] On the other hand, the photocurrent at around +1.23 V *vs.* RHE was comparable regardless of the addition of H_2O_2, implying that the CoO_x had enhanced the surface redox reactions sufficiently. However, it should be noted that the IPCE near this potential was 10%, as described previously. This suggests that the majority of the photoexcited carriers recombined in the bulk of the $SrNbO_2N$ particles or otherwise photons were not absorbed sufficiently by the $SrNbO_2N$ particle layers. Thus, $SrNbO_2N$ particles need to be designed with low defect densities and large sizes to allow for sufficient light absorption and carrier migration.

The use of flux in the synthesis of oxide precursors and in the subsequent nitridation can be proposed as a method to modify the structural properties of perovskite-type (oxy)nitrides. In our earlier studies, nitridation of $La_2Ti_2O_7$ prepared by the flux method produced single-crystalline porous $LaTiO_2N$ particles.[20] This $LaTiO_2N$, which was then modified with the CoO_x catalyst, exhibited a high quantum efficiency of 27% at 440 nm for a sacrificial oxygen evolution reaction under illumination. Flux-assisted nitridation of $La_2Ti_2O_7$ also yielded $LaTiO_2N$ with different structural properties, depending on the morphology of the precursor oxide and the ratio of the oxide and flux.[26] Doping is another possible approach. For an n-type semiconductor that is to be used as a photoanode, n-type doping can generally improve the conductivity of electrons while shortening the diffusion length of photoexcited holes and narrowing the width of depletion layers. Thus, n-type doping is expected to be effective when the electron migration from $SrNbO_2N$ particle layers to the back contact metal layer limits the PEC performance significantly. The opposite effects can be expected if the material is doped with p-type dopants. However, doping generally lowers the crystallinity of materials, which would lower electron and hole migration and possibly outweigh the positive effects of doping. An in-depth analysis of the semiconducting and structural properties of doped materials is needed.

Conclusion

$SrNbO_2N$ photoanodes fabricated by the particle transfer method exhibited a photocurrent one order of magnitude higher than those prepared *via* electrophoretic deposition and subsequent necking treatment, because of the better electronic contact between the conductive substrates and the fewer grain boundaries in $SrNbO_2N$ particle layers. $Sr_2Nb_2O_7$ was preferable to $Sr_5Nb_4O_{15}$ as a starting material for $SrNbO_2N$ using particle transfer, because the latter produced fragmented particles that could not meet the requirements of sufficient light absorption and fewer grain boundaries at the same time while being processed by the particle transfer method. CoO_x loading increased the photoanodic current on $SrNbO_2N$ photoelectrodes significantly in the positive potential region near +1.23 V *vs.* RHE. However, even after catalyst modification, a significant loss in the surface redox reaction was observed at moderately positive potentials as a result of the use of hydrogen peroxide as a sacrificial reagent. Another problem was the poor stability of the $SrNbO_2N$ photoanodes despite the catalyst modification. The development of new catalytic materials and loading methods is needed to improve the photocurrent in the moderate potential region and the stability. In addition, the IPCE of the photoanodes was limited to 10% at 400 nm even at a potential as positive as +1.23 V *vs.* RHE, where surface recombination was mostly negligible. This means that the majority of the photoexcited carriers recombined in the bulk of the $SrNbO_2N$ particles or that photons were not absorbed sufficiently by the $SrNbO_2N$ particle layers. It is a major challenge in the case of $SrNbO_2N$ to control the morphology and semiconducting properties, to allow sufficient light absorption and charge separation.

Acknowledgements

This study was supported by Grants-in-Aid for Specially Promoted Research (no. 23000009) and for Young Scientists (B) (no. 25810112), and the international exchange program of the A3 Foresight Program of the Japan Society for the Promotion of Science (JSPS) and Companhia Brasileira de Metalurgia e Mineração (CBMM). We thank them all for their support.

References

1 N. S. Lewis, *Science*, 2007, **315**, 798.
2 N. S. Lewis and D. G. Nocera, *Proc. Natl. Acad. Sci. U. S. A.*, 2006, **103**, 15729.
3 B. A. Pinaud, J. D. Benck, L. C. Seitz, A. J. Forman, Z. Chen, T. G. Deutsch, B. D. James, K. N. Baum, G. N. Baum, S. Ardo, H. Wang, E. Miller and T. F. Jaramillo, *Energy Environ. Sci.*, 2013, **6**, 1983.
4 K. Maeda and K. Domen, *J. Phys. Chem. Lett.*, 2010, **1**, 2655.
5 D. E. Scaife, *Sol. Energy*, 1980, **25**, 41.
6 K. Maeda and K. Domen, *J. Phys. Chem. C*, 2007, **111**, 7851.
7 K. Maeda, K. Teramura, D. Lu, T. Takata, N. Saito, Y. Inoue and K. Domen, *Nature*, 2006, **440**, 295.
8 X. Wang, K. Maeda, Y. Lee and K. Domen, *Chem. Phys. Lett.*, 2008, **457**, 134.
9 K. Maeda, D. Lu and K. Domen, *Chem.–Eur. J.*, 2013, **19**, 4986.
10 B. Siritanaratkul, K. Maeda, T. Hisatomi and K. Domen, *ChemSusChem*, 2011, **4**, 74.
11 K. Maeda, M. Higashi, B. Siritanaratkul, R. Abe and K. Domen, *J. Am. Chem. Soc.*, 2011, **133**, 12334.
12 M. Higashi, K. Domen and R. Abe, *J. Am. Chem. Soc.*, 2012, **134**, 6968.
13 M. Higashi, K. Domen and R. Abe, *Energy Environ. Sci.*, 2011, **4**, 4138.
14 Y. Li, L. Zhang, A. Torres-Pardo, J. M. González-Calbet, Y. Ma, P. Oleynikov, O. Terasaki, S. Asahina, M. Shima, D. Cha, L. Zhao, K. Takanabe, J. Kubota and K. Domen, *Nat. Commun.*, 2013, **4**, 2566.

15 M. Li, W. Luo, D. Cao, X. Zhao, Z. Li, T. Yu and Z. Zou, *Angew. Chem., Int. Ed.*, 2013, **52**, 1.
16 M. Higashi, K. Domen and R. Abe, *J. Am. Chem. Soc.*, 2013, **135**, 10238.
17 F. F. Abdi, L. Han, A. H. M. Smets, M. Zeman, B. Dam and R. van de Krol, *Nat. Commun.*, 2013, **4**, 2195.
18 S. D. Tilley, M. Cornuz, K. Sivula and M. l. Grätzel, *Angew. Chem., Int. Ed.*, 2010, **49**, 1.
19 T. Minegishi, N. Nishimura, J. Kubota and K. Domen, *Chem. Sci.*, 2013, **4**, 1120.
20 F. Zhang, A. Yamakata, K. Maeda, Y. Moriya, T. Takata, J. Kubota, K. Teshima, S. Oishi and K. Domen, *J. Am. Chem. Soc.*, 2012, **134**, 8348.
21 T. Hisatomi, C. Katayama, Y. Moriya, T. Minegishi, M. Katayama, H. Nishiyama, T. Yamada and K. Domen, *Energy Environ. Sci.*, 2013, **6**, 3595.
22 H. Dotan, K. Sivula, M. Grätzel, A. Rothschild and S. C. Warren, *Energy Environ. Sci.*, 2011, **4**, 958.
23 D. K. Zhong and D. R. Gamelin, *J. Am. Chem. Soc.*, 2010, **132**, 4202.
24 J. A. Seabold and K.-S. Choi, *J. Am. Chem. Soc.*, 2012, **134**, 2186.
25 T. Hisatomi, J. Kubota and K. Domen, *Chem. Soc. Rev.*, 2014, **43**, 7520.
26 A. E. Maegli, S. Pokrant, T. Hisatomi, M. Trottmann, K. Domen and A. Weidenkaff, *J. Phys. Chem. C*, 2014, **118**, 16344.

Faraday Discussions

PAPER

Artificial photosynthesis: photosensitizer/catalyst supramolecular assemblies for light driven water oxidation

Yan Gao,*[a] Lina Duan,[a] Ze Yu,[a] Xin Ding[a] and Licheng Sun*[ab]

Received 8th June 2014, Accepted 15th July 2014
DOI: 10.1039/c4fd00127c

Three new supramolecular assemblies **SA1–SA3** with different linkages between the photosensitizer and catalyst have been synthesized for light driven water oxidation. With flexible $-CH_2-CH_2-$ chains as the linkage, the assembly **SA3** displays the best performance for photocatalytic water oxidation compared with the other two assemblies, a turnover number of 34 has been obtained based on the molecular assembly **SA3** in a homogeneous system. This type of assembly connected with flexible linkages represents suitable candidates to construct photoanodes for light driven water splitting in dye sensitized photoelectrochemical devices.

1. Introduction

Utilization of solar energy to split water into molecular oxygen and hydrogen[1–3] is an attractive way to produce clean and sustainable fuel, and hopefully to reduce the emission of greenhouse gases and meet energy demands in the future. A great number of heterogeneous and homogeneous systems have been developed in past years aiming at this. In heterogeneous systems,[4–6] some significant progress has been successfully achieved on light driven water splitting using functional devices consisting of inorganic materials or molecular components. In homogeneous systems,[7–11] attempts have also been accomplished in three-components systems,[8–10,17,18] composed of a molecular catalyst,[11–16] a photosensitizer (PS) and a sacrificial electron acceptor, but only a few of these systems realize light-driven water oxidation.[16–18]

During the past few years, a series of highly efficient water oxidation catalysts with low over-potentials have been developed in our group.[11–16] Based on these catalysts, several three-component homogeneous systems have been built up, and have successfully realized light driven water oxidation. However, a considerable

[a]*State Key Laboratory of Fine Chemicals, Institute of Artificial Photosynthesis, DUT-KTH Joint Education and Research Center on Molecular Devices, Dalian University of Technology (DUT), Dalian 116024, China. E-mail: dr.gaoyan@dlut.edu.cn*

[b]*Department of Chemistry, School of Chemical Science and Engineering, KTH Royal Institute of Technology, 100 44 Stockholm, Sweden. E-mail: lichengs@kth.se*

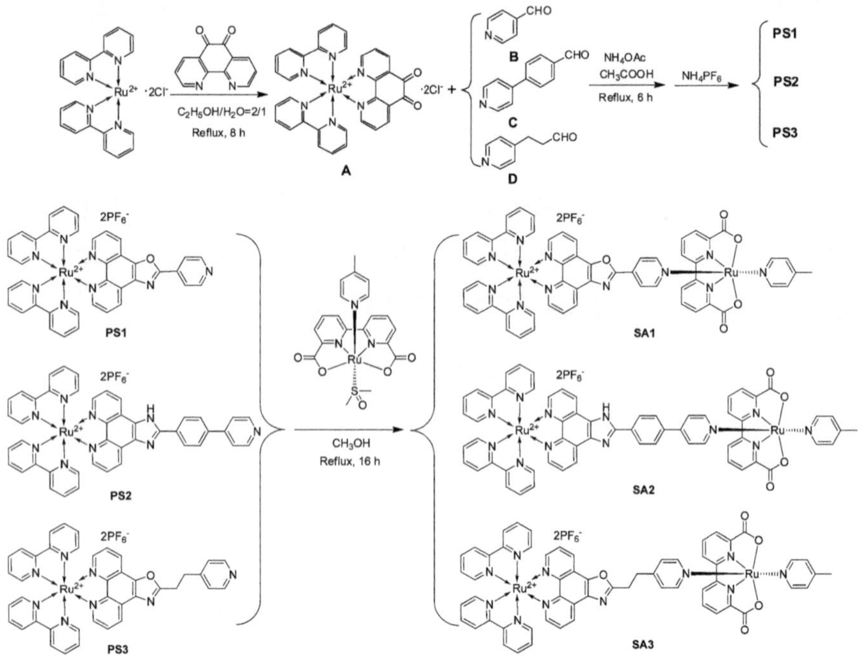

Scheme 1 Synthetic routes for preparation of assemblies **SA1**, **SA2** and **SA3**.

amount of PS is needed to keep the catalytic processes functioning in these systems. To solve this problem, a PS/catalyst supramolecular assembly (SA)[19] by coupling two PS with the Ru(bda)(pic)$_2$ (H$_2$bda = 2,2′-bipyridine-6,6′-dicarboxylic acid) catalyst has been successfully assembled and displayed good light driven water oxidation performance. Meanwhile, this type of SA[19–22] is considered to be a good model for the assembly of photoanodes in molecular devices for light driven water splitting. Whereas, further modifications of this model by coupling a single PS with the Ru(bda) catalyst should be carried out, in order to prepare a simplified TiO$_2$/PS/catalyst photoanode in the future. In this paper, we present three SAs (**SA1–SA3**) by coupling a single PS with the water oxidation catalyst Ru(bda)(pic)$_2$ through different linkages, as shown in Scheme 1. In this design, two rigid linkages and one flexible linkage are used between the PS and the catalyst, leading to different performances of these SAs on light driven water oxidation. This strategy has been proven to be very effective to enhance the photocatalytic activities and the details of the study are described in this paper.

2. Results and discussion

To construct the supramolecular assemblies, we have chosen $[Ru(bpy)_3]^{2+}$ (bpy = 2,2′-bipyridine) as the PS due to its good photophysical and photochemical properties, and Ru(bda)L$_2$ as the water oxidation catalyst unit. The synthesis started from complex **A**, as shown in Scheme 1, which was obtained through the coordination reaction of Ru(bpy)$_2$Cl$_2$ and 1,10-phenanthroline-5,6-dione in a mixture solution of ethanol–water (2 : 1, v/v). The cyclization reactions of the dione of complex **A** and the aldehyde of compounds **B–D** were carried out in the presence of excessive NH_4OAc in the reflux of acetic acid solution, leading to the formation of precursors **PS1–PS3**. The desired supramolecular assemblies **SA1–SA3** were successfully obtained, respectively, through ligand exchange reactions between the pyridyl groups in the precursors **PS1–PS3** and the DMSO in the complex Ru(II)(bda)(4-picoline)(DMSO) (DMSO = dimethylsulfoxide).

All precursors **PS1–PS3** and assemblies **SA1–SA3** were fully characterized by ^{1}H NMR spectroscopy and HRMS (high resolution mass spectrometry). The ^{1}H NMR spectra of the complexes **PS1–PS3** exhibit similar proton resonances of the bipyridine ligands. However, the proton resonances of phenanthroline ligands are found to be quite different, because of the different heterocyclic ring and the different π conjugating effects caused by the connecting linkages. A small broad peak at $\delta = 13.44$ ppm in the spectrum of the complex **PS2** is assigned to the proton resonance of H–N in the imidazole, while no peaks are found at similar positions in the spectra of the complexes **PS1** and **PS3**. These results indicate that different cyclization reactions occurred during the same synthetic process. An imidazole cycle formed on the phenanthroline ligand in complex **PS2**, while oxazole cycles formed in complexes **PS1** and **PS3**. The result could also be confirmed by comparing the proton resonances of phenanthroline ligands. A single doublet peak could be found at $\delta = 9.0$ ppm in the spectrum of complex **PS2**, while two small doublet peaks are found at around $\delta = 8.7$–8.9 ppm in the spectra of complexes **PS1** and **PS3**. The difference peaks should be attributed to the different effects of heterocyclic rings on the phenanthroline ligands. In addition, structural information is also obtained through HRMS. For complex **PS2**, the signals at $m/z = 932.1381$ and 393.5873 are assigned to the species of $[M - PF_6^-]^+$ and $[(M - 2PF_6)/2]^+$, respectively, which also fit well to complex **PS2** with imidazole involved. For complexes **PS1** and **PS3**, the corresponding signals were also obtained and fit to their structures with oxazole involved.

The ^{1}H NMR spectra of the assemblies **SA1–SA3** display more complicated proton resonances than their corresponding precursors **PS1–PS3**, because the proton resonances of the equatorial ligands and the picoline in the catalyst units are embedded in a series of aromatic resonances from the PS units. While the singlet peaks at $\delta \approx 2.26$ ppm are assigned to be the protons on the methyl group of the picoline ligands, which indicates that the SAs are successfully obtained with each catalytic unit linked to one PS unit. The cationic species with different charges in HRMS spectra could also confirm the structures of these SAs.

The cyclic voltammograms (CVs) and differential pulse voltammograms (DPVs) of the assemblies **SA1–SA3** were measured in a phosphate buffer solution (pH = 6.8), with Ag/AgCl as a reference electrode and Pt as a counter electrode, as shown in Fig. 1. The assemblies **SA1–SA3** display analogous redox peaks at *ca.* $E_{1/2} = 0.70$ V *vs.* NHE, which are attributed to the Ru^{II}–OH_2/Ru^{III}–OH_2 redox couples of the catalyst units. The onset potentials for catalytic water oxidation are found to be around $E_{1/2} = 0.95$ V *vs.* NHE. The Ru^{II}/Ru^{III} redox couples of PS units are found to be *ca.* $E_{1/2} = 1.3$ V *vs.* NHE, which is overlapping with the currents of catalytic water oxidation. Additional information of Ru^{III}/Ru^{IV} oxidation peaks from catalyst units could be found at about 1.0 V *vs.* NHE in the DPV measurements. The results are consistent with our reported works. It can also confirm that photocatalytic water oxidation could be achieved in these assemblies thermodynamically with over 300 mV as a driving force. The significantly higher catalytic current

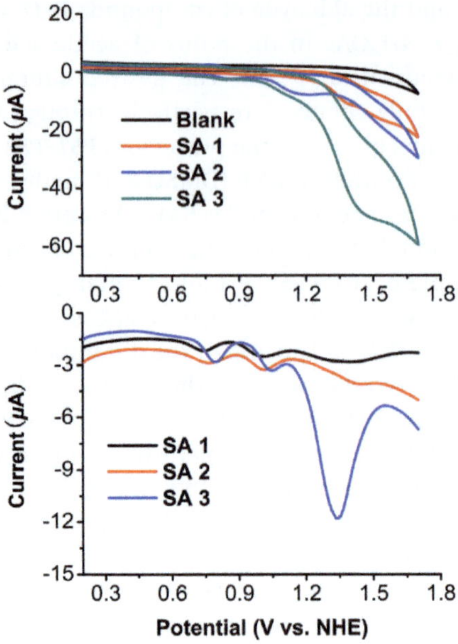

Fig. 1 Cyclic voltammograms (CVs) and differential pulse voltammograms (DPVs) of the assemblies SA1–SA3 in phosphate buffer solutions (pH = 6.8). Ag/AgCl as the reference electrode and Pt as the counter electrode, scan rate at 100 mV s^{-1} ($E_{NHE} = E_{Ag/AgCl}$ + 0.20 V).

observed for the assembly **SA3** reflects its superior performance on catalytic water oxidation, in comparison to **SA1** and **SA2**. A similar phenomenon is also observed in the photocatalytic water oxidation process that will be discussed in the following part.

The performances of the assemblies **SA1–SA3** on photocatalytic water oxidation were studied in a degassed phosphate buffer solution (pH = 6.8, 10% acetonitrile) using sodium persulfate as an electron acceptor. The experiments were performed in a 5 mL solution containing the assemblies **SA1–SA3** (5.0×10^{-5} M), and $Na_2S_2O_8$ (5.0×10^{-2} M) in a closed Schlenk bottle. Under visible light illumination ($\lambda > 400$ nm), molecular oxygen formed and the amount of generated oxygen was measured with GC as shown in Fig. 2. In the photocatalytic systems based on the assemblies **SA1–SA3**, the amount of oxygen generated is 4.0, 3.6 and 8.5 μmol, respectively. The corresponding turnover numbers (TONs) for the assemblies **SA1–SA3** are calculated to be 20, 18 and 34, respectively, according to the catalyst unit or PS unit in the assemblies. The performances of the assemblies **SA1–SA3** on photocatalytic water oxidation can be sorted in the order **SA3** > **SA1** > **SA2**. The assembly **SA3** exhibits the best performance on photocatalytic water oxidation among the SA systems studied. The difference in photocatalytic performance observed for these assemblies should be mainly attributed to the different linkages employed in the assemblies. During the photocatalytic process, the flexible $-CH_2-CH_2-$ linkage in the assembly **SA3** could most likely adjust the distance between the catalyst unit and the PS unit, which seems to be beneficial to the electron transfer and charge separation, thus improving its photocatalytic activity. On the contrary, in the assemblies **SA1** and

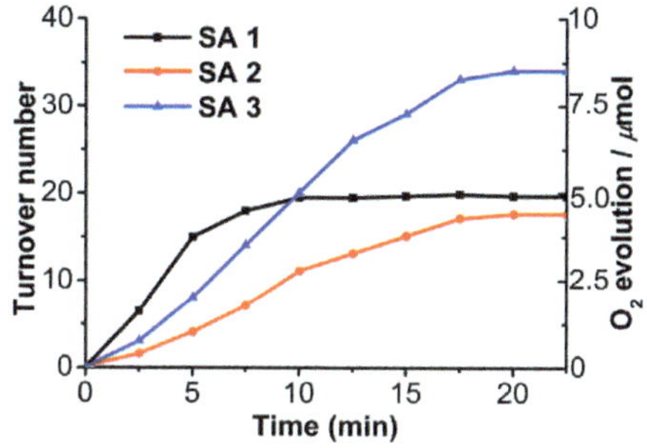

Fig. 2 Oxygen evolution measured by GC in 5 mL phosphate buffer solution (pH 6.8) employing a combined system consisting of $Na_2S_2O_8$ (5.0×10^{-2} M), **SA1** (black), **SA2** (red) and **SA3** (blue) (5.0×10^{-5} M), respectively, under identical visible light illumination ($\lambda >$ 400 nm).

SA2, the rigid linkages solidify the PS and catalyst units, which is probably detrimental to the photocatalytic activity.

More interestingly, as compared to our reported three-component photocatalytic systems,[11] the assembly **SA3** displays much better photocatalytic activity. A TON of 34 is obtained for the assembly **SA3**, which is higher than our reported assemblies with a TON of only 19.[19] Therefore, this type of assembly connected with flexible links may become more suitable to be used in the assembly of photoanodes for visible light driven water splitting in a dye-sensitized photoelectrochemical cells.

3. Conclusions

In summary, three new supramolecular assemblies **SA1–SA3** have been synthesized and fully characterized with a single PS unit covalently coupled to the water oxidation catalyst Ru(bda)(pic) unit through different linkages. The electrochemical study shows that the oxidation potential of the PS unit (*ca.* 1.3 V *vs.* NHE) is much bigger than the onset potential (*ca.* 0.95 V *vs.* NHE) of the water oxidation catalyst unit, providing a driving force of more than 300 mV. Visible light driven water oxidation was successfully achieved with these assemblies in homogeneous systems. With flexible $-CH_2-CH_2-$ chain as a linkage, the assembly **SA3** displays better performance on photocatalytic water oxidation than the other two assemblies and our earlier reported assemblies. This type of assembly connected with flexible linkages may become a suitable candidate to construct photoanodes for light driven water splitting in dye sensitized photoelectrochemical devices. Work on this direction is ongoing in our lab.

4. Experimental

Sample preparation and characterization

^1H NMR spectra were collected at 298 K using a Bruker DRX-400 instrument. Electrospray ionization mass spectra were recorded on a Q-Tof Micromass spectrometer (Manchester, England). EI-MS analyses were carried out on a gas chromatography fitted with a mass spectrometer (GC-MS 6890/5973, Agilent, USA).

UV-Vis absorption measurements were carried out on a HP 8450 spectrophotometer. The light intensity was measured by a Nova-II laser power monitor.

Electrochemistry

The electrochemical measurements were performed on a BAS-100W electrochemical potentiostat in a three-electrode system. The counter electrode was platinum wire, the reference electrode was an aqueous Ag/AgCl.

Ru(1,10-phenanthroline-5,6-dione)(bpy)$_2$Cl$_2$ (A)

Ru(II)(bpy)$_2$Cl$_2$ (2.42 g, 5 mmol), 1,10-phenanthroline-5,6-dione (0.93 g, 5 mmol) were dissolved in ethanol–water (120 mL, 2 : 1, v/v) and the solution was refluxed for 8 h under an argon atmosphere. The solvent was evaporated under reduced pressure and the obtained residue was purified by flash column chromatography with CH_2Cl_2 as the eluent, a black solid was obtained with a yield of 74.9%. ^{1}H NMR (400 MHz, DMSO-d$_6$): δ 8.85 (4H, d, $J = 8.0$ Hz), 8.54 (2H, d, $J = 8.0$ Hz), 8.21–8.16 (4H, m), 7.94 (2H, d, $J = 5.6$ Hz), 7.75–7.67 (6H, m), 7.57–7.52 (4H, m). ES-MS: $m/z = [M - Cl]^+$ 659.0519 (calcd: 659.0531).

Ru(1,10-2-(pyridin-4-yl)oxazolo[4,5-f][1,10]phenanthroline)(bpy)$_2$(PF$_6$)$_2$ (PS1)

A solution of complex **A** (1 g, 1.35 mmol), NH_4OAc (2.1 g, 27 mmol) and isonicotinaldehyde (400 μL, 4.25 mmol) in acetic acid (80 mL, 14.7 M) was refluxed for 6 h and then cooled to room temperature. The solution was neutralized with concentrated aqueous ammonia and 200 mL deionized water was added. The product was precipitated by addition of concentrated aqueous NH_4PF_6 solution, filtrated and washed with water, then dried with diethylether. After purification by flash column chromatography with CH_3OH/CH_2Cl_2 (3 : 1, v/v) as the eluent, a yellow solid was obtained with a yield of 41.8%. ^{1}H NMR (400 MHz, CD$_3$OD): δ 9.11 (1H, d, $J = 8.1$), 9.03 (1H, d, $J = 8.3$), 8.87 (2H, t, $J = 10.6$), 8.69–8.57 (6H, m), 8.35 (2H, d, $J = 5.9$), 8.21 (2H, s), 8.15 (2H, t, $J = 7.9$), 8.04 (2H, t, $J = 7.9$), 7.95–7.85 (4H, m), 7.64 (2H, t, $J = 6.6$), 7.54–7.48 (2H, m), 7.32–7.22 (2H, m). ES-MS: $m/z = [M - PF_6]^+$ 857.0934 (calcd: 857.0909). $m/z = [(M - 2PF_6)/2]^+$ 356.0644 (calcd: 356.0631).

Ru(2-(4-(pyridin-4-yl)phenyl)-1H-imidazo[4,5-f][1,10]phenanthroline)(bpy)$_2$ (PF$_6$)$_2$ (PS2)

By using similar synthetic and purified methods of **PS1**, the complex **PS2** was obtained as a yellow solid with a yield of 61.6%. ^{1}H NMR (400 MHz, CD$_3$CN): δ 13.44 (1H, brs), 9.05 (2H, dd, $J = 8.3, 1.1$), 8.77–8.59 (2H, m), 8.53 (4H, dd, $J = 14.9, 8.1$), 8.42 (2H, d, $J = 8.6$), 8.10 (2H, td, $J = 8.0, 1.5$), 8.00 (4H, ddd, $J = 9.3, 6.1, 1.4$), 7.92 (2H, d, $J = 8.5$), 7.86 (2H, d, $J = 4.9$), 7.76 (2H, dd, $J = 8.3, 5.3$), 7.70 (2H, d, $J = 6.0$), 7.67–7.57 (2H, m), 7.49–7.39 (2H, m), 7.24 (2H, t, $J = 6.5$). ES-MS: $m/z = [M - PF_6^-]^+$ 932.1381, (calcd: 932.1382), $[(M - 2PF_6)/2]^+$ 393.5873 (calcd: 393.5868).

Ru(2-(2-(pyridin-4-yl)ethyl)oxazolo[4,5-f][1,10]phenanthroline)(bpy)$_2$(PF$_6$)$_2$ (PS3)

By using similar synthetic and purified methods of **PS1**, the complex **PS3** was obtained as a yellow solid with a yield of 48.5%. ^{1}H NMR (400 MHz, CD$_3$CN): δ 8.91 (1H, d, $J = 8.3$), 8.76 (1H, d, $J = 8.3$), 8.60–8.38 (7H, m), 8.16–8.05 (4H, m),

7.99 (2H, t, $J = 7.8$), 7.81 (4H, dd, $J = 14.3$, 6.1), 7.54 (2H, t, $J = 6.2$), 7.48–7.41 (2H, m), 7.32 (2H, t, $J = 9.4$), 7.20 (2H, t, $J = 6.4$), 3.55 (2H, t, $J = 7.5$), 3.36 (2H, t, $J = 7.5$). API-ES: $m/z = [M - PF_6^-]^+$ 885.1235, (calcd: 885.1228), $[(M - 2PF_6)/2]^+$ 370.0794 (calcd: 370.0788).

PS1/catalyst supramolecular assembly (SA1)

A mixture of Ru(II)(bda) (4-picoline) (DMSO)(CH_3CN) (100 mg, 0.18 mmol), **PS1** (180.4 mg, 0.18 mmol), and dry methanol (30 mL) were refluxed for 16 h, and then cooled to room temperature. The solvent was evaporated under reduced pressure and the obtained residue was purified by flash column chromatography with CH_2Cl_2–CH_3OH (from 20 : 1 to 3 : 1, v/v) as the eluent, the product complex **SA1** was obtained as a dark brown solid with a yield of 40.3%. ^1H NMR (400 MHz, CD_3CN): δ 8.96 (1H, d, $J = 9.1$), 8.88 (1H, d, $J = 8.6$), 8.52 (5H, dd, $J = 17.0$, 8.0), 8.09 (7H, ddd, $J = 18.4$, 13.1, 7.8), 7.98 (4H, t, $J = 7.4$), 7.93–7.75 (6H, m), 7.72 (1H, d, $J = 5.1$), 7.55 (2H, dd, $J = 12.2$, 5.2), 7.48–7.41 (2H, m), 7.41–7.36 (1H, m), 7.19 (2H, t, $J = 7.2$), 7.02 (2H, d, $J = 6.3$), 2.26 (3H, s). ESI-MS: $m/z = [M - PF_6]^+$ 1294.0876 (calcd: 1294.0859), $[(M - 2PF_6)/2]^+$ 574.0605 (calcd: 574.0606).

PS2/catalyst supramolecular assembly (SA2)

By using similar synthetic and purified methods of complex **SA1**, the complex **SA2** was obtained as a dark brown solid with a yield of 58.3%. ^1H NMR (400 MHz, D_2O): δ 9.00 (2H, d, $J = 8.2$), 8.50 (4H, dd, $J = 16.5$, 8.4), 8.31 (2H, d, $J = 8.6$), 8.11–7.86 (12H, m), 7.82 (4H, t, $J = 7.1$), 7.74 (2H, d, $J = 6.5$), 7.67 (2H, dd, $J = 8.4$, 5.2), 7.53 (4H, dd, $J = 16.4$, 6.0), 7.40 (4H, s, $J = 6.8$), 7.19–7.14 (2H, m), 6.98 (2H, d, $J = 6.8$), 4.78 (1H, s), 2.26 (3H, s). ESI-MS: $m/z = [M - PF_6]^+$ 1369.1314 (calcd: 1369.1337), $[(M - 2PF_6)/2]^+$ 612.0846 (calcd: 612.0848).

PS3/catalyst supramolecular assembly (SA3)

By using similar synthetic and purified methods of complex **SA1**, the complex **SA3** was obtained as a dark brown solid with a yield of 46.6%. ^1H NMR (400 MHz, D_2O): δ 8.80 (2H, s), 8.71 (2H, s), 8.50 (4H, dd, $J = 15.8$, 8.0), 8.07 (4H, dd, $J = 18.0$, 6.4), 7.96 (4H, d, $J = 8.5$), 7.80 (8H, d, $J = 5.2$), 7.48 (2H, s), 7.45–7.34 (4H, m), 7.24 (2H, s), 7.17 (2H, s), 7.10 (2H, s), 3.40 (2H, t), 3.26 (2H, t), 2.16 (3H, s). ESI-MS: $m/z = [M - PF_6]^+$ 1322.1143 (calcd: 1322.1172), $[(M - 2PF_6)/2]^+$ 588.5744 (calcd: 588.5763).

Acknowledgements

This work was supported by the National Basic Research Program of China (973 program) (2014CB239402), the National Natural Science Foundation of China (20923006, 21003017, 21120102036, 21106015 and 91233201), the Fundamental Research Funds for the Central Universities, the Swedish Energy Agency and the K & A Wallenberg Foundation.

Notes and references

1 J. H. Alstrum-Acevedo, M. K. Brennaman and T. J. Meyer, *Inorg. Chem.*, 2005, **44**, 6802–6827.

2 N. S. Lewis and D. G. Nocera, *Proc. Natl. Acad. Sci. U. S. A.*, 2006, **103**, 15729–15735.
3 V. Balzani, A. Credi and M. Venturi, *ChemSusChem*, 2008, **1**, 26–58.
4 O. Khaselev and J. A. Turner, *Science*, 1998, **280**, 425–427.
5 A. Kudo and Y. Miseki, *Chem. Soc. Rev.*, 2009, **38**, 253–278.
6 L. Ma, J. M. Falkowski, C. Abney and W. Lin, *Nat. Chem.*, 2010, **2**, 838–846.
7 C. Wang, Z. Xie, K. E. deKrafft and W. Lin, *J. Am. Chem. Soc.*, 2011, **133**, 13445–13454.
8 M. Wang and L. Sun, *ChemSusChem*, 2010, **3**, 551–554.
9 L. Duan, A. Fischer, Y. Xu and L. Sun, *J. Am. Chem. Soc.*, 2009, **131**, 10397–10399.
10 W. Chen, F. N. Rein and R. C. Rocha, *Angew. Chem., Int. Ed.*, 2009, **121**, 9852–9855.
11 L. Duan, Y. Xu, P. Zhang, M. Wang and L. Sun, *Inorg. Chem.*, 2009, **49**, 209–215.
12 Y. Xu, A. Fischer, L. Duan, L. Tong, E. Gabrielsson, B. Åkermark and L. Sun, *Angew. Chem., Int. Ed.*, 2010, **49**, 8934–8937.
13 Y. Xu, L. Duan, L. Tong, B. Åkermark and L. Sun, *Chem. Commun.*, 2010, **46**, 6506–6508.
14 Y. Xu, G. Eilers, M. Borgström, J. Pan, M. Abrahamsson, A. Magnuson, R. Lomoth, J. Bergquist, T. Polívka and L. Sun, *Chem.–Eur. J.*, 2005, **11**, 7305–7314.
15 L. Tong, L. Duan, Y. Xu, T. Privalov and L. Sun, *Angew. Chem., Int. Ed.*, 2011, **50**, 445–449.
16 L. Li, L. Duan, Y. Xu, M. Gorlov, A. Hagfeldt and L. Sun, *Chem. Commun.*, 2010, **46**, 7307–7309.
17 G. F. Moore, J. D. Blakemore, R. L. Milot, J. F. Hull, H.-e. Song, L. Cai, C. A. Schmuttenmaer, R. H. Crabtree and G. W. Brudvig, *Energy Environ. Sci.*, 2011, **4**, 2389–2392.
18 Y. Gao, X. Ding, J. Liu, L. Wang, Z. Lu, L. Li and L. Sun, *J. Am. Chem. Soc.*, 2013, **135**, 4219–4222.
19 F. Li, Y. Jiang, B. Zhang, F. Huang, Y. Gao and L. Sun, *Angew. Chem., Int. Ed.*, 2012, **51**, 2417–2420.
20 J. J. Concepcion, J. W. Jurss, M. K. Brennaman, P. G. Hoertz, A. O. T. Patrocinio, N. Y. Murakami Iha, J. L. Templeton and T. J. Meyer, *Acc. Chem. Res.*, 2009, **42**, 1954–1965.
21 D. L. Ashford, D. J. Stewart, C. R. Glasson, R. A. Binstead, D. P. Harrison, M. R. Norris, J. J. Concepcion, Z. Fang, J. L. Templeton and T. J. Meyer, *Inorg. Chem.*, 2012, **51**, 6428–6430.
22 M. R. Norris, J. J. Concepcion, D. P. Harrison, R. A. Binstead, D. L. Ashford, Z. Fang, J. L. Templeton and T. J. Meyer, *J. Am. Chem. Soc.*, 2013, **135**, 2080–2083.

Faraday Discussions

PAPER

Inorganic core–shell assemblies for closing the artificial photosynthetic cycle

Guangbi Yuan, Anil Agiral,† Norman Pellet,‡ Wooyul Kim and Heinz Frei*

Received 11th July 2014, Accepted 11th August 2014

DOI: 10.1039/c4fd00150h

Co oxide (Co_3O_4) nanotubes are shown to act as an efficient water oxidation catalyst when driven with a visible light sensitizer (pH 7). The nanotubes form the core of a Co_3O_4–SiO_2 core–shell nanotube design for separating the carbon dioxide photoreduction from the oxygen evolution reaction. Amorphous dense phase silica of a few nanometers depth is shown to conduct protons while blocking molecular oxygen. Organic molecular wires embedded in the silica shell provide controlled charge transport between the light absorber on one side and the Co_3O_4 catalyst on the other side. Hence, the silica shell is suitable as a membrane of an assembly for closing the photosynthetic cycle on the nanometer scale under product separation.

Introduction

The use of sunlight offers a path towards replacing fossil with renewable energy resources, and single integrated artificial systems are a promising approach for achieving the very large scale on which such systems need to be made in order to have a substantial impact on transportation fuel use.[1,2] Closing the photosynthetic cycle under visible light illumination in a bias-free system has been established thus far for the reduction of protons to hydrogen (overall water splitting), primarily involving semiconductor components as light absorbers. Examples are visible light water splitting at triple junction amorphous silicon,[3,4] III–V semiconducting materials,[5] WO_3 or Fe_2O_3 photoanode/dye sensitized[6] or $BiVO_4$/a-Si tandem cells,[7] and 2-photon TiO_2/Si nanowire or n-WO_3/np$^+$Si microwire arrays.[8,9] GaN:ZnO materials functionalized with co-catalysts are an example of visible light induced overall water splitting using the semiconductor particle approach.[10] For the direct photocatalytic CO_2 reduction by H_2O, existing systems are heterogeneous and based on wide bandgap semiconductor light absorbers

Physical Biosciences Division, Lawrence Berkeley National Laboratory, University of California, Berkeley, CA 94720, USA. E-mail: HMFrei@lbl.gov; Fax: +1 510 486 7768; Tel: +1 510 486 4325

† Current address: Chevron Oronite, Richmond Technology Center, 100 Chevron Way, Richmond, CA 94801, USA.

‡ Current address: Institut des Sciences et Ingenierie Chimiques, Ecole Polytechnique Federale de Lausanne, CH-1015 Lausanne, Switzerland.

like TiO_2, $SrTiO_3$, ZnO, SiC,[11,12] or isolated Ti centers substituted in micro or mesoporous silicates.[13,14] In contrast to H_2 generating systems, light absorption for all these materials is restricted to the UV region. In the case of isolated Ti centers in mesoporous silica, CO and O_2 were shown to be the initial single photon products.[15] In addition, reports are emerging of UV-light-driven CO_2 reduction by H_2O with enhanced selectivity to specific products. Specifically, a 2-photon system, consisting of an InP photocathode functionalized by an organo Ru complex coupled to a TiO_2 particle, yields formate as the predominant product.[16] Also, MgO added to Pt loaded TiO_2 was shown to enhance selectivity towards the methane product,[17] or a thin Nafion layer on Pd-deposited TiO_2 nanoparticles was found to promote methane and ethane production in aqueous suspension.[18]

While single integrated photosystems are attractive from the standpoint of minimizing the balance of system components, which is imperative for very large scale use, a challenging requirement is that all components, from light absorbers to catalysts to separating membranes operate efficiently under the same environmental conditions. To avoid harsh pH conditions, Nature's design of closing the photosynthetic redox cycle on the length scale of nanometers is particularly relevant because it obviates the need for large amounts of added buffer or supporting electrolyte, and minimizes side reactions. Critical unmet requirements for efficient light driven reduction of CO_2 by H_2O are the closing of the photosynthetic cycle on the nanoscale using robust inorganic components, and precisely (molecularly) defined charge transport contacts between the components. Moreover, separation of the reduction chemistry from water oxidation catalysis by a proton conducting, product impermeable membrane is an unmet challenge.

To address these challenges, we have developed all-inorganic, oxo-bridged heterobinuclear units as light absorbers. The units, covalently anchored on silica nanoparticle surfaces, possess metal-to-metal charge-transfer (MMCT) absorption that reaches deep into the visible region. A dozen different oxo-bridged chromophores featuring Ti or Zr as acceptor and a first or second row transition metal as donor have been developed in our lab[19–30] and subsequently by the K. Hashimoto group.[31–33] The flexibility in selecting metal centers allows us to match the redox potential of donor or acceptor to the potential of the catalyst, which is important for converting a maximum fraction of the absorbed photon energy to chemical energy of products. The ability of most binuclear units to drive MMCT induced redox reactions reflects an unusually long lifetime of the excited charge transfer state[26] attributed to an ultrafast spin flip following optical excitation of the MMCT transition.[29] For a $TiOCr^{III}$ moiety coupled to an Ir oxide nanocluster (2 nm) on the surface of mesoporous silica, efficient visible light driven water oxidation was demonstrated.[23] Light absorbers featuring Zr as acceptor oxo-bridged to a Cu^I or Co^{II} donor center allowed the direct reduction of CO_2 to CO upon MMCT excitation of the units in mesoporous silica loaded with 1 atm of CO_2 gas.[19,28] A recently developed photodeposition method for the spatially directed assembly of a water oxidation cluster coupled to the Co donor center of a $ZrOCo^{II}$ unit allowed us to reduce CO_2 by oxidizing H_2O at such a well defined inorganic polynuclear unit for the first time.[30]

In order to accomplish the photosynthetic cycle under separation of the oxidation and reduction chemistry by a proton conducting, product impermeable membrane, we are developing core–shell nanotube assemblies as shown in the cartoon of Fig. 1. The core is a crystalline Co_3O_4 nanotube that acts as an efficient water oxidation catalyst, the shell is a dense phase oxide layer of a few nanometers depth that serves as a proton conducting, gas impermeable membrane. Its function is to separate the oxidation from the reduction chemistry, which is particularly important for the case of multistep CO_2 reduction by H_2O because of readily oxidized intermediates that are formed in the process, which otherwise would be consumed by side reactions. For tight control of electron transport across the insulating shell, hole conducting molecular wires are embedded into the separating layer. In this paper, we present crystalline Co_3O_4 nanotubes as efficient water oxidation catalysts, and nanometer-thin silica layers as proton conducting, O_2 impermeable membranes. Furthermore, efficient charge transport from light absorber on one side of the silica membrane to Co_3O_4 on the other will be presented.

Methods

Synthesis of Co_3O_4 nanotubes

Co hydroxide was precipitated by adding 10 mL 0.1 M ammonia solution to 10 mL 0.025 M $Co(NO_3)_2$. The precipitate was centrifuged (4500 rpm, 5 min) and washed a half dozen times in deionized water. After dispersing the $Co(OH)_2$ in water and adding 0.3 g $NaNO_3$, solvothermal syntheses of nanotubes were conducted in Teflon-lined stainless steel autoclaves (Parr Instrument Company, acid digestion vessel) with a capacity of 23 mL. Typical duration of the synthesis was 10 days at 250 °C. Calcination was conducted at 500 °C. Transmission electron microscopy was conducted with a Phillips CM300 instrument. Phase identification and structural analysis of nanotube samples were carried out by XRD using a Siemens D500 diffractometer equipped with Ni-filtered Cu K_α radiation ($\lambda = 1.5406$ Å). FT-IR spectra were acquired with a Bruker Vertex 70 spectrometer equipped with liquid N_2 cooled MCT detector, and FT-Raman spectra were recorded with a Bruker IFS66V spectrometer with FRA106 FT-Raman module. Dried nanotube samples were mixed with KBr into fine powders and pressed into a pellet at 5000 psi. Optical spectra of sample solutions were collected with a Shimadzu UV-2450 UV-Vis spectrophotometer.

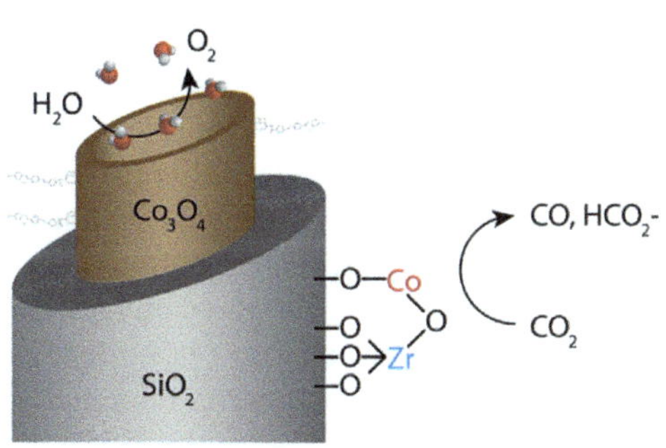

Fig. 1 Co oxide (Co_3O_4)–silica core–shell nanotube for the photoreduction of CO_2 by H_2O under separation of the evolving O_2 and reduced products.

Measurement of catalytic water oxidation activity

Photochemical water oxidation experiments were conducted using $[Ru(bpy)_3]^{2+}$ (bpy = 2,2′-bipyridine) or its carboxylated derivatives as the visible light sensitizer. Persulfate ($S_2O_8^{2-}$) served as the sacrificial electron acceptor. This sensitizer system is widely used for driving multi-electron catalysts for water oxidation in the half reaction mode.[34–36] An aqueous solution buffered at pH 7 (Na_2SiF_6–$NaHCO_3$, 3.7–70 mM) containing 13.5 mM $Na_2S_2O_8$ (Sigma-Aldrich), 1.5 mM $[Ru(bpy)_3]Cl_2 \cdot 6H_2O$ (Aldrich), and ~1 mM Co_3O_4 nanotubes was irradiated with the 476 nm emission line of an Ar ion laser (Coherent model Innova 90) at 21 °C.

Oxygen evolution during photochemical water oxidation in the aqueous phase was monitored by a Clark electrode equipped with a Pt cathode and an Ag/AgCl anode (Hansatech Oxygraph system). The electrode was assembled by adding a drop of 0.1 M KCl electrolyte on the tip of the Pt electrode and tightly covering it with a thin PTFE (Teflon) membrane, which is permeable to oxygen. Liquid phase two-point calibration was performed in which the current response was measured in an air-saturated sample solution (maximum) and in deoxygenated sample solution (minimum) at 21 °C. In a typical photochemical water oxidation experiment, 2 mL of the solution was irradiated with a 476 nm laser beam (24 mW power) which was expanded to 1 cm diameter. Prior to irradiation, the solution was purged with N_2.

For gas phase O_2 evolution experiments and mass spectrometric monitoring of the reaction, the catalyst suspension was first degassed under vacuum and filled with Ar before transfer into the reactor. The reactor, which is a 50 mL flask containing 40 mL of aqueous reactant solution, was irradiated with the 476 nm Ar laser beam (480 mW) expanded to 2 cm diameter. A 2.5 mL aliquot of gas was periodically captured from the headspace of the reactor, followed by direct injection into a quadrupole mass spectrometer (Pfeiffer model Oministar 422).

Nanosecond transient absorption spectroscopy was conducted with an Edinburgh Instruments model LP920 spectrometer equipped with a pulsed Xe probe lamp or a CW halogen lamp (for measurements > 1 ms) and a Nd:YAG laser pumped tunable Optical Paramagnetic Oscillator (Continuum Lasers model Surelite II and Surelite OPO Plus) as excitation source. The laser pulse width was 8 ns and the repetition rate was 10 Hz.

Preparation of Pt/SiO_2 electrode

Pt thin films deposited on silicon supported substrate were selected as the proton probing electrode. Briefly, a p-type Si (100) substrate (~0.005 Ω cm, Siliconquest) was cleaned in acetone, methanol and isopropanol sequentially for 10 minutes each. A 10 nm Ti layer was deposited on the silicon substrate as an adhesion layer using an e-beam evaporation system (Semicore SC600) followed by deposition of a 100 nm Pt film. After e-beam evaporation, the sample was treated in a UV ozone cleaner for 3 minutes before the electrode preparation process.

Two to eight nm deep SiO_2 films were prepared on Pt probing sample based on a modified recipe in a plasma enhanced atomic layer deposition (PEALD) system (Oxford FlexAl).[37] The SiO_2 deposition was carried out at 40 °C with bis(*t*-butylamino)silane (BTBAS, Air Products, ≥98.5%) as the silicon precursor and oxygen plasma as the oxidant. The precursor lines, carrier gas lines, and the reactor walls were all kept at 40 °C. The BTBAS precursor bubbler was also heated up to 35 °C and filled into the deposition chamber using the vapour draw method. Oxygen gas flow was held constant at 60 standard cubic centimeter per minute (sccm) throughout the deposition process. The silicon precursor exposure half cycle consisted of 2 s dosing and 10 s purging using 250 sccm high purity Ar gas. The oxygen plasma half cycle consisted of 2 s pre-plasma treatment, 3 s plasma exposure, and 5 s purging with 100 sccm N_2 and 250 sccm Ar. The applied plasma power was 250 W and was applied for 3 s during the oxygen plasma half cycle. The deposition chamber pressure was held at 80 mTorr during the BTBAS dosing and purging steps and 15 mTorr during the oxygen plasma steps. The film thicknesses were checked using UVISEL spectroscopic ellipsometer (Horiba Jobin Yvon) and confirmed with TEM measurements. The average growth rate was 1.4 Å per cycle.

The as-grown SiO_2 films were characterized by FT-IR, X-ray photoelectron spectroscopy (XPS, Kratos Axis Ultra DLD system at a takeoff angle of 0° relative to the surface normal using a monochromatized Al K_α source (1486.6 eV)), atomic force microscopy (AFM, Bruker Dimension Icon) using Bruker TAP150A probes (5 N m^{-1} spring constant; 150 kHz frequency), and TEM. After UV ozone treatment (Pt probing sample) or SiO_2 PEALD deposition, the samples were used directly for electrode fabrication. The SiO_2 coated samples were scratched at the edge to expose the Pt layer to make an electrical contact. Ohmic contact was made by fixing tinned copper wire (Conwire-811-5) with Ag epoxy (Circuitworks, CW2400) on the Pt (or SiO_2 coated Pt) side. The entire sample was sealed with non-conductive Hysol epoxy (Loctite, Hysol 615), only leaving the Pt (or SiO_2 coated Pt) surface exposed.

Electrochemical and impedance measurements

The electrochemical measurements were conducted using a CHI 609D Potentiostat/Galvanostat in a three-electrode configuration, with SiO_2 coated Pt (or Pt) as working electrode, a platinum wire as the counter electrode and a Ag/AgCl in 3 M KCl (BASi RE5B) as the reference electrode. The aqueous electrolyte solution consisted of 0.5 M sodium sulfate (≥99.0%, Sigma-Aldrich) and the solution pH was adjusted to 3.88 using sulfuric acid (95.0–98%, Sigma-Aldrich). High purity argon (Grade 4.8, Praxair) or oxygen (Grade 4.3, Praxair) gases were bubbled through the solution continuously during the electrochemical measurements. All reagents were used as-received without further purification. The current density *vs.* potential (*J*–*V*) measurements were conducted at a scan rate of 50 mV s^{-1}. All the reported potentials were converted into the potential *vs.* reversible hydrogen electrode (RHE) using the following equation.

$$E_{\mathrm{RHE}} = E_{\mathrm{measure}} + 0.059\mathrm{pH} + E^0_{\mathrm{Ag/AgCl}}$$

where E_{RHE} is the converted potential *vs.* RHE, E_{measure} is the experimentally measured applied potential against the Ag/AgCl reference electrode and $E^0_{\mathrm{Ag/AgCl}}$ is the potential at 25 °C *vs.* standard hydrogen electrode (SHE).

Results and discussion

Cobalt oxide as earth abundant water oxidation catalyst in nanotube form

We have discovered recently that nanostructured Co_3O_4 clusters in mesoporous silica SBA-15 are orders of magnitude more efficient catalysts for water oxidation under close to neutral pH conditions compared to micron size particles.[35] Driven by visible-light-generated $[Ru(bpy)_3]^{3+}$ at pH 5.8 (overpotential is 350 mV) and room temperature, the most active Co_3O_4 sample showed a turnover frequency (TOF) of 1140 O_2 molecules per second per catalyst cluster.[35] The rate was determined mass spectrometrically by probing oxygen gas that accumulated in the head space of the reaction vessel. Co_3O_4 clusters are parallel bundles of nanorods (8 nm diameter) forming a spheroidal cluster of 35 nm diameter and 50 nm length. The TOF of 1140 O_2 s^{-1} corresponds to a 1 O_2 s^{-1} nm^{-2} projected geometrical area of the catalyst cluster. For a 2-photon artificial photosystem, 150 O_2 molecules per nm^2 need to be generated at maximum solar intensity (1500 photons per s per nm^2) in order for all photons to be utilized productively.[38] Therefore, stacking of about 100 of these clusters in a silica scaffold, which is readily realized in a pressed wafer of 150 μm thickness of mesoporous silica particles at 4% Co_3O_4 loading, would be sufficient to keep up with the maximum solar flux.[39] Based on the TOF of 1140 per nanocluster, a value of 0.01 O_2 s^{-1} is calculated per surface Co center of the material (pH 5.8).[35] With the same visible light sensitization method but conducting the water oxidation at pH 7, bare surfactant-free spherical Co_3O_4 nanocrystals (4 nm diameter) were determined to have a TOF per surface Co center of 0.015 s^{-1}.[40] Therefore, crystalline Co_3O_4 is promising as a nanotube material for the proposed Co oxide–silica core–shell assembly.

Co_3O_4 nanotubes have received considerable scientific and technological attention because of their potential for various applications such as solid-state gas sensors, lithium-ion batteries, supercapacitors, and heterogeneous catalysts.[41–44] We sought to nanoengineer Co_3O_4 into high-aspect ratio nanotubes and assess the catalytic water oxidation activity. Several approaches have been developed to synthesize Co_3O_4 nanotubes such as template syntheses using chemical bath deposition, chemical vapor deposition, and atomic layer deposition and template-less synthesis using solvothermal methods, Kirkendall effect based transformation, or self-supported topotactic transformation.[41–48]

We have adopted a modified solvothermal synthesis method for preparing Co^{2+} containing oxide nanotubes first, followed by calcination to complete the transformation to Co_3O_4 nanotubes as shown in Fig. 2a–e.[45,49] The pure Co_3O_4 structure of the tubes was confirmed by characteristic XRD pattern and FT-IR bands presented in Fig. 3a and b. The prepared Co_3O_4 nanotubes typically have mean outer and inner diameters of 12 nm and 6 nm, respectively, and a tube length in the range 100–150 nm. Using the $[Ru(bpy)_3]^{2+}/S_2O_8{}^{2-}$ photosensitization system, a TOF of 0.018 s^{-1} per surface Co atom was measured (Na_2SiF_6–$NaHCO_3$ buffer solution (pH 7)) by mass spectrometric analysis of O_2 evolving in the head space (Fig. 4) combined with electrochemical probing of dissolved O_2 (Clarke electrode). This TOF value is close to our previously reported values measured for 4 nm surfactant-free Co_3O_4 nanoparticles and other spinel cobalt based nanocrystalline materials.[40] Given a footprint of 1.13 nm^2 for a 12 nm diameter vertical nanotube, a tube of 700 nm length will be able to keep up with O_2 production even at maximum solar intensity. Hence, the catalytic efficiency of the Co_3O_4 nanotube is in the proper range for use in an artificial photosynthetic assembly. It is important to note, however, that our discovery of fast and slow catalytic Co sites of Co_3O_4 by rapid-scan FT-IR spectroscopy (TOF ≥ 3 s^{-1} for the fast sites) shows that surface Co sites exist which have widely different catalytic efficiencies.[50]

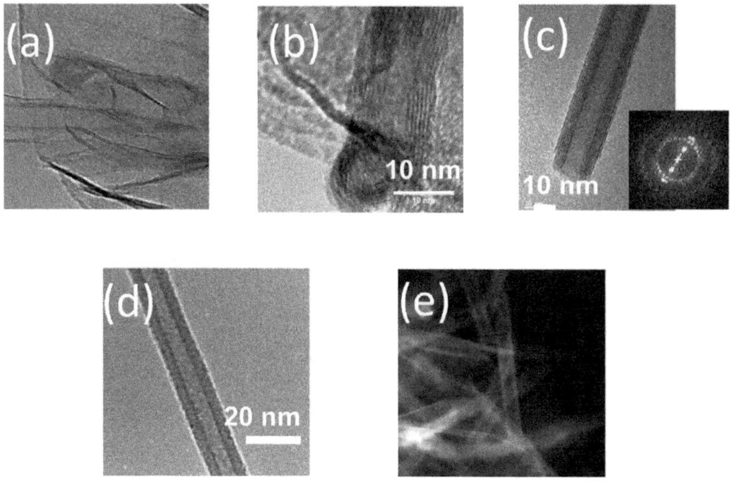

Fig. 2 Solvothermal synthesis of Co_3O_4 nanotubes. $Co(OH)_2$ sheets roll up to form open-ended nanotubes upon heating at 250 °C in a Teflon-lined autoclave (H_2O–CH_3OH mixture, $NaNO_3$), (a) after 48 h, (b) after 10 days. The $Co(OH)_2$ sheets were formed by precipitation of 0.025 M $Co(NO_3)_2$ with 0.1 M NH_4OH. Depending on nitrite concentration, heating at 250 °C gave $Co(OH)_2$ (b) or crystalline CoO nanotubes (c). (d) TEM image of Co_3O_4 nanotube formed by calcination at 500 °C. (e) Annular dark field image of Co_3O_4 nanotubes.

We note, however, that the solvothermal method for Co_3O_4 nanotube synthesis might not be suitable for subsequent asymmetrical functionalization of the nanotubes (different chemical treatment of the inside *versus* outside of the tube), which is required for the design of our artificial photosystem. To enable asymmetrical functionalization of a nanotube array, we propose to adopt a sacrificial vertical silicon nanowire array template and deposit Co_3O_4 thin films by atomic layer deposition, as demonstrated previously.[51] The silicon nanowire core serves as the blocker during the surface functionalization of molecular wires and silica shell coating, and it can subsequently be removed selectively to expose the active sites for water oxidation catalysis. The Co_3O_4 coating thickness and deposition parameters can be tuned to maximize the photocatalytic performance of the final

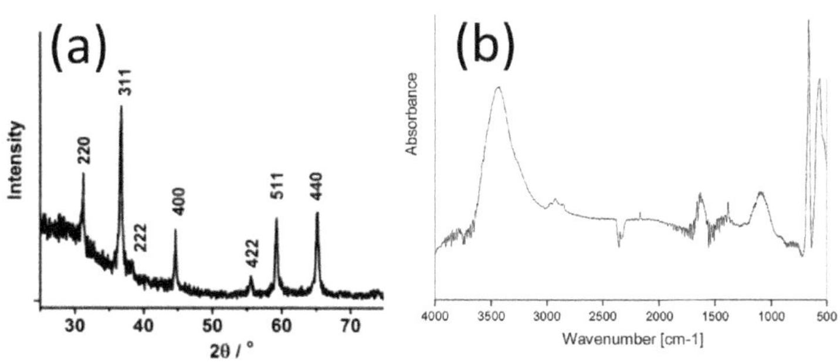

Fig. 3 (a) XRD of crystalline Co_3O_4 nanotubes. (b) FT-IR spectrum showing characteristic Co_3O_4 bands at 663 and 583 cm^{-1}.

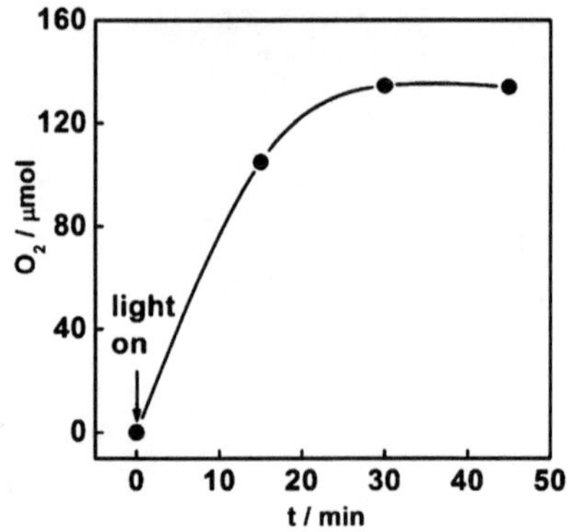

Fig. 4 O_2 evolution in the head space of the reaction vessel recorded by mass spectroscopy at mass 32.

system. We have successfully prepared vertical Co_3O_4 nanotube arrays using this approach, and deposition of the silica layers with embedded molecular wires (described below) is in progress in our lab.

Nanoscale silica layer as a proton-conducting, gas blocking membrane

Amorphous pore-free silica layers a few nanometer thick hold promise as robust materials for separating the visible light absorber and reductive chemistry from the metal oxide catalyst for water oxidation (Fig. 1). Silica is known to transmit protons across thin layers[52–54] while blocking methanol.[54] We anticipate that such a thin SiO_2 layer also blocks O_2 efficiently. These are the critical permeability requirements of a membrane for a nanoscale photosystem for CO_2 reduction by H_2O. Other types of inorganic oxide layers of nanometer depth exhibit similar permeability properties. For example, Domen demonstrated that Cr_2O_3 or lanthanoid oxides transmit protons but block O_2 when used as protective layers of a noble metal proton reduction catalyst.[55,56] Hwang observed selectivity for proton *versus* O_2 permeability in the case of 1 nm NiO overlayers.[57] Hundreds of nanometer thick films of silica have been demonstrated as proton exchange membranes in microfabricated fuel cells.[54] Furthermore, derivatized silicate compounds that exhibit enhanced proton conduction property are proposed as promising candidate for the next generation intermediate temperature solid-state fuel cells.[52]

We have synthesized 2–4 nm silica thin films using plasma enhanced atomic layer deposition method and investigated both the proton conduction and gas blocking properties of these films by electrochemical measurements. Our approach consists of the preparation of nanometer thin SiO_2 layers on a Pt probing electrode and monitoring of the proton adsorption/desorption and oxygen reduction reactions by cyclic voltammetry. When the SiO_2 coated Pt working electrode was immersed in acidic, argon saturated solution (pH 4), we observed the reduction current of protons to Pt–H indicating that protons permeate through the silica layer and reach the Pt surface. When oxygen is introduced to the solution, we should measure minimal oxygen reduction current due to the silica barrier. Since we aim for CO_2 reduction reaction, the hydrogen permeability is not considered here. Notably, all experiments were carried out at room temperature, which is different from proton conductivity measurements in fuel cell studies. Various thicknesses of SiO_2 thin films were deposited on top of the Pt of the Pt/Si working electrode. The applied number of cycles allowed the growth of silica layers with precise depth. Fig. 5a shows a cross sectional high resolution TEM of a 3.8 ± 0.3 nm thickness, and a root-mean-square surface roughness of 1.93 Å determined by AFM (Fig. 5b). All films with thickness equal or greater than 3.8 nm were shown to grow uniformly on the Pt surface without cracks, particles or pinholes.

Before starting electrochemical measurements, all electrodes were cycled between 0.05 V to 1.15 V *vs.* RHE at 0.1 V s^{-1} under argon atmosphere to remove any impurities and stabilize the electrode. The sweep potential was limited below 1.15 V to prevent alterations of the Pt surface, such as high potential induced oxidation and reconstruction.[58,59] Cathodic sweeps in O_2-free aqueous solution showed a broad peak due to proton to H (Pt–H) reduction centered at 0.23 V RHE (Fig. 6). Because of the polycrystalline nature of Pt prepared by e-beam evaporation, the proton adsorption/desorption wave measured here represents a combination of proton adsorption/desorption peaks corresponding to different Pt crystal planes, such as (110), (100), (111). Interestingly, the proton adsorption current densities were significantly increased on the 3.8 nm SiO_2 coated Pt electrode, which suggests that the SiO_2 coating activates the proton adsorption sites on the Pt surface. Control experiments confirmed that a simple plasma treatment (identical SiO_2 growth conditions without using Si precursor) did not activate the Pt surface for proton adsorption.

The proton adsorption wave was reproducible over several hours of CV scans, confirming that there is sustained proton transport. At the same time, the 3.8 nm silica layer serves as a good O_2 diffusion barrier since no O_2 reduction current could be observed when conducting the CV scans for an O_2 saturated solution (Fig. 7). Hence, silica layers of this thickness and larger are proton conducting, O_2 impermeable membranes. Measurement of the SiO_2 thickness dependence of the overpotential for proton reduction and proton conductivity measurements are our

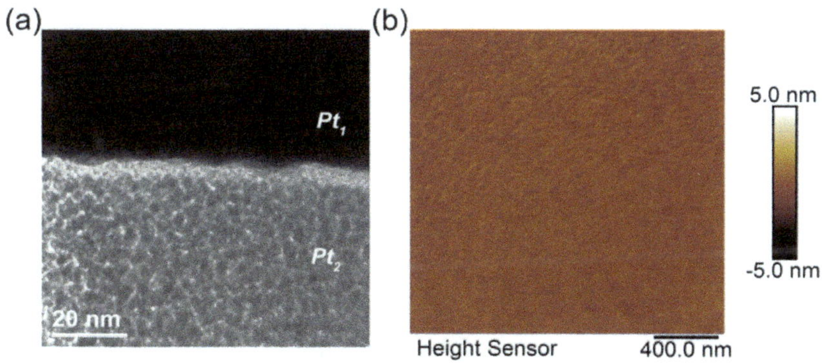

Fig. 5 (a) Cross-sectional TEM image of SiO_2 coated Pt sample. Pt_1 is the evaporated Pt film as the contrast background. Pt_2 is the 100 nm Pt film prepared by e-beam evaporation. SiO_2 film depth is 3.8 ± 0.3 nm. The surface roughness stems from the deposited Pt film. (b) AFM image of the SiO_2 deposited on Si substrate.

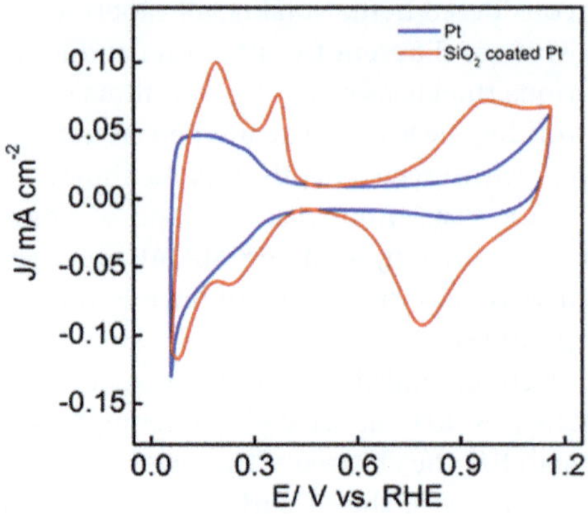

Fig. 6 Cyclic voltammetry of Pt (blue) and 3.8 nm SiO_2 coated Pt electrodes (red) at 298 K in 0.5 M aqueous Na_2SO_4 solution. The solution was adjusted to pH 4 using H_2SO_4 and Ar saturated solution. The scan rate was 50 mV s^{-1}.

next steps to further evaluate the potential of using nanometer thin silica film for separation membrane in the artificial photosynthesis system.

Embedded molecular wires for controlled electron transport across the silica membrane

In addition to transmitting protons while separating photosynthetic products, a critical requirement for photosynthetic efficiency is tightly controlled transport of electrons (holes) from the light absorber to the Co_3O_4 catalyst across the membrane.

Our approach for controlled charge transfer is to embed organic molecular wires with appropriate electronic properties into the silica layer and covalently attach them to the catalyst surface on one side, and to a light absorber on the other. If proof of concept can be established, such a composite membrane could

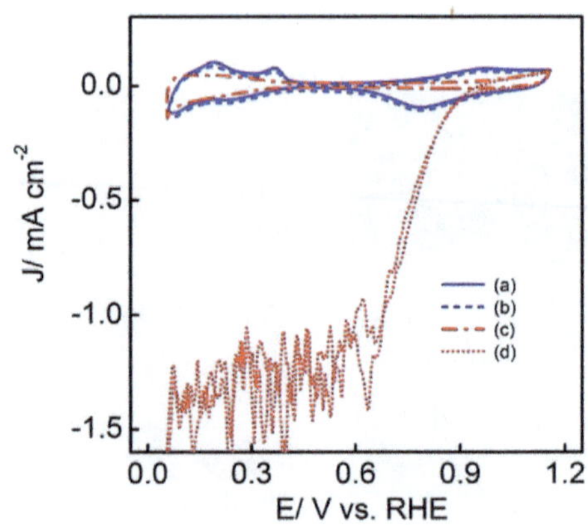

Fig. 7 Cyclic voltammetry for 3.8 nm SiO_2 coated Pt electrodes under Ar (a) and oxygen (b) bubbling. Pt control electrodes under Ar (c) and oxygen (d) bubbling. All the measurements were carried out at 298 K in 0.5 M aqueous Na_2SO_4 solution. The solution was adjusted to pH 4 using H_2SO_4, the scan rate was 50 mV s^{-1}.

be used in Co_3O_4–SiO_2 core–shell nanotubes, which have the proper geometry for CO_2 reduction by H_2O under product separation.

We have synthesized *p*-oligo(phenylenevinylene) molecules with 3 aryl units (1,3-di((*E*)-styryl) benzene, abbrev. PV3). These molecules are well established hole conductors and have the potential of the HOMO (highest occupied molecular orbital) properly positioned between the potential of a Ru bipyridyl sensitizer and the upper valence band position of Co_3O_4 nanoparticles (Fig. 8c).[60] The molecules were functionalized on one end with tridentate $(HOCH_2)_3C$-groups or carboxyl groups for covalent anchoring on the Co_3O_4 particle surface, and sulfonate groups on the other end (Fig. 8a).[61] The tripodal anchors combined with the negative charge of the SO_3^- groups (surface repelling) were selected in order to impose vertical (radial) arrangement of the wire molecules[62] on the Co oxide surface (Fig. 8b). Covalent anchoring of the wire molecules on the Co_3O_4 particle surface was conducted with a peptide coupling agent. The wire molecule has sufficient potential ($E^o = 1.46$ V *vs.* NHE, Fig. 8c) to drive a Co_3O_4 catalyst for water oxidation since even unfunctionalized $[Ru(bpy)_3]^{3+}$ with a potential of just 1.26 V has been shown to induce water oxidation at Co_3O_4 nanoparticles efficiently.[35,50] For visible light induced hole injection into the PV3 molecule, $[Ru(bpy(CO_2CH_3)_2)_3]^{2+}$ was used, which has a redox potential of 1.86 V (NHE). The LUMO of PV3 is situated close to 0.8 V above the reduction potential of the excited sensitizer (Fig. 8c), thus rendering electron flow in the undesired direction (towards Co_3O_4 catalyst) inefficient. For proof of concept of charge transport across silica embedded molecular wires, a simple spherical core–shell particles (4 nm) system was chosen rather than the more complex nanotube construct. The integrity of the PV3 structure, its electronic properties and level of surface loading (12 PV3 nm^{-2}

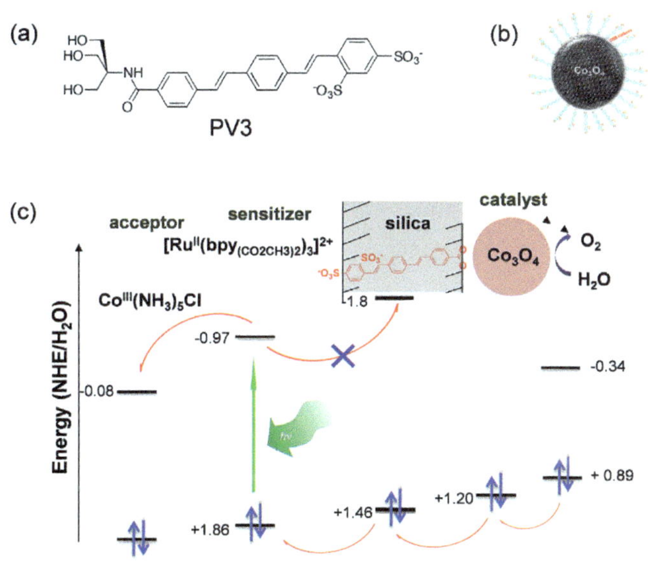

Fig. 8 (a) Molecular wire with 3 aryl units and tripodal anchoring group (PV3). (b) Covalently anchored PV3 wires on Co_3O_4 surface. (c) Energetics of the HOMO and LUMO of the coupled sensitizer, molecular wire and water oxidation catalyst of Co_3O_4–SiO_2 core–shell assembly with embedded PV3 molecular wire.

surface of the Co_3O_4 particle) were established by UV-Vis, FT-IR and FT-Raman spectroscopy.[61]

Upon excitation of $[Ru(bpy(CO_2CH_3)_2)_3]^{2+}$ in the presence of $[Co(NH_3)_5Cl]^{2+}$ acceptor and Co_3O_4 particles covered with anchored PV3 (symbolized by Co_3O_4_PV3) suspended in aqueous solution (pH 7) with a 450 nm nanosecond laser pulse (20 mJ per pulse), hole injection into anchored wire molecules was directly observed by the recovery of the bleach of the $[Ru(bpy(CO_2CH_3)_2)_3]^{2+}$ absorption at 470 nm with a time constant of 10.4 ± 0.2 μs, as shown in Fig. 9a. The sequence of charge transfer steps was

$$\left[Ru\left(bpy(CO_2CH_3)_2\right)_3\right]^{2+} + [Co(NH_3)_5Cl]^{2+} \xrightarrow{h\nu} \left[Ru\left(bpy(CO_2CH_3)_2\right)_3\right]^{3+} + Co^{2+} + 5NH_3 + Cl^-$$

$$[Ru(bpy(CO_2CH_3)_2)_3]^{3+} + Co_3O_4\_PV3 \rightarrow [Ru(bpy(CO_2CH_3)_2)_3]^{2+} + Co_3O_4\_PV3^+$$

As a control, the supernatant was excited with a 450 nm laser pulse in a separate experiment to find out whether any dissolved PV3 wire molecules could have contributed to the observed decay. As can be seen from the grey trace of Fig. 9a, the $[Ru(bpy(CO_2CH_3)_2)_3]^{2+}$ bleach persists and no decay is observed on

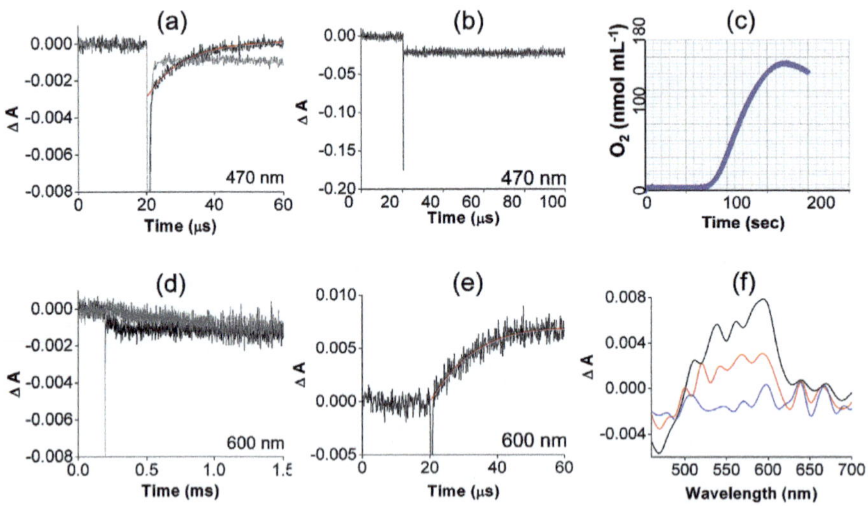

Fig. 9 (a) Recovery of transient bleach of $[Ru(bpy(CO_2CH_3)_2)_3]^{2+}$ at 470 nm due to hole injection of $[Ru(bpy(CO_2CH_3)_2)_3]^{3+}$ into attached molecular wires. The grey trace confirms that no reduction of $[Ru(bpy(CO_2CH_3)_2)_3]^{3+}$ occurs in the absence of Co_3O_4_PV3 particles. (b) Hole injection from oxidized sensitizer into bare Co_3O_4 particles is slow on the microsecond time scale. (c) Electrochemical detection of O_2 upon sensitization of Co_3O_4_PV3 particles. (d) Hole in PV3 attached to Co_3O_4 is too short lived to allow detection. (e) Rise and (d) spectrum at 600 nm of transient hole on PV3 attached to SiO_2 particle.

the microsecond time scale, indicating that such a process is not responsible for the observed recovery of the reduced sensitizer. Furthermore, the reduced $[Ru(bpy(CO_2CH_3)_2)_3]^{2+}$ species is completely regenerated within tens of μs (Fig. 9a). Because direct hole transfer from the oxidized sensitizer to Co_3O_4 particles is several orders of magnitude slower process (Fig. 9b) with a time constant of 25 ms,[63] holes are injected into Co_3O_4 catalyst particles exclusively *via* anchored wire molecules. As shown in the electrochemical measurement of Fig. 9c, evolution of O_2 in the same Co_3O_4_PV3 suspension containing Ru sensitizer and Co acceptor is readily observed upon visible light illumination. Therefore, holes injected into the attached PV3 molecules are transferred to the Co_3O_4 particle where they induce catalytic water oxidation. The characteristic transient absorption of the hole residing on the PV3 at 600 nm is not observed (Fig. 9d), indicating that transfer of the charge to the Co_3O_4 catalyst is too fast. On the other hand, the $PV3^+$ absorption at 600 nm is detected when the wire molecules are covalently attached to SiO_2 nanoparticles, as shown in Fig. 9e and f. Silica is an insulator and has no electronic states that could accept the charge. Comparison of the transient signal amplitude at 600 nm for PV3 on SiO_2 particles and the noise of the kinetic trace for PV3 on Co_3O_4 allowed us to estimate an upper limit for the hole residence time in the case of Co_3O_4_PV3. The transfer rate from PV3 to Co_3O_4 is found to be $7 \times 10^5\ s^{-1}$ or faster.[61] Therefore, covalently anchored PV3 molecules offer an approach for tightly controlled, efficient charge transport between visible light absorber and Co_3O_4 catalyst.

In the next step, we developed a modified method based on the original solvothermal approach by Stoeber[64] for casting a dense phase amorphous silica shell around the Co_3O_4 attached wire molecules.[40] Conformal coating of 4 nm Co_3O_4 catalyst particles by a 2 nm silica shell was demonstrated by HR TEM imaging and EELS measurement (Fig. 10a and b). FT-Raman spectra revealed an intact casting procedure (Fig. 10c), which was further corroborated by FT-IR and UV-Vis spectroscopy.[40] Using the same photosensitization method for suspensions of the resulting Co oxide core–silica shell particles with embedded wires, as discussed above for Co_3O_4_PV3 particles, nanosecond transient absorption measurements of the bleach recovery of the reduced $[Ru(bpy(CO_2CH_3)_2)_3]^{2+}$ sensitizer indicated diffusion controlled hole transfer from the light absorber across the silica shell to the Co_3O_4 core (Fig. 11). No fast recovery of the 470 nm band of $[Ru(bpy(CO_2CH_3)_2)_3]^{2+}$ was observed when conducting the same experiments with Co_3O_4–SiO_2 particles without embedded molecular wires, as shown by the grey trace of Fig. 11b. Using the same method described above, an upper limit of a few microseconds was inferred for the transit time of the hole across the embedded PV3 molecule to the Co_3O_4 particle. The result was corroborated by monitoring of the Ru oxidation state of the sensitizer by the bpy ligand vibrational modes in light-on/light-off experiments using core–shell particles. The technique used was rapid-scan FT-IR spectroscopy in the attenuated total reflection mode, which allowed the direct observation of the reduction of the oxidized sensitizer $[Ru(bpy(CO_2CH_3)_2)_3]^{3+}$ in the presence of core–shell particles with embedded PV3 wires. No reduction of $[Ru(bpy(CO_2CH_3)_2)_3]^{3+}$ was observed for particles with no wire molecules in the shell.[40] We conclude that the hole-conducting wire molecules cast into the dense phase silica shell provide a controlled, fast charge transfer path through the product-separating SiO_2 nanolayer between the visible light sensitizer and the Co oxide catalyst. Furthermore, the silica minimizes oxidative damage of the organic material.

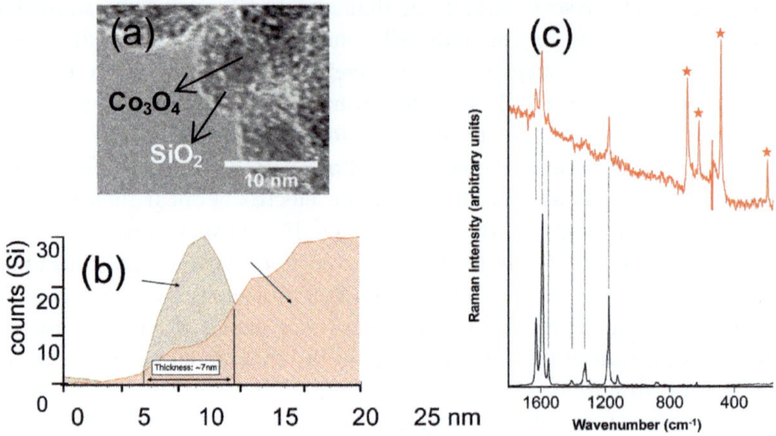

Fig. 10 Co_3O_4–SiO_2 core–shell particles with intact PV3 molecular wires embedded in a silica shell. (a) TEM. (b) EELS showing silica shell (brown) and Co oxide core (red). (c) FT-Raman spectrum of core–shell particle (top) agrees with the spectrum of pure PV3 wires (bottom). Bands with asterisk are Co_3O_4.

Conclusions

In this work, we have demonstrated Co_3O_4 nanotubes as efficient multi-electron catalysts for water oxidation and found that amorphous silica layers of a few nanometers depth function as proton conducting, O_2 impermeable membranes.

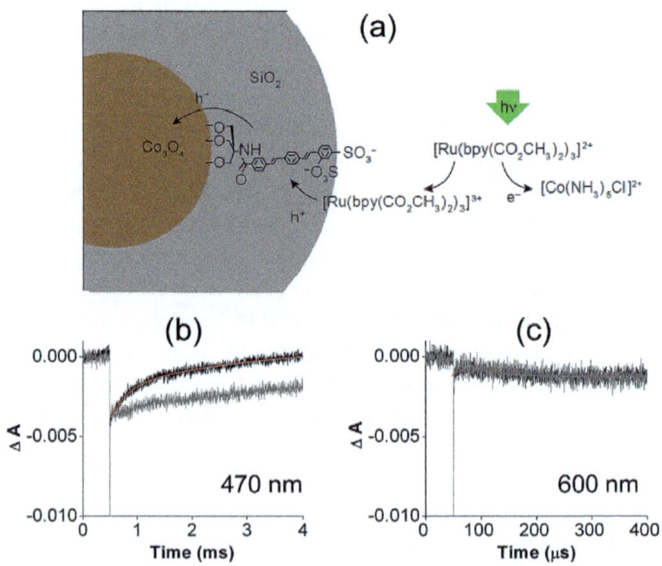

Fig. 11 (a) Transient absorption spectroscopy for monitoring visible light sensitized hole injection into silica embedded molecular wires. (b) Recovery of reduced $[Ru(bpy(CO_2CH_3)_2)_3]^{2+}$ sensitizer upon hole transfer to molecular wire. (c) No transient hole on the PV3 wire observed, indicating fast charge transfer to Co_3O_4 particle.

Combined with our recently established solvothermal method of casting hole conducting molecular wires into silica for controlled charge transport from light absorber on one side of the SiO_2 barrier to the catalyst on the other, these results provide the basis for developing Co_3O_4–SiO_2 core–shell nanotubes as units for closing the cycle of photocatalytic CO_2 reduction by H_2O on the nanometer scale. A large array of vertically arranged core–shell nanotubes, with CO_2 conversion taking place in the space between the tubes, would accomplish separation of O_2 evolution from carbon dioxide reduction on the macroscale.

Achieving this goal poses interesting challenges in terms of materials synthesis as well as functional characterization of the assembly. Specifically, methods need to be developed for the selective anchoring of charge conducting molecular wires (PVn, n between 4 and 6 (n is the number of aryl units)) on the outside of the nanotube while keeping the inner surface of the Co_3O_4 tube free for water oxidation catalysis. An approach currently explored in our lab utilizes a silicon nanowire array as a sacrificial template for growing Co oxide nanotubes, with the silicon wires blocking access to PVn molecules during anchoring on the outer catalyst surface. Furthermore, tuning of the proton transport rates of the silica membrane in order to match the H^+ generation rate during water oxidation catalysis might become important for maintaining photocatalytic efficiency. Transition metal or phosphorous doping is known to strongly affect the proton flux through silicate membranes of micro fuel cells, an approach that could be explored during casting of the silica around the Co oxide nanotubes by either solvothermal or plasma enhanced ALD methods. A further synthetic challenge is precise charge transport contact between the donor center of the heterobinuclear light absorbers[19–33] and the embedded molecular wires when anchoring the units on the outer silica surface of the core–shell nanotube.

Acknowledgements

This work was supported by the Director, Office of Science, Office of Basic Energy Sciences, Division of Chemical, Geological and Biosciences of the U. S. Department of Energy under Contract no. DE-AC02-05CH11231. Portions of this work (plasma enhanced atomic layer deposition, e-beam evaporation, ellipsometry) were performed as a User Project at The Molecular Foundry, Lawrence Berkeley National Laboratory, which is supported by the Office of Science, Office of Basic Energy Sciences. The authors acknowledge the support of the National Center for Electron Microscopy, Lawrence Berkeley National Laboratory, which is supported by the U.S. Department of Energy. A. Agiral acknowledges the Netherland Organization for Scientific Research (NWO) for a Rubicon Fellowship. We acknowledge Prof. Han Sen Soo, Nanyang University of Technology, Singapore, for his research contribution at the early stages of the work reported in this paper.

References

1 N. S. Lewis and D. G. Nocera, *Proc. Natl. Acad. Sci. U. S. A.*, 2006, **103**, 15729.
2 T. Faunce, S. Styring, M. R. Wasielewski, G. W. Brudvig, A. W. Rutherford, J. Messinger, A. F. Lee, C. L. Hill, H. de Groot, M. Fontecave, D. R. McFarlane, B. Hankamer, D. G. Nocera, D. M. Tiede, H. Dau, W. Hillier, L. Wang and R. Amal, *Energy Environ. Sci.*, 2013, **6**, 1074.

3 R. E. Rocheleau, E. L. Miller and A. Misra, *Energy Fuels*, 1998, **12**, 3.
4 S. Y. Reece, J. A. Hamel, K. Sung, T. D. Jarvi, H. J. Esswein, J. J. H. Pijpers and D. G. Nocera, *Science*, 2011, **334**, 645.
5 O. Khaselev and J. A. Turner, *Science*, 1998, **280**, 425.
6 J. Brillet, J. H. Yum, M. Cornuz, T. Hisatomi, R. Solarska, J. Augustynski, M. Gratzel and K. Sivula, *Nat. Photonics*, 2012, **6**, 824.
7 F. F. Abdi, L. Han, A. H. M. Smets, M. Zeman, B. Dam and R. van de Krol, *Nat. Commun.*, 2013, **4**, 1.
8 C. Liu, J. Tang, H. M. Chen, B. Liu and P. Yang, *Nano Lett.*, 2013, **13**, 2989.
9 M. R. Shaner, K. T. Fountaine, S. Ardo, R. H. Coridan, H. A. Atwater and N. S. Lewis, *Energy Environ. Sci.*, 2014, **7**, 779.
10 K. Maeda, K. Teramura, D. Lu, T. Takata, N. Saito, Y. Inoue and K. Domen, *Nature*, 2006, **440**, 295.
11 J. C. Hemminger, R. Carr and G. A. Somorjai, *Chem. Phys. Lett.*, 1978, **57**, 100.
12 T. Inoue, T. A. Fujishima, S. Konishi and K. Honda, *Nature*, 1979, **277**, 637.
13 M. Anpo, H. Yamashita, K. Ikeue, Y. Fujii, S. G. Zhang, Y. Ichihashi, D. R. Park, Y. Suzuki, K. Koyano and T. Tatsumi, *Catal. Today*, 1998, **44**, 327.
14 M. Anpo and M. Takeuchi, *J. Catal.*, 2003, **216**, 505.
15 W. Lin, H. Han and H. Frei, *J. Phys. Chem. B*, 2004, **108**, 18269.
16 S. Sato, T. Arai, T. Morikawa, K. Uemura, T. M. Suzuki, H. Tanaka and T. Kajino, *J. Am. Chem. Soc.*, 2011, **133**, 15240.
17 S. Xie, Y. Wang, Q. Zhang, W. Fan, W. Deng and Y. Wang, *Chem. Commun.*, 2013, **49**, 2451.
18 W. Kim, T. Seok and W. Choi, *Energy Environ. Sci.*, 2012, **5**, 6066.
19 W. Lin and H. Frei, *J. Am. Chem. Soc.*, 2005, **127**, 1610.
20 W. Lin and H. Frei, *J. Phys. Chem. B*, 2005, **109**, 4929.
21 H. Han and H. Frei, *Microporous Mesoporous Mater.*, 2007, **103**, 265.
22 H. Han and H. Frei, *J. Phys. Chem. C*, 2008, **112**, 8391.
23 H. Han and H. Frei, *J. Phys. Chem. C*, 2008, **112**, 16156.
24 X. Wu, W. W. Weare and H. Frei, *Dalton Trans.*, 2009, 10114.
25 H. Frei, *Chimia*, 2009, **63**, 721.
26 T. Cuk, W. W. Weare and H. Frei, *J. Phys. Chem. C*, 2010, **114**, 9167.
27 H. S. Soo, M. L. Macnaughtan, W. W. Weare, J. Yano and H. Frei, *J. Phys. Chem. C*, 2011, **115**, 24893.
28 M. L. Macnaughtan, H. S. Soo and H. Frei, *J. Phys. Chem. C*, 2014, **118**, 7874.
29 B. A. McClure and H. Frei, *J. Phys. Chem. C*, 2014, **118**, 11601.
30 W. Kim, B. A. McClure, G. Yuan and H. Frei, *J. Am. Chem. Soc.*, 2014, **136**, 11034.
31 A. Okamoto, R. Nakamura, H. Osawa and K. Hashimoto, *Langmuir*, 2008, **24**, 7011.
32 A. Okamoto, R. Nakamura, H. Osawa and K. Hashimoto, *J. Phys. Chem. C*, 2008, **112**, 19777.
33 R. Nakamura, A. Okamoto, H. Osawa, H. Irie and K. Hashimoto, *J. Am. Chem. Soc.*, 2007, **129**, 9596.
34 A. Harriman, I. J. Pickering, J. M. Thomas and P. A. Christensen, *J. Chem. Soc., Faraday Trans. 1*, 1988, **84**, 2795.
35 F. Jiao and H. Frei, *Angew. Chem., Int. Ed.*, 2009, **48**, 1841.
36 M. Hara, J. T. Lean and T. E. Mallouk, *Chem. Mater.*, 2001, **13**, 4668.

37 G. Dingemans, C. A. A. van Helvoirt, M. C. M. van de Sanden and W. M. M. Kessel, *ECS Trans.*, 2011, **35**, 191.
38 F. Jiao and H. Frei, *Energy Environ. Sci.*, 2010, **3**, 1018.
39 http://www.rredc.nrel.gov/solar/spectra/am1.5.
40 A. Agiral, H. S. Soo and H. Frei, *Chem. Mater.*, 2013, **25**, 2264.
41 W. Y. Li, L. N. Xu and J. Chen, *Adv. Funct. Mater.*, 2005, **15**, 851.
42 J. Xu, L. Gao, J. Y. Cao, W. C. Wang and Z. D. Chen, *Electrochim. Acta*, 2010, **56**, 732.
43 Z. Y. Fei, S. C. He, L. Li, W. J. Ji and C. T. Au, *Chem. Commun.*, 2012, **48**, 853.
44 C. Y. Lee, K. Y. Lee and P. Schmuki, *Angew. Chem., Int. Ed.*, 2013, **52**, 2077.
45 L. H. Zhuo, J. C. Ge, L. H. Cao and B. Tang, *Cryst. Growth Des.*, 2009, **9**, 1.
46 T. Li, S. Yang, L. Huang, B. Gu and Y. Du, *Nanotechnology*, 2004, **15**, 1479.
47 X. P. Shen, H. J. Miao, H. Zhao and Z. Xu, *Appl. Phys. A: Mater. Sci. Process.*, 2008, **91**, 47.
48 C. C. Li, X. M. Yin, Q. H. Li, L. B. Chen and T. H. Wang, *Chem.–Eur. J.*, 2011, **17**, 1596.
49 N. Pellet, Master thesis, Ecole Polytechnique Federale de Lausanne, 2012.
50 M. Zhang, M. de Respinis and H. Frei, *Nat. Chem.*, 2014, **6**, 362.
51 Y. J. Lin, G. B. Yuan, S. Sheehan, S. Zhou and D. W. Wang, *Energy Environ. Sci.*, 2011, **4**, 4862.
52 Y. Aoki, H. Habazaki, S. Nagata, A. Nakao, T. Kunitake and S. Yamaguchi, *J. Am. Chem. Soc.*, 2011, **133**, 3471.
53 S. Prakash, W. E. Mustain, S. H. Park and P. A. Kohl, *J. Power Sources*, 2008, **175**, 91.
54 C. W. Moore, J. Li and P. A. Kohl, *J. Electrochem. Soc.*, 2005, **152**, A1606.
55 M. Yoshida, K. Takanabe, K. Maeda, A. Ishikawa, J. Kubota, Y. Sakata, Y. Ikezawa and K. Domen, *J. Phys. Chem. C*, 2009, **113**, 10151.
56 M. Yoshida, K. Maeda, D. L. Lu, J. Kubota and K. Domen, *J. Phys. Chem. C*, 2013, **117**, 14000.
57 H. Husin, W. N. Su, C. J. Pan, J. Y. Liu, J. Rick, S. C. Yang, W. T. Chuang, H. S. Sheu and B. J. Hwang, *Int. J. Hydrogen Energy*, 2013, **38**, 13529.
58 S. K. Zecevic, J. S. Wainright, M. H. Litt, S. Lj. Gojkovic and R. F. Savinell, *J. Electrochem. Soc.*, 1997, **144**, 2973.
59 L. Liu, R. Viswanathan, R. Liu and E. S. Smotkin, *Electrochem. Solid-State Lett.*, 1999, **1**, 123.
60 W. B. Davis, W. A. Svec, M. A. Ratner and M. R. Wasielewski, *Nature*, 1998, **396**, 554.
61 H. S. Soo, A. Agiral, A. Bachmeier and H. Frei, *J. Am. Chem. Soc.*, 2012, **134**, 17104.
62 P. G. Hoertz, R. A. Carlisle, G. J. Meyer, D. Wang, P. Piotrowiak and E. Galloppini, *Nano Lett.*, 2003, **3**, 325.
63 M. De Respinis, Master thesis, Dept. of Physics, Technical University of Denmark, 2011.
64 W. Stoeber, W. Fink and E. Bohn, *J. Colloid Interface Sci.*, 1968, **26**, 62.

Faraday Discussions

PAPER

Mesoscopic photosystems for solar light harvesting and conversion: facile and reversible transformation of metal-halide perovskites

Hauke Arne Harms,[a] Nicolas Tétreault,[a] Norman Pellet,[a] Michaël Bensimon[b] and Michael Grätzel*[a]

Received 20th August 2014, Accepted 2nd September 2014
DOI: 10.1039/c4fd00160e

Recently, hybrid organic–inorganic metal halide perovskites have gained prominence as potent light harvesters in thin film solid-state photovoltaics. In particular the solar-to-electric power conversion efficiency (PCE) of devices using $CH_3NH_3PbI_3$ as sensitizer has increased from 3 to 20.1% within only a few years. This key material can be prepared by solution processing from PbI_2 and CH_3NH_3I in one step or by sequential deposition. In the latter case an electron capturing support such as TiO_2 is first covered with PbI_2, which upon exposure to a CH_3NH_3I solution is converted to the perovskite. Here we apply for the first time quartz crystal microbalance (QCMD) measurements in conjunction with X-ray diffraction and scanning electron microscopy to analyse the dynamics of the conversion of PbI_2 to $CH_3NH_3PbI_3$. Employing 200 nm thick PbI_2 films as substrates we discover that the CH_3NH_3I insertion in the PbI_2 is reversible, with the extraction into the solvent isopropanol occurring on the same time scale of seconds as the intercalation process. This offers an explanation for the strikingly rapid and facile exchange of halide ions in $CH_3NH_3PbX_3$ by solution processing at room temperature.

1 Introduction

The application of hybrid organic–inorganic methylammonium metal halide perovskites in photovoltaic cells has gained importance over the last few years, mainly due to the surge in published power conversion efficiency from 3.8% in 2009 up to 20.1% certified efficiency in 2014.[1–8] The concept of using a perovskite material in the device architecture of a solid-state dye sensitized solar cell can be seen as a result of the search for inorganic absorber materials other than quantum dots or Sb_2S_3, that can fulfil a similar function as the molecular

[a]*Laboratory for Photonics and Interfaces (LPI), Faculty of Basic Science, EPFL, Lausanne, Switzerland. E-mail: michael.graetzel@epfl.ch; Fax: +41 (0)21 6936100; Tel: +41 (0)21 6933112*

[b]*Central Environmental Laboratory (CEL), Environmental Engineering Institute (IIE), EPFL, Lausanne, Switzerland. E-mail: michael.bensimon@epfl.ch; Fax: +41 (0)21 6936330; Tel: +41 (0)21 6932331*

absorber in a dye sensitized solar cell. Methylammonium lead iodide has proven the most successful material yet, and size estimations of the constituting lattice by Mashiyama *et al.* as well as by Mitzi *et al.* suggest that it forms more stable crystals than other methylammonium lead halides.[9] Methylammonium tin halide perovskites are less stable than their lead counterparts due to the facile oxidation of Sn^{2+} to Sn^{4+}.

One route to high-efficiency devices employs a sequential deposition of a PbI_2 film that is subsequently converted to $CH_3NH_3PbI_3$ perovskite by incorporation of methylammonium iodide (CH_3NH_3I or MAI) from solution. This route, which often gives better morphological control over the resulting film than a one-step deposition of the perovskite, was first successfully demonstrated on a mesoporous titania film by Burschka *et al.*,[6] attaining efficiencies of 15% and more recently over 17% [J.-H. Im, I.-H. Jang, N. Pellet,[2] M. Grätzel and N. G. Park, *Nat. Nanotechnol.*, 2014, **9** (11), 927–932].

The sequential deposition has been adapted to the conversion of PbI_2 films on compact titania and zinc oxide films, which are solution-processed at low temperatures and yield power conversion efficiencies of 13.7% and 15.7%, respectively.[10,11]

1.1 Perovskite structure

Fig. 1 illustrates the crystal structure of the ideal perovskite. The ideal ABX_3 compound crystallises in a cubic unit cell with the larger cation A in its centre, and the smaller cation B located at the corners. The smaller cation B is octahedrally coordinated to the anions X that lie at the centre of the edges of the cubic unit cell. In inorganic–organic hybrid perovskites, there is often a competition between the 3-dimensional perovskite structure ABX_3, and the 2-dimensional perovskite structure A_2BX_4 because of the size difference between the organic cation A and the inorganic cation B. The 2-dimensional perovskite consists of corner-sharing BX_6 octahedra that are separated by a double layer of organic cations A. This layered structure does in principle impose no limitations on the size of the organic cation A.

The synthesis of phase-pure organic–inorganic hybrid perovskites can be difficult, and methylammonium along with formamidinium is one of the few organic cations that readily crystallises in a 3-dimensional perovskite structure

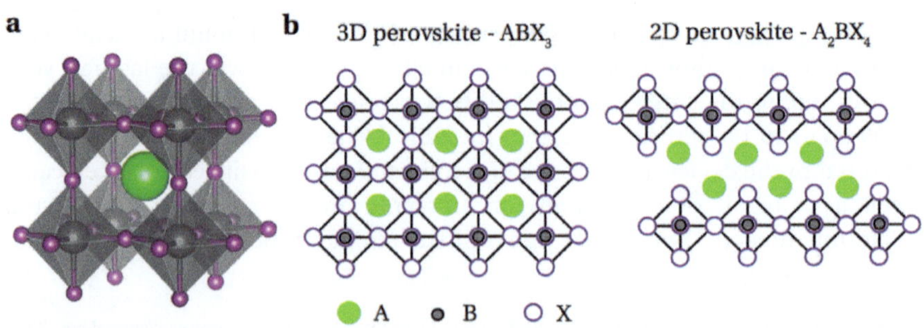

Fig. 1 A perspective view of the perovskite ABX_3 crystal structure is shown in part (a). Projected views of the 3-dimensional ABX_3 perovskite and the 2-dimensional A_2BX_4 perovskite crystal structure are given in (b). Graphic from Burschka.[12]

when used in a lead-halide or tin-halide compound. Methylammonium tin halide perovskites are less stable than their lead counterparts because of the facile oxidation of Sn^{2+} to Sn^{4+}. Regarding the different methylammonium lead-halide perovskites, the iodide compound $CH_3NH_3PbI_3$ has attracted the most interest for photovoltaic applications due its strong light absorption covering all the visible range.

1.2 Methylammonium lead iodide perovskites for photovoltaic applications and their synthesis by conversion from lead-iodide salt

The direct bandgap of $CH_3NH_3PbI_3$ is 1.54 eV which corresponds to an absorption onset at approximately 800 nm. In a typical solar cell configuration, the perovskite absorber is contacted by TiO_2 on one side, acting as an electron-selective contact and as an n-type conductor. The other side of the perovskite is contacted with a hole transport material acting as a hole-selective contact, usually spiro-MeOTAD (2,2′,7,7′-tetrakis(*N*,*N*-di-*para*-methoxyphenyl-amine)-9-9′-spiro-bifluorene). Fig. 2 shows a simplified energy level diagram for this configuration on a scale against vacuum. The valence band of the $CH_3NH_3PbI_3$ was measured to be −5.4 eV by ultraviolet photoelectron spectroscopy.[3] Considering the optical bandgap of 1.54 eV, the conduction band edge lies around −3.9 eV. If we use the published value of −4.0 eV for the TiO_2 conduction band, and of −5.1 eV for the valence band of spiro-MeOTAD, we obtain an offset of 0.1 eV between the conduction bands of the perovskite and the TiO_2, and 0.3 eV between the valence band of the perovskite and the spiro-MeOTAD. The valence band and conduction band of the $CH_3NH_3PbI_3$ perovskite are formed only by the orbitals of Pb and I, whereas the organic cation has no direct contribution. However the size of the organic cation influences the structure of the perovskite crystal and thereby the overlap between the orbitals of Pb and I, affecting its optoelectronic properties.

The crystal phase of $CH_3NH_3PbI_3$ depends on temperature, and three different phases have been observed.[13] The ideal cubic perovskite phase prevails at temperatures above 54 °C, and allows for rapid reorientation of the methylammonium cation that can occupy 24 different disordered states.[14] Between 54 °C

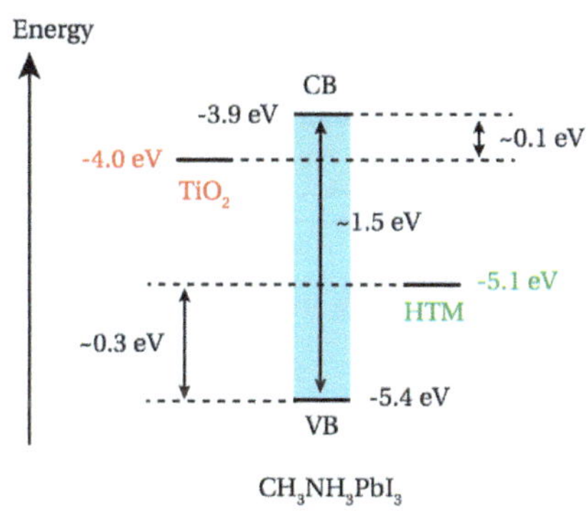

Fig. 2 Simplified energy diagram, illustrating the conduction and valence band of the perovskite as well as the conduction band of TiO_2 and the valence band of the hole transport material spiro-MeOTAD. All energy levels are given in eV against vacuum level. Graphic from Burschka.[12]

and −112 °C, the PbI_2 octahedra undergo a tilting around the *c*-axis of the crystal and thereby reduce the structure to a tetragonal phase, limiting the number of disordered states of the methylammonium cation to 8 possibilities. Below −112 °C, the $CH_3NH_3PbI_3$ crystallises in an orthorhombic phase, and the alignment of the methylammonium anion is fixed. For the $CH_3NH_3PbI_3$ material, no 2-dimensional phase of the form A_2BX_4 has been observed.

For a single-step deposition methods, the perovskite is usually processed by dissolving two precursors, CH_3NH_3I and PbI_2, in polar organic solvents like γ-butyrolactone or *N*,*N*-dimethylformamide (DMF). The $CH_3NH_3PbI_3$ perovskite readily crystallises at room temperature upon evaporation of the solvent. However, it can be difficult to control the film morphology or the infiltration of mesoporous films when crystallising the perovskite material directly from solution. In order to achieve a better control of film morphology, Burschka *et al.* further developed a previously described method of sequential deposition.[6,17] First, a PbI_2 film is deposited by spin-casting a concentrated PbI_2 solution in DMF which subsequently is exposed to a solution of methylammonium iodide in a solvent such as isopropanol that does not dissolve the PbI_2. The process is illustrated in Fig. 3.

Two-step-depositions which entail a conversion reaction are often used to achieve a certain morphology. One example from literature is the conversion of II–V semiconductor nanostructures to III–V materials by ion exchange reactions whilst maintaining the initial architecture.[17] In those examples, the thermodynamic driving force was the difference in bulk lattice energy between the different compounds. Here, the starting material PbI_2 is itself a well-studied semiconductor. PbI_2 crystallises in layers of edge-sharing PbI_6 octahedra, which lead to strong intralayer bonding in a hexagonal crystal lattice, but only weak interlayer bonding, which allows for the formation of different polytypes for the stacking of layers. Fig. 4 is an illustration of the crystal structure of PbI_2 in the most prevalent layer stacking of the 2H polytype. The weak interlayer bonding seems to allow for rapid intercalation of guest molecules into the PbI_2 crystal. The conversion results into a two-fold increase of volume per formula, as can be estimated from the lattice parameters for PbI_2 and $CH_3NH_3PbI_3$ that are listed in Table 1. The

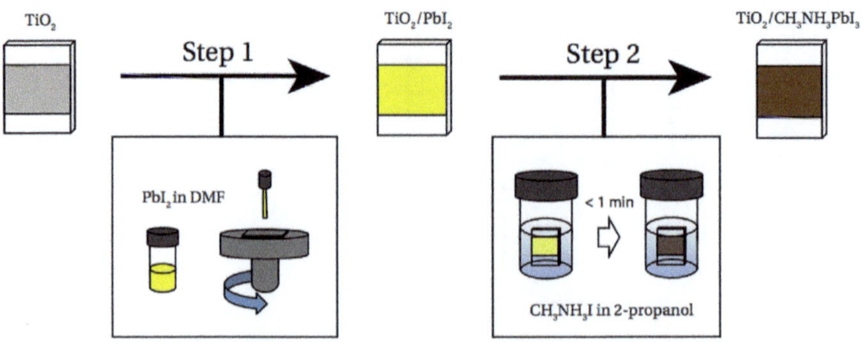

Fig. 3 Schematic illustration of the sequential deposition technique in two steps. First, a PbI_2 film is spin-cast from a solution of DMF. Subsequently, the film is immersed in a solution of CH_3NH_3I in isopropanol for less than one minute. In our experimental work, a pre-wetting with isopropanol has been applied prior to the second step. From Burschka.[12]

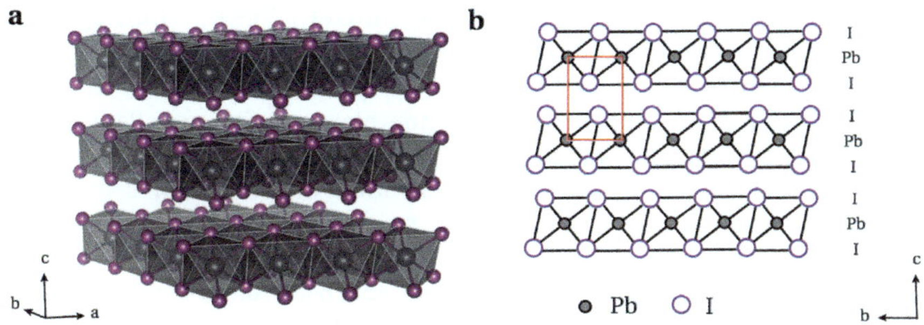

Fig. 4 Illustration of the layered PbI_2 crystal structure in the hexagonal 2H polytype. (a) is a perspective view and (b) shows a projection along the *a*-axis. From Burschka.[12]

literature values of the density are 6.16 g cm^{-3} for PbI_2 and 4.22 g cm^{-3} for $CH_3NH_3PbI_3$.

1.3 Sample preparation and characterisation

The sample preparation aimed for producing perovskite films by sequential deposition, as illustrated in Fig. 3 in the previous section. We decided to use PbI_2 and perovskite films on an underlayer of flat, compact TiO_2 as a model system. The details of PbI_2 deposition and conversion to the perovskite were adapted from the procedure used by Yella *et al.*[10]

Compact TiO_2 underlayer. Prior to deposition of the mesoporous TiO_2 film, the QCM-D sensor or the silicon substrate is cleaned, and an underlayer of thin, compact TiO_2 of approximately 20 nm thickness is deposited. One method applied was atomic layer deposition (ALD) from a tetrakis(dimethylamino)titanium precursor at 150 °C, as described in ref. 18, with subsequent annealing for 3 hours at 420 °C. Alternatively, the underlayer was applied by spray-pyrolysis of an ethanolic solution of di-iso-propoxy titanium-bis(acetylacetone) with an added amount of acetylacetone, using oxygen as a carrier gas for the aerosol,[19–21] and heating the substrate at 420 °C during deposition.

Table 1 Comparison of the crystal lattice parameters of PbI_2 and the tetragonal phase of $CH_3NH_3PbI_3$. Adapted from,[12] based on literature data[13,15,16]

		PbI_2	$CH_3NH_3PbI_3$
Crystal system		Trigonal	Tetragonal
Space group		$P3m1$	$I4/mcm$
Lattice parameters/Å	a	4.558	8.874
	b	4.558	8.874
	c	6.986	12.67
Formula for unit cell volume		$a^2c\sin(60°)$	a^2c
Unit cell volume/Å^3	V	125.7	997.9
Formula units per cell	Z	1	4
Volume per formula unit	V/Z	125.7	249.5
Molecular weight/g mol^{-1}	M	461.0	620.0
Density/g cm^{-1}	ρ	6.16	4.22

Deposition of the PbI_2 film. The PbI_2 was dissolved in *N*,*N*-dimethylformamide at a concentration of 600 mg ml^{-1} by heating the solution to 120 °C under occasional stirring. Subsequently, the solution was cooled down to 70 °C. For deposition of an approximately 200 nm thick PbI_2 film on a flat, compact TiO_2 underlayer, 50 μl of solution were applied to the sample by spin-casting at 6000 RPM for 30 s, using an acceleration of 4500 RPM s^{-1}, and by repeating this procedure once. The entire preparation was carried out in a dry-air box.

QCM-D measurement of the conversion and back-conversion. For QCM-D measurements, the sensor with the PbI_2 film was transferred in a closed sample box with dry silica gel from the drybox to the QCM-D setup, where humidity was kept under 5% for most of the time. After mounting the sensor, the flow cell was flushed with pure isopropanol and signal acquisition was re-started soon after the filling of the chamber. After approximately 10 min of rinsing with isopropanol, the liquid handling system was switched to the solution of 8 mg ml^{-1} CH_3NH_3I in isopropanol, which had been prepared previously in a dry box. In order to achieve a rapid exchange of solvent inside the liquid handling system without creating a concentration gradient, the solvent was flushed in at the maximum pump speed for about 30 s. The maximum pump speed is approximately 800 μl min^{-1}. The same procedure was applied when changing back to pure isopropanol. The usual nominal pump speed during the experiment was 100 μl min^{-1}. The experiment in Section 2.2 will show that increasing this regular nominal pump speed to 500 μl min^{-1} does not significantly affect the reaction kinetics in our experiment. We can therefore assume that the kinetics monitored here are not limited by depletion of species from or their accumulation in the solvent.

Humidity control on the QCM-D setup. The technical details of the QCM-D E4 machine and the setup of the flow cell have been explained previously in ref. 22. In order to keep humidity low, but maintain easy access to the liquid handling system, the QCM-D machine was placed in a plexiglas box that was flushed with large currents of dry air and had a large hatch for manipulations. At 35% ambient humidity in the lab, the humidity inside the box was below 5% when the hatch was closed. Upon opening the hatch for manipulation, the humidity would jump up to 20%, but go back down to 5% within 30 s after closing the hatch.

Samples for SEM and XRD. The films for SEM imaging were prepared on Si wafer substrates, that had been coated with a TiO_2 underlayer just as the QCM-D sensors by ALD or by spray-pyrolysis, and the PbI_2 film was deposited in the same way as above unless otherwise mentioned. Sample preparation then continued in a dry air box in analogy to those in the QCM-D experiment: pre-wetting in isopropanol for 10 min, approximately 50 s of immersion in 0.8 μl min^{-1} CH_3NH_3I in isopropanol, and rinsing with isopropanol for 5 min. Different samples were taken out of the process at each stage of interest; *e.g.* after conversion to the perovskite, solvent was removed by spin-coating at 4000 RPM for 30 s and subsequent drying at 70 °C on a hotplate. The samples were stored in dry air for less then 24 hours and transferred to the SEM facility in a closed box filled with dry air and silica gel. After SEM imaging, the same samples were further used for XRD analysis.

2 Reversible interconversion of PbI_2 and MAI to $CH_3NH_3PbI_3$

2.1 Quantitative analysis by QCM-D

A QCM-D measurement of the conversion of PbI_2 and MAI to $CH_3NH_3PbI_3$ perovskite produces a strong and well reproducible signal. Fig. 5 shows experimental data, starting from a 200 nm thick PbI_2 film that is rinsed with plain isopropanol. During the rinse, which corresponds to pre-wetting, a small but continuous up-drift of approximately +40 Hz in 10 min is observed. This is insignificant compared to the overall mass of the PbI_2 film, which corresponds to *ca.* 6600 Hz in frequency shift, and elemental analysis of the outflow suggest that small amounts of iodide are dissolved, but no lead. After $t = 720$ s the film is exposed to 8 mg ml^{-1} CH_3NH_3I in isopropanol, resulting in a stark drop in frequency by -1000 Hz. This is caused by the incorporation of CH_3NH_3I into the PbI_2, resulting in a large mass increase. During the exposure to CH_3NH_3I for more than 60 s, we observe a small linear decrease in frequency, possibly corresponding to a much slower, continued incorporation of CH_3NH_3I. The XRD analysis in Section 2.4 will reveal that we do indeed form the $CH_3NH_3PbI_3$ perovskite, but that about half of the perovskite remains unconverted after 60 s of CH_3NH_3I exposure under these particular conditions. Concomitant to the steep drop in frequency, the dissipation shifts up by 2.0×10^{-5}, correlating with a change in film morphology, as SEM images in Section 2.3 will show. At $t = 800$ s, we start rinsing again with pure isopropanol until the end of the displayed measurement. Almost instantly, the frequency starts to shift upwards in an approximately linear manner until after 180 s, it reaches a plateau at -70 Hz, close to its initial value. The dissipation passes through a maximum of 5.0×10^{-5}, before returning to a value of 1.8×10^{-5}. The linear increase in frequency reflects the dissolution of a species which is identified below by XRD and ICP-MS to be CH_3NH_3I .The peculiar behaviour of the dissipation correlates again with a change in film morphology as

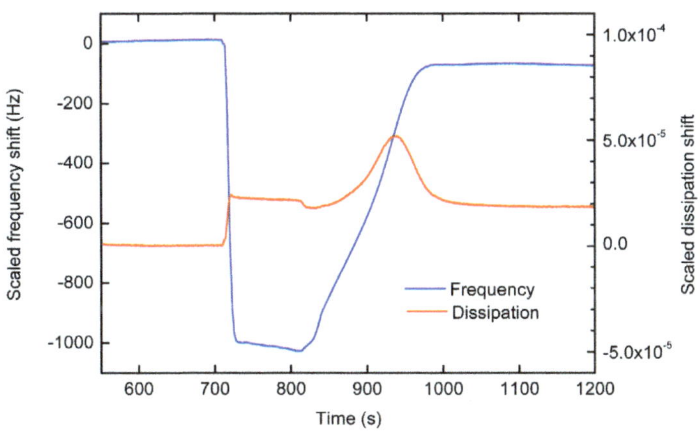

Fig. 5 QCM-D signal of the conversion of a PbI_2 film to $CH_3NH_3PbI_3$ perovskite by exposure to a CH_3NH_3I solution in isopropanol, and the subsequent back-conversion by exposure to isopropanol.

observed by SEM in Section 2.3 and 4. This shift in dissipation will be addressed along with the effect of PbI_2 film morphology on the QCM-D signals in Section 2.5.

The conversion and back-conversion reaction can be repeated several times on the same film, and the PbI_2 itself can be subsequently dissolved in *N*,*N*-dimethylformamide (DMF), as demonstrated in Fig. 6, leading to an increase in frequency of +6600 Hz, keeping isopropanol as a reference. This allows us to estimate the quantity of PbI_2 in the film and compare it to the amount of CH_3NH_3I incorporated during the transformation to the $CH_3NH_3PbI_3$ perovskite.

To estimate the amount of incorporated CH_3NH_3I, we note from the QCM-D measurement in Fig. 5 that the decrease in frequency is approximately −1100 Hz, which corresponds to a change in Sauerbrey mass of $+19.25 \times 10^{-6}$ g cm^{-2}. If we consider that the initial PbI_2 is porous and assume that the $CH_3NH_3PbI_3$ perovskite is not, then we should account for the amount of isopropanol adding to the wet mass of the initial PbI_2 film, which will be discussed in detail in Section 2.5. The amount of trapped isopropanol is 0.97×10^{-6} g cm^{-2}, and the true mass of the incorporated CH_3NH_3I should thus be approximately 20.22×10^{-6} g cm^{-2}. Using the molecular weight of CH_3NH_3I = 59 g mol^{-1}, we obtain an amount of 0.127×10^{-6} mol cm^{-2} of incorporated CH_3NH_3I after conversion to the $CH_3NH_3PbI_3$ perovskite for this type of preparation. The dissolution of the back-converted PbI_2 in DMF solvent does result in an increase of frequency by +6600 Hz, which corresponds to a Sauerbrey mass of $+115.50 \times 10^{-6}$ g cm^{-2}. The isopropanol trapped inside the porous, back-converted PbI_2 amounts to an area mass of approximately 4.63×10^{-6} g cm^{-2}, and thus the dry mass of the PbI_2 film should be around 110.87×10^{-6} g cm^{-2}. Assuming that the PbI_2 is stoichiometric, we use the molar mass of $MW_{PbI_2} = 461$ g mol^{-1} and derive the amount of PbI_2 in the film to be 0.240×10^{-6} mol cm^{-2}. The molar ratio between PbI_2 and CH_3NH_3I after conversion to the perovskite is thus approximately (2 : 1) PbI_2 : CH_3NH_3I, and not (1 : 1), as expected for complete conversion. This ratio indicates that only half of the PbI_2 has been transformed to $CH_3NH_3PbI_3$ during the initial exposure to CH_3NH_3I. This finding is in very good agreement with XRD

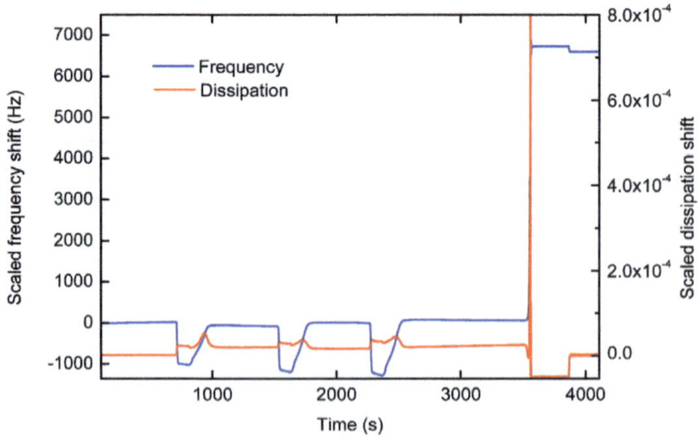

Fig. 6 QCM-D signal of the repetitive conversion of a PbI_2 film to $CH_3NH_3PbI_3$ perovskite, as well as the final dissolution of the PbI_2 in DMF. Subsequent rinsing with isopropanol is in order to account for effects of the bulk liquid on the QCM-D signal.

data presented in Section 2.4. The incomplete conversion on the flat, compact TiO_2 underlayer is in agreement with the general experimental experience, that rapid complete conversion from liquid phase is only achieved for PbI_2 films that are deposited on either mesoporous or very rough compact metal oxides.[6]

We note that the correction made for the liquid trapped inside the pores is smaller than the variation that we observe between different samples made by the same procedure. Initial CH_3NH_3I incorporation produces frequency shifts between −950 Hz and −1200 Hz, and PbI_2 dissolution leads to upshifts between +6000 Hz and +7000 Hz. These variations are probably due to variations in the morphology and thickness of the PbI_2 film, and we have observed occasional irregularities of the dissolution of PbI_2 in the DMF solution prior to spin-casting. We have further observed durations of the back-conversion process that range from 150 s to 250 s. We have carried out the above QCM-D experiment also with flat layers of TiO_2 deposited by ALD or by spray pyrolysis, and we have found no systematic difference in the QCM-D signals for the conversion of the PbI_2 or its re-dissolution.

2.2 Elemental analysis of the outflow by ICP-MS

The QCM-D measurement used for collecting the outflow is shown in Fig. 7. This measurement was performed at a higher nominal pump speed of 500 μl min^{-1} which is five times faster than the usual 100 μl min^{-1}. The increased flow allowed us to avoid intermixing and uncontrolled dilution inside the tubing of the outflow. This was necessary because dissolution processes on the film take place within less than 2 min, and the specific sample had to be taken from the outflow accordingly. Apart from the advantage in technical handling, this measurement shows that the QCM-D results are approximately the same for 100 μl min^{-1} and 500 μl min^{-1} nominal pump speed. Our conversion and back conversion reactions are therefore not limited by depletion or accumulation of chemical species

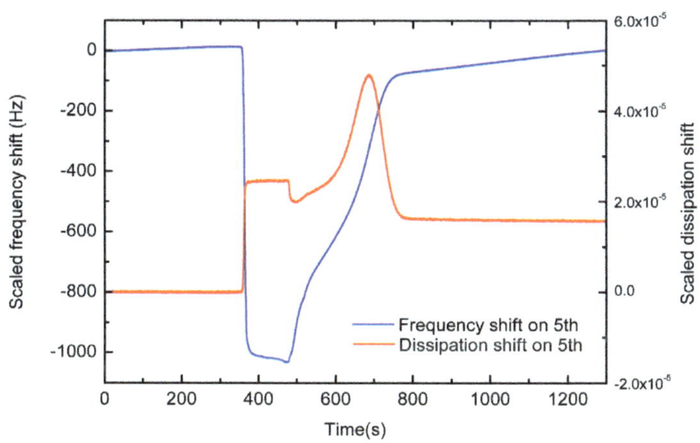

Fig. 7 QCM-D signal of the conversion of a PbI_2 film to $CH_3NH_3PbI_3$ perovskite by exposure to a 10 mg ml^{-1} CH_3NH_3I solution in isopropanol, and the subsequent back-conversion by exposure to isopropanol. The outflow from this measurement was used for elemental analysis by ICP-MS; the nominal pump speed was 500 μl min^{-1}.

in the flow chamber, which might influence reaction kinetics at very low flow speeds. The measurement dates from an earlier period, using 10 mg ml^{-1} CH_3NH_3I in isopropanol.

In Table 2 we present the results of the ICP-MS and the approximate concentrations of lead and iodide in the outflow, collected at a nominal pump speed of 500 μl min^{-1}. The actual ICP-MS measurement was performed on samples diluted with an aqueous solution of 1% nitric acid, in order to obtain measured concentrations in the range of 5 ppb to 500 ppb where precision of the ICP-MS setup is best. The quantitative analysis of lead is very precise (±1 ppb) whereas that of iodide bears a larger error of ±30 ppb.

The outflow during the rinsing of the initial PbI_2 contained small traces of iodide, nominally 670 ppb, but almost no lead was detected (6 ppb). The outflow during exposure to the CH_3NH_3I solution has not been measured due to its excessive iodide content. Upon rinsing of the perovskite with pure isopropanol, the outflow during the significant mass loss, from $t = 850$ s to $t = 1000$ s in Fig. 5, was collected separately from the outflow after the signals had reached a plateau. The solution collected during the mass loss was of a high concentration of iodide, nominally 2.8×10^5 ppb, but only a very small amount of lead (35 ppb). The outflow during the plateau seemed to contain some iodide, nominally 1220 ppb, and very little lead (23 ppb), which is very similar to the outflow from rinsing the initial PbI_2 film.

2.3 Morphology by SEM

The morphology of the films was investigated by scanning electron microscopy acquiring high-resolution images of the top view and of the cross-section. As described in Section 1.4, the films for SEM imaging were prepared on Si wafer substrates, that had been coated with approximately 20 nm TiO_2 by ALD and subsequently annealed for 30 min at 420 °C, so the underlayer was the same as for the QCM-D sensors. The PbI_2 film was prepared the same way as on the on the QCM-D sensors, according to the procedure used by Yella *et al.*[10]

Table 2 Concentrations of lead and iodide in the outflow during conversion and back-conversion obtained by inductive coupling plasma mass spectrometry (ICP-MS). As explained in the text, the concentration of iodide bears a large error, and numbers are understood as nominal values

	Measured concentration on diluted sample		Dilution factor	Estimated concentration in outflow	
	c (I)	c (Pb)		c (I)	c (Pb)
Initial rinsing with isopropanol	67 ppb	0.6 ppb	×10	670 ppb	6 ppb
Mass loss from $t = 500$ s to $t = 750$ s	20 224 ppb	2.5 ppb	×14	2.8×10^5 ppb	35 ppb
Rinsing with isopropanol on the plateau from $t = 850$ s onwards	122 ppb	2.3 ppb	×10	1220 ppb	23 ppb
Final rinsing with DMF	27 ppb	48.5 ppb	>125	—	—

The samples were then prepared in a dry air box in analogy to those in the QCM-D experiment: pre-wetting in isopropanol for 5 min to 10 min, approximately 50 s of immersion in 0.8 mg ml^{-1} CH_3NH_3I in isopropanol, and rinsing with isopropanol for 5 min. Samples were collected at the stage of interest, *e.g.* initially and after conversion to the perovskite. Solvent was removed by spin-coating at 4000 RPM for 30 s and subsequent drying at 70 °C on a hotplate.

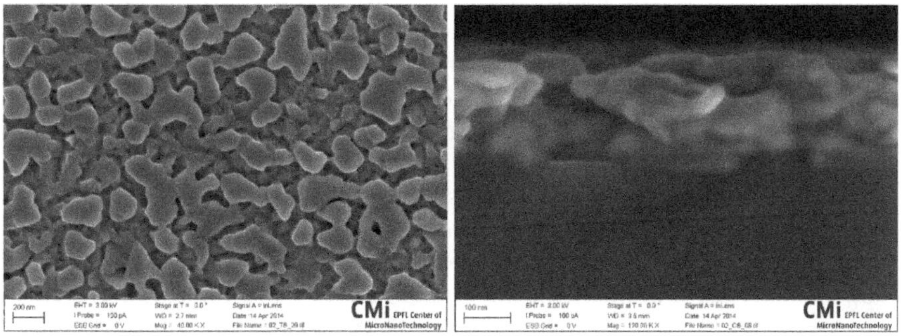

(a) PbI_2 prior to conversion, top view 40 kX. (b) PbI_2 prior to conversion, cross-sectional view 120 kX.

(c) $CH_3NH_3PbI_3$ after conversion, top view 40 kX. (d) $CH_3NH_3PbI_3$ after conversion, cross-sectional view 120 kX.

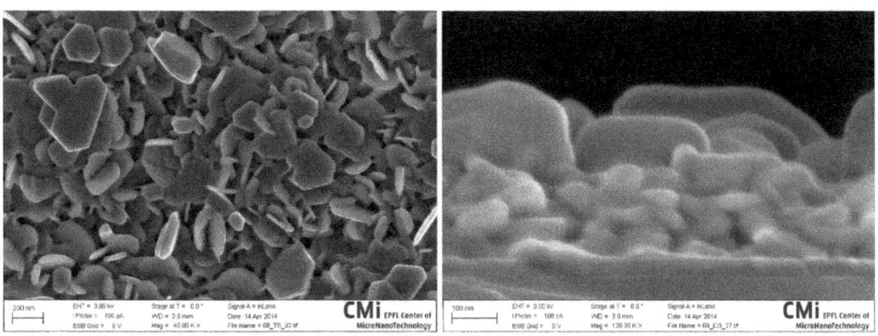

(e) PbI_2 after back-conversion, top view 40 kX. (f) PbI_2 after back-conversion, cross-sectional view 120 kX.

Fig. 8 SEM images of the film morphology at the major steps of initial, pre-wet PbI_2 (a and b), after conversion to $CH_3NH_3PbI_3$ perovskite by exposure to CH_3NH_3I for 50 s (c and d), and after back-conversion to PbI_2 by rinsing with plain isopropanol for 300 s (e and f).

Fig. 8 shows SEM images of the initial PbI_2 film, after conversion to the $CH_3NH_3PbI_3$ perovskite, and after back-conversion to PbI_2. The top view of the initial PbI_2 film in Fig. 8(a) shows features between 100 nm and 200 nm, and the flat appearance on the top is a result of the double-spin-coating process. The side view in Fig. 8(b) confirms the existence of up to 200 nm large pores, and the locally homogenous film thickness of slightly less than 200 nm. Fig. 8(c) and (d) show the top and cross-sectional view of the film after conversion to the $CH_3NH_3PbI_3$ perovskite by 50 s long immersion in a solution of 0.8 mg ml^{-1} CH_3NH_3I in isopropanol. The SEM images show homogeneously distributed crystallites of cubic appearance, which have lateral dimensions of up to 500 nm and extend to a maximum height of about 300 nm from the surface of the TiO_2 underlayer. Although the cubic crystallites form large protrusions, the film does not show any visible pores. The last set of Fig. 8(e) and (f) show the PbI_2 film after back-conversion from the perovskite by rinsing for more than 5 min with pure isopropanol. The top view (e) reveals a homogenous distribution of up to 200 nm large platelets of hexagonal shape. Most of these platelets have a thickness of approximately 20 nm, though some of them are up to 80 nm thick. The cross-sectional view (f) suggests that the hexagonal platelets lie on top of a layer of smaller particles of a less regular shape. The overall thickness of the layer on this sample is about 350 nm.

The film morphology of the initial PbI_2 and the $CH_3NH_3PbI_3$ perovskite correspond very well to previously published results. Since the back conversion has never been studied explicitly, the PbI_2 film after back-conversion is noteworthy for its pronounced morphology change compared to the initial PbI_2 film. The shape of the particles points to a re-crystallisation process that leads to a larger crystallite size. This is confirmed by XRD in Section 2.4.

2.4 Structural analysis by XRD

The same samples that had been used for SEM imaging in the previous Section 2.3 were subsequently taken for crystallographic analysis by X-ray diffraction (XRD). Fig. 9 shows the spectra of the initial, pre-wet PbI_2, of the film after (partial) conversion to the $CH_3NH_3PbI_3$ perovskite by 50 s exposure to CH_3NH_3I, and of the film after back-conversion to PbI_2 by rinsing with pure isopropanol for more than 5 min. Furthermore, the spectra of a film during back-conversion is shown. In this case, the back-conversion was stopped by starting the drying process of spin-coating and heating after 150 s of rinsing the perovskite film in pure isopropanol.

An overview of the XRD spectra from $2\theta = 10°$ to $2\theta = 60°$ is shown in Fig. 9(a). Fig. 9(b) shows a close-up of the PbI_2-related peak at 12.7° and the $CH_3NH_3PbI_3$ perovskite-related peak at 14.2°, which are exemplary for all PbI_2 and perovskite peaks. The initial PbI_2 peak at 12.7° has an intensity of 640 counts. After conversion to the perovskite, the PbI_2 peak at 12.7° decreases to half of its intensity (360 counts), indicating that only part of the crystalline PbI_2 material has been converted to the $CH_3NH_3PbI_3$ perovskite, which corresponds to the peak at 14.2°. We note that the perovskite peak is actually split into one lower peak at 14.1°, and a second, slightly higher peak at 14.2°. After 120 s of rinsing the perovskite with isopropanol, the back-conversion is not yet complete, as we know from the QCM-D measurements in the beginning of this Section 2.1. However, the according XRD spectrum shows no more perovskite peaks, but a PbI_2 peak at

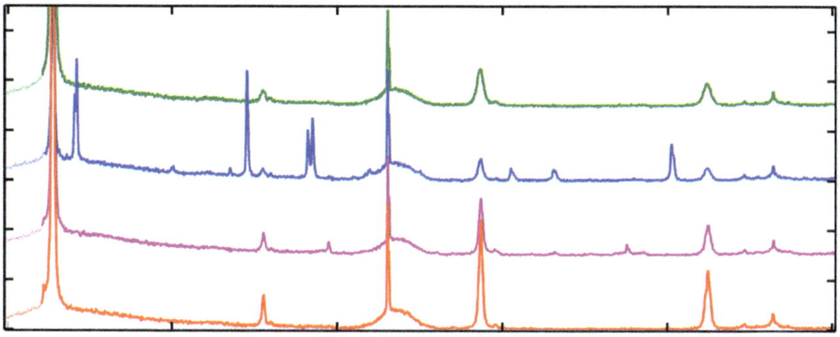

(a) Complete XRD spectra.

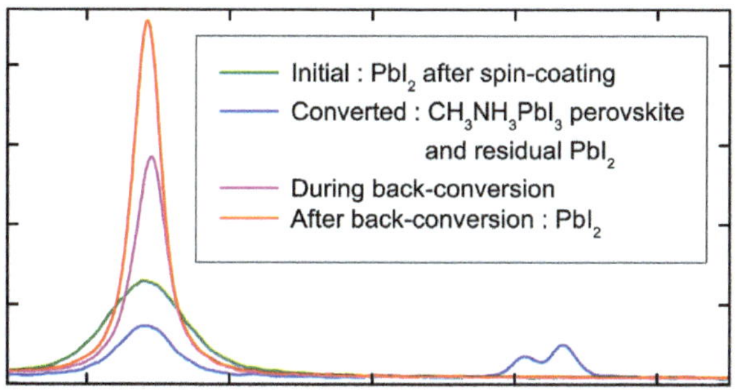

(b) Details of the PbI_2-related peak at 12.7° and the CH_3NH_3 PbI_3 perovskite peat at 14.2°.

Fig. 9 XRD spectra of the PbI_2 film prior to conversion, after conversion to the $CH_3NH_3PbI_3$ perovskite, during back-conversion, and after completion of the back-conversion to PbI_2.

12.7°, which is of a higher intensity (1420 counts) than the peak of the initial PbI_2. We note that in the complete spectrum in Fig. 9(a), this sample shows two low-intensity peaks at diffraction angles of 29.45° and 47.51°. These two peaks do not exist in any of the other samples in this Figure, and they indicate that the back-conversion may proceed *via* a different crystalline phase. After having rinsed the perovskite for more than 5 min with isopropanol and thus completing the back-conversion to PbI_2, the PbI_2 peak at 12.7° has the highest intensity of all peaks (2270 counts), and the most narrow line width.

This XRD measurement gives important information on the crystal structure, and the different species present. The emergence of the perovskite peak upon CH_3NH_3I exposure, and the decrease of the PbI_2 peak confirm that we have indeed the desired conversion reaction. The fact that the PbI_2 peak does not disappear proves that the conversion is incomplete, though the reaction kinetics have reached a quasi-plateau as seen by earlier QCM-D analysis. Upon back-conversion, the perovskite structure seems to disintegrate before the CH_3NH_3I has actually left the material. Back-conversion leads to a significant increase in intensity of the PbI_2 peak and to a narrowing of its width, indicating a significant increase in crystallite size upon back-conversion. This correlates very well with the morphology observed by SEM in the previous Section 2.3. We note that the area under the PbI_2 peak is significantly larger after back-conversion than it was on the initial film. Small variations could be related to inconsistent film thickness between the different samples or to the XRD machine. However, the increase in peak area may be an indication that the initial PbI_2 film has a rather amorphous character with a very small average crystallite size, which would result into a much lower peak intensity and area than for larger crystallites.

2.5 Effect of PbI_2 morphology on QCM-D signals

Frequency shift, trapped liquid and effective porosity. Theses measurements aim to understand the influence the morphology of the PbI_2 film on the QCM-D frequency shift and dissipation shift. The porous character leads to a certain amount of liquid trapped inside the film, adding to its wet mass and thereby increasing the measured frequency shift. We furthermore observe an influence of the morphology on the dissipation, indicating a significant drag at the interface between the porous film and the overlying bulk liquid.

We have carried out a QCM-D measurement while changing between two solvents of different density and viscosity, similar to the analysis previously shown on thin mesoporous TiO_2 films.[23] Our choice of solvents was isopropanol ($\rho = 0.781$ g cm^{-3}, $\eta = 2.040$ mPa s) and chlorobenzene ($\rho = 1.106$ g cm^{-3}, $\eta = 0.753$ mPa s), since they do not dissolve PbI_2. The according measurement is shown in Fig. 10: the green line shows the QCM-D response on a flat, compact TiO_2 film as a reference, which corresponds to the changes in bulk viscosity and density of the overlying liquid, and should correspond to the description by Kanazawa and Gordon:[24–26]

$$\Delta f = {f_0}^{3/2}\sqrt{\frac{\eta\rho}{\pi\mu_q\rho_q}} \text{ and } \Delta D = \frac{1}{t_q\rho_q}\sqrt{\frac{\eta\rho}{\pi f_0}}.$$

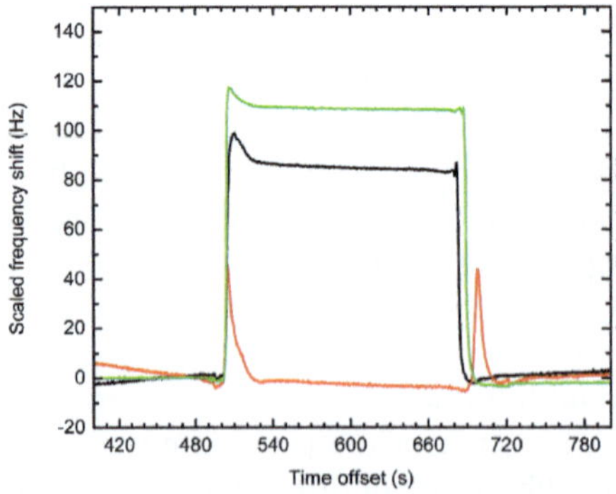

(a) Frequency shift.

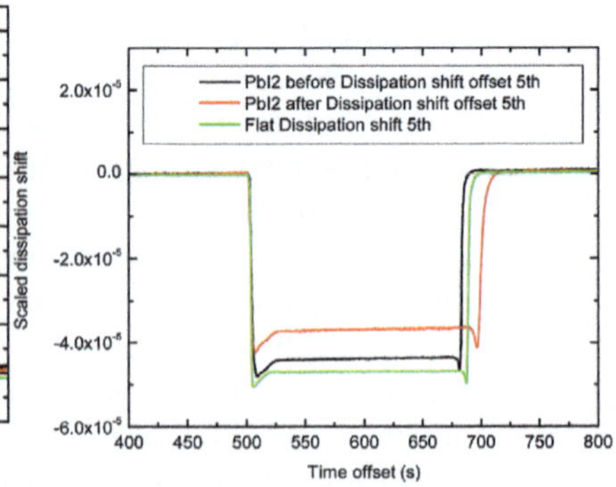

(b) Dissipation shift.

Fig. 10 QCM-D data for solvent exchange from isopropanol to chlorobenzene and back to isopropanol. Comparing a PbI_2 film prior to conversion (black), the same PbI_2 film after back-conversion (red), and a flat, compact TiO_2 film as a reference (green).

η and ρ are the viscosity and the density of the liquid, and μ_q and t_q are the shear modulus and the transverse sound velocity of the AT-cut quartz. The black line shows the QCM-D response on the initial PbI_2 film deposited by spin-casting prior to conversion. The red line shows QCM-D shifts of frequency and dissipation on the same PbI_2 film after conversion to the $CH_3NH_3PbI_3$ perovskite and back-conversion to PbI_2. The data are offset in time and y-value in order to facilitate a direct comparison.

Table 3 summarises the shifts of frequency and dissipation upon changing from isopropanol to chlorobenzene displayed in Fig. 10 at offset time $t = 560$ s. The last two columns show the differences $\delta f = \Delta f_{PbI_2} - \Delta f_{flat,TiO_2}$ and $\delta D = \Delta D_{PbI_2} - \Delta D_{flat,TiO_2}$ relative to the flat film. We observe a small difference in dissipation shift δD when comparing the dissipation shift on flat film to each of the dissipation shifts on the porous PbI_2. The difference in dissipation shift is $\delta D = +3 \times 10^{-6}$ for the PbI_2 film prior to conversion, and $\delta D = +10 \times 10^{-6}$ for the PbI_2 film after conversion to the $CH_3NH_3PbI_3$ perovskite and back-conversion to PbI_2.

However, for a very rough analysis, we will neglect the shift of dissipation and apply the Sauerbrey equation in order to obtain the approximate amount of trapped liquid from the frequency shift. As a rule of thumb, the Sauerbrey equation is still valid if the dissipation shift in 10^{-6} does not exceed 10% of the value of the frequency shift in Hz. This is the case in the above measurement, and we will discuss the dissipation shift at the end of this section. This rule of thumb, which relates the unit-less dissipation to the frequency shift in Hz, is best understood by recalling that QCM-D measures the properties of a resonance. The frequency shift is the shift of the (central) resonance frequency upon mass loading. The dissipation is proportional to the width of the resonance divided by its absolute frequency, $D = 1/Q = 2\Gamma/f_0 + 3 \times 10^{-6}$, ($D$: dissipation, Q: quality-factor, Γ: full width at half maximum in Hz, f_0: resonance frequency in Hz). The above rule of thumb expresses that the change in the width of the resonance has to be small compared to the shift of the resonance frequency.

The difference in frequency shift δf_{PbI_2} should be primarily related to the change in mass given by the product of the volume of trapped liquid $V_{liq.}$, and the difference in volumetric density ($\rho_{chlorobenzene} - \rho_{isopropanol}$) between chlorobenzene and isopropanol, described by the Sauerbrey equation:

$$-C \times \delta f = \delta m = V_{liq.} \times (\rho_{chlorobenzene} - \rho_{isopropanol}),$$

with the Sauerbrey constant $C = f_0/q_t t_q \approx 17.5$ ng cm^{-2} Hz^{-1}. δm and $V_{liq.}$ are mass and volume per area. Resolving this equation gives the volume of liquid

Table 3 Summarising shifts of frequency and dissipation upon changing from isopropanol to chlorobenzene. Sample surfaces are flat, compact TiO_2, porous initial PbI_2 before conversion, and porous PbI_2 after back-conversion. Values of Δf and ΔD are taken from measurements displayed in Fig. 10 at offset time $t = 560$ s. The last column shows the differences $\delta f = \Delta f_{PbI_2} - \Delta f_{flat,TiO_2}$ and δD relative to the flat film

	$\Delta_{flat\ TiO_2}$	$\Delta_{PbI_2,\ before}$	$\Delta_{PbI_2,\ after}$	$\delta_{PbI_2,\ before}$	$\delta_{PbI_2,\ after}$
Frequency shift Δf, δf in Hz	+109	+89	−1	−23	−110
Dissipation shift ΔD, δD	-47×10^6	-44×10^6	-37×10^6	$+3 \times 10^6$	$+10 \times 10^6$

trapped inside the initial PbI_2 film prior to conversion, and the volume of liquid trapped inside the PbI_2 film after one back-conversion: $V_{liq.}(PbI_2, before) = 1.24 \times 10^{-12}$ m^3 cm^{-2} and $V_{liq.}(PbI_2, after) = 5.92 \times 10^{-12}$ m^3 cm^{-2}.

On the initial PbI_2 film, the trapped isopropanol thus accounts for $V_{liq.}(PbI_2, before) \times \rho_{isopropanol} = 967$ ng cm^{-2} of its wet mass. On the PbI_2 film after back-conversion and before complete dissolution in DMF, the trapped isopropanol accounts for $V_{liq.}(PbI_2, before) \times \rho_{isopropanol} = 4626$ ng cm^{-2} of its wet mass. The latter can be used to calculate the dry mass of the adsorbed PbI_2 by dissolution of the entire film, which gives a measure of its wet mass. We did this in Section 2.1 and obtained a dry area mass of 110.92×10^{-6} g cm^{-2} for the PbI_2 film. Taking into account the density of PbI2, $\rho_{PbI_2} = 6.160$ g cm^{-3}, we can obtain a PbI_2 volume per area of 18.01×10^{-12} m^3 cm^{-2}. For the PbI_2 film after one back-conversion from the perovskite, we measured a trapped liquid volume of 5.92×10^{-12} m^3 cm^{-2}, giving an effective porosity of 24.7%. Prior to conversion, the volume of trapped liquid is approximately 1.24×10^{-12} m^3 cm^{-2}. This leads us to estimate the effective porosity of the initial, spin-cast PbI_2 film to be only around 6.4%.

Dissipation shift and viscoelastic behaviour. Evaluating the difference in dissipation shift observed in the measurement above (Fig. 10) is less straightforward. In order to obtain a general overview, we recall the QCM-D measurement of the conversion from PbI_2 to the perovskite and its back-conversion, as displayed in Fig. 5.

During conversion to the perovskite, the dissipation went up to approximately 25×10^{-6}, including a small contribution from changes in bulk viscosity, which produced 5×10^{-6} dissipation shift on a flat, non-reacting control sample (not shown). During back-conversion, dissipation goes through a peak of approximately 50×10^{-6}, and finally falls back to a value that is 18×10^{-6} higher than on the initial PbI_2 film. Upon dissolution of the PbI_2 film, the dissipation goes approximately back to the same value as the initial PbI_2 film (dissipation shift 0). The SEM picture in Sections 2.3 show, that the change of dissipation clearly correlates with a change in film morphology.

The change in dissipation may be due to a change in the way the surface of the film couples to the overlying bulk liquid. In a simplified picture, the different morphology will produce a different drag in the liquid, which leads to a change in dissipation. This would require to introduce a form factor describing the interface morphology into the equations of Kanazawa and Gordon which describe the coupling to the overlying bulk liquid:

$$\Delta f = {f_0}^{3/2}\sqrt{\frac{\eta\rho}{\pi\mu_q\rho_q}} \text{ and } \Delta D = \frac{1}{t_q\rho_q}\sqrt{\frac{\eta\rho}{\pi f_0}}.$$

We do not attempt to provide a quantitative description of this correlation yet, but we can try to exclude other possible sources of a shift in dissipation. A common reason for an increase in dissipation would be a softening of the adsorbed film. This is not the case, since the PbI_2 film and the perovskite itself are rigid and they should be well interconnected. It would be possible that the attachment of the PbI_2 film to the TiO_2 underlayer is weakened upon the first conversion and thereby causes an upshift in dissipation. However, we think this is unlikely because the conversion was incomplete in this experiment, and the SEM pictures in Section 2.3 indicate that some of the PbI_2 close to the TiO_2 substrate does not change its morphology.

Another possible reason for the frequency shift would be turbulent drag due to the flow inside the flow chamber. A turbulent flow can be described by the Reynolds number Re, which is usually defined as a dimensionless number

$$\mathrm{Re} = \frac{\rho v}{\eta}\hat{D}.$$

ρ and η are the density and the viscosity of the liquid, and v is a mean velocity of the liquid. $\hat{D}$ is a form factor of a length unit m, which describes the flow around an object, for example a tube, a sphere, or a rough surface. Below a certain Reynolds number, the flow is laminar, whereas it becomes turbulent above a certain Reynolds number. Increased turbulent flow may lead to increased dissipation in the QCM-D signal as shown by Johannsmann *et al.*[27] However, we do not see any significant change in the dissipation when either stopping the flow, or when pumping with a nominal pump speed of 500 μl instead of the usual 100 μl as shown in Section 2.2, which changes the velocity of the liquid v and thus will influence the turbulent drag.

The quantitative impact of the morphology on QCM-D signals is not understood in detail, and we therefore refrain from further analysis of this phenomena, since its influence on shifts of frequency and dissipation appears to be relatively small compared to the major shifts of frequency due to mass change, which we observe during the conversion and back-conversion reaction.

Concerning the dissipation shift, the most remarkable feature is that during the first conversion, the dissipation shifts up by 20×10^{-6} and then returns approximately to this value after the second and the third cycle of conversion and back-conversion. As explained in Section 2.5, the dissipation shift correlates directly with large changes in film morphology, although we cannot explain the correct mechanism yet. Furthermore, there is an influence of the bulk viscosity and density, which are slightly higher during the exposure to the CH_3NH_3I solution than for the pure isopropanol and account for a shift of approximately $+5 \times 10^{-6}$ on a flat film.

3 Conclusions

Whereas the conversion process from the PbI_2 to the $CH_3NH_3PbI_3$ perovskite has been studied previously by UV-Vis absorption and emission techniques such as XRD and SEM, the back-conversion has not been observed so far. From the data collected until now we can already draw significant insight into this intriguing process. The results presented have been collected on a thin film with an initial PbI_2 thickness of 200 nm but the results should be of general validity, in particular for the mesoscopic systems used in perovskite photovoltaics.

The most evident conclusion from the above results is, that the conversion reaction of PbI_2 to $CH_3NH_3PbI_3$ perovskite is reversible by extraction of the CH_3NH_3I from the perovskite into the solvent. This runs counter to the prevailing contention that the entire perovskite disintegrates in contact with polar solvents.

The implications are numerous. First, the dissolution of the CH_3NH_3I from the perovskite in a moderately polar solvent like isopropanol implies that the same should hold true for any polar solvent, including water. Since CH_3NH_3I is volatile (the solid sublimes under ambient conditions), this suggests one possible decomposition pathway in humid environments. Strongly polar solvents will equally dissolve parts of the PbI_2.

In general, the reversibility of the conversion reaction should hold true for any lead-halide perovskite, the ease of the transformation being controlled by the interplay of the respective PbX_2 salt and CH_3NH_3X in solution. This also suggests, that a readily formed perovskite of one halide may possibly be converted directly to the perovskite of another halide, simply by exposing to a CH_3NH_3X solution of the second halide. This corresponds to an ion-exchange reaction whose extent is controlled by the mass action law. Importantly we have confirmed recently that a strikingly rapid and facile exchange of halide ions in $CH_3NH_3PbX_3$ can indeed be achieved by solution processing at room temperature [Pellet *et al.*, submitted].

The SEM study in Section 2.3 has shown that the morphology of the PbI_2 film is affected by the conversion and back-conversion, resulting in a PbI_2 film that consists of larger PbI_2 crystallites with improved crystallinity The QCM-D results in Section 3 further indicate that this does increase the amount of incorporated CH_3NH_3I, as well as the size of perovskite crystallites. This effect can be used to direct the morphology in a desired way and improve the overall amount of converted material on flat films.

Beyond the back-conversion, the QCM-D measurements also shed some light on the kinetics of the conversion process from PbI_2 to $CH_3NH_3PbI_3$. There clearly is a rapid initial conversion, which occurs in the first 30 s of the exposure to CH_3NH_3I on a pre-wet sample. This correlates well with optical measurements.[6] However, the QCM-D results suggest that during continued CH_3NH_3I exposure, there is additional CH_3NH_3I incorporated into the film, but at a much slower rate. We recall that the SEM images in Section 3 and 4 indicate that only the upper part of the film is converted to perovskite crystallites, whereas the bottom part resembles the morphology of PbI_2 crystallites. This suggests that the perovskite layer forms a compact cover, impairing the access of CH_3NH_3I to the underlying PbI_2.

References

1 A. Kojima, K. Teshima, Y. Shirai and T. Miyasaka, *J. Am. Chem. Soc.*, 2009, **131**, 6050–6051.

2 J.-H. Im, C.-R. Lee, J.-W. Lee, S.-W. Park and N.-G. Park, *Nanoscale*, 2011, **3**, 4088–4093.

3 H.-S. Kim, C.-R. Lee, J.-H. Im, K.-B. Lee, T. Moehl, A. Marchioro, S.-J. Moon, R. Humphry-Baker, J.-H. Yum, J. E. Moser, M. Grätzel and N.-G. Park, *Sci. Rep.*, 2012, **2**, 1–7.

4 M. M. Lee, J. Teuscher, T. Miyasaka, T. N. Murakami and H. J. Snaith, *Science*, 2012, **338**, 643–647.

5 L. Etgar, P. Gao, Z. Xue and P. Qin, *J. Am. Chem. Soc.*, 2012, **134**, 17396–17399.

6 J. Burschka, N. Pellet, S.-J. Moon, R. Humphry-Baker, P. Gao, M. K. Nazeeruddin and M. Grätzel, *Nature*, 2013, **499**, 316–319.

7 S. Ryu, J. H. Noh, N. J. Jeon, Y. C. Kim, W. S. Yang, J. Seo and S. Il Seok, *Energy Environ. Sci.*, 2014, **7**, 2614–2618.

8 NREL, *Best research-cell efficiencies*, NREL, 2014.

9 H. Mashiyama, Y. Kurihara and T. Azetsu, *J. Korean Phys. Soc.*, 1998, **32**, 156–158.

10 A. Yella, L.-P. Heiniger, P. Gao, M. K. Nazeeruddin and M. Grätzel, *Nano Lett.*, 2014, **14**, 2591–2596.

11 D. Liu and T. L. Kelly, *Nat. Photonics*, 2013, **8**, 133–138.

12 J. Burschka, PhD Thesis, École polytechnique fédérale de Lausanne (EPFL), "High performance solid state mesoscopic solar cells", 2013.

13 T. Baikie, Y. Fang, J. M. Kadro, M. Schreyer, F. Wei, S. G. Mhaisalkar, M. Graetzel and T. J. White, *J. Mater. Chem. A*, 2013, **1**, 5628–5641.

14 Y. Kawamura, H. Mashiyama and K. Hasebe, *J. Phys. Soc. Jpn.*, 2002, **71**, 1694–1697.

15 D. Weber, *Z. Naturforsch., B: J. Chem. Sci.*, 1978, **33**, 1443–1445.

16 S. E. Derenzo, M. J. Weber and M. K. Klintenberg, *Nucl. Instrum. Methods Phys. Res., Sect. A*, 2002, **486**, 214–219.

17 K. Liang, D. B. Mitzi and M. T. Prikas, *Chem. Mater.*, 1998, **10**, 403–411.

18 N. Tétreault, L.-P. Heiniger, M. Stefik, P. L. Labouchère, É. Arsenault, N. K. Nazeeruddin, G. A. Ozin and M. Grätzel, *ECS Trans.*, 2011, **41**, 303–314.

19 F. Matteocci, G. Mincuzzi, F. Giordano, A. Capasso, E. Artuso, C. Barolo, G. Viscardi, T. M. Brown, A. Reale and A. Di Carlo, *Org. Electron.*, 2013, **14**, 1882–1890.

20 L. Kavan and M. Grätzel, *Electrochemistry*, 1995, **40**, 643–652.

21 H. J. Snaith and M. Grätzel, *Adv. Mater.*, 2006, **18**, 1910–1914.

22 H. A. Harms, N. Tétreault, V. Gusak, B. Kasemo and M. Grätzel, *Phys. Chem. Chem. Phys.*, 2012, **14**, 9037–9040.

23 H. Harms, Chapter 4, PhD Thesis,Ecole Polytechnique Fédérale de Lausanne, 2014.

24 K. K. Kanazawa and J. Gordon, *Anal. Chim. Acta*, 1985, **175**, 99–105.

25 K. K. Kanazawa and J. G. Gordon, *Anal. Chem.*, 1985, **57**, 1770–1771.

26 M. Rodahl, *Sens. Actuators, B*, 1996, **37**, 111–116.

27 D. Johannsmann, I. Reviakine and R. P. Richter, *Anal. Chem.*, 2009, **81**, 8167–8176.

Faraday Discussions

PAPER

Liquid phase deposition of TiO_2 nanolayer affords $CH_3NH_3PbI_3$/nanocarbon solar cells with high open-circuit voltage†

Haining Chen,‡[a] Zhanhua Wei,‡[a] Keyou Yan,[a] Ya Yi,[b] Jiannong Wang[b] and Shihe Yang*[a]

Received 23rd July 2014, Accepted 18th August 2014
DOI: 10.1039/c4fd00155a

Hybrid organic/inorganic perovskite solar cells are attracting intense attention and further developments largely hinge on understanding the fundamental issues involved in the cell operation. In this paper, a liquid phase deposition (LPD) method is developed to design and grow a TiO_2 nanolayer at room temperature for carbon-based perovskite solar cells. The TiO_2 nanolayer grown on FTO glass is compact but polycrystalline consisting of tiny anatase TiO_2 nanocrystals intimately stacked together. By directly exploiting this TiO_2 nanolayer in a solar cell of TiO_2 nanolayer/$CH_3NH_3PbI_3$/nanocarbon, we have achieved a V_{oc} as high as 1.07 V, the highest value reported so far for hole transporter-free $CH_3NH_3PbI_3$ solar cells. This is rationalized by the slower electron injection and longer electron lifetime due to the TiO_2 nanolayer, which enhances the electron accumulation in $CH_3NH_3PbI_3$ and consequently the V_{oc}. By employing a rutile TiO_2 nanorod (NR) array as a base structure for the LPD-TiO_2 nanolayer to support the $CH_3NH_3PbI_3$ layer, the photocurrent density is considerably increased without obviously compromising the V_{oc} (1.01 V). As a result, the power conversion efficiency is boosted from 3.67% to 8.61%. More elaborate engineering of the TiO_2 nanolayer by LPD in conjunction with judicious interfacing with other components has the potential to achieve higher performances for this type of solar cell.

1 Introduction

Hybrid organic/inorganic perovskite solar cells have the potential to offer high conversion efficiencies while permitting low-cost solution based manufacturing. In the last five years, the efficiencies of perovskite solar cells have skyrocketed

[a]*Department of Chemistry, The Hong Kong University of Science and Technology, Clear Water Bay, Kowloon, Hong Kong, China. E-mail: chsyang@ust.hk*

[b]*Department of Physics, The Hong Kong University of Science and Technology, Clear Water Bay, Kowloon, Hong Kong, China*

† Electronic supplementary information (ESI) available: Supplementary SEM images, Raman shift, $(\alpha h\nu)^2$–$h\nu$ plot and UPS spectrum. See DOI: 10.1039/c4fd00155a

‡ These authors contributed equally to this work.

from the initial 3% to 17.9%,[1–11] and over 20% efficiencies are expected to be achieved in the near future with the progress in basic understanding of their working mechanism and the further development of more controllable fabrication processes.[4,12–14] Furthermore, through integration with silicon panels to create tandem cells, obtaining much higher efficiencies may be possible. Therefore, perovskite solar cells are predicted as the "next big thing" in photovoltaics.

Hybrid organic/inorganic perovskite solar cells typically consist of TiO_2/perovskite/hole transport material (HTM)/noble metal electrode. However, other cell structures are exploited and developed due to the excellent photovoltaic properties of perovskite materials ($CH_3NH_3PbI_3$ and $CH_3NH_3PbI_{3-x}Cl_x$), including strong solar absorption,[2,3,7] ambipolar charge transport,[4,13,15] and long balanced carrier diffusion lengths (100–1000 nm).[16,17] Among those new cell structures, HTM-free perovskite solar cells have become more and more promising, which avoid using unstable and/or expensive organic HTMs (mainly spiro-OMeTAD).[18–27] Very recently, carbon-based perovskite solar cells stand out for their high efficiencies (a certified efficiency of 12.8%) and the obvious advantages of the carbon electrode, such as material abundance, low cost, superior stability and printable fabrication processes.[24–28]

Regardless of the development of different cell structures, a TiO_2 nanolayer (dense layer and/or mesoporous layer) is still needed for high efficiency perovskite solar cells. In dye-sensitized solar cells (DSSCs), the TiO_2 nanolayer accepts injected electrons and is the unique electron transport pathway. However, since perovskite itself is an efficient electron transporter, the TiO_2 nanolayer is not a unique electron transport pathway in perovskite solar cells.[13,15] And it has been proposed that partial electron accumulation and transport in the perovskite is favorable for obtaining a higher open-circuit voltage (V_{oc}) (due to the higher conduction band edge of perovskite than that of TiO_2) and thus a higher efficiency.[3,29,30]

Herein, we report the growth of a TiO_2 nanolayer by the liquid phase deposition (LPD) method[31,32] at room temperature for $CH_3NH_3PbI_3$/nanocarbon solar cells. In contrast to the commonly used TiO_2 nanolayer, this unique LPD-TiO_2 nanolayer could slow electron injection and prolong electron lifetime, which as a result enhanced the electron accumulation in $CH_3NH_3PbI_3$ and led to a considerably higher V_{oc} of 1.07 V, the highest value for HTM-free $CH_3NH_3PbI_3$ solar cells. We have further used a TiO_2 nanorod (NR) array as a base structure for this LPD-TiO_2 nanolayer to support the $CH_3NH_3PbI_3$ photoactive layer, and achieved a significantly boosted photocurrent density without obviously compromising V_{oc} (1.01 V), promoting the cell efficiency from 3.67% to 8.61%.

2 Experimental

2.1 Preparation

Preparation of LPD-TiO_2 nanolayers. To prepare TiO_2 nanolayers by the LPD method,[31,32] the reaction solutions were firstly prepared by dissolving 0.05 M $(NH_4)_2TiF_6$ and 0.15 M boric acid (H_3BO_3) in distilled water. Then, FTO glasses were placed in the reaction solution with the FTO thin film sides facing down. The reaction was carried out at room temperature and the thickness of the TiO_2 nanolayers was controlled by varying the reaction duration.

TiO_2 NR array as a base structure for LPD-TiO_2 nanolayers. The TiO_2 NRs were grown on FTO glass using the hydrothermal method. Briefly, FTO glass was immersed in the Teflon-lined stainless steel autoclave (100 mL capacity) containing 0.6 mL of titanium(IV) butoxide, 25 mL of deionized (DI) water and 25 mL of concentrated HCl (36–38%).[8,33] The hydrothermal process was carried out by putting the autoclave into an oven and heating at 180 °C for 3 h. The autoclave was then removed from the oven and cooled down to room temperature naturally. The samples were taken out and washed with DI water, and finally annealed at 500 °C for 1 h in air.

To deposit a LPD-TiO_2 nanolayer on the FTO glass/TiO_2 NR array, the TiO_2 NR array was immersed in the LPD solutions with the TiO_2 NR array side facing down.

Fabrication of $CH_3NH_3PbI_3$/nanocarbon solar cells. First, the TiO_2 nanolayers were pre-heated to 100 °C. Then, a *N*,*N*-dimethylformamide solution containing 460 mg mL^{-1} PbI_2, which was pre-heated to 100 °C, was spin coated onto the TiO_2 nanolayer-coated FTO glasses at 2500 rpm for 30 s.[6] After being dried in air, the samples were heated up to 100 °C and left for 15 min to crystallize the PbI_2. Finally, the samples were cooled naturally to room temperature again.

The deposition of the nanocarbon electrode was carried out by spin coating a chlorobenzene suspension of 15 mg mL^{-1} nanocarbon particles onto the top of TiO_2/PbI_2. Then, the samples were dried at room temperature and heated at 100 °C for 5 min.

Conversion of the PbI_2 layer to $CH_3NH_3PbI_3$ was performed by spin coating a 2-propanol solution containing 10 mg mL^{-1} CH_3NH_3I onto the top of TiO_2/PbI_2/nanocarbon at 100 rpm for 20 s.[6] Finally, the samples were gently dried under a flow of dry air and then heated to 100 °C and left for 15 min to produce the $CH_3NH_3PbI_3$/nanocarbon solar cells.

For measurement, another piece of FTO glass, serving as an assistant collecting electrode, was clamped with the above $CH_3NH_3PbI_3$/nanocarbon component for solar cell performance evaluation.

Preparation of USP-TiO_2 nanolayer. A commonly used TiO_2 dense layer for control experiments was deposited on FTO by a home-made ultrasonic spray deposition (USP) system at 500 °C using a titanium diisopropoxide bis(acetylacetonate) solution (0.05 M, ethanol) as the precursor and air as the carrier gas.

2.2 Characterization

The morphology was evaluated using scanning electron microscopy (SEM, JEOL6700F) at an accelerating voltage of 5 kV. X-ray diffraction (XRD) patterns were recorded on a Philips PW-1830 X-ray diffractometer with Cu Kα radiation ($K = 0.15418$ nm). Diffuse reflectance UV-vis absorption spectra were recorded on a Perkin-Elmer UV-vis spectrophotometer (model Lambda 20). The XPS spectra were measured on a Perkin-Elmer model PHI 5600 XPS system with a resolution of 0.3–0.5 eV from a monochromated aluminium anode X-ray source with Mo Kα radiation (1486.6 eV). Ultraviolet photoelectron spectroscopy (UPS) measurements were performed on a photoelectron spectrometer (Kratos Analytical, AXIS-ultra-DLD) with He(I) excitation of 21.22 eV and pass energy of 5 eV.

The Mott–Schottky (M–S) plot was acquired using a CHI 660A electrochemical analyzer in a three-electrode system. The LPD-TiO_2 nanolayer, an Ag/AgCl electrode and a Pt wire served as the working, reference and counter electrodes, respectively. The electrolyte used for this measurement contained 1 M KOH and the frequency was set at 3 kHz.

2.3 Photovoltaic measurements

The solar light simulator (Newport solar simulator, model number 6255, 150 W Xe lamp, AM 1.5 global filter) was calibrated to 1 sun (100 mW cm^{-2}) using a silicon reference solar cell equipped with a KG-5 filter. A black metal shading mask (0.09 cm^2) was used to define the illumination area of the cells to be tested and tightly packaged to prevent light incident from the edges. Current density–voltage (*J*–*V*) characteristic curves and open circuit voltage decay (OCVD) curves were recorded on an IM6x electrochemical workstation (ZAHNER-Elektrik GmbH & Co., KG, Germany). Incident-photo-to-current conversion efficiency (IPCE) was recorded using an IPCE kit developed by ZAHNER-Elektrik in AC mode with a frequency of 1 Hz. For time-resolved photoluminescence (TRPL) measurements, a tunable Ti:sapphire femtosecond-pulsed laser was used as the excitation light source, with an excitation wavelength of 400 nm and an incident light intensity of 1 W cm^{-2}. A Hamamatsu C5680-04 streak camera was used for the TRPL experiments. The TRPL measurements were performed under vacuum.

3 Results and discussion

3.1 The formation of the LPD-TiO_2 nanolayer and its influence on $CH_3NH_3PbI_3$/nanocarbon solar cells

To prepare LPD-TiO_2 nanolayers, we used an aqueous solution containing 0.05 M $(NH_4)_2TiF_6$ and 0.15 M H_3BO_3. The LPD of metal oxide nanolayers was performed *via* a chemical equilibrium reaction between a metal fluoro-complex and a metal oxide in an aqueous solution.[31,32] Specifically, the deposition reactions of LPD-TiO_2 nanolayers can be expressed as:

$$TiF_6^{2-}\ (aq) + 2H_2O \rightarrow TiO_2\ (s) + 4HF\ (aq) + 2F^-\ (aq) \tag{1}$$

$$H_3BO_3\ (aq) + 4\ HF\ (aq) \rightarrow BF_4^-\ (aq) + 3H_2O\ (l) + H^+\ (aq) \tag{2}$$

Interestingly, H_3BO_3 here can serve as an F^- scavenger, which consumes the F^- ions (eqn (2)) and thus accelerates the hydrolysis reaction (eqn (1)). In this way, LPD at low temperature[31,32] could produce partially crystalline TiO_2, which could be used to control the electron injection by modulating the TiO_2/perovskite interface.

SEM was firstly applied to evaluate the morphology of the LPD-TiO_2 nanolayers. As shown in Fig. 1(A), although the surface feature of FTO glass could still be observed, the FTO substrate becomes rough in nanoscale due to the coverage of tiny nanoparticles after LPD for 6 h. The cross sectional SEM image in Fig. 1(A1) shows that the film thickness is about 50 nm. As the LPD duration increases to 12 h, the surface features of FTO glass could hardly be observed, demonstrating the formation of a relatively flat LPD-TiO_2 nanolayer. The size of the tiny nanoparticles seems to be unchanged, suggesting that generating a new crystal nucleus instead of the continued growth of old crystals mainly promotes the growth of the LPD-TiO_2 nanolayer, which is favorable for producing a compact

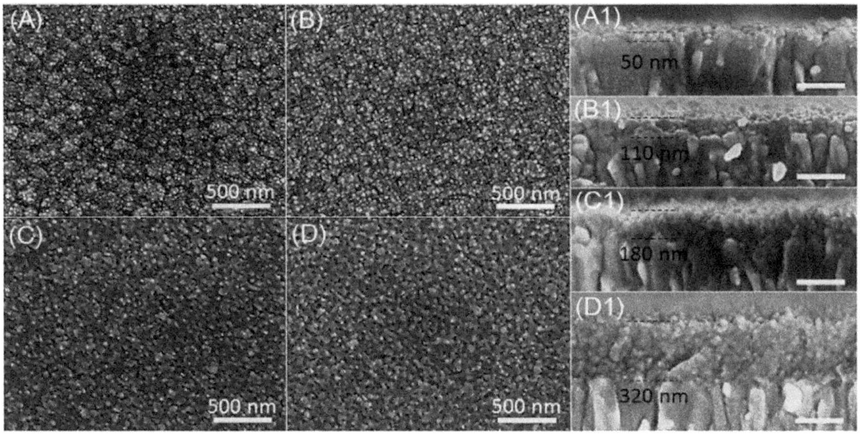

Fig. 1 Morphology evaluation of the LPD-TiO_2 nanolayer synthesised with different LPD durations. Top-view SEM images: (A) 6 h, (B) 12 h, (C) 18 h and (D) 36 h. (A1), (B1), (C1) and (D1) are the corresponding cross sectional SEM images. Scale bar: 200 nm.

TiO_2 nanolayer. As indicated in Fig. 1(B1), the LPD-TiO_2 nanolayer very compactly covers the FTO glass with the thickness increasing to about 110 nm.

When the LPD duration was extended to 18 h, the size of the tiny nanoparticles is still very small (Fig. 1(C)), but small pores with the size similar to that of the tiny particles are homogeneously dispersed on the surface. As depicted in Fig. 1(C1), the film thickness is increased to about 180 nm and the film is still intimately covered on the surface of the FTO glass. Further extending the LPD duration to 36 h resulted in a slight increase in size for both the tiny nanoparticles and the pores (Fig. 1(D)) with the film thickness increasing to about 320 nm (Fig. 1(D1)). As indicated in the SEM images and observed during experiments, the LPD-TiO_2 nanolayers show excellent adherence to the FTO glass, which could not be destroyed by ultrasonic agitation.

XRD patterns and Raman spectra were recorded to evaluate the composition and structural phase of the LPD-TiO_2 nanolayers. As shown in Fig. 2(A), in addition to the characteristic diffraction peaks of FTO glass, other peaks at 25.0° and 47.8° are observed, which can be well indexed to the (101) and (200) planes of anatase TiO_2, respectively. Similarly, as indicated in Fig. 2(B), the peaks at 166 cm^{-1}, 501 cm^{-1} and 623 cm^{-1} can be assigned to the three Raman-active modes of E_g, A_{1g} and E_g of anatase TiO_2,[34–38] respectively. However, compared with the corresponding Raman shift data in the literature,[34–36] obvious blue shifts occurs, plausibly due to the much smaller size of the anatase TiO_2 crystals prepared by LPD.[38] According to the relation between the position of the E_g mode (ω) and the diameter of TiO_2 nanocrystals (L),[38]

$$\omega = 100(1/L^{1.5}) + \omega_0 \tag{3}$$

where $\omega_0 = 142\ cm^{-1}$, the diameter of the anatase TiO_2 nanocrystals should be only about 3 nm. Together, both the XRD and Raman shift results demonstrate that the LPD-TiO_2 nanolayer is in the anatase phase and composed of tiny

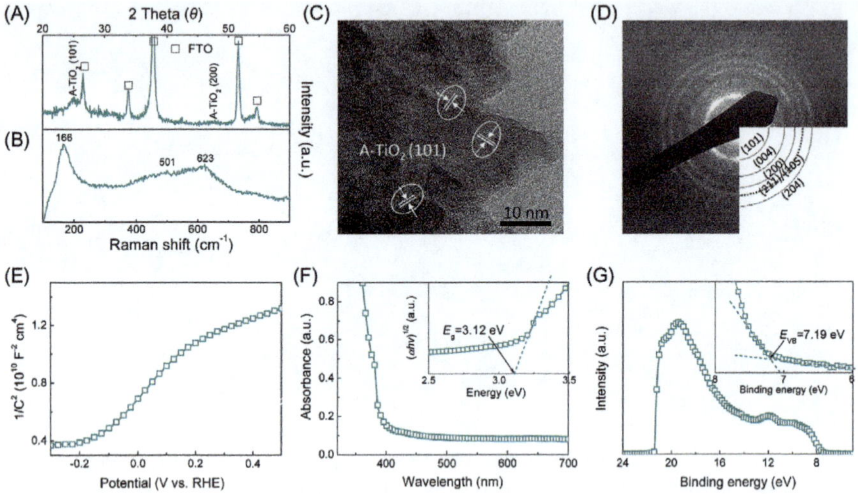

Fig. 2 Composition and structural phase of the LPD-TiO_2 nanolayer: (A) XRD pattern, (B) Raman spectrum, (C) HRTEM images, and (D) electron diffraction pattern. Semiconducting properties of the LPD-TiO_2 nanolayer: (E) M–S plot, (F) UV-vis absorption spectrum, and (G) UPS spectrum. Inset in (F) is the $(\alpha hv)^{1/2}$–hv plot and inset in (G) is the magnified region between of the UPS spectrum, 6–8 eV.

nanocrystals. Both the peak intensities of the XRD pattern and Raman shift are low, again suggesting the low crystallinity of the anatase TiO_2 nanolayers.

TEM was also applied to characterize the LPD-TiO_2 nanolayers. It can be readily seen in Fig. 2(C) that the LPD-TiO_2 nanolayers are polycrystalline. As indicated by the white line circles, the nanocrystal diameter is about several nanometers, consistent with the Raman shift results. Note that these tiny nanocrystals are compactly stacked together, making it possible to form a uniform film. The observed electron diffraction pattern (Fig. 2(D)) can be readily indexed to the (101), (004), (200), (211), (105) and (204) planes of anatase TiO_2, confirming the polycrystalline nature of anatase TiO_2 nanolayer prepared by LPD.

To study the semiconducting properties of the LPD-TiO_2 nanolayer, Mott–Schottky (M–S) plot was first measured. As shown in Fig. 2(E), the M–S plot shows a positive slope, indicating an n-type semiconductor behavior. The free charge carrier density in the LPD-TiO_2 nanolayer could be calculated using eqn (4):[39]

$$N_c = (2/e_0\varepsilon\varepsilon_0)[d(1/C^2)/dV]^{-1} \tag{4}$$

where e_0 is the electron charge, ε the dielectric constant of anatase TiO_2, ε_0 the permittivity of vacuum, N_c the carrier density and V the applied potential. N_c for the LPD-TiO_2 nanolayer is calculated to be about 1.27×10^{20} cm^{-3}. This high N_c will be favorable for electron transport in, for example, solar cells.

From the UV-vis absorption spectrum in Fig. 2(F), the absorption threshold of the LPD-TiO_2 nanolayer is obtained at about 390 nm, corresponding to the electronic transition from the valence band (VB) to the conduction band (CB) of anatase TiO_2. The $(\alpha hv)^{1/2}$–hv plot in the inset of Fig. 2(F) indicates that the band gap (E_g) values of the LPD-TiO_2 nanolayer is about 3.12 eV, consistent with the typical value for anatase TiO_2. The ultraviolet photoelectron spectroscopy (UPS) result in Fig. 2(G) indicates that the VB edge for the LPD-TiO_2 nanolayer is about −7.19 eV with reference to the vacuum level. The CB edge of the LPD-TiO_2 nanolayer is calculated to be about −4.07 eV using the relation: $E_g = E_{CB} - E_{VB}$, which is lower than the CB edge of $CH_3NH_3PbI_3$. Therefore, the energy band structure of the LPD-TiO_2 nanolayer is suitable for electron injection in the $CH_3NH_3PbI_3$/nanocarbon solar cells.

To fabricate the $CH_3NH_3PbI_3$/nanocarbon solar cells based on the LPD-TiO_2 nanolayer, the processes illustrated in Fig. 3(A) are applied. Firstly, the *N*,*N*-dimethylformamide solution containing 460 mg mL^{-1} PbI_2 was spin coated onto the LPD-TiO_2 nanolayer.[6] After crystallization, a vivid PbI_2 layer composed of sheet-like nanocrystals is deposited on the LPD-TiO_2 nanolayer, with a thickness of about 300 nm (Fig. S1(A)†). Nanocarbon electrode was coated on the top of FTO/TiO_2/PbI_2 by spin coating a chlorobenzene solution containing 15 mg mL^{-1} nanocarbon particles. After drying, a porous nanocarbon layer tens of micrometers thick is formed on the top of the PbI_2 layer with nanocarbon nanoparticles of about 30–80 nm in diameter, as shown in Fig. S1(B) and (C).† Then, the 2-propanol solution containing 10 mg mL^{-1} CH_3NH_3I was spin coated on the top of FTO/TiO_2/PbI_2/nanocarbon.[6] Due to the porous structure of nanocarbon electrode, the CH_3NH_3I solution could penetrate through the nanocarbon electrode to reach and react with the PbI_2 layer to generate $CH_3NH_3PbI_3$. As a result, the PbI_2 layer would be converted to $CH_3NH_3PbI_3$ and this conversion was further promoted after heating at 100 °C for 15 min.

Fig. 3(B) represents the cross sectional SEM image of the typical TiO_2/$CH_3NH_3PbI_3$/nanocarbon solar cells, showing a clear layer-by-layer structure. The cubic crystal could be vividly observed at the $CH_3NH_3PbI_3$ interlayer,

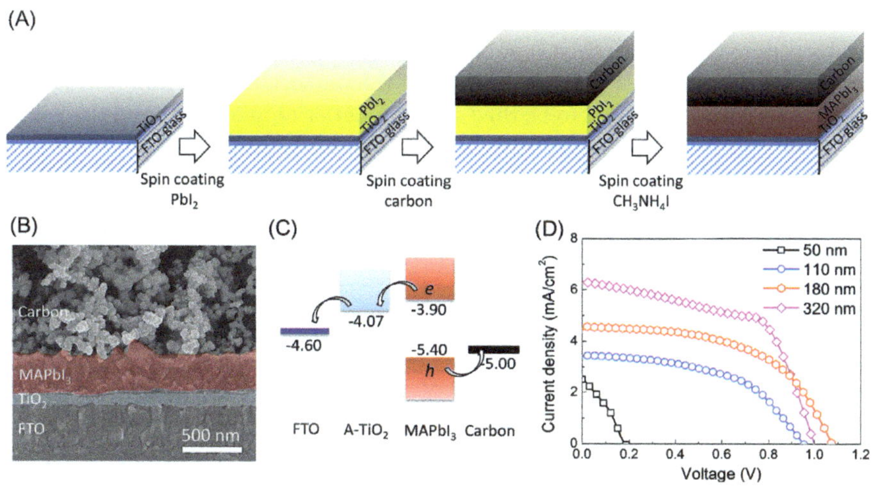

Fig. 3 (A) Schematic illustration of the preparation processes of the $CH_3NH_3PbI_3$/nanocarbon solar cells based on the LPD-TiO_2 nanolayer. (B) Cross sectional SEM images and (C) energy level diagram and charge transfer processes of the $CH_3NH_3PbI_3$/nanocarbon solar cells. And (D) *J*–*V* curves of the $CH_3NH_3PbI_3$/nanocarbon solar cells based on LPD-TiO_2 nanolayers with different thicknesses.

implying the high quality of $CH_3NH_3PbI_3$ crystals prepared by our preparation strategy. As expected, the thickness of the $CH_3NH_3PbI_3$ interlayer is increased to about 400 nm in comparison with that of PbI_2 layer because the volume would be increased after the conversion of PbI_2 to $CH_3NH_3PbI_3$. As a result, through a careful observation, a fraction of the bottom nanocarbon layer attached onto the PbI_2 layer is embedded in the $CH_3NH_3PbI_3$ layer, forming a very intimate contact between $CH_3NH_3PbI_3$ and the nanocarbon layers favorable for hole extraction in operating solar cells.

The energy level diagram and charge transfer processes in the $CH_3NH_3PbI_3$/nanocarbon solar cells are illustrated in Fig. 3(C). When $CH_3NH_3PbI_3$ is excited by light illumination, electrons and holes are generated on the CB and VB, respectively. Since the CB of the LPD-TiO_2 nanolayer is located at about −4.07 eV, electron injection will take place at the TiO_2/$CH_3NH_3PbI_3$ interface from the CB (−3.90 eV) of $CH_3NH_3PbI_3$ to the CB of TiO_2, and these electrons will be finally collected at the FTO glass. In parallel, the holes are extracted by the nanocarbon electrode through the $CH_3NH_3PbI_3$/nanocarbon interface from the VB (−5.40 eV) of $CH_3NH_3PbI_3$ to the nanocarbon electrode (Fermi level: −5.0 eV).[24]

Fig. 3(D) exhibits the J–V curves of the $CH_3NH_3PbI_3$/nanocarbon solar cells based on the LPD-TiO_2 nanolayers with different thicknesses, and Table 1 lists the corresponding photovoltaic parameters. For comparison, the solar cells based on bare FTO glass were also prepared, and their performance was so poor and unstable that their J–V curve could not be obtained. Comparatively, the LPD-TiO_2 nanolayer about 50 nm thick allowed us to obtain reproducible J–V curves, but the V_{oc} is only about 0.18 V and the J_{sc} is also very low, resulting in a low η at about 0.18%. As the film thickness is increased to 110 nm, the cell performance is obviously improved with a V_{oc} of 0.98 V and a J_{sc} of 3.44 mA cm^{-2}, amounting to a η of 1.67%.

Further thickening of the film to 180 nm continued to improve the cell performance with a V_{oc} as high as 1.07 V, which is the highest reported V_{oc} for HTM-free $CH_3NH_3PbI_3$ solar cells including carbon and Au-based $CH_3NH_3PbI_3$ solar cells.[18,19,21–27] Since the CB of LPD-TiO_2 nanolayer is located at −4.07 eV, the maximum obtainable V_{oc} is about 0.93 V if the highest electron and lowest hole energy levels are assumed to be the CB of the LPD-TiO_2 nanolayer and the Fermi

Table 1 Photovoltaic parameters obtained from the $CH_3NH_3PbI_3$/nanocarbon solar cells based on the LPD-TiO_2 nanolayers with different LPD durations (or film thicknesses) on FTO glass and FTO glass/TiO_2 NRs

Substrate	LPD duration (film thickness)	V_{oc} (V)	J_{sc} (mA cm^{-2})	FF	η (%)
FTO glass	6 h (50 nm)	0.18	2.50	0.40	0.18
	12 h (110 nm)	0.95	3.44	0.51	1.67
	18 h (180 nm)	1.07	4.54	0.54	2.62
	36 h (320 nm)	1.00	6.28	0.58	3.67
FTO glass/TiO_2 NRs	0 h	0.54	9.35	0.27	1.36
	2 h	0.67	12.98	0.35	3.04
	5 h	1.01	14.20	0.60	8.61
	9 h	0.89	15.74	0.49	6.86

level of the nanocarbon (−5.0 eV), respectively. The ultrahigh V_{oc} (1.07 V) we obtained further verifies the proposition of carrier accumulation in $CH_3NH_3PbI_3$.[4,13,15] What is more, J_{sc} and FF are also increased to 4.54 mA cm^{-2} and 0.54, respectively, leading to a η of 2.62%. Finally, when the film thickness is increased to 320 nm, V_{oc} is slightly decreased to 1.00 V, but J_{sc} continued to increase to 6.28 mA cm^{-2}, corresponding to a η up to 3.67%. It seems that the level of electron injection from perovskite to TiO_2 plays an important role is dictating the V_{oc} and J_{sc}.

3.2 Insight into the influence of the LPD-TiO_2 nanolayer on the cell performance

To gain a deeper insight into the performance enhancement mechanism of the LPD-TiO_2 nanolayer, a commonly used dense anatase TiO_2 nanolayer with high crystallinity (Fig. S2†) for perovskite solar cells was prepared by ultrasonic spray pyrolysis (USP). For meaningful comparison, the thicknesses of the two TiO_2 nanolayers were controlled at about 100 nm. As shown in Fig. 4(A), the current densities for both solar cells are not very high since they are compact and thin. But it can still be seen that the USP-TiO_2 nanolayer based solar cells achieve a higher J_{sc} (5.12 mA cm^{-2}) than the LPD-TiO_2 nanolayer based ones (3.02 mA cm^{-2}). Whereas the latter achieves a considerably higher V_{oc} (0.94 V) and FF (0.56) than the former (V_{oc} = 0.78 V and FF = 0.31). As a result, the former achieves a η of 1.63%, which is higher than the latter one (1.37%).

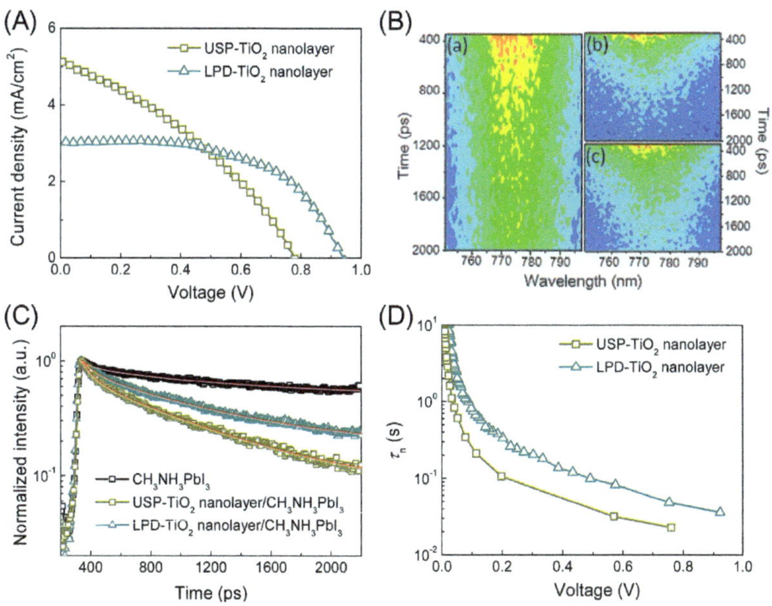

Fig. 4 Performance comparison between the USP-TiO_2 nanolayer and LPD-TiO_2 nanolayer. (A) J–V curves of $CH_3NH_3PbI_3$/nanocarbon solar cells; (B) TRPL streak camera images of (a) $CH_3NH_3PbI_3$, (b) USP-TiO_2 nanolayer/$CH_3NH_3PbI_3$ and (c) LPD-TiO_2 nanolayer/$CH_3NH_3PbI_3$; (C) Fitted TRPL with biexponential decay traces; and (D) τ_n–V_{oc} curves of $CH_3NH_3PbI_3$/nanocarbon solar cells.

To pin down the origin of the performance difference between these two TiO_2 nanolayers, TRPL was first performed to investigate the electron injection rate from $CH_3NH_3PbI_3$ to TiO_2 nanolayers. Fig. 4(B) presents the streak camera images of TRPL, which map the PL decays at different wavelengths and provide an overview of the transient spectral behaviors. As can be easily observed, bare $CH_3NH_3PbI_3$ exhibits the slowest PL decays, while accelerated PL decays occur to the two TiO_2 nanolayers presumably due to the electron injection. More interestingly, the PL decay is slower for $CH_3NH_3PbI_3$ on the LPD-TiO_2 nanolayer than that on the USP-TiO_2 nanolayer, suggesting the slower electron injection of the former.

The PL decay curves extracted from Fig. 4(B) are displayed in Fig. 4(C). The PL lifetime (τ), which is defined as the time when the PL intensity decreases to $1/e$ of the original value, is calculated for the $CH_3NH_3PbI_3$ on the LPD-TiO_2 nanolayer to be about 880 ps, which is lower than that on the USP-TiO_2 nanolayer, about 515 ps. The slower electron injection could partially account for the lower J_{sc} for the LPD-TiO_2 nanolayer. But more importantly, this slower electron injection is expected to accumulate electrons in $CH_3NH_3PbI_3$, causing an increase in V_{oc} if recombination loss in the solar cells is not significant.

Further information on the recombination loss in the solar cells can be obtained from the electron lifetime (τ_n) measured by the OCVD technique. OCVD curves were recorded immediately after the interruption of light illumination and τ_n was determined according to eqn (5):[40]

$$\tau_n = k_B T/q(\mathrm{d}V_{oc}/\mathrm{d}t)^{-1} \tag{5}$$

where k_B is the Boltzmann constant, T the absolute temperature and q the positive elementary charge. As shown in Fig. 4(D), the τ_n for the $CH_3NH_3PbI_3$/nanocarbon solar cells based on the LPD-TiO_2 nanolayer is about 2–4 times longer than that based on the USP-TiO_2 nanolayer in the whole potential range, implying a slower recombination reaction in the former solar cells than in the latter, in keeping with the higher V_{oc} and FF.

Conclusively, the slower electron injection and longer electron lifetime in the LPD-TiO_2 nanolayer based solar cells are held responsible for their increased V_{oc} and FF compared with those in USP-TiO_2 nanolayer based solar cells. The lower crystallinity and/or the smaller size of the TiO_2 nanocrystals in the LPD-TiO_2 nanolayer may account for the slower electron injection, while the slower electron recombination hints at a better interface contact between the LPD-TiO_2 nanolayer and $CH_3NH_3PbI_3$. Clearly, further investigations are needed to better understand the detailed causality involved. Nevertheless, it has been clearly demonstrated that the LPD-TiO_2 nanolayer holds promise for designing perovskite solar cells with high V_{oc} and FF.

Though the ultrahigh V_{oc} was achieved for the $CH_3NH_3PbI_3$/nanocarbon solar cells based on the LPD-TiO_2 nanolayers, the photocurrent densities were low, which limited the power conversion efficiencies. The low current density should be attributed to the fact that all these LPD-TiO_2 nanolayers are flat and compact, which makes it difficult to achieve a homogenous coverage of a sufficiently thick and effective $CH_3NH_3PbI_3$ layer through solution-based processes.[15,41,42] This low quality $CH_3NH_3PbI_3$ layer would lead to a poor light absorption of $CH_3NH_3PbI_3$/nanocarbon solar cells. To resolve this problem, we made an attempt to employ

TiO_2 NR array grown on FTO glass as a base structure for the LPD-TiO_2 nanolayers to support a high quality $CH_3NH_3PbI_3$ layer as described below.

3.3 LPD-TiO_2 nanolayers on a rutile NR array for $CH_3NH_3PbI_3$/nanocarbon solar cells

To grow the TiO_2 NR array, a typical hydrothermal method was applied with FTO thin film as the seed layer. After 3 h of hydrothermal treatment, TiO_2 NRs with an average length of 600 nm were grown on FTO glass, as shown in Fig. 5(A). The diameter of TiO_2 NRs ranges from 40 to 80 nm. After LPD for 2 h, the surface of the TiO_2 NRs becomes rough (Fig. 5(B)) in comparison with those in Fig. 5(A), indicating the secondary growth of TiO_2 on the TiO_2 NRs. Besides, the exposed surface of FTO glass between TiO_2 NRs is also covered with a new growth layer. As the LPD duration was extended to 5 h, the surface of the TiO_2 NRs becomes more rough, resulting in an obvious increase in diameter (Fig. 5(C)), and the overlayer on the exposed surface of FTO glass between the TiO_2 NRs is also thickened. On further increasing the LPD duration to 9 h, an obvious nanobranch structure appears to be formed, and the space between TiO_2 NRs is reduced accompanied by the crowded roots of the TiO_2 NRs (Fig. 5(D)).

Next, XRD patterns and Raman spectra were acquired to evaluate the composition of the samples before and after LPD. For bare TiO_2 NRs, in addition to the characteristic diffraction peaks of FTO glass (Fig. 5(E)), the diffraction peaks at 27.3°, 36.0°, 41.2°, 51.6°, 54.4° and 62.7° are indexed to (110), (101), (111), (211), (220) and (002) planes of rutile TiO_2, respectively. After LPD, two new diffraction peaks at 25.0° and 47.9° that could be indexed to the (101) and (200)

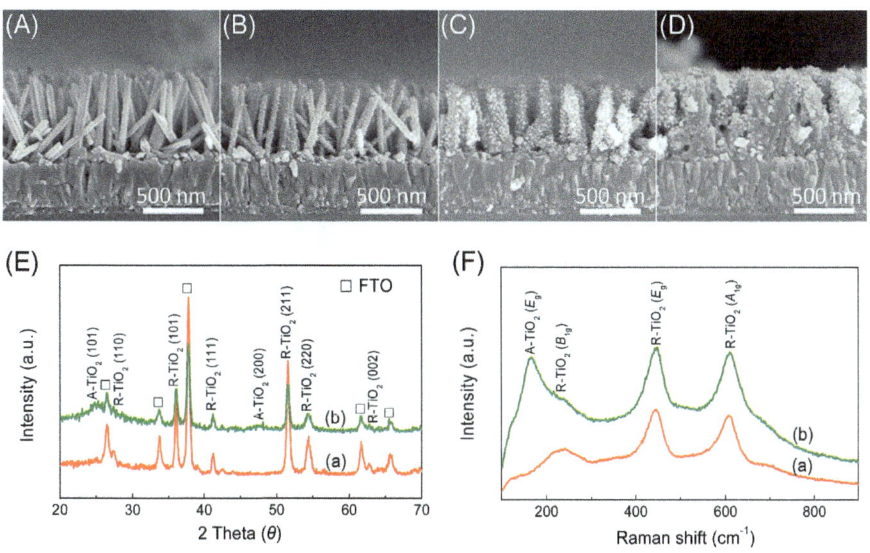

Fig. 5 Cross-sectional SEM images of the TiO_2 NRs/LPD-TiO_2 nanolayers with different LPD durations: (A) 0 h, (B) 2 h, (C) 5 h and (D) 9 h. Phase structure and composition characterizations of (a) pristine TiO_2 NRs and (b) TiO_2 NRs/LPD-TiO_2 nanolayer: (E) XRD patterns and (F) Raman spectra.

planes of anatase TiO_2, respectively, are observed, implying that anatase TiO_2 was deposited on the rutile TiO_2 NRs after LPD.

The Raman spectra in Fig. 5(F) indicate that only the peaks at 238 cm^{-1}, 445 cm^{-1} and 608 cm^{-1} are observed for bare TiO_2 NRs, consistent with the three Raman-active modes of B_{1g}, E_g and A_{1g} of rutile TiO_2,[43,44] respectively, confirming the rutile phase of TiO_2 NRs. After LPD, a new and obvious peak accompanying those of rutile TiO_2 is detected at 166 cm^{-1}, which can be easily assigned to the E_g mode of anatase TiO_2. Therefore, both XRD and Raman shift results corroborate that the secondary growth through LPD forms the anatase TiO_2-coated layer on the rutile TiO_2 NR array.

To tap into the structural features of the TiO_2 NRs/LPD-TiO_2 nanolayer, TEM was further applied. For the bare TiO_2 NRs, the surface is smooth as shown in Fig. 6(A). The HRTEM in Fig. 6(A1) demonstrates the single-crystalline structure of the TiO_2 NRs with the growth orientation along the [001] direction. After LPD for 2 h, a new outer layer with a thickness of about several nanometers is deposited, which coarsens the surface of the TiO_2 NRs. The HRTEM image in Fig. 6(B1) demonstrates that the outlined fringes are consistent with the (101) planes of anatase TiO_2, further confirming the deposition of an anatase TiO_2 polycrystalline nanolayer on the surface of the rutile TiO_2 NRs. The TEM image in Fig. 6(C) indicates that the outer anatase TiO_2 nanolayer thickens up to 30–40 nm as the LPD duration extends to 5 h. The electron diffraction pattern in the inset of Fig. 6(C) indicates that in addition to the diffraction spots of single rutile TiO_2, the polycrystalline diffraction rings corresponding to the (101), (004) and (200) planes of anatase TiO_2 are also detected. Therefore, anatase TiO_2 polycrystalline nanolayer is deposited on the single rutile TiO_2 NRs after LPD. The HRTEM image in Fig. 6(C1) indicates that the size of the anatase TiO_2 nanocrystals is about several nanometers and they stack compactly on each other.

On the basis of the SEM, XRD, Raman shift and TEM results, it can be concluded that the nature of the TiO_2 nanolayer prepared by LPD is barely changed when the substrate is changed from FTO glass to FTO glass/TiO_2 NRs,

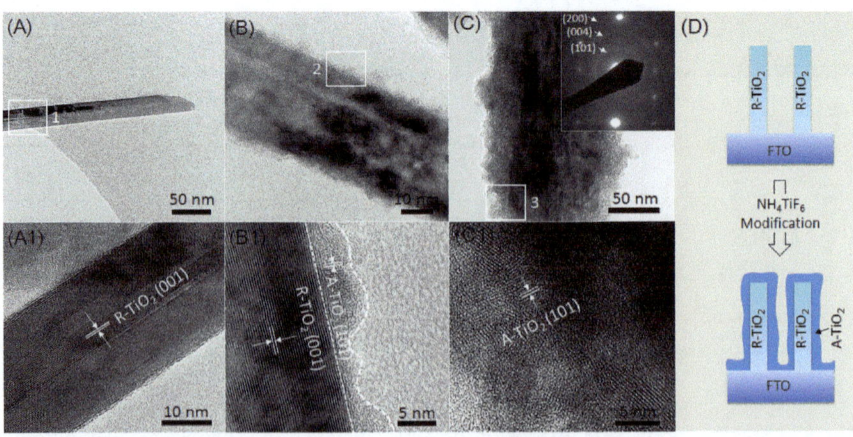

Fig. 6 TEM images of TiO_2 NRs/LPD-TiO_2 nanolayers with different LPD durations: (A) 0 h, (B) 2 h and (C) 5 h. (A1), (B1) and (C1) represent the corresponding HRTEM images. (D) Schematic illustrating the structure and composition of the TiO_2 NRs before and after LPD.

suggesting that the LPD method can be applied to deposit TiO_2 nanolayers on various substrates with similar material properties. To sum up, the structure and composition of the FTO glass/TiO_2 NRs before and after LPD can be illustrated in Fig. 6(D).

To fabricate $CH_3NH_3PbI_3$/nanocarbon solar cells based on the TiO_2 NRs/LPD-TiO_2 nanolayers, the same fabrication processes to those based on LPD-TiO_2 nanolayers were applied. The cross sectional SEM image of the solar cells in Fig. 7(A) exhibits the vivid three-layer structure. The interlayer becomes compact in comparison with the cross sectional morphology of the TiO_2 NRs/LPD-TiO_2 nanolayers in Fig. 5 due to the penetration and deposition of $CH_3NH_3PbI_3$ between TiO_2 NRs. The thickness of the interlayer is increased to 800 nm in comparison with that of the bare TiO_2 NRs/LPD-TiO_2 nanolayer (Fig. 5), suggesting the deposition of a top layer of $CH_3NH_3PbI_3$ on the TiO_2 NRs. The scheme in Fig. 7(B) clearly illustrates the cross sectional structure of the $CH_3NH_3PbI_3$/nanocarbon solar cells based on the TiO_2 NRs/LPD-TiO_2 nanolayers.

The energy level diagram and charge transfer processes in the $CH_3NH_3PbI_3$/nanocarbon solar cells based on TiO_2 NRs/LPD-TiO_2 nanolayers are illustrated in Fig. 7(C). Photogenerated electrons on the CB of $CH_3NH_3PbI_3$ will firstly be injected into the CB of anatase TiO_2 (−4.07 eV) and then transferred to the CB of rutile TiO_2 NRs (−4.10 eV) (see Fig. S4†). Finally, the electrons, after being transported through the single-crystalline rutile TiO_2 NRs, will be collected at the FTO glass. Meanwhile, the holes in the CB of $CH_3NH_3PbI_3$ will be extracted by the nanocarbon electrode.

Fig. 7(D) presents the *J–V* curves of the $CH_3NH_3PbI_3$/nanocarbon solar cells based on the TiO_2 NRs/LPD-TiO_2 nanolayers with different LPD durations and

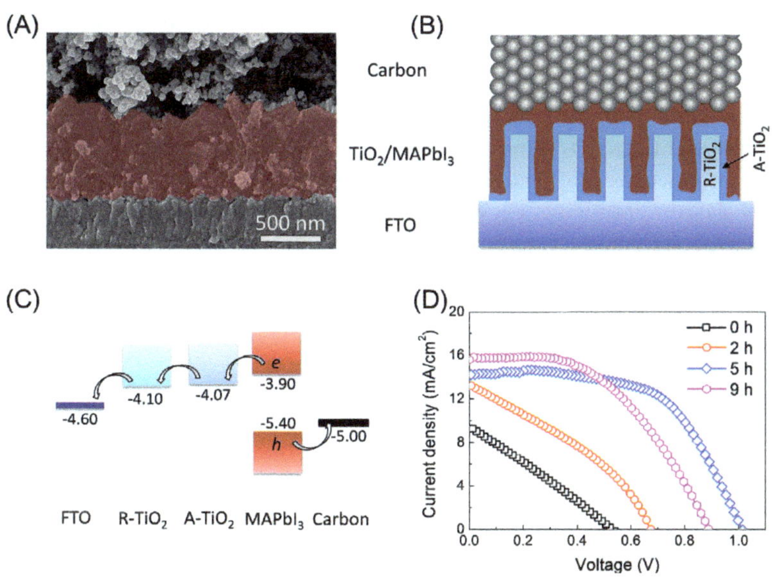

Fig. 7 (A) Cross sectional SEM image, (B) cross sectional structure scheme, and (C) energy level diagram of the $CH_3NH_3PbI_3$/nanocarbon solar cells based on the TiO_2 NRs/LPD-TiO_2 nanolayers. (D) *J–V* curves of the $CH_3NH_3PbI_3$/nanocarbon solar cells based on the TiO_2 NRs/LPD-TiO_2 nanolayers with different LPD durations.

Table 1 lists the corresponding photovoltaic parameters. The solar cells based on bare TiO_2 NRs exhibit a considerably higher $J_{sc} = 9.35$ mA cm^{-2} than those based on LPD-TiO_2 nanolayers, which should be due to the deposition of the thicker effective $CH_3NH_3PbI_3$ photoactive layer. However, both V_{oc} and FF are low, 0.54 V and 0.27, respectively, implying the serious recombination at the exposed FTO glass/nanocarbon and/or the rutile TiO_2/nanocarbon interfaces. As a result, only a low η (1.36%) is obtained.

The LPD-TiO_2 nanolayer on the exposed FTO glass and the TiO_2 NR array appear to effectively suppress charge recombination at the above mentioned interfaces. In addition, electron accumulation in $CH_3NH_3PbI_3$ will also be enhanced. As a result, the V_{oc}, J_{sc}, FF and η are all obviously improved after deposition of the LPD-TiO_2 nanolayer. At the optimized LPD duration of 5 h, η as high as 8.61% was obtained accompanied by high J_{sc}, V_{oc} and FF of 14.20 mA cm^{-2}, 1.01 V and 0.60, respectively.

Therefore, by using the rutile TiO_2 NR array as a base structure for the LPD-TiO_2 nanolayer to support a relatively thick effective $CH_3NH_3PbI_3$ photoactive layer, we are able to achieve a considerably increased J_{sc} without obviously compromising V_{oc} (1.01 V) and hence produce a significantly improved η.

4 Conclusion

To control electron injection in carbon-based $CH_3NH_3PbI_3$ solar cells, we have prepared a TiO_2 nanolayer by the LPD method at room temperature. The TiO_2 nanolayer is compact and composed of tiny anatase TiO_2 nanocrystals intimately stacked together. Directly applying the TiO_2 nanolayer onto FTO glass to $CH_3NH_3PbI_3$/nanocarbon solar cells produced a V_{oc} as high as 1.07 V, which constitutes the highest value for HTM-free $CH_3NH_3PbI_3$ solar cells. Comparatively, the commonly used TiO_2 nanolayer could only achieve a low V_{oc} of 0.78 V. We have shown that the considerably higher V_{oc} for the solar cells based on the LPD-TiO_2 nanolayer is due to a slower electron injection and longer electron lifetime, which enhanced the electron accumulation in $CH_3NH_3PbI_3$. The application of the LPD-TiO_2 nanolayer to a rutile TiO_2 NR array was found to significantly enhance the photocurrent density of the corresponding solar cells without obviously compromising V_{oc} (1.01 V), upgrading the η from 3.67% to 8.61%. Higher performance is anticipated for the perovskite solar cells when the LPD-TiO_2 nanolayer is better interfaced with other components of the devices.

Acknowledgements

This work was supported by the HK-RGC General Research Funds (GRF no. HKUST 606511 and 604410).

References

1 A. Kojima, K. Teshima, Y. Shirai and T. Miyasaka, *J. Am. Chem. Soc.*, 2009, **131**, 6050.

2 H. S. Kim, C. R. Lee, J. H. Im, K. B. Lee, T. Moehl, A. Marchioro, S. J. Moon, R. Humphry-Baker, J. H. Yum, J. E. Moser, M. Gratzel and N. G. Park, *Sci. Rep.*, 2012, **2**, 591.

3 M. M. Lee, J. Teuscher, T. Miyasaka, T. N. Murakami and H. J. Snaith, *Science*, 2012, **338**, 643.
4 M. A. Green, A. Ho-Baillie and H. J. Snaith, *Nat. Photonics*, 2014, **8**, 506.
5 H.-S. Kim, S. H. Im and N.-G. Park, *J. Phys. Chem. C*, 2014, **118**, 5615.
6 J. Burschka, N. Pellet, S. J. Moon, R. Humphry-Baker, P. Gao, M. K. Nazeeruddin and M. Gratzel, *Nature*, 2013, **499**, 316.
7 J. H. Im, C. R. Lee, J. W. Lee, S. W. Park and N. G. Park, *Nanoscale*, 2011, **3**, 4088.
8 J. Qiu, Y. Qiu, K. Yan, M. Zhong, C. Mu, H. Yan and S. Yang, *Nanoscale*, 2013, **5**, 3245.
9 Z. L. Zhu, J. A. Ma, Z. L. Wang, C. Mu, Z. T. Fan, L. L. Du, Y. Bai, L. Z. Fan, H. Yan, D. L. Phillips and S. H. Yang, *J. Am. Chem. Soc.*, 2014, **136**, 3760.
10 M. Liu, M. B. Johnston and H. J. Snaith, *Nature*, 2013, **501**, 395.
11 N. J. Jeon, J. H. Noh, Y. C. Kim, W. S. Yang, S. Ryu and S. I. Seok, *Nat. Mater.*, 2014, **13**, 897–903.
12 N.-G. Park, *J. Phys. Chem. Lett.*, 2013, **4**, 2423.
13 H. J. Snaith, *J. Phys. Chem. Lett.*, 2013, **4**, 3623.
14 S. Kazim, M. K. Nazeeruddin, M. Gratzel and S. Ahmad, *Angew. Chem.*, 2014, **53**, 2812.
15 T. C. Sum and N. Mathews, *Energy Environ. Sci.*, 2014, **7**, 2518.
16 S. D. Stranks, G. E. Eperon, G. Grancini, C. Menelaou, M. J. Alcocer, T. Leijtens, L. M. Herz, A. Petrozza and H. J. Snaith, *Science*, 2013, **342**, 341.
17 G. Xing, N. Mathews, S. Sun, S. S. Lim, Y. M. Lam, M. Gratzel, S. Mhaisalkar and T. C. Sum, *Science*, 2013, **342**, 344.
18 L. Etgar, P. Gao, Z. S. Xue, Q. Peng, A. K. Chandiran, B. Liu, M. K. Nazeeruddin and M. Gratzel, *J. Am. Chem. Soc.*, 2012, **134**, 17396.
19 W. Abu Laban and L. Etgar, *Energy Environ. Sci.*, 2013, **6**, 3249.
20 S. Aharon, B. E. Cohen and L. Etgar, *J. Phys. Chem. C*, 2014, **118**, 17160.
21 S. Aharon, S. Gamliel, B. El Cohen and L. Etgar, *Phys. Chem. Chem. Phys.*, 2014, **16**, 10512.
22 J. Shi, J. Dong, S. Lv, Y. Xu, L. Zhu, J. Xiao, X. Xu, H. Wu, D. Li, Y. Luo and Q. Meng, *Appl. Phys. Lett.*, 2014, **104**, 063901.
23 J. Shi, Y. Luo, H. Wei, J. Luo, J. Dong, S. Lv, J. Xiao, Y. Xu, L. Zhu, X. Xu, H. Wu, D. Li and Q. Meng, *ACS Appl. Mater. Interfaces*, 2014, **6**, 9711.
24 Z. Ku, Y. Rong, M. Xu, T. Liu and H. Han, *Sci. Rep.*, 2013, **3**, 3132.
25 Z. Li, S. A. Kulkarni, P. P. Boix, E. Shi, A. Cao, K. Fu, S. K. Batabyal, J. Zhang, Q. Xiong, L. H. Wong, N. Mathews and S. G. Mhaisalkar, *ACS Nano*, 2014, **8**, 6797.
26 Y. Rong, Z. Ku, A. Mei, T. Liu, M. Xu, S. Ko, X. Li and H. Han, *J. Phys. Chem. Lett.*, 2014, **5**, 2160.
27 Z. Wei, K. Yan, H. Chen, Y. Yi, T. Zhang, X. long, J. Li, L. Zhang, J. Wang and S. Yang, *Energy Environ. Sci.*, 2014, **7**, 3326.
28 A. Mei, X. Li, L. Liu, Z. Ku, T. Liu, Y. Rong, M. Xu, M. Hu, J. Chen, Y. Yang, M. Grätzel and H. Han, *Science*, 2014, **345**, 295.
29 D. Bi, S.-J. Moon, L. Häggman, G. Boschloo, L. Yang, E. M. Johansson, M. K. Nazeeruddin, M. Grätzel and A. Hagfeldt, *RSC Adv.*, 2013, **3**, 18762.
30 H.-S. Kim, I. Mora-Sero, V. Gonzalez-Pedro, F. Fabregat-Santiago, E. J. Juarez-Perez, N.-G. Park and J. Bisquert, *Nat. Commun.*, 2013, **4**, 2242.
31 S. Deki, Y. Aoi, O. Hiroi and A. Kajinami, *Chem. Lett.*, 1996, 433.

32 S. Deki, N. Yoshida, Y. Hiroe, K. Akamatsu, M. Mizuhata and A. Kajinami, *Solid State Ionics*, 2002, **151**, 1.

33 H. Chen, Z. Wei, K. Yan, Y. Bai and S. Yang, *J. Phys. Chem. Lett.*, 2014, **5**, 2890.

34 T. Ohsaka, *J. Phys. Soc. Jpn.*, 1980, **48**, 1661.

35 W. F. Zhang, Y. L. He, M. S. Zhang, Z. Yin and Q. Chen, *J. Phys. D: Appl. Phys.*, 2000, **33**, 912.

36 T. Ohsaka, F. Izumi and Y. Fujiki, *J. Raman Spectrosc.*, 1978, **7**, 321.

37 J. C. Parker and R. W. Siegel, *J. Mater. Res.*, 1990, **5**, 1246.

38 A. S. Pottier, S. Cassaignon, C. Chaneac, F. Villain, E. Tronc and J. P. Jolivet, *J. Mater. Chem.*, 2003, **13**, 877.

39 M. G. Walter, E. L. Warren, J. R. McKone, S. W. Boettcher, Q. X. Mi, E. A. Santori and N. S. Lewis, *Chem. Rev.*, 2010, **110**, 6446.

40 A. Zaban, M. Greenshtein and J. Bisquert, *ChemPhysChem*, 2003, **4**, 859.

41 G. E. Eperon, V. M. Burlakov, P. Docampo, A. Goriely and H. J. Snaith, *Adv. Funct. Mater.*, 2014, **24**, 151.

42 A. Dualeh, N. Tétreault, T. Moehl, P. Gao, M. K. Nazeeruddin and M. Grätzel, *Adv. Funct. Mater.*, 2014, **24**, 3250.

43 J. H. Nicola, C. A. Brunharoto, M. Abramovich and C. E. T. Concalvesdasila, *J. Raman Spectrosc.*, 1979, **8**, 32.

44 G. Liu, Z. G. Chen, C. L. Dong, Y. N. Zhao, F. Li, G. Q. Lu and H. M. Cheng, *J. Phys. Chem. B*, 2006, **110**, 20823.

Faraday Discussions

PAPER

Effect of double blocking layers at TiO_2/Sb_2S_3 and Sb_2S_3/spiro-MeOTAD interfaces on photovoltaic performance

Hyun-Woo Kang, Jin-Wook Lee and Nam-Gyu Park*

Received 6th June 2014, Accepted 10th July 2014
DOI: 10.1039/c4fd00126e

The effect of double blocking layers on the Sb_2S_3-sensitized all solid state solar cell are investigated. Thin layers of ZrO_2 (blocking layer 1, BL1) and ZnS (blocking layer 2, BL2) are introduced at the TiO_2/Sb_2S_3 and Sb_2S_3/hole transporting material (HTM) interfaces. The presence of BL1 is found to improve the open-circuit voltage (V_{oc}) from 457 mV to 618 mV, whereas BL2 improves mainly short-circuit current density (J_{sc}) from 11.1 mA cm^{-2} to 14.0 mA cm^{-2}. Transient photovoltage confirms that the BL1 efficiently blocks charge recombination, responsible for the V_{oc} enhancement, whereas the BL2 has little effect on suppression of charge recombination. Surface modification of Sb_2S_3 by BL2, on the other hand, leads to recovery of Sb_2S_3 from the surface oxidized Sb_2O_5, which affects the J_{sc} increment. The present results suggest that the TiO_2/Sb_2S_3 interface is the main pathway for charge recombination rather than the Sb_2S_3/HTM interface. Double blocking layers enhanced the power conversion efficiency by 30%.

1. Introduction

Since the first report on the solid-state sensitized mesoscopic solar cell in 1998,[1] it has been regarded as a platform technology for an inorganic–organic hybrid type p–i–n junction structure. As a sensitizer for this device, diverse materials can be used such as molecular dyes or inorganic semiconductors in the form of quantum dots. The sensitizer is deposited on the mesoporous TiO_2 surface, which is in contact with a hole transporting material (HTM) such as 2,2′,7,7′-tetrakis(*N*,*N*-*p*-dimethoxy-phenylamino)-9,9′-spirobifluorene (spiro-MeOTAD). Photo-excitation of the sensitizer produces electron–hole excitons that are subsequently separated in the respective electron and hole materials. The photovoltage is determined by the difference in the quasi-Fermi level between the electron- and hole-transporting solids.

The first solid-state sensitized solar cell empolyed a conventional organometallic dye, coded N719, which showed a power conversion efficiency (PCE) of less

School of Chemical Engineering and Department of Energy Science, Sungkyunkwan University, Suwon 440-746, Korea. E-mail: npark@skku.edu; Fax: +82-31-290-7272; Tel: +82-31-290-7241

than 1%.[1] Synthesis of organic dyes with high absorption coefficients[2–5] and formulation of spiro-MeOTAD solution[6] have boosted PCEs up to 7%. In the solid-state sensitized solar cell, the mesoporous TiO_2 layer thickness is limited to less than 2 μm because of the limited hole diffusion length of HTM materials, which reduces the total surface area for the adsorbing sensitizer. Inorganic semiconductors have been adopted in the thin film TiO_2 based solid state sensitized structure because of a higher absorption coefficient than molecular dyes. Among the inorganic sensitizers,[7] Sb_2S_3 demonstrated the highest PCEs of around 6%,[8–10] which is in part due to the high absorption coefficient of $\sim 10^5\ cm^{-1}$ in the visible region.[11,12] More recently, organolead iodide perovskite materials demonstrated PCE higher than 15% and approaching 20%.[13]

Despite the advantages of Sb_2S_3 such as high absorption coefficients, long-term stability and tunable band gap, it suffers from low photovoltage. The open-circuit voltages (V_{oc}) of the Sb_2S_3-sensitized solid state devices were reported to be much lower than the theoretical voltage.[7] This problem is often found not only from Sb_2S_3 but also from most chalcogenide quantum dots.[14–16] The low V_{oc} could be attributed to a significant recombination at TiO_2/Sb_2S_3 interfaces and/or Sb_2S_3/HTM interfaces. It was reported that surface treatment of Sb_2S_3 deposited TiO_2 with a 1-decyl phosphonic acid (DPA) improved photovoltaic performance mainly due to longer electron lifetime after the DPA surface treatment.[17] Surface treatment of TiO_2 with various insulating oxides in the TiO_2/Sb_2S_3/CuSCN configuration was reported to improve PCE from 2.8% to 4.1% because of surface passivation against recombination of the injected electrons in TiO_2.[18] The previous observations indicate that both interfaces could provide the recombination pathways. However, it has not been determined whether both interfaces contribute equally to the charge recombination or if there is a major recombination pathway. Further improvement can be expected if the exact role of each interface is clarified in Sb_2S_3-sensitized solid state solar cells.

We report here on the effect of double blocking layers at TiO_2/Sb_2S_3 and Sb_2S_3/HTM interfaces in the Sb_2S_3-sensitized solid state solar cell on photovoltaic performance to determine a major recombination pathway. ZrO_2 is introduced between TiO_2 and Sb_2S_3, which is denoted as a blocking layer 1 (BL1) and ZnS is introduced between Sb_2S_3 and spiro-MeOTAD, which is denoted as blocking layer 2 (BL2). Investigation of the role of each blocking layer's photovoltaic performance is investigated for the TiO_2/BL2/Sb_2S_3/BL2/spiro-MeOTAD configuration with double blocking layers. ZrO_2 is selected as BL1 due to a higher conduction band edge[19,20] than TiO_2, and its chemical compatibility to TiO_2. ZnS is selected as BL2 due to a lower valence band edge[21] than HTM, and its chemical compatibility to Sb_2S_3.

2. Experimental

2.1. Preparation of mesoporous TiO_2 film

50 nm-sized anatase TiO_2 nanoparticles were synthesized using 20 nm-sized TiO_2 nanoparticles as a seed. The seed TiO_2 nanoparticles were synthesized by autoclaving a hydrolyzed solution of titanium(IV) isoproxide (97%, Aldrich) at 230 °C for 12 h.[22] To prepare nonaqueous TiO_2 paste, water in the TiO_2 colloid solution was replaced with ethanol. The ethanol solution of TiO_2 nanoparticle was mixed with terpineol (Aldrich, 99.5%), ethyl cellulose (Aldrich, 46 cp) and lauric acid

(Fluka, 96%) and ethanol was evaporated to get viscous TiO_2 paste. The paste was further treated using a three-roll-mill apparatus to disperse the nanoparticles sufficiently. The nominal composition of TiO_2 : terpineol : ethylcellulose : lauric acid was about 1.25 : 6 : 0.3 : 0.1. Fluorine-doped tin oxide (FTO) glass (Pilkington, TEC-8, 8 Ω sq^{-1}) was cleaned by UV/ozone treatment and sonication with ethanol. The FTO surface was covered with a thin compact TiO_2 layer prior to depositing the nanocrystalline TiO_2 paste by spin-coating the 0.15 M 1-butanol solution of titanium diisopropoxide bis(acetylacetonate) (75 wt%, Aldrich) at a stepwise spin rate of 700 rpm for 8 s $\rightarrow$ 1000 rpm for 10 s $\rightarrow$ 2000 rpm for 40 s, followed by drying at 125 °C for 5 min. The spin-coating and drying process was repeated using a 0.3 M titanium diisopropoxide bis(acetylacetonate) solution. The compact layer was finally annealed at 500 °C for 15 min. The nanocrystalline TiO_2 paste was deposited on the compact blocking layer coated FTO glass using screen printing technique and subsequently annealed at 550 °C for 1 h in a muffle furnace to obtain a mesoporous TiO_2 film. The thickness of the annealed mesoporous TiO_2 layer was measured to be 1.5 ± 0.2 μm by an alpha-step IQ surface profiler (KAL Tencor). The annealed TiO_2 film was immersed in the 20 mM $TiCl_4$ (Aldrich) aqueous solution at 70 °C for 10 min. The substrates were washed with distilled water and dried under air flow, which was followed by annealing at 500 °C for 30 min.

2.2. Formation of a blocking layer on TiO_2 with ZrO_2 (BL1)

A 2-propanol solution of 0.04 M zirconium(IV) butoxide (Sigma-Aldrich, 80 wt% in 1-butanol) was prepared under an argon atmosphere and used as the precursor solution to form the blocking layer on TiO_2 surface.[23] 20 μL of the precursor solution was spin-coated on the annealed mesoporous TiO_2 film at a spin rate of 4000 rpm for 20 s. The film was then dried under N_2 flow. After drying at room temperature, the film was annealed at 500 °C for 20 min in air. The thickness was controlled by varying the precursor concentration and the number of spin coating/annealing cycles.[24]

2.3. Sb_2S_3 deposition

The Sb_2S_3 layer was deposited by chemical bath deposition (CBD) method.[18] Mesoporous TiO_2 film pre-treated with UV/ozone for 30 min was dipped in the bath containing 1.3 g $SbCl_3$ (Aldrich, 99.99%) in 5 mL acetone (Samchun, 99.9%), 50 mL of 1 M $Na_2S_2O_3$ (Fluka) and 145 mL of distilled water for 70 min at 10 °C. The as-deposited films were annealed at 330 °C for 30 min under Ar atmosphere, where the orange color turned to dark brown after annealing. The Sb_2S_3 deposited substrate was cooled down to 300 °C in Ar atmosphere and then quenched at 300 °C in air.

2.4. Formation of blocking layer on Sb_2S_3 with ZnS (BL2)

Precursor solutions for ZnS were prepared by mixing 0.02 M ZnCl (0.01363 g, Sigma-Aldrich, reagent grade, >98%) for Zn^{2+} source and 0.04 M thiourea (0.01522 g, Sigma-Aldrich, ACS reagent, >99%) for S^{2-} source in deionized water (5 mL) under an Ar atmosphere. 30 μL of precursor solution was spin-coated onto the Sb_2S_3 coated TiO_2 film at a spin rate of 4000 rpm for 20 s and subsequently dried on a hot plate at 200 °C for 10 min.

2.5. Fabrication of solid-state solar cells

Solid-state Sb_2S_3 sensitized solar cells were fabricated by infiltration of spiro-MeOTAD into the Sb_2S_3 adsorbed mesoporous TiO_2 layer. The composition of HTM was 0.08 M spiro-MeOTAD, 0.198 M 4-*tert*-butylpyridine and 64 mM Li salt as described elsewhere.[25] 30 μL of HTM solution was spin-coated onto the Sb_2S_3 coated TiO_2 film at 0 rpm for 20 s and 2500 rpm for 30 s. The counter electrode was deposited by thermal evaporation of silver under 1.0×10^{-5} Torr at 7.0–8.0 Å s^{-1}. The thickness of Ag was adjusted to 200 nm. The active area of the device was 0.125 cm^2.

2.6. Characterizations

The absorption spectra of the film were recorded by a UV-vis spectrophotometer (Agilent 8453) in the wavelength range 300 nm to 900 nm. Photocurrent and voltage were measured using a solar simulator (Oriel Sol 3A class AAA) equipped with 450 W Xenon lamp (Newport 6279NS) and a Keithley 2400 source meter. The NREL-calibrated Si solar cell with KG-2 filter was used to adjust light intensity to one sun illumination (100 mW cm^{-2}). While measuring photocurrent and voltage, the cell was covered with an aperture mask. External quantum efficiency (EQE) was measured using a specially designed EQE system (PV measurement Inc.). A 75 W xenon lamp (USHIO, Japan) was used as a white light source to generate a monochromatic beam. Calibration was conducted using NIST-calibrated photodiode G425 as a standard. EQE data were collected at DC mode. The deposited blocking layers of ZrO_2 and ZnS were investigated using a high-resolution transmission electron microscope (HR-TEM, JEOL JEM-2100F) at an acceleration voltage of 200 kV. Energy dispersive X-ray (EDX) spectra were obtained using a field-emission scanning electron microscope (FE-SEM, JEOL JSM-7600F) to determine the presence of the blocking layers. The time constants for charge recombination were obtained from the transient photovoltage. The cells were probed with a weak laser pulse at 532 nm superimposed on a relatively large, background (bias) illumination at 680 nm. The details on the transient setup and data collection are described elsewhere.[22]

3. Results and discussion

Fig. 1 shows HR-TEM images before and after surface modifications of TiO_2 and Sb_2S_3 with ZrO_2 and ZnS. Compared to the bare TiO_2 (Fig. 1a), thin layers are seen on the TiO_2 surface after surface modification. It is clear from the enlarged TEM images that ZrO_2 are conformally coated on the TiO_2 surface with thickness of about 1 nm (Fig. 1b). A lattice fringe appears on the surface of TiO_2, which is different from that of the core TiO_2 particles as can be seen in the inset of Fig. 1b. This indicates that the surface deposited ZrO_2 has crystallinity to some extent. For the case of Sb_2S_3 in Fig. 1c, a thin layer of Sb_2S_3 forms on the TiO_2 surface with thickness ranging between 2.3 and 4.7 nm. After deposition of ZnS on the Sb_2S_3 layer, a rather conformal coating is made with thickness of about 4.8 nm, which is, however, not obviously distinguished from Sb_2S_3 (Fig. 1d). The lack of a distinct boundary between Sb_2S_3 and ZnS is probably related to the similar bond distance between Sb–S and Zn–S due to similar cation radius between Zn^{2+}

(74 pm) and Sb^{3+} (76 pm).[26] The deposition of blocking layers is further clarified by EDX analysis.

The EDX spectra are obtained with the TiO_2/ZrO_2 and $TiO_2/Sb_2S_3/ZnS$ samples in Fig. 2. Zr is presented in the TiO_2/ZrO_2 sample, which indicates that the thin layer as observed from TEM measurement is confirmed to be ZrO_2. Zn is also detected along with Sb and S, which implies that the thin layer on the TiO_2 surface (Fig. 1d) is composed of both ZnS and Sb_2S_3.

3.1. Effect of blocking layer 1 (ZrO_2) on TiO_2

The effect of ZrO_2 (blocking layer 1, BL1) on photovoltaic performance is first investigated. *J–V* curves and photovoltaic parameters are compared before and after BL1 addition in Fig. 3 and Table 1. The open-circuit voltage (V_{oc}) is significantly improved from 0.457 V to 0.618 V after introduction of BL1 onto the TiO_2 surface, while little changes in short-circuit current density (J_{sc}) and fill factor (FF) are observed. As a consequence of the 35% increment in V_{oc}, the PCE increases from 2.87% to 3.82%. The improvement of V_{oc} indicates that the Fermi level of TiO_2 is higher for the surface modified TiO_2 compared to the bare TiO_2, possibly due to an improved electron lifetime by suppression in charge recombination. It can be postulated that the ZrO_2 layer blocks the recombination of the injected electrons in TiO_2. Transient photovoltage decay is compared before and after the introduction of ZrO_2 thin layer on TiO_2 surface in order to investigate the recombination behavior.

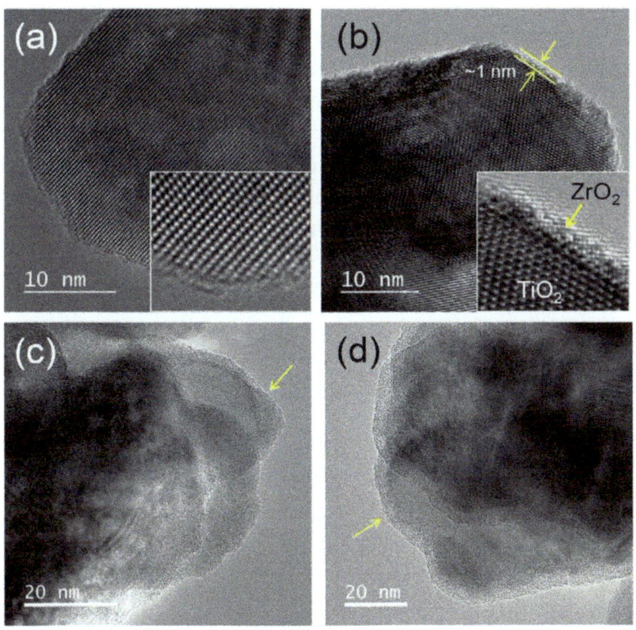

Fig. 1 HR-TEM images of (a) bare TiO_2, (b) ZrO_2 coated TiO_2 (TiO_2/ZrO_2), (c) Sb_2S_3 coated TiO_2 (TiO_2/Sb_2S_3) and (d) ZnS coated TiO_2/Sb_2S_3 ($TiO_2/Sb_2S_3/ZnS$). Arrows in (c) and (d) indicate thin layers on the TiO_2 surface.

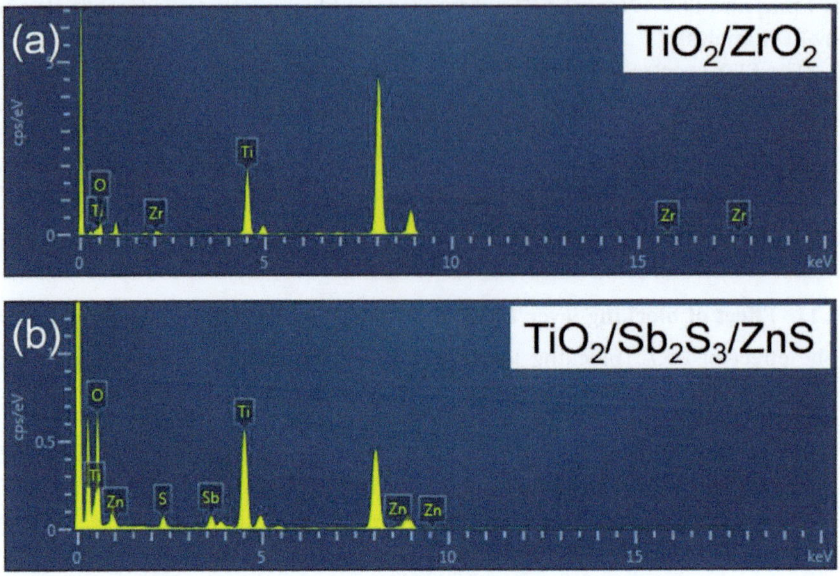

Fig. 2 EDX spectra for (a) TiO_2/ZrO_2 and (b) $TiO_2/Sb_2S_3/ZnS$.

Fig. 4 shows transient photovoltage profiles of Sb_2S_3-sensitized TiO_2 with and without BL1. A slower decay is observed for the TiO_2 with BL1 (Fig. 4b) compared to the TiO_2 without BL1 (Fig. 4a), indicative of slower charge recombination in the presence of BL1. The time constant for charge recombination is estimated to be twice as long for the TiO_2 with BL1 ($\tau = 7.73 \times 10^{-4}$ s) than for the bare TiO_2 ($\tau = 3.88 \times 10^{-4}$ s). It is thus obvious that the ZrO_2 layer effectively blocks charge recombination, which is responsible for the higher V_{oc}.

Before investigating the role of the second blocking layer, the effect of the BL1 layer thickness is investigated. Fig. 5 shows the dependence of photovoltaic parameters on the zirconia precursor concentration. The growth of ZrO_2 layer thickness can be controlled by either the precursor concentration or the number of coating cycles.[24] We vary the precursor concentration to control the ZrO_2

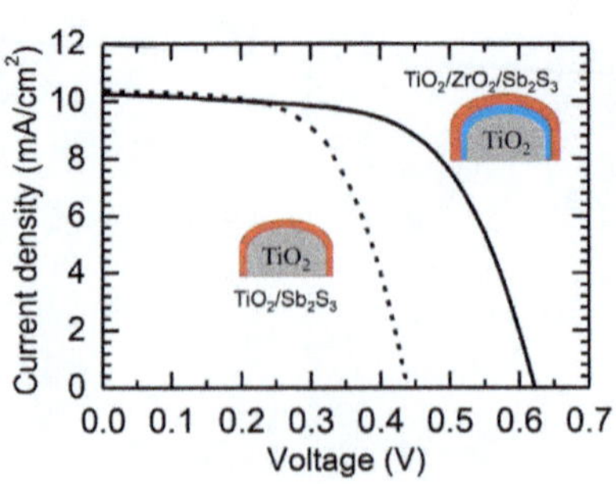

Fig. 3 Current density–voltage curves of the Sb_2S_3-sensitized solid state solar cells employing mesoporous TiO_2 layer with (solid line) and without (dotted line) the surface ZrO_2 layer.

Table 1 Short-circuit photocurrent density (J_{sc}), open-circuit voltage (V_{oc}), FF and PCE of Sb_2S_3-sensitized TiO_2 with and without surface ZrO_2. The thickness of the mesoporous TiO_2 layer was 1.5 ± 0.2 μm. The data were obtained under AM 1.5G one sun illumination (100 mW cm^{-2})

	J_{sc} (mA cm^{-2})	V_{oc} (V)	FF	PCE (%)
Bare	10.1 ± 0.60	0.457 ± 0.02	0.621 ± 0.01	2.87 ± 0.16
w/ZrO_2 (BL1)	10.3 ± 0.22	0.618 ± 0.01	0.611 ± 0.02	3.82 ± 0.12

thickness instead of using the non-reproducible coating cycle method. When the precursor concentration is 0.02, J_{sc} slightly increases but starts to decrease as the concentration is higher than 0.02 M. Similar tendency is observed for FF. However, V_{oc} increases continuously with increasing concentration upto 0.04 M. Due to the relatively large increment of V_{oc} from 0.45 V to 0.61 V, PCE reaches maximum at 0.04 M. This means that efficiency is mainly dependent on open circuit voltage, which confirms again the role of BL1. For the further study on double blocking layers, we select the precursor concentration of 0.04 M for BL1.

3.2. Effect of blocking layer 2 (ZnS) on Sb_2S_3

The effect of ZnS (blocking layer 2, BL2) on photovoltaic performance is investigated. In Fig. 6, J–V curves are compared before and after introduction of BL2 on Sb_2S_3, both J_{sc} and V_{oc} are improved after surface treatment of Sb_2S_3 with ZnS. The photovoltaic parameters are listed in Table 2. The increment of J_{sc} from 11.1 mA cm^{-2} to 14.0 mA cm^{-2}, corresponding to a 26% increment, is relatively more pronounced than the V_{oc} from 0.512 V to 0.545 V (6.4% increment). IPCE spectra in Fig. 6b confirm that photocurrent increases at entire wavelength, where the calculated current density values based on the IPCE data are in good agreement

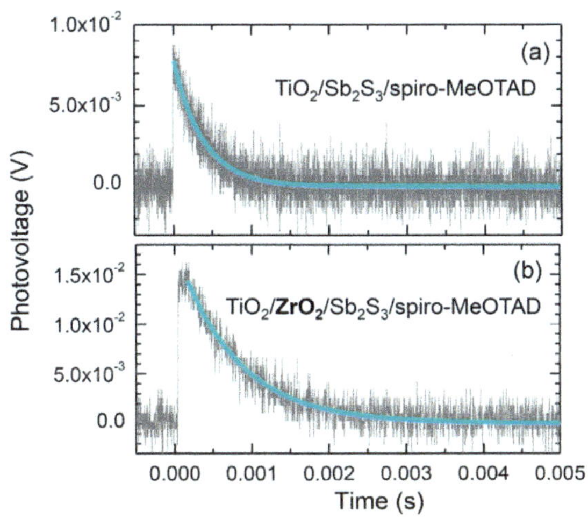

Fig. 4 Transient photovoltage profile of the Sb_2S_3-sensitized TiO_2 (a) without and (b) with ZrO_2 (BL1). Light blue solid lines represent the best fit results.

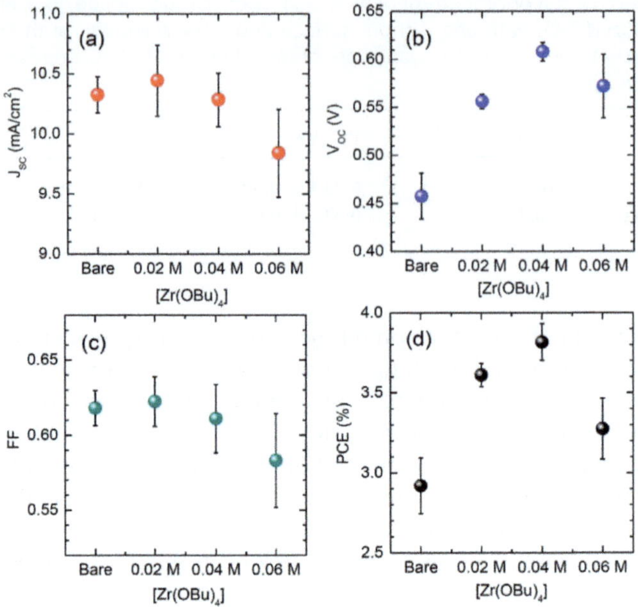

Fig. 5 Effect of the precursor ($Zr(OBu)_4$) concentration for the ZrO_2 (BL1) layer on photovoltaic parameters of (a) J_{sc}, (b) V_{oc}, (c) FF and (d) PCE.

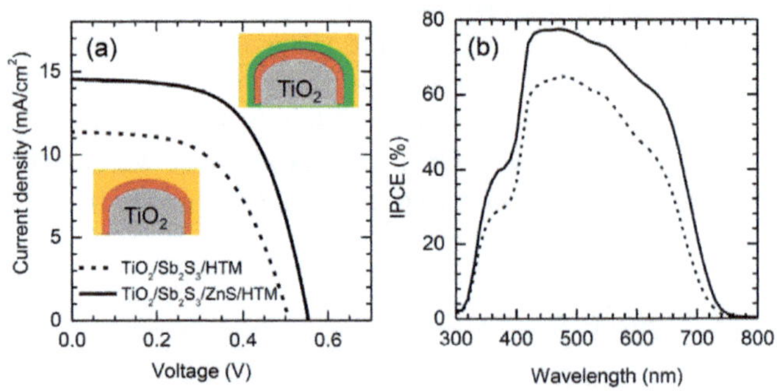

Fig. 6 (a) J–V curves and (b) IPCE spectra of Sb_2S_3-sensitized solid state solar cells before and after surface treatment of Sb_2S_3 with ZnS.

with the observed J_{sc} values. Due to the fact that the presence of BL2 improves mainly J_{sc}, the average PCE is improved from 2.99% to 4.55%. Since there is no noticeable enhancement in V_{oc} after the introduction of BL2 on Sb_2S_3, it can be said that the interface between Sb_2S_3 and spiro-MeOTAD is less critical for charge recombination compared to the TiO_2/Sb_2S_3 interface.

Transient photovoltage confirms that the surface treatment of Sb_2S_3 hardly retards the recombination kinetics as can be seen in Fig. 7. Time constants obtained from the best fit are 4.35×10^{-4} s and 4.78×10^{-4} s before and after BL2, respectively. Little difference in the time constant underlines that recombination at the Sb_2S_3/HTM interface can be negligible. A slightly improved V_{oc} is

Table 2 Short-circuit photocurrent density (J_{sc}), open-circuit voltage (V_{oc}), FF and PCE of Sb_2S_3-sensitized solid state solar cells before and after surface treatment of Sb_2S_3 with ZnS

	J_{sc} (mA cm^{-2})	V_{oc} (V)	FF	PCE (%)
Bare	11.1 ± 1.14	0.512 ± 0.01	0.527 ± 0.06	2.99 ± 0.42
w/ZnS (BL2)	14.0 ± 0.40	0.545 ± 0.01	0.557 ± 0.03	4.25 ± 0.31

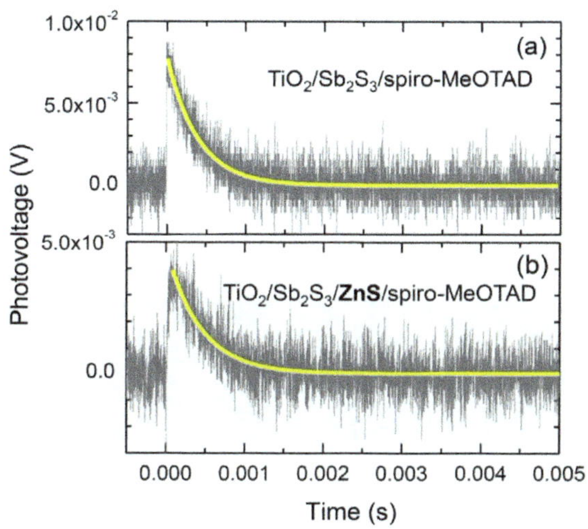

Fig. 7 Transient photovoltage profiles of the Sb_2S_3-sensitized solid state solar cells (a) before and (b) after surface treatment of Sb_2S_3 with ZnS.

therefore likely to be related to the increased photocurrent, associated with the increased density of the injected electrons, rather than the recombination issue.

To understand the basis for the increased J_{sc} by the surface modification of Sb_2S_3 with ZnS, UV-vis absorbance is measured and compared before and after BL2 in Fig. 8. Absorbance is found to increase after BL2, where the degree of increment in absorbance becomes significant as wavelength decreases from 700 nm to 500, which agrees well with the tendency in the IPCE improvement as observed in Fig. 6b. Since Sb_2S_3 deposited electrode was quenched in air at 300 °C, Sb_2S_3 surface could be oxidized to Sb_2O_5.[27] Surface treatment of Sb_2S_3 with ZnS may lead to the reduction of the oxidized layer on the Sb_2S_3 surface. If this process is involved, the absolute amount of Sb_2S_3 will increase after ZnS treatment. To clarify this possibility, elemental analysis is conducted.

Fig. 9 shows EDX elemental mapping images of the TiO_2/Sb_2S_3 and TiO_2/Sb_2S_3/ZnS particles, and in Table 3 atomic percentage of each component is listed. Little change in the atomic percentage of Sb is observed, while the sulfur atomic percentage is significantly increased after ZnS coating. When taking the S content in ZnS into consideration, the ratio of S to Sb is estimated and found to be largely increased from *ca.* 9–40% after ZnS coating. In addition, oxygen is relatively reduced after ZnS coating. Elemental analysis implies that the surface Sb_2O_5 is in part reduced to Sb_2S_3 by ZnS coating, which eventually leads to an increase in the

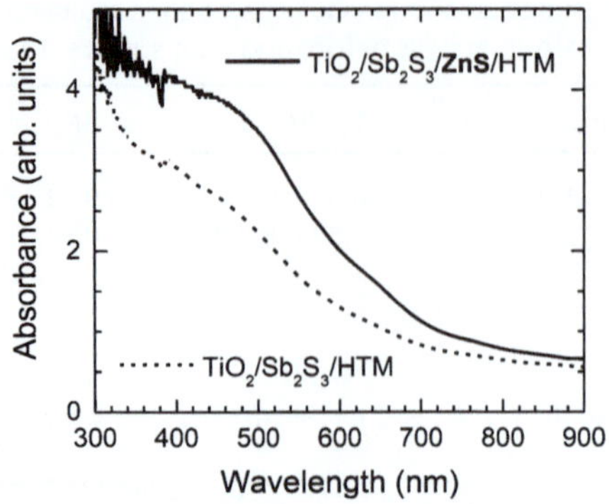

Fig. 8 Absorbance spectra of the ZnS (BL2)-free TiO_2/Sb_2S_3/HTM electrode (dotted line) and the BL2 contained TiO_2/Sb_2S_3/BL2/HTM electrode (solid line).

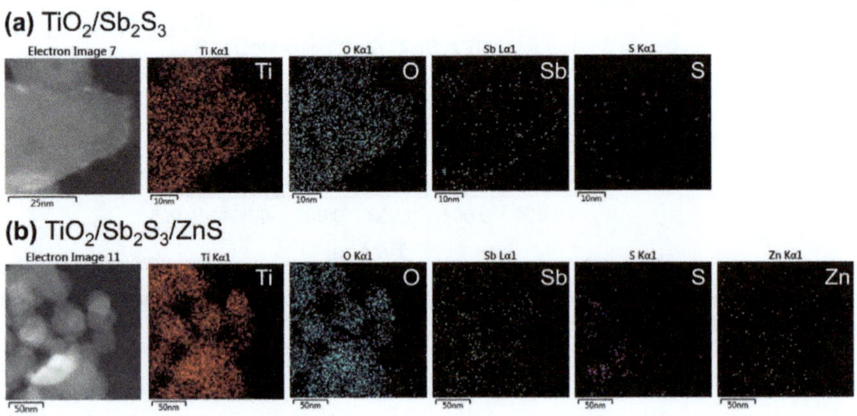

Fig. 9 Elemental mapping images of TEM micrographs for (a) TiO_2/Sb_2S_3 and (b) TiO_2/Sb_2S_3/ZnS.

amount of the adsorbed Sb_2S_3 on TiO_2. The increased J_{sc} is thus attributed to the increased amount of Sb_2S_3 by a surface reduction mechanism as depicted schematically in Fig. 10.

3.3. Effect of double blocking layers

Based on the understanding of the main role of each blocking layer, we finally investigate effect of double blocking layers on photovoltaic performance. TiO_2 surface is first covered conformally with the 1 nm-thick ZrO_2 layer, which is followed by coating Sb_2S_3 and then Sb_2S_3 surface is finally modified by ZnS. As can be seen in Fig. 11, J_{sc} and V_{oc} are substantially improved after the introduction of double blocking layers. Changes in photovoltaic parameters are listed in Table 4, where J_{sc} increases by 14%, V_{oc} by 13% and as a result PCE increases by 30%. From the results on the double blocking layers, it is proven that the increased V_{oc} is due to the suppression of charge recombination at the TiO_2/Sb_2S_3 interface by

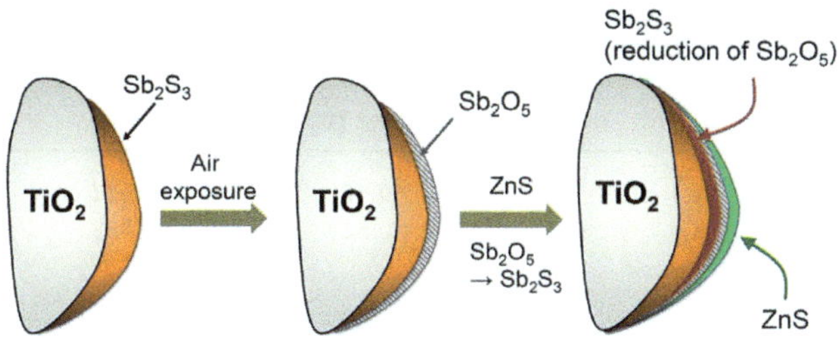

Fig. 10 Schematic diagram of the surface reduction mechanism. The Sb_2O_5 forms on the Sb_2S_3 surface due to quenching at 300 °C in air, which is in part reduced to Sb_2S_3 by ZnS coating.

Table 3 Elemental analysis before and after ZnS coating on Sb_2S_3

Element	wt% (w/o ZnS)	wt% (w/ZnS)	at% (w/o ZnS)	at% (of w/ZnS)
O	45.70	39.08	73.38	67.26
S	0.19	0.94	0.15	0.81
Ti	46.25	52.45	24.81	30.15
Zn	—	0.36	—	0.15
Sb	7.86	7.18	1.66	1.63
Total	100.00	100.00	100.00	100.00

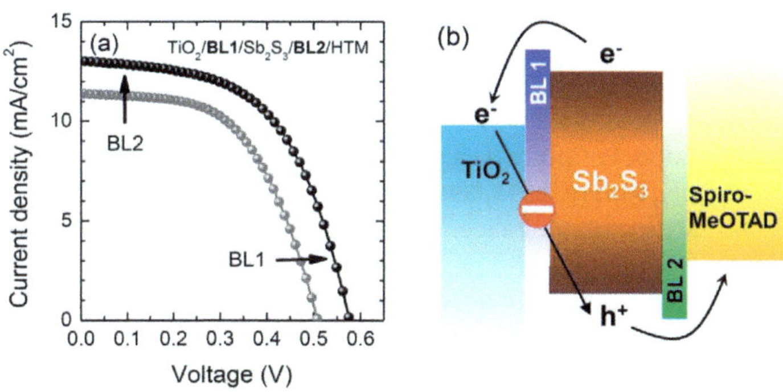

Fig. 11 (a) J–V curves and (b) charge pathway in the Sb_2S_3-sensitized solid state solar cell employing BL1 (ZrO_2) and BL2 (ZnS). BL1 and BL2 play a role in increasing V_{oc} by blocking charge recombination and in increasing J_{sc} by recovering Sb_2S_3 from the surface oxidized layer, respectively.

ZrO_2 blocking layer and the increased J_{sc} is ascribed to the partial reduction of Sb_2O_5 formed on the Sb_2S_3 surface by the ZnS coating (Fig. 11b).

4. Conclusions

In this study, we investigated the major recombination pathway in Sb_2S_3-sensitized all solid state solar cells by inserting thin layers at TiO_2/Sb_2S_3 and Sb_2S_3/HTM

Table 4 Short-circuit photocurrent density (J_{sc}), open-circuit voltage (V_{oc}), FF and PCE of Sb_2S_3-sensitized solid state solar cells before and after employing BL1 (ZrO_2) and BL2 (ZnS)

Device ID	J_{sc} (mA cm^{-2})	V_{oc} (V)	FF	PCE (%)
Bare	11.4	0.509	0.551	3.20
BL1/BL2	13.0	0.577	0.553	4.16

interfaces. We found that V_{oc} loss occurred mainly at the TiO_2/Sb_2S_3 interface because of charge recombination. Thus BL1 at this interface efficiently blocked charge recombination, leading to a significant improvement in V_{oc}. On the other hand, charge recombination was not noticeable at the Sb_2S_3/HTM interface. Instead, the partially oxidized surface at Sb_2S_3 could be recovered by BL2, leading to an increase in J_{sc}. According to the present studies, we can draw an important conclusion that photo-injected electrons in TiO_2 prefer to recombine with oxidized Sb_2S_3 compared to the holes separated to HTM.

Acknowledgements

This work was supported by the National Research Foundation of Korea (NRF) grants funded by the Ministry of Science, ICT & Future Planning (MSIP) of Korea under contract no. NRF-2010-0014992, NRF-2012M1A2A2671721, NRF-2012M3A7B4049986 (Nano Material Technology Development Program) and NRF-2012M3A6A7054861 (Global Frontier R&D Program on Center for Multiscale Energy System).

References

1 U. Bach, D. Lupo, P. Comte, J. E. Moser, F. Weissörtel, J. Salbeck, H. Spreitzer and M. Grätzel, *Nature*, 1998, **395**, 583–585.
2 L. Schmidt-Mende, S. M. Zakeeruddin and M. Grätzel, *Appl. Phys. Lett.*, 2005, **86**, 013504.
3 L. Schmidt-Mende, U. Bach, R. Humphry-Baker, T. Horiuchi, H. Miura, S. Ito, S. Uchida and M. Grätzel, *Adv. Mater.*, 2005, **17**, 813–815.
4 M. Wang, S.-J. Moon, M. Xu, K. Chittibabu, P. Wang, N.-L. Cevey-Ha, R. Humphry-Baker, S. M. Zakeeruddin and M. Grätzel, *Small*, 2010, **6**, 319–324.
5 N. Cai, S.-J. Moon, L. Cevey-Ha, T. Moehl, R. Humphry-Baker, P. Wang, S. M. Zakeeruddin and M. Grätzel, *Nano Lett.*, 2011, **11**, 1452–1456.
6 J. Burschka, A. Dualeh, F. Kessler, E. Baranoff, N.-L. Cevey-Ha, C. Yi, M. K. Nazeeruddin and M. Grätzel, *J. Am. Chem. Soc.*, 2011, **133**, 18042–18045.
7 J. H. Rhee, C.-C. Chung and E. W.-G. Diau, *NPG Asia Mater.*, 2013, **5**, e68.
8 J. A. Chang, J. H. Rhee, S. H. Im, Y. H. Lee, H.-J. Kim and S. I. Seok, *Nano Lett.*, 2010, **10**, 2609–2612.
9 S. H. Im, C.-S. Lim, J. A. Chang, Y. H. Lee, N. Maiti, H.-j. Kim, M. K. Nazeeruddin, M. Grätzel and S. I. Seok, *Nano Lett.*, 2011, **11**, 4789–4793.
10 J. A. Chang, S. H. Im, Y. H. Lee, H.-j. Kim, C.-S. Lim, J. H. Heo and S. I. Seok, *Nano Lett.*, 2012, **12**, 1863–1867.
11 M. Y. Versavel and J. A. Haber, *Thin Solid Films*, 2007, **515**, 7171–7176.

12 N. Tigau, V. Ciupina, G. Prodan, G. I. Rusu, C. Gheorghies and E. Vasile, *J. Optoelectron. Adv. Mater.*, 2004, **6**, 211–217.
13 N.-G. Park, *J. Phys. Chem. Lett.*, 2013, **4**, 2423–2429.
14 N. Zhao, T. P. Osedach, L.-Y. Chang, S. M. Geyer, D. Wanger, M. T. Binda, A. C. Arango, M. G. Bawendi and V. Bulovic, *ACS Nano*, 2010, **4**, 3743–3752.
15 Y.-L. Lee and Y.-S. Lo, *Adv. Funct. Mater.*, 2009, **19**, 604–609.
16 Y.-L. Lee, B.-M. Huang and H.-T. Chien, *Chem. Mater.*, 2008, **20**, 6903–6905.
17 T. Fukumoto, T. Moehl, Y. Niwa, M. K. Nazeeruddin, M. Grätzel and L. Etgar, *Adv. Energy Mater.*, 2013, **3**, 29–33.
18 K. Tsujimoto, D.-C. Nguyen, S. Ito, H. Nishino, H. Matsuyoshi, A. Konno, G. R. A. Kumara and K. Tennakone, *J. Phys. Chem. C*, 2012, **116**, 13465–13471.
19 B. Neppolian, Q. Wang, H. Yamashita and H. Choi, *Appl. Catal., A*, 2007, **333**, 264–271.
20 A. K. Chandiran, M. K. Nazeeruddin and M. Grätzel, *Adv. Funct. Mater.*, 2014, **24**, 1615–1623.
21 P. E. Lippens and M. Lannoo, *Phys. Rev. B: Condens. Matter Mater. Phys.*, 1989, **39**, 10935–10942.
22 M.-J. Kim, C.-R. Lee, W.-S. Jeong, J.-H. Im, T. I. Ryu and N.-G. Park, *J. Phys. Chem. C*, 2010, **114**, 19849–19852.
23 J. Y. Kim, S. Lee, J. H. Noh, H. S. Jung and K. S. Hong, *J. Electroceram.*, 2009, **23**, 422–425.
24 E. Palomares, J. N. Clifford, S. A. Haque, T. Lutz and J. R. Durrant, *J. Am. Chem. Soc.*, 2003, **125**, 475–482.
25 H.-S. Kim, J.-W. Lee, N. Yantara, P. P. Boix, S. A. Kulkarni, S. Mhaisalkar, M. Grätzel and N.-G. Park, *Nano Lett.*, 2013, **13**, 2412–2417.
26 R. D. Shannon, *Acta Crystallogr., Sect. A: Cryst. Phys., Diffr., Theor. Gen. Crystallogr.*, 1976, **32**, 751–767.
27 E. L. Gui, A. M. Kang, S. S. Pramana, N. Yantara, N. Mathews and S. Mhaisalkar, *J. Electrochem. Soc.*, 2012, **159**, B247–B250.

Faraday Discussions

PAPER

Mesoporous perovskite solar cells: material composition, charge-carrier dynamics, and device characteristics

Yixin Zhao, Alexandre M. Nardes and Kai Zhu*

Received 9th June 2014, Accepted 14th July 2014

DOI: 10.1039/c4fd00128a

We report on our investigations on charge transport and recombination in TiO_2-based mesoporous solar cells using PbI_2 and various perovskite compositions, including $CH_3NH_3PbI_3$, $CH_3NH_3PbI_2Br$, $CH_3NH_3PbIBr_2$, and $CH_3NH_3PbBr_3$. The mesoporous TiO_2 film is about 650 nm thick. Electron microscopy measurements show that no perovskite capping layer is formed on the top surface of the TiO_2 film. Intensity-modulated photocurrent/photovoltage spectroscopies show that the electron diffusion coefficient and recombination lifetime are governed by the underlying mesoporous TiO_2 film and thus do not depend on the perovskite composition. However, replacing the perovskite absorber with PbI_2 leads to a diffusion coefficient that is about a factor of 5 slower than that in perovskite-based devices. We also find that $TiCl_4$ treatment of the mesoporous TiO_2 film prior to device fabrication substantially reduces the charge recombination kinetics in mesoporous perovskite solar cells.

Introduction

Organometallic halide perovskites (ABX_3) are a special class of organic–inorganic layered materials that generally comprise an inorganic framework of corner-sharing BX_6 octahedra, where B is usually a divalent metal cation (*e.g.*, Pb^{2+} and Sn^{2+}) and X is a halogen (I^-, Cl^-, Br^-). The A^+ is a small organic cation (*e.g.*, $CH_3NH_3^+$) to neutralize the total charge. These perovskite materials have recently emerged as a revolutionary class of light-absorbers for solar conversion application. A representative material composition of this absorber family is $CH_3NH_3PbI_3$ (or $MAPbI_3$). Solar cells using this absorber have demonstrated impressive progress in photovoltaic (PV) cell performance from 3.8% in 2009[1] to a certified 17.9% in 2014.[2] An open-circuit voltage of 1.13 V has been demonstrated with a 1.55 eV bandgap perovskite absorber.[3] Halide perovskites consist of Earth-abundant elements and have the capability of bandgap tuning by tailoring their inorganic and organic components.[4,5] Moreover, high-efficiency perovskite solar

Chemical and Materials Science Center, National Renewable Energy Laboratory, 15013 Denver West Parkway, Golden, CO 80401, USA. E-mail: Kai.Zhu@nrel.gov

cells can be made by low-cost solution processing, which is ideal for high-throughput scale-up, such as roll-to-roll processing. Despite the rapid progress in device performance, understanding the structural/electronic properties of halide perovskites and the device physics of perovskite-based solar cells is still in its infancy.

Three types of device architectures are normally used for halide perovskites. The first type is adapted from solid-state dye-sensitized solar cells (DSSCs), in which the perovskites simply replace the dye molecules to absorb light and 2,2′,7,7′-tetrakis-(*N*,*N*-di-*p*-methoxyphenyl-amine)-9,9′-spirobifluorene (spiro-MeOTAD) is normally used as the hole transport material (HTM).[6] The device also works well in the absence of the HTM layer.[7] The second type is a planar structure, in which a thin layer of perovskite is sandwiched between the electron- and hole-contact layers such as TiO_2 and spiro-MeOTAD, respectively.[8] The third type is a mesoporous-planar hybrid structure consisting of a TiO_2 mesoporous layer (filled with perovskites) and a perovskite capping layer. The hybrid structure is generally considered as the mesoporous structure. However, the structure and device characteristics between the hybrid and the mesoporous solar cell architectures differ significantly.[9] These different device architectures with various contact materials have attracted enormous worldwide attention during the past two years.[3,4,10–22]

In the present study, we focus on the mesoporous perovskite solar cells based on 650 nm-thick TiO_2 films. Different perovskite compositions are examined, including $MAPbI_3$, $MAPbI_2Br$, $MAPbIBr_2$, and $MAPbBr_3$. Charge transport and recombination properties are examined by intensity-modulated photocurrent/photovoltage spectroscopies (IMPS/IMVS). Electron microscopy measurements show that no perovskite capping layer is formed. We find that the electron diffusion coefficient and the recombination lifetime do not depend on the perovskite composition and are governed by the underlying mesoporous TiO_2 film. Increasing the amount of Br content in the perovskite composition is found to shift the absorption edge and spectral response toward shorter wavelengths, as a result of the larger bandgap. Replacing the perovskite absorber with PbI_2 significantly lowers the device performance. Transport in a PbI_2-based cell is about a factor of 5 slower than that in perovskite-based devices. $TiCl_4$ treatment of the mesoporous TiO_2 film prior to device fabrication is found to substantially reduce the charge recombination kinetics.

Experimental

CH_3NH_3I was synthesized by reacting methylamine (33 wt% ethanol solution) and hydroiodic acid (57 wt% in water, Aldrich) at the molar ratio of 1.2 : 1 in an ice bath for 2 h with stirring followed by vacuum drying and cleaning with ethyl acetate. CH_3NH_3Br was synthesized by reacting methylamine (33 wt% ethanol solution) and 47 wt% hydrobromide acid at the molar ratio of 1.2 : 1 in an ice bath for 2 h with stirring followed by vacuum drying and cleaning with acetonitrile. To make the PbI_2 precursor solution, 1.000 g PbI_2 was dissolved in 2.75 g dimethylformamide (DMF). Four compositions of perovskite solutions were made: the $CH_3NH_3PbI_3$ precursor solution consisted of 0.693 g PbI_2 and 0.239 g CH_3NH_3I in 2.75 g DMF; the $CH_3NH_3PbI_2Br$ precursor solution consisted of 0.693 g PbI_2 and 0.169 g CH_3NH_3Br in 2.6 g DMF; the $CH_3NH_3PbIBr_2$ precursor solution consisted of 0.551 g $PbBr_2$ and 0.239 g CH_3NH_3I in 2.6 g DMF; and the $CH_3NH_3PbBr_3$ precursor solution was made of 0.551 g $PbBr_2$ and 0.169 g CH_3NH_3Br in 2.8 g DMF.

The fluorine-doped transparent conducting SnO_2-coated glass substrate FTO (TEC15, Hartford, USA) was pre-patterned by etching with Zn powder and 25 wt% HCl solution for about 2 min. The patterned FTO substrate was then cleaned by soaking in a base bath (5 wt% NaOH alcohol) for 16 h followed by rinsing with deionized (DI) water and ethanol. The cleaned FTO substrate was subsequently coated with a compact layer of TiO_2 by spray pyrolysis using 0.2 M Ti(IV)bis(ethyl acetoacetate)-diisopropoxide in 1-butanol solution at 450 °C, followed by annealing at 450 °C for 1 h. The TiO_2 paste based on 20 nm-sized TiO_2 nanoparticles was screen-printed using a screen with a 3 μm emulsion thickness on the compact TiO_2 layer on the patterned FTO substrates as described previously.[23] The printed mesoporous TiO_2 film was annealed at 500 °C for 30 min. The average thickness of the mesoporous TiO_2 film was around 650 nm as determined using an Alpha-Step 500 surface profiler. The annealed TiO_2 films were then soaked in 0.04 M $TiCl_4$ solution at 70 °C for 30 min, followed by rinsing with DI water and ethanol, and finally dried under N_2. The $TiCl_4$-treated TiO_2 films were annealed again at 500 °C for 30 min before use for perovskite deposition. For comparison purposes, some mesoporous TiO_2 films were not treated with $TiCl_4$.

Solar cells were fabricated in ambient conditions (unless stated otherwise) as detailed below. The perovskite and PbI_2 precursor solutions were spin-coated onto the 650 nm-thick mesoporous TiO_2 films on FTO at 2500 rpm for 30 s, followed by annealing on a hotplate in air at 100 °C for about 5 min. An HTM solution was spin-coated on the perovskite-covered TiO_2 electrodes at 4000 rpm for 30 s. The HTM solution consists of 0.1 M 2,2′,7,7′-tetrakis-(*N*,*N*-di-*p*-methoxy-phenyl-amine)-9,9′-spirobifluorene (spiro-MeOTAD), 0.035 M bis(trifluoromethane)sulfonimide lithium salt (Li-TFSi), and 0.12 M 4-*tert*-butylpyridine (tBP) in chlorobenzene–acetonitrile (10 : 1, v/v) solution. Finally, a 150 nm-thick Ag layer was deposited on the HTM layer by thermal evaporation. The active area of each device was about 0.25 cm^2.

The crystal structures of the perovskite and PbI_2 films were characterized by X-ray diffraction (XRD, Rigaku D/Max 2200 diffractometer with Cu K_α radiation). The UV-vis absorption spectra of these perovskite films were measured using a Cary-6000i spectrophotometer. The photocurrent–voltage (J–V) characteristics of the perovskite and PbI_2 solar cells were measured with a Keithley 2400 source meter under simulated AM 1.5G illumination (100 mW cm^{-2}; Oriel Sol3A Class AAA Solar Simulator), from open circuit to short circuit with a scan rate of 0.02 V s^{-1}. Less than 5% J–V hysteresis was observed when scanned from short circuit to open circuit. Charge transport and recombination properties of the mesostructured perovskite cells were measured by IMPS/IMVS as described previously.[24]

Results and discussion

Fig. 1 shows the typical top view of scanning electron microscopy (SEM) images of four different perovskites (*i.e.*, $MAPbI_3$, $MAPbI_2Br$, $MAPbIBr_2$, and $MAPbBr_3$) deposited on 650 nm-thick mesoporous TiO_2 films. The SEM images of the mesoporous TiO_2 film and PbI_2 deposited on the TiO_2 film are also given for comparison. The PbI_2, $MAPbI_3$, and $MAPbI_2Br$ films show essentially the same

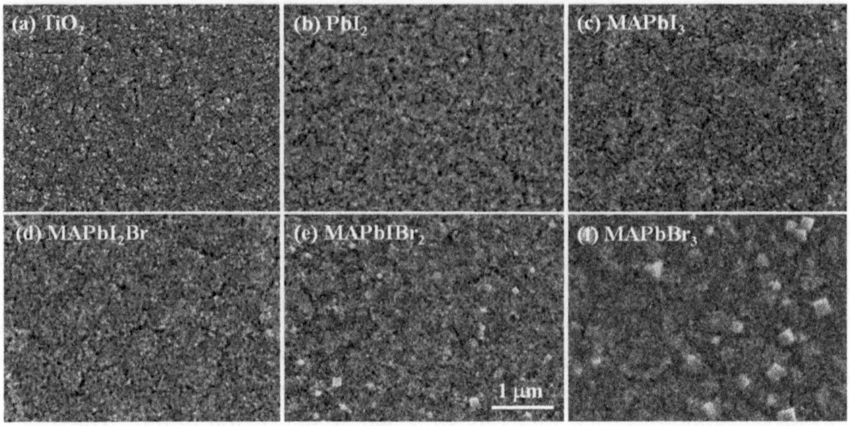

Fig. 1 Typical top view of SEM images of (a) a mesoporous TiO_2 film and various absorbers, including (b) PbI_2, (c) $MAPbI_3$, (d) $MAPbI_2Br$, (e) $MAPbIBr_2$, and (f) $MAPbBr_3$ deposited on 650 nm-thick mesoporous TiO_2 films.

morphology as the underlying mesoporous TiO_2 films. For $MAPbIBr_2$, some cubic crystallites of perovskites are visible on the top surface of the TiO_2 film. The crystallites become larger for the all-bromide perovskite ($MAPbBr_3$) film. This is likely caused by the fact that the $MAPbI_3$ has a tetragonal perovskite phase whereas the $MAPbIBr_2$ and $MAPbBr_3$ have a cubic perovskite phase. Despite some isolated perovskite crystallites on the top surface of the TiO_2 films, the overall morphology adopted by the Br-containing perovskite films is still primarily the mesoporous structure of the underlying TiO_2 film. These top-view SEM images suggest that the perovskite film morphology in this study can be properly described as a mesoporous structure, as described in the Introduction. This observation is further confirmed with the cross-sectional view of the SEM images shown in Fig. 2. No obvious difference is observed for the microstructures of mesoporous TiO_2/absorbers (PbI_2 and perovskite) films, which is consistent with the conclusion that the morphology of mesoporous films is dominated by the underlying mesoporous TiO_2 films.

Fig. 3a compares the ultraviolet-visible (UV-vis) absorption spectra of PbI_2 and the perovskites with four different compositions as indicated. The absorption

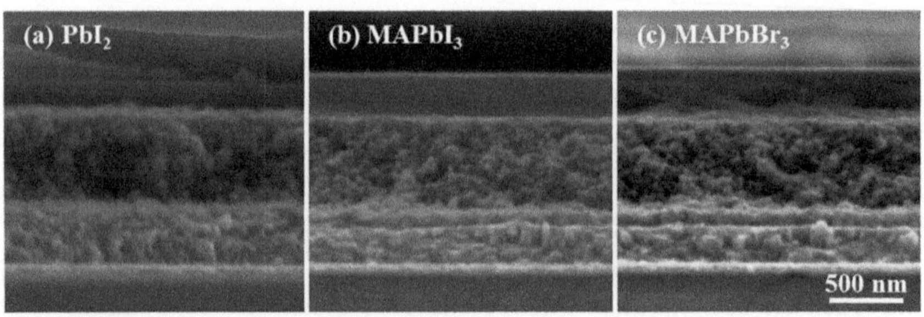

Fig. 2 Typical cross-sectional SEM images of (a) PbI_2, (b) $MAPbI_3$, and (c) $MAPbBr_3$ deposited on 650 nm-thick mesoporous TiO_2 films.

spectrum of $MAPbI_3$ increases sharply as the wavelength decreases from about 800 to 750 nm. The absorption continues to increase (relatively slowly) with further decreasing wavelength. The absorption spectrum of $MAPbI_3$ agrees with our previous studies.[23] The optical bandgap determined from the absorption spectrum is about 1.55 eV for $MAPbI_3$. As the amount of Br in the perovskite composition is increased, the absorption edge of the perovskite film shifts towards shorter wavelength. In this case, the bandgap of the Br-containing perovskites is increased to about 1.75 eV for $MAPbI_2Br$, 2.02 eV for $MAPbIBr_2$, and 2.30 eV for $MAPbBr_3$ as compared to $MAPbI_3$. The bandgap values of these perovskites based on mixed I and Br are consistent with previous reports where perovskites were prepared by using the mixed $MAPbI_3$ and $MAPbBr_3$ precursors.[4] PbI_2 shows a strong absorption at a wavelength shorter than 520 nm, corresponding to a bandgap of 2.4 eV. Fig. 3b compares the main characteristic X-ray diffraction (XRD) peak of the four perovskites and the PbI_2 film to illustrate the difference between these samples. PbI_2 shows the characteristic peak at 12.58°, corresponding to the reflection from (001) planes. For $MAPbI_3$, the main perovskite (110) diffraction peak appears at 14.08°. This diffraction peak shifts to about 14.36° for $MAPbI_2Br$, to 14.68° for $MAPbIBr_2$, and to 15° for $MAPbBr_3$. This shift toward a higher degree angle is a result of the decreased lattice spacing induced by the substitution of larger I atoms with the smaller Br atoms.[4]

Charge transport and recombination times in mesostructured cells using different perovskite absorbers are investigated by IMPS and IMVS, respectively. Both IMPS and IMVS involve small-amplitude modulation of the photocarrier density around steady-state (under illumination) during the photocurrent and photovoltage measurements, respectively.[25] Specifically, IMPS measures the modulated photocurrent response to the modulation of illumination light intensity (usually) at short circuit, whereas IMVS detects the photovoltage response to the modulation of light intensity (usually) at open circuit. Alternative techniques include impedance,[26] time-of-flight,[27] and open-circuit voltage decay.[28] Fig. 4 shows the typical Nyquist plots of IMPS and IMVS response curves for the $MAPbI_3$-based solar cell. Other perovskite solar cells and PbI_2-based devices show similar response curves, *i.e.*, all IMPS and IMVS plots display a semicircle in the complex plane. The charge transport (or collection) time can be estimated from

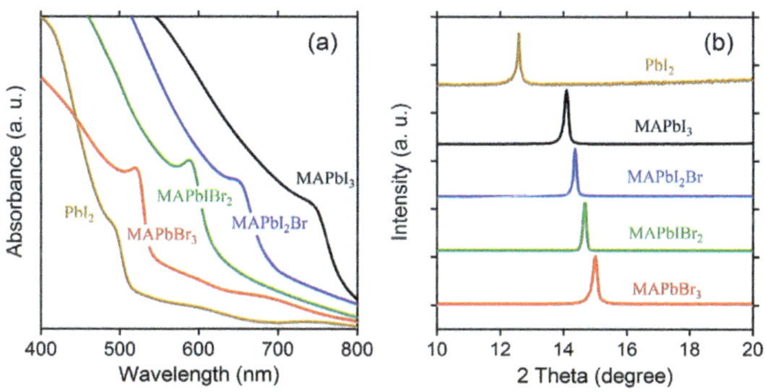

Fig. 3 (a) UV-vis absorption spectra and (b) XRD patterns of PbI_2 and perovskites with different compositions (as indicated) deposited on 650 nm-thick mesoporous TiO_2 films.

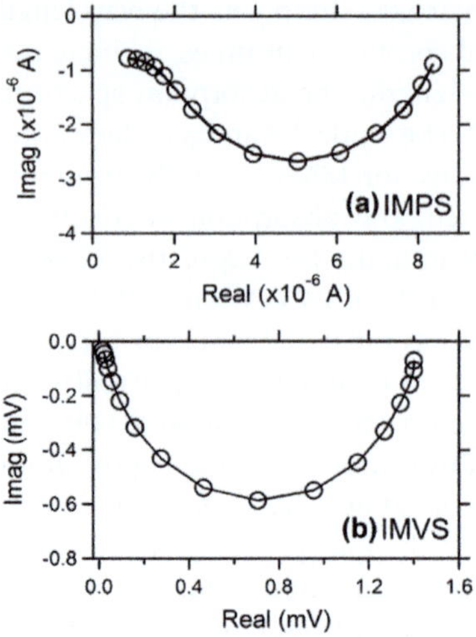

Fig. 4 Typical (a) IMPS and (b) IMVS responses in the complex plane of a perovskite solar cell based on a 650 nm-thick mesoporous TiO_2 film.

the expression: $\tau_c = 1/2\pi f_c$, where f_c is the characteristic frequency at the minimum of the IMPS imaginary component. From the transport time τ_c and the film thickness L, the electron diffusion coefficient D can be calculated according to the expression $D = L^2/2.35\tau_c$.[29] Similarly, the recombination time can be determined from $\tau_r = 1/2\pi f_r$, where f_r is the characteristic frequency at the minimum of the IMVS imaginary component.

Fig. 5a shows the effect of perovskite composition on the diffusion coefficient D as a function of photoelectron density (n). The average thickness for the mesoporous TiO_2 film is about 650 nm. The transport results of the PbI_2-based solar cell are also shown as a comparison. All perovskite cells and the PbI_2-based device exhibit the same power-law dependence of diffusion coefficient on the photoelectron density. For dye-sensitized solar cells (DSSCs), such non-linear dependence is usually described as $D \propto N_{tot}{}^{(1/3)-(1/\alpha)}n^{(1/\alpha)-1}$, where N_{tot} is the total trap

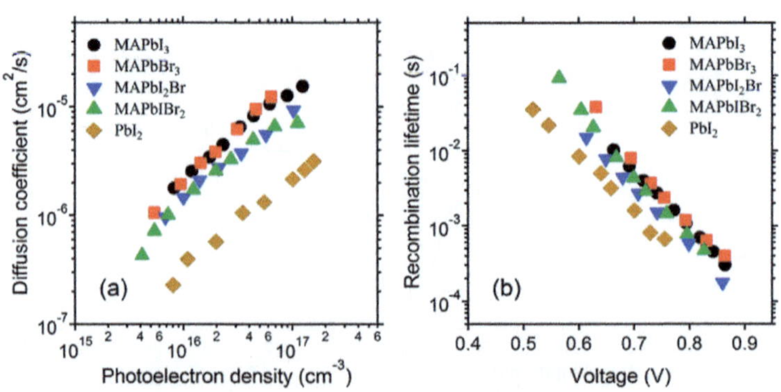

Fig. 5 Comparison of (a) electron diffusion coefficient and (b) recombination lifetime of mesoporous solar cells based on PbI_2 and various perovskite absorbers (as indicated).

density and α is usually between 0 and 1. This non-linear electron density dependence of the electron diffusion coefficient is attributed to the electrons undergoing multiple trapping and detrapping through a network of trap states across the mesoporous TiO_2 electrode film.[25] The parameter α is related to the distribution of these trap states.[30] Best fits of the data to the power-law dependence ($D \propto n^{(1/\alpha)-1}$) yield $\alpha = 0.50 \pm 0.05$ for all samples, including the PbI_2-based device. When compared at a fixed electron density (*e.g.*, 10^{16} cm^{-3}), all perovskite solar cells with different perovskite compositions exhibit similar D values within a factor of two, suggesting that transport is dominated by the properties of the TiO_2 film rather than by those of the perovskite layer. The trap distribution is thus not strongly influenced by the composition of perovskites coated on the TiO_2 surface. This observation confirms our previous finding that the transport property of $MAPbI_3$-sensitized cells is the same as that of the conventional solid-state DSSCs, in which the electron transport is limited by the trap states associated with TiO_2.[23] It is worth noting that a recent paper shows that the transport of mesostructured perovskite solar cells depends on the concentration of the perovskite precursors.[31] When a relatively high concentration of perovskite precursor solution is used, electrons could also conduct through the perovskite layer. However, when a high-concentration precursor is used, the device architecture has presumably transitioned from the sensitized mesoporous structure to the mesoporous-planar hybrid solar cell structure, which consists of a mesoporous layer (filled with perovskites) and a perovskite capping layer. In this situation, the charge carrier dynamics are likely determined by perovskite rather than TiO_2.[9] This is different from the film morphology discussed in this study. In contrast to the perovskite solar cells, the PbI_2-based device shows a much lower diffusion efficient (about a factor of 5 lower than that of the perovskite solar cells), which implies a much higher trap density (N_{tot}) of the PbI_2-based cell than that of perovskite solar cells. A recent study has shown that for $MAPbI_3$ only shallow traps are likely to be formed because of their low formation energies.[15] In contrast, both shallow- and deep-level defects could form in PbI_2 films.[32] Thus, it appears valuable to eliminate the PbI_2 phase from the perovskite materials for the development of perovskite solar cells. This finding has direct implications for device manufacturing.

Fig. 5b shows the effect of perovskite composition on the electron recombination lifetime in mesoporous perovskite solar cells. The voltage-dependence of the recombination lifetime is commonly seen for DSSCs and is associated with faster recombination at higher electron density. The voltage dependence of the recombination kinetics in DSSCs has been explained in detail previously.[26] Our recent study shows that the recombination kinetics in mesoporous $MAPbI_3$ cells are comparable to those in mesoporous solid-state DSSCs using standard Z907 dye.[23] Consistent with this observation, Fig. 5b shows that, for a given voltage, there is essentially no difference in recombination lifetime among the mesoporous perovskite cells with different perovskite compositions. This implies that for mesoporous perovskite cells, the recombination kinetics are dominated by recombination between the electrons in TiO_2 and the holes in spiro-MeOTAD. In contrast to Fig. 5b, a recent study[33] using impedance spectroscopy has shown that increasing the Br content of the perovskite composition significantly reduces the recombination resistance and leads to an improved photovoltage for perovskite solar cells. This previous study, however, used a 250 nm-thick mesoporous TiO_2

film and a higher precursor concentration. Although the film/cell morphology was not shown, it is likely that a mesoporous-planar hybrid solar cell structure (*i.e.*, a mesoporous bottom layer plus a planar top capping layer) was actually formed rather than the mesoporous structured investigated in our study. Finally, we find that the recombination lifetime for PbI_2-based device follows the same voltage dependence but is slightly smaller than for the perovskite cells. This is likely caused by the higher trap density as discussed above.

Fig. 6 shows the effect of $TiCl_4$ treatment (on a 650 nm-thick mesoporous TiO_2 film) upon the diffusion coefficient and recombination lifetime for mesoporous perovskite cells. The data for the $TiCl_4$-treated samples are taken from Fig. 5. The $TiCl_4$ treatment of mesoporous TiO_2 films is a standard procedure to improve the device performance of dye-sensitized solar cells. The improvement has been attributed to various factors, including improvement of the connection between TiO_2 nanoparticles,[34] enhancement of the surface roughness of TiO_2,[35] passivation of surface traps,[36] and movement of the TiO_2 conduction band position.[37] While these studies focused on DSSCs based on a liquid iodide/triiodide electrolyte, a more recent study has shown for spiro-MeOTAD-based solid-state DSSCs that the main effect of $TiCl_4$ treatment is to reduce the recombination kinetics associated with a subtle change of the film morphology.[38] Similar to the DSSC fabrication procedure, $TiCl_4$ treatment has also been used for preparing perovskite solar cells.[6,7,39–42] However, its impact on charge transport and recombination for perovskite solar cells has not been reported. In this study, $MAPbI_3$ and $MAPbBr_3$ are used as examples to understand the $TiCl_4$ treatment effect. We find that the charge transport is not affected by $TiCl_4$ treatment for mesoporous perovskite solar cells (Fig. 6a), whereas the recombination lifetimes for cells based on $TiCl_4$-treated TiO_2 films are about a factor of 4–5 longer than those without $TiCl_4$ treatment (Fig. 6b). These results are consistent with the recent study on solid-state DSSCs as mentioned above,[38] suggesting that $TiCl_4$ treatment can reduce the recombination kinetics for mesoporous perovskite solar cells. It is well-known that $TiCl_4$ treatment usually increases the surface roughness of mesoporous TiO_2 films. The larger surface roughness presumably improves the coating/contact of the perovskites on the TiO_2 surface, thus reducing the back-recombination between electrons in the TiO_2 network and holes in the spiro-MeOTAD layer.

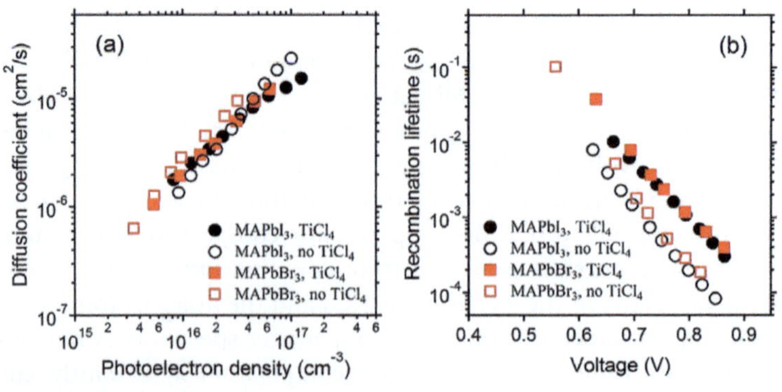

Fig. 6 Effect of $TiCl_4$ treatment on (a) electron diffusion coefficient and (b) recombination lifetime for mesoporous perovskite solar cells.

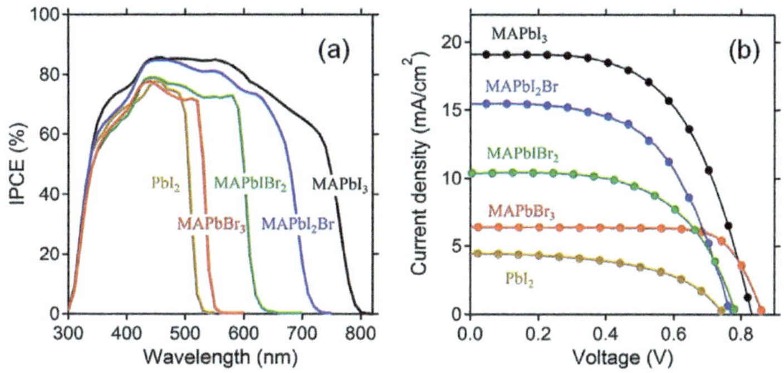

Fig. 7 Comparison of (a) IPCE and (b) J–V characteristics of perovskite ($MAPbI_3$, $MAPbI_2Br$, $MAPbIBr_2$, and $MAPbBr_3$) and PbI_2 solar cells based on 650 nm-thick mesoporous TiO_2 films treated with $TiCl_4$.

Table 1 Effect of the absorber composition on the short-circuit photocurrent density J_{sc}, open-circuit voltage V_{oc}, fill factor FF, and conversion efficiency η of mesoporous solar cells

Cell type	J_{sc} (mA cm^{-2})	V_{oc} (V)	FF	η (%)
$MAPbI_3$	19.09	0.835	0.575	9.16
$MAPbI_2Br$	15.39	0.773	0.567	6.75
$MAPbIBr_2$	10.35	0.785	0.583	4.74
$MAPbBr_3$	6.35	0.857	0.757	4.12
PbI_2	4.42	0.747	0.527	1.74

Fig. 7a shows the incident photon-to-current efficiency (IPCE) spectra for the perovskite ($MAPbI_3$, $MAPbI_2Br$, $MAPbIBr_2$, and $MAPbBr_3$) and PbI_2-based mesoporous solar cells. The long-wavelength onset of the IPCE spectrum of perovskite solar cell shifts to lower wavelengths with increasing Br content, which is in good agreement with the absorption spectrum for each absorber material (Fig. 2a). Fig. 7b compares the J–V characteristics of mesoporous solar cells using the same set of absorbers as used in Fig. 7a. The short-circuit photocurrent density (J_{sc}) value of these solar cells increases in the order of $PbI_2 < MAPbBr_3 < MAPbIBr_2 < MAPbI_2Br < MAPbI_3$, which is consistent with the changes of their respective IPCE spectra. In contrast to the significant and systematic changes of the J_{sc} values, the open-circuit voltages (V_{oc}) of these devices are near 0.8 V. This suggests that the voltage for the mesoporous TiO_2-based perovskite solar cell is limited by the conduction band level and density of states of TiO_2. When ZrO_2 or Al_2O_3 is used as the scaffold, a higher photovoltage is usually observed,[33,43] which presumably results from fewer trap states in these oxides than in TiO_2. The photovoltaic (PV) parameters (J_{sc}, V_{oc}, and fill factor FF and conversion efficiency η) of these devices are shown in Table 1. It is clear that the PbI_2-based device shows a much lower efficiency with a poorer fill factor than the perovskite solar cells. Since several approaches for making perovskites could lead to excess PbI_2 in the film, it is important to remove PbI_2 to achieve good perovskite cells. Moreover, analysis of a recent report[44] shows that the conduction band edge of PbI_2 is about 0.5 eV above that of $MAPbI_3$. Thus, if PbI_2 exists between perovskite and TiO_2, it could act as a barrier for charge injection from perovskite into TiO_2, which further limits the solar conversion efficiency.

Conclusions

In conclusion, we have investigated the influence of perovskite composition upon charge transport and recombination in mesoporous perovskite solar cells based on 650 nm-thick TiO_2 films. Four perovskite compositions were examined, including $MAPbI_3$, $MAPbI_2Br$, $MAPbIBr_2$, and $MAPbBr_3$. Electron microscopy measurements indicate that the cell architecture in this study can be properly described as a mesoporous structure inasmuch that no perovskite capping layer is found on the top surface of the TiO_2 film. Increasing the amount of Br content in the perovskite composition shifts the absorption edge and spectral response toward a shorter wavelength. The electron diffusion coefficient and recombination lifetime in the mesoporous perovskite cells do not depend on the perovskite composition for the samples investigated in this study. However, we find that transport in the PbI_2-based device is significantly slower than that in the perovskite solar cells. Finally, $TiCl_4$ treatment of the mesoporous TiO_2 film prior to device fabrication is found to reduce the charge recombination kinetics substantially.

Acknowledgements

We acknowledge the support by the U.S. Department of Energy/National Renewable Energy Laboratory's Laboratory Directed Research and Development (LDRD) program under Contract no. DE-AC36-08GO28308. K.Z. acknowledges the support on the charge transport and recombination studies by the Division of Chemical Sciences, Geosciences, and Biosciences, Office of Basic Energy Sciences, U.S. Department of Energy, under contract no. DE-AC36-08GO28308 with the National Renewable Energy Laboratory.

References

1 A. Kojima, K. Teshima, Y. Shirai and T. Miyasaka, *J. Am. Chem. Soc.*, 2009, **131**, 6050–6051.

2 NREL, NREL's latest PV chart on record cell efficiencies, http://www.nrel.gov/ncpv/images/efficiency_chart.jpg.

3 M. M. Lee, J. Teuscher, T. Miyasaka, T. N. Murakami and H. J. Snaith, *Science*, 2012, **338**, 643–647.

4 J. H. Noh, S. H. Im, J. H. Heo, T. N. Mandal and S. I. Seok, *Nano Lett.*, 2013, **13**, 1764–1769.

5 G. E. Eperon, S. D. Stranks, C. Menelaou, M. B. Johnston, L. M. Herz and H. J. Snaith, *Energy Environ. Sci.*, 2014, **7**, 982–988.

6 H.-S. Kim, C.-R. Lee, J.-H. Im, K.-B. Lee, T. Moehl, A. Marchioro, S.-J. Moon, R. Humphry-Baker, J.-H. Yum, J. E. Moser, M. Grätzel and N.-G. Park, *Sci. Rep.*, 2012, **2**, 591.

7 L. Etgar, P. Gao, Z. Xue, Q. Peng, A. K. Chandiran, B. Liu, M. K. Nazeeruddin and M. Grätzel, *J. Am. Chem. Soc.*, 2012, **134**, 17396–17399.

8 M. Liu, M. B. Johnston and H. J. Snaith, *Nature*, 2013, **501**, 395–398.
9 S. Ryu, J. H. Noh, N. J. Jeon, Y. C. Kim, W. S. Yang, J. W. Seo and S. I. Seok, *Energy Environ. Sci.*, 2014, **7**, 2614–2618.
10 J.-H. Im, C.-R. Lee, J.-W. Lee, S.-W. Park and N.-G. Park, *Nanoscale*, 2011, **3**, 4088–4093.
11 B. Cai, Y. Xing, Z. Yang, W.-H. Zhang and J. Qiu, *Energy Environ. Sci.*, 2013, **6**, 1480–1485.
12 J. Qiu, Y. Qiu, K. Yan, M. Zhong, C. Mu, H. Yan and S. Yang, *Nanoscale*, 2013, **5**, 3245–3248.
13 D. Bi, L. Yang, G. Boschloo, A. Hagfeldt and E. M. J. Johansson, *J. Phys. Chem. Lett.*, 2013, **4**, 1532–1536.
14 E. Edri, S. Kirmayer, D. Cahen and G. Hodes, *J. Phys. Chem. Lett.*, 2013, **4**, 897–902.
15 W. J. Yin, T. T. Shi and Y. F. Yan, *Appl. Phys. Lett.*, 2014, **104**, 063903.
16 S. Colella, E. Mosconi, P. Fedeli, A. Listorti, F. Gazza, F. Orlandi, P. Ferro, T. Besagni, A. Rizzo, G. Calestani, G. Gigli, F. De Angelis and R. Mosca, *Chem. Mater.*, 2013, **25**, 4613–4618.
17 P. Docampo, J. M. Ball, M. Darwich, G. E. Eperon and H. J. Snaith, *Nat. Commun.*, 2013, **4**, 2761.
18 J. You, Z. Hong, Y. Yang, Q. Chen, M. Cai, T.-B. Song, C.-C. Chen, S. Lu, Y. Liu and H. Zhou, *ACS Nano*, 2014, **8**, 1674–1680.
19 J. Y. Jeng, Y. F. Chiang, M. H. Lee, S. R. Peng, T. F. Guo, P. Chen and T. C. Wen, *Adv. Mater.*, 2013, **25**, 3727–3732.
20 J. A. Christians, R. C. M. Fung and P. V. Kamat, *J. Am. Chem. Soc.*, 2014, **136**, 758–764.
21 P. Qin, S. Tanaka, S. Ito, N. Tetreault, K. Manabe, H. Nishino, M. K. Nazeeruddin and M. Grätzel, *Nat. Commun.*, 2014, **5**, 3834.
22 Y. Zhao and K. Zhu, *J. Phys. Chem. Lett.*, 2013, **4**, 2880–2884.
23 Y. Zhao, A. M. Nardes and K. Zhu, *J. Phys. Chem. Lett.*, 2014, **5**, 490–494.
24 K. Zhu, N. Kopidakis, N. R. Neale, J. van de Lagemaat and A. J. Frank, *J. Phys. Chem. B*, 2006, **110**, 25174–25180.
25 A. J. Frank, N. Kopidakis and J. van de Lagemaat, *Coord. Chem. Rev.*, 2004, **248**, 1165–1179.
26 Q. Wang, S. Ito, M. Grätzel, F. Fabregat-Santiago, I. Mora-Sero, J. Bisquert, T. Bessho and H. Imai, *J. Phys. Chem. B*, 2006, **110**, 25210–25221.
27 N. Kopidakis, E. A. Schiff, N. G. Park, J. van de Lagemaat and A. J. Frank, *J. Phys. Chem. B*, 2000, **104**, 3930–3936.
28 A. Zaban, M. Greenshtein and J. Bisquert, *ChemPhysChem*, 2003, **4**, 859–864.
29 J. van de Lagemaat and A. J. Frank, *J. Phys. Chem. B*, 2001, **105**, 11194–11205.
30 J. van de Lagemaat, K. Zhu, K. D. Benkstein and A. J. Frank, *Inorg. Chim. Acta*, 2008, **361**, 620–626.
31 T. Leijtens, B. Lauber, G. E. Eperon, S. D. Stranks and H. J. Snaith, *J. Phys. Chem. Lett.*, 2014, **5**, 1096–1102.
32 E. Lifshitz, L. Bykov, M. Yassen and Z. ChenEsterlit, *Chem. Phys. Lett.*, 1997, **273**, 381–388.
33 B. Suarez, V. Gonzalez-Pedro, T. S. Ripolles, R. S. Sanchez, L. Otero and I. Mora-Sero, *J. Phys. Chem. Lett.*, 2014, **5**, 1628–1635.
34 S. Kambe, S. Nakade, Y. Wada, T. Kitamura and S. Yanagida, *J. Mater. Chem.*, 2002, **12**, 723–728.

35 S. Ito, P. Liska, P. Comte, R. L. Charvet, P. Pechy, U. Bach, L. Schmidt-Mende, S. M. Zakeeruddin, A. Kay, M. K. Nazeeruddin and M. Grätzel, *Chem. Commun.*, 2005, 4351–4353.

36 S. W. Lee, K. S. Ahn, K. Zhu, N. R. Neale and A. J. Frank, *J. Phys. Chem. C*, 2012, **116**, 21285–21290.

37 P. M. Sommeling, B. C. O'Regan, R. R. Haswell, H. J. P. Smit, N. J. Bakker, J. J. T. Smits, J. M. Kroon and J. A. M. van Roosmalen, *J. Phys. Chem. B*, 2006, **110**, 19191–19197.

38 A. Marchioro, A. Dualeh, A. Punzi, M. Grätzel and J. E. Moser, *J. Phys. Chem. C*, 2012, **116**, 26721–26727.

39 J. H. Heo, S. H. Im, J. H. Noh, T. N. Mandal, C. S. Lim, J. A. Chang, Y. H. Lee, H. J. Kim, A. Sarkar, M. K. Nazeeruddin, M. Grätzel and S. I. Seok, *Nat. Photonics*, 2013, **7**, 486–492.

40 J. Burschka, N. Pellet, S. J. Moon, R. Humphry-Baker, P. Gao, M. K. Nazeeruddin and M. Grätzel, *Nature*, 2013, **499**, 316–319.

41 J.-W. Lee, T.-Y. Lee, P. J. Yoo, M. Grätzel, S. Mhaisalkar and N.-G. Park, *J. Mater. Chem. A*, 2014, **2**, 9251–9259.

42 Y. Zhao and K. Zhu, *J. Phys. Chem. C*, 2014, **118**, 9412–9418.

43 D. Bi, S.-J. Moon, L. Haggman, G. Boschloo, L. Yang, E. M. J. Johansson, M. K. Nazeeruddin, M. Grätzel and A. Hagfeldt, *RSC Adv.*, 2013, **3**, 18762–18766.

44 R. Lindblad, D. Bi, B.-w. Park, J. Oscarsson, M. Gorgoi, H. Siegbahn, M. Odelius, E. M. J. Johansson and H. Rensmo, *J. Phys. Chem. Lett.*, 2014, **5**, 648–653.

Faraday Discussions

DISCUSSIONS

Solar cells and photocatalytic systems: general discussion

Joachim Maier, Pengfei Liu, Xile Hu, Galen Stucky, Shi-Gang Sun, Liang Chen, Wei Han, Dong Ryeol Whang, Gang Chen, Chao Zhan, Jian He, Lee Cronin, Shihe Yang, Erwin Reisner, Lulu Long, Andrew Mount, Ram Seshadri, Xudong Yang, Fuping Pan, Hong Liu, Zhonghua Li, Zhongqun Tian, Heinz Frei, Jingshan Luo, Fei Li, Pingwu Du, Baohua Wang, Yu Li, Yicheng Zhao, Licheng Sun, Takashi Hisatomi, Can Li, Kai Zhu, Nam-Gyu Park, Huiqiong Wang and Michael Grätzel

DOI: 10.1039/C5FD90002F

Xile Hu opened the discussion of the paper by Can Li: When you change the photoanode to a more porous structure, is the higher photocurrent due to a higher degree of light absorption, or is it due to better charge transfer kinetics?

Can Li responded: Both could be possible, but more light absorption should be considered when it is really a porous material.

Lee Cronin remarked: How stable are the plate-like WO_3? Could the be used at pH 1 or below? Does the morphology change?

Can Li replied: WO_3 is quite stable under acid conditions, and the morphology does not change evidently.

Pingwu Du commented: Could the enhancement of photocurrent on a nano-structured (plate-like) tungsten oxide photoanode result from an increase of the surface area, rather than its catalytic activity?

Can Li replied: Compared with the compact WO_3 film, the plate-like WO_3 photoanodes show more electrochemical active surface area, increased light absorption and lower electrode/electrolyte interface charge-transfer resistance. So, the improved photocurrent cannot be singly attributed to the changing surface area.

Pengfei Liu asked: According to the EIS figure, the plate-like WO_3 has a much smaller R_{ct}, which also improve the photocurrent. Accordingly, how will the morphology influence the charge transfer resistance?

Can Li responded: The morphology influences the charge separation mainly *via* the driving force of charge separation between the different facets.

Fei Li said: What is the difference between the non-redox metals used for the etching step?

Can Li replied: The morphologies of the resulting electrodes can be different.

Shihe Yang communicated: (1) The electrochemically active surface area of plate-like WO_3 is twice as large as the sputtered compact WO_3 film, and its charge-transfer resistance is also much lower than that of the compact WO_3 film. Why is the photocurrent density generated by the plate-like WO_3 sample just a little higher than that of the compact WO_3 film? (2) What is actually the main reason for the enhanced performance for the plate-like WO_3 sample? The increase in specific surface area or the improvement in catalytic activity? (3) Is there possible doping, amorphization or disordering due to etching, which could change the electronic properties of WO_3?

Can Li communicated in reply: (1) This is possibly due to the poorly ordered assembly on the electrode. (2) One of the actual reasons for the increase in specific facets is due to the charge separation between different facets. (3) It is possible that the disordering on the surface could change the catalytic properties.

Jingshan Luo opened the discussion of the paper by Erwin Reisner: Regarding the *J–V* curve, there seems to be a peak in the curve of photosystem II; is it an oxidation peak?

Erwin Reisner replied: A small peak is observed in the *J–V* curve at +0.95 V *vs.* RHE. However, the standard deviation for this measurement (20%) makes it difficult to interpret this feature with confidence. Photodegradation of PSII or film-loss of PSII are possible explanations for the peak.

Chao Zhan addressed Can Li and Erwin Reisner: What is the concentration of PSII on the surface? I think the TOF will decrease with increasing concentration of PSII. The low stability of the ITO/PSII system may be caused by the weak interation between the substrate and PSII. This is very common in organic–inorganic system. How can this be improved?

Can Li answered: The concentration of PSII on the surface is difficult to measure because it is a weak adsorption, or physical adsorption under a dynamic state. For our case (not in this work), we tried to employ self assembly of PSII on the surface to improve the interaction.

Erwin Reisner continued: The geometrical surface density of adsorbed PSII on ITO is approximately 19 pmol cm^{-2}. We used the total amount of PSII adsorbed on ITO to calculate the TOF, which is therefore a conservative estimate. An increasing amount of PSII could result in an apparently lower TOF because of the multi-layer formation of PSII and a consequently higher fraction of electroinactive PSII on the ITO electrode. We can increase the stability of PSII on ITO through covalent binding of PSII to the ITO surface (see M. Kato, T. Cardona, A. W. Rutherford and E. Reisner, *J. Am. Chem. Soc.*, 2013, **135**, 10610–10613).

Andrew Mount remarked: In Fig. 3b(i) it was stated that it is unclear if the response actually peaks near $E = 1$ V, given the standard deviation of the measurements (which I presume is the s.d. for repeated measurements on different electrodes). It is also stated in the paper that this system is relatively unstable, with a lifetime of around 4 min. This opens up the possibility that if the peak were real, it could be produced by a decrease in photoactivity with time. Have the authors made or considered repeated measurements of these photocurrents in this E range on the same electrode, stepping both to increasing and decreasing values of E, to gain insight into these effects and potentially obtain standard deviations of successive measurements on the same electrode?

Erwin Reisner responded: The standard deviation of 20% for the PSII–ITO trace in Fig. 3b(i) was determined for repeated measurements on different electrodes. The large error makes it difficult to interpret with confidence the small decrease in photocurrent from 1.0 to 1.2 V. We can speculate that the decrease may be due to photo-degradation or film loss of the enzyme during the measurement. We have not stepped back and forth between the potentials to investigate the small decrease in more detail.

Xile Hu commented: What is the energy of the hole release from PSII? Do you think that the I–V curve of PSII is influenced by different recombination rates at different potentials?

Erwin Reisner responded: The hole is not 'released' by PSII, it is 'consumed' and provides the MnCa-cluster with an approximately 350 mV driving force to oxidise water to O_2. We cannot exclude a dependence of recombination events on immobilised PSII on the applied potential and it would be interesting to investigate this in more detail.

Heinz Frei asked: Is there a small subset of PSII systems on your electrodes running much faster than average, while others might be very slow or not contributing at all?

Erwin Reisner responded: PSII was immobilised in a random orientation on mesoporous ITO in this study. Thus, only PSII in the correct orientation on ITO (*i.e.* electroactive enzyme) contributes to interfacial electron transfer, whereas all of the other PSII (electroinactive enzyme) does not. It is therefore likely that a small subset of PSII is responsible for the majority of the photocurrent and the remaining PSII shows only small or no photoactivity.

Pingwu Du asked: Why do you need the titanium for deposition (in the WO_3-Ti/Ni system)? As for the photoanode functionalized with PSII, what happened to the stability during catalysis?

Erwin Reisner responded: We employed the molecular precursor of $[Ti_2(OEt)_9(NiCl)]_2$ to generate hydrolytically a protecting TiO_2 and a NiO_x

electrocatalyst layer on WO_3. TiO_2 is an alkaline resistant material and can partly protect WO_3 from direct contact with the alkaline solution, thereby enhancing the stability of WO_3. In addition, TiO_2 can decrease the rate of charge recombination by serving as a charge separation layer for transferring holes from the WO_3 to the catalyst. The photoactivity of the PSII–ITO photoanode decreases readily during light exposure and catalysis. The photo-degradation is due to the intrinsic photo-instability of PSII during irradiation and the fragility of the PSII–ITO interface.

Lee Cronin commented: Have you tried to pattern some electrode surfaces using nano imprint lithography (or similar) and bind the PSII into the cavities? This may allow the redox behaviour to be modified as a function of the architectures on the electrode.

Erwin Reisner responded: We have developed improved electrode materials for PSII and other enzymes over the last couple of years and we observe a substantially improved integration and activity of the proteins within tailor-made 3D architectures (unpublished work).

Andrew Mount communicated: Could you use scanning electrochemical microscopy to interrogate those quinones in your PSII layer that are reduced but are not in close enough contact with the electrode for electron transfer to occur?

Erwin Reisner communicated in reply: We considered using SECM for the study of electroactive PSII on the ITO surface by following the release of O_2. I think that it would be easier to follow catalytically generated O_2 than reduced quinones.

Andrew Mount communicated: Can you orient directionally PSII with respect to the surface of the electrode by directional control/functionalisation of the surface deposition?

Erwin Reisner communicated in reply: Yes, we have improved the orientation of PSII on ITO by exploiting electrostatic interactions between a self-assembled monolayer of negatively charged carboxylates with the positive dipole on the electron acceptor side of PSII (see M. Kato, T. Cardona, A. W. Rutherford, E. Reisner, *J. Am. Chem. Soc.*, 2013, **135**, 10610–10613).

Andrew Mount communicated: Does/could loading the PSII electrode film with more quinone mediators lead to enhanced electrode communication?

Erwin Reisner communicated in reply: Yes, we were able to improve the photocurrent response of PSII electrodes with excess quinones as electron mediator.

Zhonghua Li opened the discussion of the paper by Takashi Hisatomi: Can you compare films produced by different methods at the same thickness?

Takashi Hisatomi responded: Photoelectrodes with fewer grain boundaries and a better electric contact to a back conductive layer will show higher photocurrent if the thicknesses of the particulate photocatalyst layers are the same.

Therefore, the use of single-crystalline particles would be ideal. The optimum thickness (or the particle size) would be dependent on various factors, such as the depletion layer widths at the working conditions, minority carrier diffusion lengths, electric conductivities and absorption coefficients.

Zhonghua Li commented: Why do you choose H_2O_2 as electrolyte?

Takashi Hisatomi responded: H_2O_2 was used because it had been proven to be applicable to hematite photoanodes and it was actually effective for $SrNbO_2N$. One could also use sulfite ions, iodide ions, ferrocyanide ions or methanol as hole trapping reagents.

Chao Zhan addressed Takashi Hisatomi: In the *I–V* curve (Fig. 2) we can find transient currents, especially in the low bias, and the transient current induced by light-on is larger than that for light-off. Also, for the steady-state current, in the low voltage bias it is decreasing and in the high bias it is increasing – why is this? I think we can perhaps find more information from the transient and steady current, just like the time scale.

Takashi Hisatomi replied: A transient photoanodic current is observed in the negative potential region (0.3-0.9 V *vs.* RHE) when the potential is swept from negative to positive as shown in Fig. 2. This kind of photoanodic response is not observed when the potential is maintained or swept from positive to negative. Similar behavior is observed commonly for photoanodes modified with Co-containing catalysts and is probably due to the transient oxidation of the Co-containing species. The photoanodic current in the negative potential region is attributable to a transient oxidation process of the CoO_x catalyst. On the other hand, the photoanodic current in the positive potential region is mainly attributable to continuous water oxidation, as the Faradaic efficiency of approximately 90% for water oxidation was observed. Therefore, the photocurrent increases as the potential becomes more positive. It is possible to extract the capacitance and the time scale using a more time-resolved system. Please refer to Le Formal *et al.*, *J. Phys. Chem. C*, 2012, **116**, 26707 and *J. Am. Chem. Soc.*, 2014, **136**, 2564.

Jingshan Luo commented: Regarding the particle transfer method, How reproducible is it? Is there any corrosion of the metal contacts if they are not fully covered by the active materials, and does the exposed metal surface have an impact on the dark current?

Takashi Hisatomi answered: The particle transfer method is mostly reproducible and also applicable to fabrication of photoelectrodes of other particulate semiconductors (*e.g.*, *Chem. Sci.*, 2013, **4**, 1120–1124; *Energy Environ. Sci.*, 2014, 7, 2239–2242; *J. Phys. Chem. C*, 2014, **118**, 16386–16392). Sputtering conditions of metal films, such as the temperature of a sample substrate and the pressure of a chamber, should be chosen carefully. I also think that metal layers should be sufficiently thicker than the roughness of particle layers so that they can be a continuous film.

Pengfei Liu commented: Although the electrode fabricated by the particle transfer method exhibits much improved photocurrent compared with that made of EPD, the current is still not large enough. Will electrodes prepared through *in situ*-growth method be more suitable for PEC?

Takashi Hisatomi replied: *In situ*-growth methods may be suitable for some oxide photoelectrodes but probably not for $SrNbO_2N$ and other (oxy)nitride materials. $SrNbO_2N$ was prepared by nitridation under an NH_3 flow at 1173 K in this study. Many conductive substrates would not be stable in such reactive conditions. Therefore, $SrNbO_2N$ powder was firstly prepared and processed into photoelectrodes by the particle transfer method. The advantage of the particle transfer method over the conventional electrophoretic deposition method was described in the main text.

Shihe Yang opened a general discussion by addressing Can Li and Erwin Reisner: Regarding the integration of artificial and natural photosystems, how far can one go considering that the natural photosystem has a huge arsenal of support, such as self-healing mechanisms, but the artificial system is very primitive?

Can Li answered: Yes, natural photosynthesis has many merits including self-healing, and the artificial system is still quite primitive, but the artificial system based on inorganic compounds is durable and able to harvest more solar light.

Erwin Reisner responded: Most artificial photosynthetic systems indeed lack efficient self-repair mechanisms. This issue is recognised by the community and should receive more attention in the future. We have not yet addressed the photodamage issue in our PSII–ITO assembly.

Shihe Yang continued: Effective use of photosystem II is being pursued actively, as we see from the discussion. This mainly enhances the internal quantum efficiency. In fact, there are other biological systems, such as those with photonic structures (*e.g.*, in peacock feathers), which may be used to harvest sunlight more efficiently. Is it possible to integrate these systems? What are the prospects in this direction?

Can Li answered: This is a good idea, I think it is possible to integrate these systems. The prospects in this direction may be more concerned with understanding before developing more stable and efficient photosystems.

Erwin Reisner replied: Yes, photonic structures can be used to enhance light absorption and, consequently, external quantum efficiencies. We are working on this approach.

Heinz Frei addressed Can Li: In the PSII assembly you used, is the light protection function (*i.e.* non-photochemical quenching (NPQ)) included? The achievable photocatalytic efficiency would be limited if NPQ is operative, while durability would be improved.

Can Li responded: You are right: the photocatalytoc efficiency is limited if NPQ is operative, while for our current system the activity is still far from the saturation, so the durability seems good.

Zhongqun Tian continued: The three speakers have demonstrated new and important progress, which is very impressive. Natural systems always use weak bond interactions to more easily and timely adjust active reaction sites for the best selectivity and yield. I just wonder if the combination of your catalytic systems with soft matter could get close to natural systems with higher efficiency?

Can Li responded: It is possible to achieve a more efficient system by building up a hybrid system of natural photosynthesis and a soft matter system.

Lee Cronin addressed Erwin Reisner: Can you imagine applying flux to a material/molecular system as it is assembled to make a non-equilibrium material, *e.g.* playing rock music to optimise the efficiency of a polymer solar cell?! This requires that you set up a dissipative flux during the assembly of the system. Why not shine light on your next interface when you are making it?

Erwin Reisner responded: We are interested in non-equilibrium and dynamic self-assembly of interfaces, in particular driven by solar flux. I can imagine to use it as a tool to assemble photoactive materials. I think that the assembly can also occur *in situ* during irradiation from a pre-organised equilibrium state.

Fei Li asked: How did you deposit the nickel oxide? Is it the same material as Nocera's Ni-Bi?

Erwin Reisner answered: We believe that the NiO_x is comparable to Nocera's Ni-Bi. Nano$WO_3|TiO_2|NiO_x$ was prepared by spin coating and hydrolytic decomposition of the molecular single source precursor $[Ti_2(OEt)_9(NiCl)]_2$ on nanostructured WO_3. NiO_x is an active water oxidation catalyst in basic borate solution (Bi) and acts as electro-catalyst driven by photo-generated holes from the valence band of WO_3.

Dong Ryeol Whang opened the discussion of the paper by Licheng Sun: Fig. 2 from your paper tells us that SA3 shows the highest turnover numbers (TON), as you mentioned in the talk. However, in terms of turnover frequency (TOF), SA1 is the most efficient catalyst. Can you explain why there exists the kind of trade-off between TON and TOF?

Licheng Sun replied: Both the photosensitizer and the catalyst units in SA1 and SA3 are the same; the difference in chemical structure between these two assemblies is the bridging ligands. In SA1 the bridging ligand is rigid while in SA3 the bridging ligand is flexible. Different bridging ligands might have a big influence on the forwards electron transfer from the catalyst unit to the photogenerated oxidized form of the sensitizer and the charge recombination. These factors may cause the initial rate (TOF) of light driven water oxidation to be faster in the case of SA1 and slower in the case of SA3, but with a lower TON for SA1 and a higher TON for SA3. Electron transfer studies on these assemblies by time-resolved spectroscopic techniques are in progress.

Chao Zhan asked: As we know, a conjugated system is better for charge transfer, but SA3 is better than SA1 and SA2. Why is this? Can we imagine that the better charge conduction also leads to easier charge recombination, or the time scale of the charge separation transfer and the reaction should match together? Perhaps some time related experiments are needed, using different flexible linkages like carbon? Can the linkages be designed just allowing one-direction charge transfer like in a p–n junction?

Licheng Sun replied: Typically when you have a conjugated linker between the photosensitizer and the catalyst, the electron transfer is fast, however the charge recombination is also fast; this might be the case for SA1. When you have a non-conjugated linker between the photosensitizer and the catalyst, the electron transfer is slow, however the charge recombination could also become slow; this might be the case for SA3. Indeed, a fast forward electron transfer and a slow charge recombination is the key for a long lived charge separated state, which in turn can speed up the catalysis, in our case the catalytic water oxidation.

Fuping Pan enquired: The performance of photocatalytic water oxidation can be increased by using supramolecular assemblies. Can much large molecular assemblies be built to improve the performance?

Licheng Sun replied: Yes, large molecular assemblies can be built to improve the performance of catalytic water oxidation. We are preparing a supramolecular assembly where a photosensitizer tethered with a water oxidation catalyst can be immobilized on the surface of nanostructured TiO_2, aiming to create a functional device for light-driven water oxidation at the photoanode and hydrogen generation at the cathode.

Lee Cronin asked: Is it possible to devise a catalyst system that can run for trillions of turnovers? What time scale/TON/TOF can you realistically aim for?

Licheng Sun answered: First, when you design an efficient catalyst, you need a certain catalytic rate (TOF), the higher the better. However, the TOF is limited by the nature of metal ion, ligand and experimental conditions. As for water oxidation, a catalyst with a TOF of more than 300 s^{-1} is comparable with the OEC in photosystem II. Secondly, once you have a catalyst with high TOF, such as 300 s^{-1}, how stable (TON) the catalyst is becomes very important. It is not impossible to have a catalyst that can run for trillions of turnovers, such a robust catalyst is desirable for practical application in functional devices.

Erwin Reisner opened the discussion of the paper by Heinz Frei by asking: Do you have any information on the ideal length and inner-diameter of your Co oxide nanotubes with respect to efficient water oxidation and product diffusion?

Heinz Frei responded: We have quantified thus far the (average) catalytic activity per Co surface center for nanotubes with a 12 nm inner diameter and

150 nm length, but not yet for other sizes and aspect ratios. Planned photocatalytic experiments with complete nanotube arrays with different tube diameters, aspect ratios and spacing between the nanotubes, which can be done by varying the structure of the Si template, will allow us to optimize the geometry of the nanotubes and the array in terms of photocatalytic yield.

Joachim Maier asked: Is it known how the protons move in silica? Perhaps by proton hopping from an internal OH^--group to regular oxygen?

Heinz Frei replied: Yes, proton conduction through silica has been reported (ref. 52–54), and our electrochemical experiments confirm both the proton conductivity and blocking of O_2 (at silica layer thickness greater than 3 nm). A hopping mechanism involving internal silanol and siloxy groups is most likely, but I am not aware of direct spectroscopic evidence for the proposed mechanism.

Joachim Maier remarked: Is the proton motion in silica the same as in proton conducting perovskites (phonon assisted proton transfer from internal OH^- to O^{2-})?

Heinz Frei answered: Proton diffusion through thin films of dense phase amorphous silica is thought to occur by proton release from internal silanol groups. I am not aware of reports on direct spectroscopic observations of the process.

Xile Hu asked: You want to make an all inorganic device, but your hole transporter is an organic molecule. How stable is it?

Heinz Frei replied: Because we are not aware of (desirable) all-inorganic hole conducting molecular wires with the HOMO at a suitable potential for transporting the hole from the light absorber to the catalyst at minimal loss of energy, organic OPPV wires seem the best approach at this time. These molecules are used in electronics applications (although not in a reactive environment), and in our system cast into silica. The latter should provide adequate protection against oxidative damage, but we have not yet investigated the durability on longer time scales.

Xile Hu enquired: Why do you have a potential gradient across your silica membrane?

Heinz Frei responded: A potential gradient is needed across the embedded organic molecular wires in order for the transient hole generated on the visible light absorber to transfer spontaneously across the wire (hence the silica layer) and inject into the Co_3O_4 catalyst nanotube. Therefore, there needs to be a modest potential decrease (shift to less positive potential) from the HOMO of the light absorber to the HOMO of Co_3O_4.

Xile Hu asked: Does your device work at all pH values? Does the pH make a difference to its performance?

Heinz Frei answered: We have conducted photocatalytic experiments so far at pH values between 4 and 8. On the water oxidation side, rates increase with increasing pH, as expected. However, our goal is to run the system with gaseous H_2O and CO_2, not in an aqueous solution environment. We have demonstrated with a vapor phase system the photoreduction of CO_2 by H_2O, but not yet for a system that features silica as separating membrane.

Lulu Long asked: What is the role of the silica layer? If we use a nanometer carbon layer instead of the silica layer, what will happen? Is the role of carbon layer similar to the silica layer?

Heinz Frei replied: The dense (pore-free) silica layer functions as a proton conducting, gas impermeable membrane for the separation of the water oxidation reaction from the light absorber and the CO_2 reduction catalyst on the opposite side. O_2 impermeability and proton transmission has been demonstrated, and silica is known to block desired CO_2 reduction products such as methanol (silica is permeable for hydrogen, but this is not a desired product of our system). Since the membrane needs to be an insulator (the electrons are transported through embedded molecular wires with precisely tuned electronic properties); while carbon layers are semiconducting or behave like a metal, carbon layers are not suitable as membranes in this construct. Also, carbon layers may not have significant proton permeability although it might be possible to change this property through functionalization of the carbon.

Gang Chen asked: How big is the inner diameter of your core-shell nanotube? Will the small diameter of the inner tube prevent water from entering it due to surface tension, so the water oxidation reaction will not be efficient?

Heinz Frei replied: The inner diameter of the Co_3O_4 nanotube is determined by the diameter of the silicon nanowires that are used as sacrificial template to grow the nanotubes (by ALD). There is flexibility in selecting Si nanowire thicknesses and, therefore, in the choice of inner nanotube diameter (from tens to hundreds of nm). Our results of efficient water oxidation at Co_3O_4 nanoclusters grown inside the channels of mesoporous silica SBA-15 (Ref. 35) show that for nano-channels as narrow as 8 nm, access of water to the catalyst surface is not to be impeded.

Zhongqun Tian added: You have shown some very interesting approaches to artificial photosynthesis. You have designed some good artificial systems based on understanding of natural mechanisms. What is the best way to further improve the efficiency? Natural systems are very smart, as their catalytic sites change the structure dynamically for better activity during multi-step procedures. The surrounding molecules can assist each step collaboratively. For artificial photosynthesis, how can we design rationally and optimize dynamic activity in the supramolecular assemblies?

Heinz Frei responded: The primary task for improving the efficiency of complete artificial photosystems is the improvement of the yield and the energetics of charge transport between light absorbers, catalysts, and half reaction units. Equally urgent for efficiency improvement is to accomplish separation of the water oxidation catalysis from the reduction chemistry by a proton conducting, molecule-impermeable membrane, particularly in the case of carbon dioxide reduction, in order to minimize cross reactions. As the natural system shows, the dynamic involvement of the protein environment can play a key role for achieving high efficiency or selectivity of photosynthetic transformations. Some aspects, for example the tyrosine charge transport relay between the P680 light absorber and water oxidation catalyst of Photosystem II have already been successfully implemented as a design principle in artificial photosystems. Rational design of dynamic involvement of the surrounding matrix will likely be most fruitful for soft matter and molecular systems, and understanding of biological systems at the most detailed level is a powerful approach for progress in this direction. Dynamic involvement of the surrounding matrix may be harder to design rationally in the case of inorganic assemblies, but other opportunities exist for this approach, as recent discoveries illustrate. For example, nanostructured inorganic surfaces offer extremely large numbers of sites for storing transferred charge in the form of chemical change and for accomplishing subsequent catalytic transformations. Here, the sheer number of available active sites relaxes substantially constraints regarding catalytic rates of individual sites. New phenomena for the kinetic control of charge transport in all-inorganic systems are emerging that can be exploited for efficiency enhancement of photosynthetic assemblies. Progress here towards rational design of efficient assemblies requires discovery coupled with the advancement of theory and functional/atomic level characterization tools.

Licheng Sun continued: Nature indeed has created a very smart oxygen evolving complex (OEC) in photosystem II, and water oxidation occurs with the help of surrounding molecules. When we design a molecular catalyst for water oxidation, we should also consider the influence of the second and third coordination spheres, such as hydrogen bonding networks. In our Ru water oxidation catalysts with the bda ligand, the carboxylate can play such a role.

Shihe Yang communicated: Since the ionicity of SiO_2 is low, proton conductivity is expected to be low. What is the actual magnitude? Is it possible through the atomic diffusion of H, which seems to be a stable species in SiO_2?

Heinz Frei communicated in reply: We have measured electrochemically a proton flux of 2.5 protons s^{-1} nm^{-2} (pH 4). For our core–shell nanotube design, this is in the proper range in order for proton transport not be become rate-limiting at maximum solar intensity. If higher proton flux is needed modest doping by phosphorous or transition metals, which can be applied during the atomic layer deposition process, will be used to enhance the proton conductivity (for artificial photosynthesis, proton permeability through membranes does not need to exceed 10^{-4} Siemens per cm). This approach has been successfully demonstrated for silicate films used in micro fuel cell applications, although these membranes are much thicker and the flux requirements much higher. The mechanism involves proton transport across the silica, presumably by hopping along internal siloxy and silanol groups. Silica is permeable for hydrogen, but in our experiment the hydrogen is produced only after proton transport through the silica layer at the silica–Pt electrode interface.

Baohua Wang opened the discussion of the paper by Michael Grätzel: Is the film in an intermediate state when PbI_2 is treated with CH_3NH_2I solution so it can turn back in 2-propanol?

Michael Grätzel replied: The film is not in an intermediate state as the XRD measurements show that tetragonal crystals of $CH_3NH_3PbI_3$ perovskite are formed. In polar solvent the reaction is reversible and the CH_3NH_3I is extracted from the perovskite resulting in the formation of PbI_2.

Baohua Wang said: Does the perovskite film decompose in humid air? We noticed that the film will turn transparent if exposed in air for several days.

Michael Grätzel answered: Yes the perovskite film decomposes in humid air. First a colorless H_2O adduct is formed and then the CH_3NH_3I ions are extracted by water resulting in the formation of PbI_2, as shown by XRD analysis.

Ram Seshadri commented: Is it not true that the same considerations that make the halides easy to form into films, including the high mobility of ions in and out of the structures, as you described, are also considerations that might lead to decreased robustness of the materials?

Michael Grätzel answered: Your thoughts are correct. For example metal halide perovskites are very sensitive to humidity as the water leaches out the CH_3NH_3I ions leaving behind PbI_2. This sensitivity to moisture is one of the main caveats of metal halide perovskite solar cells.

Galen Stucky remarked: What is the configuration of the printed device that you made to test the performance stability of the perovskite?

Michael Grätzel responded: The printed device has a triple layer structure comprising a mesoscopic TiO_2 film supported by the transparent conducting oxide (fluorine doped tin oxide) acting as an electron selective conduct. This is covered by a mesoscopic ZrO_2 film which in turn is contacted by a mesoporous carbon film acting as a back contact. The perovskite is infiltrated by drop casting into the mesoporous network. More details can be found in the following publication: A. Mei, X. Li, L.Liu, Z. Ku, T.Liu, Y. Rong, M. Xu, M.Hu, J. Chen, Yi. Yang, M.Grätzel, and H. Han, A hole-conductor free, fully printable mesoscopic perovskite solar cell with high stability, *Science*, 2014, **345**, 295.

Galen Stucky asked: What is the approximate mesoscopic pore diameter?

Michael Grätzel responded: For perovskite solar cells the appropriate mesoscopic pore diameter is in the 30–50 nm range.

Yicheng Zhao commented: As Professor Grätzel told us, the PbI_2 can not be fully reacted into perovskite in the planar PbI_2. Why are there some papers reporting high efficiency of the planar heterojunction structures by using a two-step method? Isn't this contradictory?

Michael Grätzel answered: You are that right several papers reported the formation of planar heterojunctions by the sequential conversion method from 'planar' PbI_2 films. However solution-deposited PbI_2 films are often porous, as shown in our paper. Also, more drastic steps such as heating for longer periods of time at 100 °C are often applied to drive the conversion to completion in such a case. An example for such a method has recently been published by the Huang group in *Nature Communications*. (DOI: 10.1038/ncomms6784).

Shihe Yang communicated: (1) While in QCM-D, the frequency change is often associated with the mass change of the film, how do we understand the dissipation factor of the perovskite film? Can the QCM-D study provide some direct information about the conversion kinetics? (2) Have you used the back converted PbI_2 to prepare perovskite solar cells? And if yes, how about their performance? (3) How about the reversible transformation for the metal-halide perovskites in other solvents, such as water, ethanol, methanol and so on? Do you have any suggestions?

Michael Grätzel communicated in reply: (1) You are correct a mass change does can be directly linked to the change in the resonance frequency of the quartz microcrystal by the Sauerbrei equation which is given in section 2.5 of our paper. By contrast, the dissipation is related to the damping of the amplitude of the oscillation and is proportional to the width of the resonance signal divided by the absolute frequency at which it occurs. Thus relation is given in the same section. Thus, if during the oscillation, the crystal covered with PbI_2 or perovskite dragging along solvent of the structure supported by the quartz crystal itself is soft, then this drag produces a stronger dissipation signal than in its absence. Yes: the conversion kinetics are directly reflected in the rate of mass uptake or release. (2) We are currently performing such PV studies and will submit a report soon for publication. (3) These studies are very important. For example treating $CH_3NH_3PbI_3$ with water will leach out the perovskite leaving behind PbI_2 which is a carcinogenic compound. Any polar solvent that only dissolves poorly the Pb_{I2} would lend itself to such studies.

Joachim Maier opened the discussion of the paper by Nam-Gyu Park: What was the indication for ferroelectricity? Such polarization features are induced by ion conduction and stoichiometry polarization as a consequence of blocking the ions by the interfaces. Indeed, ion conductivity and such effects were confirmed by us (Yang, Gregori, Pellet, Grätzel and Maier, submitted).

Nam-Gyu Park answered: Yes, ion conduction could be one of the bases for ferroelectricity. According to impedance spectroscopy studies, we also observed a very high dielectric constant at very low frequency, which seems to be related to orientational polarization due to permanent dipoles from the methylammonium ion. Interfacial polarization cannot be excluded because charge accumulation is possible in perovskites under an external field, which may be related to ion conduction.

Lee Cronin asked: You emphasise the need for efficiency, but I wonder if this is the most important issue. Isn't scalability more important than efficiency? Will you run out of raw materials if you don't consider the ability to scale up?

Nam-Gyu Park answered: Yes, scalability is also important. However, low-cost high efficiency is very important in solar cell research and development in terms of low-cost electricity supply. Secondly, once we achieve such a goal, we will have to move on to the scalability issue, where effective processing methods should be developed.

Liang Chen addressed Michael Grätzel and Nam-Gyu Park: $CH_3NH_3PbI_3$ is a good material for perovskite solar cells, but Pb is a very poisonous metal. Sn is viewed as a good alternative for Pb. However, $CH_3NH_3SnI_3$ has a worse stability not only for moisture but also for oxygen. How can we solve these problems? Is Sn the best alternative for Pb, but what about other metal elements?

Nam-Gyu Park replied: Absolutely you are right. For the moment, Sn seems to be best alternative to Pb when you consider tolerance-factor calculations. We are trying to find effective ways to protect from oxygen and moisture, as well as stabilize Sn^{2+}.

Michael Grätzel also replied: You are entirely correct that Pb is a very poisonous material and Sn analogues for $CH_3NH_3PbI_3$ are very unstable in air and are also very problematic for the environment. Hence we have to extend our research to other inorganic pigments that are more benign. Using mesoscopic architectures will be of great importance to address the issue of short carrier diffusion lengths.

Shihe Yang communicated: (1) The three control samples (bare) are supposed to have the same configuration: TiO_2/Sb_2S_3/HTM. However, their *J–V* data are different. Why is this the case? (2) Why is the performance of BL1/BL2 is poorer than W/ZnS (BL2)? One would expect it to be better. (3) Have you done similar work on perovskite solar cells? What do you expect?

Nam-Gyu Park communicated in reply: (1) A slight difference in *J–V* property for the three control samples is due to different batches. In each batch, it is difficult to have exactly the same *J–V* data although the same process was used for each batch. (2) We actually expected the performance of BL1/BL2 to be better than that of only BL2 or BL1. However, we found that the process of Sb_2S_3 deposition was not variable because the annealing and quenching processes were difficult to control. However, we would like to compare the BL1/BL2 sample with the control sample prepared at the same batch, to see what happened. We observed reproducibly higher J_{sc} and V_{oc} from BL1/BL2 compared to those of the control sample. (3) We are doing the similar work, but the perovskite solar cell structure is somewhat different from Sb_2S_3 QD cell. We are trying to use blocking layers in perovskite solar cell for a different purpose.

Joachim Maier opened the discussion of the paper by Kai Zhu: You found the nature of the perovskite not to be decisive for transport parameters, yet you found a big difference between PbI_2 and perovskite in that respect. What is your explanation?

Kai Zhu answered: It is commonly accepted that transport in dye-sensitized solar cells is limited by trapping defects at the TiO_2 surface. We have shown that the transport mechanism in mesoporous-TiO_2-based perovskite solar cells is similar to that in the dye cells. The density of the defects affects strongly the transport rate. Thus, the observed difference of transport between perovskites and PbI_2 is likely to be caused by the different densities of defects created at the TiO_2/perovskite or TiO_2/PbI_2 interfaces. For example, it has been shown that for perovskites only shallow traps are likely to be formed because of their low formation energies. However, the defect chemistry/physics may be very different for PbI_2.

Baohua Wang commented: If the charge diffusion length of $CH_3NH_3PbI_{3-x}Br_x$ is much longer than the pore size in the porous TiO_2 film, can the technique used be able to tell the difference between these perovskite materials?

Kai Zhu answered: Yes. For example, there is a clear difference on transport between perovskites and PbI_2, as shown in Fig. 5. Similar transport properties among all perovskite-based mesoporous cells suggest that the transport-limiting defects are similar among all perovskite based cells.

Hong Liu said: When you make $MAPbX_3$ with X occupied by different halogen atoms, how do you quantitatively and correctly measure the content of each element in the perovskite lattice? Can you determine their chemical bonding situation in the perovskite lattice and distinguish them from other forms of existence, *e.g.*, just as free ions or molecules?

Kai Zhu replied: We used energy dispersive X-ray (EDX) analysis to determine the relative ratio between I, Br, and Cl in the perovskite films. At this point, we cannot determine their chemical bonding situation in the perovskite lattice.

Shihe Yang communicated: (1) The authors found that the PbI_2-based device showed a much lower diffusion coefficient (about a factor of 5 lower than that of the $MAPbI_3$ solar cells), so the PbI_2 phase should be eliminated from the perovskite absorber. However, according to another paper published by Yang Yang (*Nano Lett.*, 2014, **14**, 4158–4163), a certain degree of PbI_2 passivation helps to reduce the recombination between the electrons from perovskite and holes from HTM. How do you rationalize it regarding the PbI_2 role in the perovskite solar cell in more detail? (2) What is the cell performance difference before and after the $TiCl_4$ treatment? What are the causes?

Kai Zhu communicated in reply: (1) This is still a subject under debate. In our lab, we found it to be critical to remove PbI_2 in order to achieve devices with reasonable performance, and it is true for both mesoporous and planar perovskite solar cells. However, the exact role of PbI_2 could depend on the properties of the perovskite materials (*e.g.*, morphology, crystallinity, grain size, and other defect properties) and the relative amount of PbI_2 in the perovskite film. This could vary a lot from different labs and should depend on the exact perovskite growth conditions. For example, Yang Yang's *Nano Lett.* paper also shows that device performance is reduced when there is too much PbI_2. (2) the device performance is better with $TiCl_4$ treatment for mesoporous perovskite cells. As we have shown in the paper, $TiCl_4$ treatment increases the recombination lifetime for about a factor of 4–5 for mesoporous-TiO_2-based perovskite solar cells. It is known that $TiCl_4$ solution treatment can enhance the surface roughness of TiO_2 electrodes, which could improve the contact between perovskites and TiO_2 to improve the quality of coated perovskite, and consequently, suppresses the back-electron-transfer from TiO_2 to spiro-MeOTAD.

Yu Li opened the discussion of the paper by Shihe Yang: When you perform TRPL with the streak camera, the LPD method actually shows a slower decay than the USP counterpart. However, the recombination lifetime between them is not significantly different. How do you make sure that charge accumulation has been enhanced?

Shihe Yang responded: If the injection is slowed down and the recombination lifetime is not much different (in our case it is prolonged), more electrons will surely be accumulated in the perovskite layer. This will raise the electron quasi-Fermi level Fn.

Fuping Pan remarked: How about the binding force between TiO_2 and FTO? Can the thickness of TiO_2 impact the performance of the solar cells? If so, what is optimal value of TiO_2 thickness?

Shihe Yang responded: Since the interaction between TiO_2 and FTO has both ionic and covalent character, it is relatively strong. There should be an optimum TiO_2 thickness for solar cell performance. If it is too thin, charge recombination would be strong, but if it is too thick, it would be transport-limited. The optimal thickness has been summarized in the paper.

Zhonghua Li commented: According to your paper, holes will be transferred from the titania to the perovskite? As a result, the stability will be decreased. Why choose titania?

Shihe Yang answered: We did not state that holes are transferred from titania to perovskite. Rather, we suggested that electron injection is modulated by the titania nanolayer. We chose titania because it can be structurally and electronically controlled and it is compatible with the substrate and easy to use to form uniform films.

Zhonghua Li asked: How about the stability of this material compared to that without titania?

Shihe Yang responded: The stability is not much relevant here with or without titania. The main issue here is how the titania layer affects charge injection and recombination.

Yicheng Zhao began a general discussion by addressing Michael Grätzel: Because of the hysteresis phenomenon in perovskite solar cells, do you think if there is a need to set up a new measurement standard to testify our solar cells, like the steady-state *I–V* curves?

Michael Grätzel answered: Yes, ultimately one would need to measure the power produced by the solar cell in a steady state mode if the hysteresis is so severe that if cannot be removed by slowing down the scanning speed.

Liang Chen addressed Michael Grätzel and Nam-Gyu Park: 2D hybrid perovskite material often has a large band gap, but it is more stable to moisture than 3D hybrid perovskites such as $CH_3NH_3PbI_3$. I wonder whether a 2D hybrid perovskite material is a good choice for solar cells if decreasing the band gap. Are there any other problems for 2D hybrid perovskite solar cells?

Nam-Gyu Park replied: This is excellent question. Yes, if the band gap of a 2D perovskite is comparable to that of a 3D one, this will be expected to be better in terms of stability. However, I am not quite sure whether or not we can still have a high absorption coefficient from a 2D perovskite.

Michael Grätzel continued: Apart from the larger band gap, there is indeed a potential problem that you raise from the anisotropy of the electrical conductivity which is excellent within the plane of the 2D perovskite, but poor in the vertical direction. Hence, if the 2D crystals lie flat on the current collectors the electronic transport across these layers would be impaired resulting in a low fill factor and small carrier collection efficiencies.

Xudong Yang asked: Why is the hysteresis so different for different types of perovskite solar cells? For example, there's always a large hysteresis in TiO_2/perovskite/spiro-OMeTAD based solar cells, while it is much smaller in PCBM/perovskite/PEDOT:PSS based cells. How can we reduce the large hysteresis in the first case?

Michael Grätzel responded: The cause for the hysteresis is still under active investigation. There is evidence that ions are moving under the effect of the electric field that is generated by light-induced charge separation in the perovskite. Other researchers have ascribed this to deep trapping-states present at the interface of the perovskite with the electron collecting contact. In a recent paper published in *Nature Communications* Professor Huang's group ascribes the decrease in the hysteresis observed for the PCBM/perovskite/PEDOT:PSS solar cells to a smaller level of defect states at the PCBM/perovskite interface (DOI: 10.1038/ncomms6784).

Jian He commented: In Fig. 8 of your paper (a) and (e) show the PbI_2 film morphology, before converting to $CH_3NH_3PbI_3$ and after back-conversion respectively. Before converting to $CH_3NH_3PbI_3$, PbI_2 slices lie on the substrate horizontally. However, after back-conversion, many PbI_2 slices become vertical. What is the reason for this change? Does this change lead to a more porous surface and a larger surface area?

Michael Grätzel answered: You are correct that the PbI_2 slices are more randomly oriented after back-conversion. We attribute this to the large volume expansion on converting the PbI_2 to $CH_3NH_3PbI_3$. When the $CH_3NH_3PbI_3$ is back-converted the volume of the solute contracts leaving porous space behind. Indeed, we report in section 2.5 of our paper that the porosity of the PbI_2 film increases from 6.4 to 24.7% upon one cycle of conversion to $CH_3NH_3PbI_3$ and back conversion to PbI_2.

Zhongqun Tian asked: Recently you published a very interesting paper in *Science* on water photolysis at 12.3% efficiency *via* perovskite photovoltaics and Earth-abundant catalysts. Could you comment on the advantages and disadvantages of this new combination of perovskite photovoltaics and electrocatalysis on water splitting?

Michael Grätzel communicated in reply: The advantage of the system is that it employs only low-cost earth abundant materials that can be applied by solution processing to achieve a solar-to-hydrogen conversion efficiency of 12.3%, which is substantially higher than that achieved with the combination of silicon solar cells with water electrolyzer. However the perovskite cells employed in this work were unsealed and need to be encapsulated appropriately to stabilize their power output.

Zhongqun Tian addressed Shihe Yang, Nam-Gyu Park and Michael Grätzel: You all have shown that perovskite solar cells are very promising as one of the next-generation energy devices. For the further development, it's important to have an *in situ* tool to charactarize the materials, especially their interfaces. What could be the best tools or combined techniques in terms of good spatial and temporal resolution?

Shihe Yang answered: Perovskite solar cells as a system involve important structures and dynamics over a broad range of spatial and time scales. A single technique cannot solve all these problems. So far transient absorption, time-resolved luminescence, electron microscopy, scanning probe microscopy, impedance spectroscopy, *etc.* have been used. New techniques need be developed to probe more details about the local structure and dynamics, such as molecular orientation, ferroelectric domain evolution, ionic movement, *etc.* The probes may be modern sources of photons, electrons, photoelectrons, *etc.* with high spatial, temporal and spectral resolution. This is quite challenging.

Nam-Gyu Park continued: High resolution TEM may be helpful to figure out the interfacial structure. For charge separation, photoluminesence spectroscopy may be helpful. Electron-beam-induced current (EBIC) combined with SEM can also reveal useful information on interfaces.

Michael Grätzel concluded: Indeed characterizing the interfaces of the materials is crucial for the further development of perovskite solar cells. For spatial resolution there has been, fortunately, a huge progress in electron microscopy and scanning technologies that allow atomic resolution and hence detection of defect states. To investigate the pore filling of mesoporous system tomographic

HREM looks particularly promising. However the temporal resolution pulsed laser techniques combined *e.g.* with IR, SERS and terahertz spectroscopy look extremely interesting.

Wei Han addressed Shihe Yang, Kai Zhi, Nam-Gyu Park and Michael Grätzel: What are the advantages of inorganic solar cell materials over organic polymer solar cell materials?

Shihe Yang answered: In general, inorganic materials are more stable than organic materials. They also have higher mobilities. As for the organohalide perovskites, they are actually organic/inorganic hybrid materials and thus can combine advantages of both. There is also room for improving solar cell perfromance by modulating the organic components.

Kai Zhu continued: Inorganic solar cell materials are generally believed to be more robust than organic solar cell materials.

Nam-Gyu Park added: Inorganic light absorbers usually have higher absorption coefficients, one or two orders of magnitude higher than organic or polymeric light absorbers.

Michael Grätzel concluded: The organic polymer solar cells suffer from low efficiency and problems with stability. They are sensitive to water and oxygen and would need to be hermetically sealed to maintain their power output. Also they have a tendency to undergo phase separation, degrading their performance over time, even if they are kept in the dark in a drawer. Finally the polymers that produce the highest efficiencies (*ca.* 10%) are very difficult and very expensive to synthesize. Most of these disadvantages are not shared by inorganic or dye sensitized solar cells.

Shi-Gang Sun continued: Recently, the investigation of perovskite materials for solar cells has progressed very fast. As we heard in your presentation, the PCE even reached 28%, which seems not yet to be the limit. With such rapid development, what are your predictions for the future applications for perovskite solar celcomparisonarison with silicon solar cells?

Michael Grätzel answered: The certified solar to electric power conversion efficiency of perovskite solar cells (PCE) measured under standard air mass 1.5 reporting conditions has reached 20.1%, not 28%. The PCE of 28% refers to an unpublished value obtained with a tandem configuration with silicon. Such tandem cells will certainly be of great interest in the future and may become one of the most interesting future applications of perovskite solar cells.

Faraday Discussions

PAPER

Data-driven discovery of energy materials: efficient $BaM_2Si_3O_{10}$: Eu^{2+} (M = Sc, Lu) phosphors for application in solid state white lighting

Jakoah Brgoch,*[ab] Kathryn Hasz,[b] Kristin A. Denault,[bc] Christopher K. H. Borg,[b] Alexander A. Mikhailovsky[d] and Ram Seshadri[abcd]

Received 6th June 2014, Accepted 18th July 2014

DOI: 10.1039/c4fd00125g

In developing phosphors for application in solid state lighting, it is advantageous to target structures from databases with highly condensed polyhedral networks that produce rigid host compounds. Rigidity limits channels for non-radiative decay that will decrease the luminescence quantum yield. $BaM_2Si_3O_{10}$ (M = Sc, Lu) follows this design criterion and is studied here as an efficient Eu^{2+}-based phosphor. M = Sc^{3+} and Lu^{3+} compounds with Eu^{2+} substitution were prepared and characterized using synchrotron X-ray powder diffraction and photoluminescence spectroscopy. Substitution with Eu^{2+} according to $Ba_{1-x}Eu_xSc_2Si_3O_{10}$ and $Ba_{1-x}Eu_xLu_2Si_3O_{10}$ results in UV-to-blue and UV-to-blue-green phosphors, respectively. Interestingly, substitution with Eu^{2+} in the Lu^{3+} containing material produces two emission peaks at low temperature and with 365 nm excitation, as allowed by the two substitution sites. The photoluminescence of the Sc^{3+} compound is robust at high temperature, decreasing by only 25% of its room temperature intensity at 503 K, while the Lu-analogue suffers a large drop (75%) from its room temperature intensity. The decrease in emission intensity is explained as stemming from charge transfer quenching due to the short distances separating the luminescent centers on the Lu^{3+} substitution site. The correlation between structure and optical response in these two compounds indicates that even though the structures are three-dimensionally connected, high symmetry is required to prevent structural distortions that could impact photoluminescence.

1 Introduction

Advances in lighting technology provide an opportunity to replace conventional incandescent and fluorescent bulbs with solid state lighting devices. Solid state

[a]*Mitsubishi Chemical Center for Advanced Materials, University of California, Santa Barbara, CA, 93106, USA. E-mail: jrbrgoch@mrl.ucsb.edu*

[b]*Materials Research Laboratory, University of California, Santa Barbara, CA 93106, USA*

[c]*Materials Department, University of California, Santa Barbara, CA 93106, USA*

[d]*Department of Chemistry and Biochemistry, University of California, Santa Barbara, CA 93106, USA*

white lights require less energy, have longer lifetimes, and are free of toxic heavy metals, among other advantages.[1,2] These devices generally consist of a light-emitting diode (LED) and an inorganic phosphor that partially down-converts the monochromatic LED emission. The LED and phosphor are blended in the correct ratio, giving rise to the observed white light.[3] The current search for novel phosphors has recently centered on identifying compounds that contain highly rigid bonding networks.[4–6] This concept is classically employed to understand photoluminescence efficiency in molecular chemistry[7] as well as in inorganic–organic frameworks.[8] Extending this principle to bulk inorganic compounds should aid in the identification of new, highly efficient phosphors. Structures with the most rigid bonding networks tend to emerge in high symmetry structures with dense polyhedral packing.[6] Thus, preparing materials that have a large degree of connectivity makes it possible to limit non-radiative relaxation by suppressing phonons. Additionally, the photoluminescence quantum yield (Φ) can be influenced by the symmetry of a structure. Crystal structures with higher symmetry are not prone to structural relaxation (or distortion) due to chemical substitution or electronic excitation, which can produce soft-phonon modes ultimately affecting Φ. Previous research has successfully applied this approach to understand the exceptional efficiency of two phosphors, $YAG:Ce^{3+}$ (ref. 4) and $La_3Si_6N_{11}:Ce^{3+}$.[5] These compounds contain high symmetry and dense polyhedral networks that yield a rigid structure and drive their excellent Φ. This concept also provides an interpretation of the unique temperature-dependent photoluminescence in the orthosilicate solid solution, $(Sr_xBa_{2-x})SiO_4:Eu^{2+}$.[9] Moreover, a number of novel phosphors have been identified based on the principle of structural rigidity, including $A_3RE(BO_3)_3:Ce^{3+}$ (A = Ba, Sr; RE = Y, La),[10] $NaY_9Si_6O_{26}:Ce^{3+}$, $Na_3YSi_2O_7:Ce^{3+}$,[11] and $(Ba_{1-x}Sr_x)Al_2Si_2O_8:Eu^{2+}$.[12]

Continuing the search for structure-types with a dense packing of polyhedra has revealed $BaM_2Si_3O_{10}$ (M = Sc, Lu, Y, Gd),[13,14] which contains fully connected SiO_4 and MO_6 polyhedral units. Although a variety of M atoms have been identified in this structure-type, the optical properties of only the yttrium analogue have been reported. Substituting a variety of luminescent centers in $BaY_2Si_3O_{10}$ has resulted in emission in the blue (Ce^{3+}), green (Tb^{3+}), and red (Eu^{3+}) regions of the visible spectrum.[15] Moreover, this phosphor host can be co-doped with Ce^{3+} and Tb^{3+} to tune the emission color between blue and green.[16] Preparing M = Sc or Lu analogues as phosphors, with the smaller ionic radii of the M atoms, is ideal for maintaining the highly connected crystal structure while enhancing structural rigidity following Coulomb's law for ionic bonding. Moreover, decreasing the unit cell volume should enhance the crystal field splitting of the Eu^{2+} 5d orbitals, providing a desired red-shift in the emission spectrum.

Thus, we present two new Eu^{2+}-substituted silicate phosphors, $Ba_{1-x}Eu_xM_2Si_3O_{10}$ (M = Sc, Lu), that are capable of efficiently down-converting a UV-LED emission into the blue and blue-green regions, respectively, of the visible spectrum. Their preparation was achieved using high temperature solid state methods and the products were characterized using synchrotron X-ray powder diffraction and optical property measurements. The Φ of these two compounds is reported as a function of the Eu^{2+} concentration, along with their temperature- and time-dependent optical properties. From the in-depth property analysis of these two compounds, we are able to further our understanding of the structure–

property relationships in inorganic phosphors that are necessary for the next generation of solid state white lighting.

2 Experimental

2.1 Sample preparation

Polycrystalline samples of $Ba_{1-x}Eu_xM_2Si_3O_{10}$ (M = Sc, Lu; x = 0.005, 0.01, 0.02, 0.03, 0.04, 0.05, 0.06) were prepared using traditional high temperature solid state methods. Starting materials ($BaCO_3$, Cerac, 99.9%; Eu_2O_3, Aldrich, 99.9%; Lu_2O_3, Materion, 99.9%; Sc_2O_3, Stanford Materials, 99.99%; SiO_2, Johnson-Matthey, 99.9%) were weighed out in desired stoichiometric ratios and thoroughly ground with an agate mortar and pestle. The molar ratio of Ba^{2+} to Eu^{2+} was systematically varied to explore the effect of Eu^{2+} concentration on the optical properties. The mixtures were pressed into pellets and placed in alumina crucibles (3 mm; AdValue Technology) with sacrificial powder of the same composition to prevent contamination between the pellet and the crucible. The pellets were fired in a tube furnace under reducing conditions (5% H_2/95% Ar). $Ba_{1-x}Eu_xSc_2Si_3O_{10}$ (x = 0.005, 0.01, 0.02, 0.03, 0.04, 0.05, 0.06) was fired at 1375 °C for 10 h whereas $Ba_{1-x}Eu_xLu_2Si_3O_{10}$ (x = 0.005, 0.01, 0.02, 0.03, 0.04, 0.05, 0.06) was fired at 1475 °C for 5 h. Both reactions employed heating and cooling rates of 3 °C min^{-1}. The resulting compounds were ground with an agate mortar and pestle for characterization.

2.2 Structure and optical properties

Structural properties were probed using synchrotron X-ray powder diffraction (XRD) collected at beamline 11-BM at the Advanced Photon Source, Argonne National Laboratory[17] using a calibrated wavelength of 0.413837 Å. The room temperature diffraction data were analyzed using the Rietveld method within the General Structure Analysis System (GSAS).[18,19] The background was handled using a 9-term shifted-Chebyshev function, while the peak shapes were handled using a pseudo-Voigt function, with Finger–Cox–Jephcoat asymmetry to correct for axial divergence at low angle.

Room temperature photoluminescence spectra and quantum yields were measured with a Horiba Fluoromax-4 fluorometer using a 150 W xenon arc lamp for excitation and a 150 mm Spectralon™-coated Quantum-Φ integrating sphere attachment. Each of the samples was thoroughly mixed with silicon resin (GE Silicones, RTV615) and applied to a quartz substrate (Chemglass) for measurement. The Φ was calculated based on the method of de Mello *et al.*[20]

Temperature- and time-dependent measurements of the photoluminescence were collected at 77 K, 295 K, and 503 K using a home-built fluorimeter incorporating a cryostat for low temperature measurements and a heating stage for high temperature data collection. Lifetime measurements were carried out using the Time-Correlated Single Photon Counting (TCSPC) technique[21] as described previously.[9] Here, 200 fs excitation pulses with a wavelength of 365 nm were generated by doubling the fundamental frequency of femtosecond Ti : sapphire laser (Coherent Mira 900) pulses in a commercial optical harmonic generator (Inrad). The laser repetition rate was reduced to 200 kHz by a home-made acousto-optical pulse picker to avoid saturation of the phosphor. The TCSPC system is equipped with an ultrafast microchannel plate photomultiplier tube detector (Hamamatsu R3809U-51) and electronics board (Becker & Hickl SPC-630), and has an instrument response time of about 65 ps. The triggering signal for the TCSPC board was generated by sending a small fraction of the laser beam onto a fast (400 MHz bandwidth) Si photodiode (Thorlabs Inc.). Fluorescence signals were dispersed using an Acton Research SPC-500 monochromator after passing through a pump blocking, long-wavelength, autofluorescence-free interference filter (Omega Filters, ALP series). The monochromator was equipped with a CCD camera (Roper Scientific PIXIS-400), allowing the monitoring of the time-averaged fluorescence spectrum. Luminescence transients were not deconvolved from the instrument response function since their characteristic time constants were much longer than the width of the system response to the excitation pulse.

2.3 Density functional theory

All calculations were carried out using the Vienna *ab initio* Simulation Package (VASP; version 5.2),[22,23] which employs the projector augmented wave method (PAW) of Blöchl[24] as adapted by Kresse and Joubert.[25] Initial relaxation of the atomic positions was performed to ensure that the compounds are in their electronic ground state. The elastic tensors were then determined from the stress–strain relationship of six finite distortions of the crystal[26] using displacements of ± 0.015 Å. Exchange and correlation were described by the Perdew–Burke–Ernzerhof generalized gradient approximation (PBE).[27] The energy cut-off of the plane wave basis set was 650 eV and approximately 100 *k*-points (determined using a Gamma grid) were used. The energy convergence criterion was set to 1×10^{-6} eV per formula unit to ensure accurate electronic convergence. The Debye temperature (Θ_D) was estimated from the VASP-calculated elastic constants (C_{ij}) as described previously.[6]

3 Results and discussion

Materials with a high Θ_D are ideal phosphor hosts due to the presence of high-energy phonon modes that limit non-radiative recombination. The highest Θ_D values arise with connected polyhedral units that produce rigidly bound structural units. As a result, the Θ_D for the three-dimensionally connected $BaM_2Si_3O_{10}$ (M = Sc, Lu) host was calculated prior to synthesis, ensuring the potential of this compound. The resulting Debye temperatures of 543 K and 500 K for M = Sc and Lu, respectively, surpass related compounds by nearly 100 K. For example, $Ba_9Y_2Si_6O_{24}$, which crystallizes with two-dimensional layers of YO_6 octahedra that share corners with SiO_4 tetrahedra, has a Θ_D of 409 K.[28] Thus, based on the higher Θ_D and more rigid bonding network, $BaM_2Si_3O_{10}$ (M = Sc, Lu) is expected to have improved optical properties compared to related systems.

$Ba_{1-x}Eu_xM_2Si_3O_{10}$ (M = Sc, Lu) was prepared using high temperature solid state methods and the structure refined against synchrotron X-ray powder diffraction data. The compounds both crystallize in space group $P2_1/m$ (no. 11) with the $BaGd_2Si_3O_{10}$ structure-type, as illustrated in Fig. 1a. As shown in Fig. 1b and c, the compounds are nearly phase pure with the small unidentified impurity peaks likely corresponding to alternate silicate phases. Using the known Gd-based structure type as an initial structural model, the lattice parameters and atomic positions were refined using the Rietveld method and are presented in Table 1 and Table 2.

This structure contains Ba^{2+} cations on a mirror plane that are coordinated by 10 neighboring oxygen atoms with an average Ba–O bond length of 2.9(2) Å for both compounds, in agreement with related silicates.[28,29] The structure also contains distorted SiO_4 tetrahedra that form corner-sharing zig-zag trimers. There are two crystallographically independent Si^{4+} cations with average Si–O

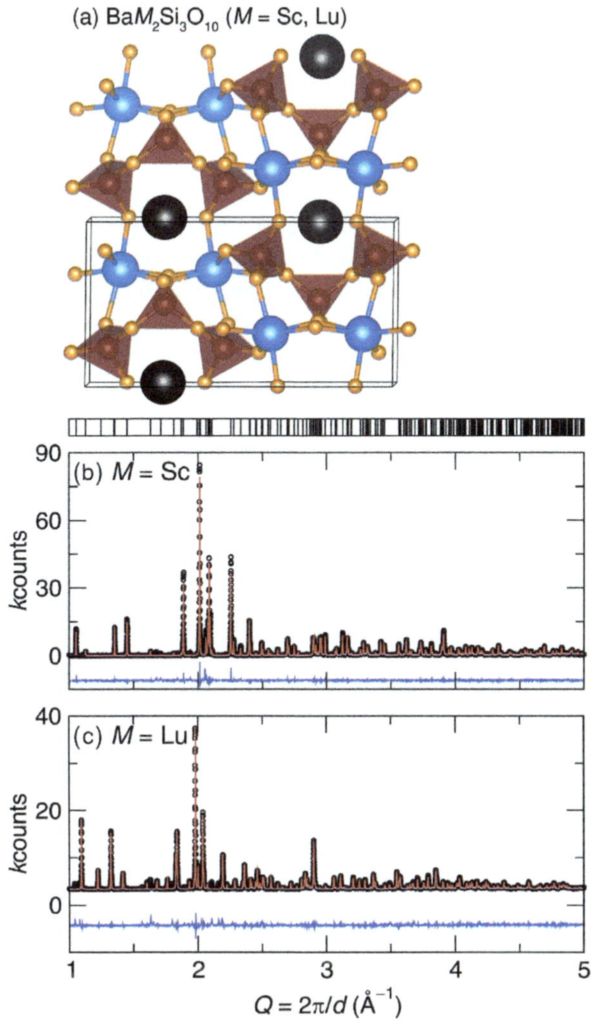

Fig. 1 (a) The crystal structure of $Ba_{1-x}M_2Si_3O_{10}$ illustrated near the [100] direction with the SiO_4 tetrahedra (red) and MO_6 distorted octahedra (blue) highlighted; the Ba is illustrated in black. The synchrotron X-ray powder diffraction data (black circles) were refined using the Rietveld method (fit in red) for (b) $Ba_{0.99}Eu_{0.01}Sc_2Si_3O_{10}$ and (c) $Ba_{0.97}Eu_{0.03}$-$Lu_2Si_3O_{10}$. The difference between the data and the Rietveld fit is shown in blue.

Table 1 Crystal structure refinement data for the isostructural $Ba_{0.99}Eu_{0.01}Sc_2Si_3O_{10}$ and $Ba_{0.97}Eu_{0.03}Lu_2Si_3O_{10}$, determined using the Rietveld method against synchrotron X-ray powder diffraction data

	M = Sc	M = Lu
Radiation type, λ (Å)	Synchrotron X-ray (11-BM), 0.413837	
2θ range (degree)	2.5–50	
Temperature (K)	295	
Crystal system	Monoclinic	
Space group; Z	$P2_1/m$ (no. 11); 2	
Lattice parameters	$a = 5.27359(1)$ Å	$a = 5.35430(1)$ Å
	$b = 11.92596(4)$ Å	$b = 12.1205(1)$ Å
	$c = 6.59206(2)$ Å	$c = 6.78250(1)$ Å
	$\beta = 107.0428(2)°$	$\beta = 106.6045(3)°$
Volume (Å^3)	396.386(1)	421.820(1)
R_p (%)	11.38	13.37
R_{wp} (%)	16.35	20.21

bond lengths of 1.57(9) Å (Si1) and 1.62(8) Å (Si2) in the Lu^{3+} structure. These SiO_4 trimers are connected *via* two highly distorted, edge-sharing MO_6 octahedra. When M = Sc^{3+}, the average Sc–O bond length is 2.16(8) Å, whereas the Lu–O bond length is 0.12 Å longer. The M–M bond length separating the dimers is 3.531(1) Å and 3.635(1) Å for Sc^{3+} and Lu^{3+}, respectively. Due to the connectivity of the MO_6 dimers and SiO_4 trimers, the structure bears the desirable three-dimensional bonding network, which gives rise to its high calculated Θ_D.

Incorporating Sc^{3+} or Lu^{3+} as M in the compound leads to moderate changes in the refined crystal structure due to their respective ionic radii. For example, the MO_6 polyhedral volume increases from 12.42 Å^3 to 14.84 Å^3 going from the smaller 6-coordinate Sc^{3+} ($r = 0.745$ Å) to the larger Lu^{3+} ($r = 0.861$ Å) cation.[30] The increase in polyhedral volume is directly proportional to the observed 6% increase in unit cell volume. Additional structural modifications also occur that may not be anticipated. For example, the MO_6 octahedra are more distorted when M = Sc^{3+} relative to M = Lu^{3+}, as highlighted by their respective quadratic elongations ($\langle\lambda\rangle$, eqn (1)).[31] Here, $\langle\lambda\rangle$ is defined by l_0, the center-to-vertex distance of a regular polyhedron of the same volume, and l_i, the distance from the central atom to the ith coordinating atom.

$$\langle\lambda\rangle = \frac{1}{n}\sum_{i=1}^{n}\left(\frac{l_i}{l_0}\right)^2 \tag{1}$$

From the calculation of these polyhedron parameters, provided in Table 3, the ScO_6 octahedron is nearly ideal with a $\langle\lambda\rangle$ approaching 1, whereas the structure with the LuO_6 shows a large distortion of the octahedron with $\langle\lambda\rangle$ increasing dramatically. These changes surely stem from the ionic size of the M atom. Because the crystal structure is three-dimensionally connected, the substitution by Lu^{3+} can not be solely accommodated by an increase in the unit cell volume. A distortion of the local environment must also occur. Regardless, M does not affect

Table 2 Refined atomic coordinates and equivalent isotropic displacement parameters of $Ba_{0.99}Eu_{0.01}Sc_2Si_3O_{10}$ and $Ba_{0.97}Eu_{0.03}Lu_2Si_3O_{10}$, determined by Rietveld refinement of powder synchrotron X-ray diffraction data collected at room temperature. The concentration of Eu^{2+} (x) was excluded from the refinement due to the low substitution level

Atom	Wyck.	x	y	z	U_{eq} (Å^2)
$Ba_{0.99}Eu_{0.01}Sc_2Si_3O_{10}$					
Ba1	2e	0.7599(1)	$\frac{1}{4}$	0.02665(8)	0.0072(1)
Sc1	4f	0.1526(2)	0.10197(9)	0.6753(1)	0.0067(3)
Si1	2e	0.5741(5)	$\frac{1}{4}$	0.4907(4)	0.0050(5)
Si2	4f	0.2996(3)	0.0541(1)	0.2089(2)	0.0005(4)
O1	2e	0.3412(9)	$\frac{1}{4}$	0.6011(7)	0.011(1)
O2	2e	0.8680(9)	$\frac{1}{4}$	0.6286(7)	0.002(1)
O3	4f	0.5418(6)	0.1429(2)	0.3310(5)	0.010(1)
O4	4f	0.4500(7)	0.9419(2)	0.2025(5)	0.007(1)
O5	4f	0.1490(6)	0.1119(2)	0.9899(5)	0.010(1)
O6	4f	0.1044(7)	0.0422(2)	0.3499(5)	0.008(1)
$Ba_{0.97}Eu_{0.03}Lu_2Si_3O_{10}$					
Ba1	2e	0.7606(2)	$\frac{1}{4}$	0.0202(2)	0.0094(3)
Lu1	4f	0.1542(1)	0.10006(5)	0.68249(9)	0.0014(1)
Si1	2e	0.584(1)	$\frac{1}{4}$	0.4923(8)	0.009(1)
Si2	4f	0.3043(8)	0.0578(3)	0.2115(6)	0.003(1)
O1	2e	0.378(2)	$\frac{1}{4}$	0.599(1)	0.006(3)
O2	2e	0.858(2)	$\frac{1}{4}$	0.619(1)	0.008(3)
O3	4f	0.547(1)	0.1445(6)	0.332(1)	0.007(2)
O4	4f	0.442(1)	0.9420(6)	0.195(1)	0.006(2)
O5	4f	0.153(1)	0.1271(7)	0.003(1)	0.007(2)
O6	4f	0.106(1)	0.0564(6)	0.340(1)	0.005(2)

BaO_{10}, with the polyhedral volume changing by only 0.11 Å^3 between the two compounds. This is further illustrated by the small change of the Ba^{2+} polyhedral distortion index (D),[32] defined by eqn (2), where l_i is the distance from the central atom to the ith coordinating atom and l_{av} is the average bond length:

Table 3 Changes in polyhedral volume and distortion of the cation sites for $Ba_{0.99}Eu_{0.01}Sc_2Si_3O_{10}$ and $Ba_{0.97}Eu_{0.03}Lu_2Si_3O_{10}$

	M = Sc	M = Lu
MO_6 poly. vol.	12.42 Å	14.84 Å
$\langle\lambda\rangle$	1.005	1.048
σ^2 (deg.^{2})	177.68	151.88
BaO_{10} poly. vol. (Å^3)	49.54	50.05
BaO_{10} distortion index	0.0699	0.0668

$$D = \frac{1}{n}\sum_{i=1}^{n}\frac{|l_i - l_{av}|}{l_{av}} \quad (2)$$

Substituting Eu^{2+} for Ba^{2+} sites following $Ba_{1-x}Eu_xM_2Si_3O_{10}$ (M = Sc, Lu) is ideal due to their isovalency. By varying the concentration of Eu^{2+} ($x = 0.005, 0.01, 0.02, 0.03, 0.04, 0.05, 0.06$) the optimal quantum yield was identified for both compositions. As shown in Fig. 2, the Φ value for M = Sc is negligible at 0.5% Eu^{2+} substitution but dramatically increases to ≈35% at 1% substitution, followed by a decrease with the further addition of rare-earth ions. Lu^{3+} achieves a maximum Φ of 24% at 3% Eu^{2+}. The moderate quantum yields reported here could likely be improved through the inclusion of a flux during preparation or additional post processing to control the particle size and morphology. Nevertheless, the Φ values measured for these compounds are lower than expected based on their high Debye temperatures.

The room temperature excitation and emission spectra for $Ba_{1-x}Eu_xM_2Si_3O_{10}$ (M = Sc, Lu) with the optimal Eu^{2+} concentration are presented in Fig. 3. Both compounds have broad excitation spectra ($\lambda_{em} = 430$ nm) extending from <250 to 400 nm with a $\lambda_{max} \approx 315$ nm. Additionally, there is a shoulder present in both of the excitation spectra at ≈360 nm. In both compounds, the emission peaks are broad (FWHM > 75 nm) arising from the Eu^{2+} 5d → 4f ($^8H_j \rightarrow {}^8S_{7/2}$) transitions. Specifically, exciting $Ba_{0.99}Eu_{0.01}Sc_2Si_3O_{10}$ at 260 nm or 315 nm produces a single emission peak as fitted by a Gaussian function centered at approximately 440 nm. The emission spectrum of $Ba_{0.97}Eu_{0.03}Lu_2Si_3O_{10}$ (λ_{ex} = 260 nm and 315 nm) can also be fitted by a single Gaussian function and is slightly blue shifted compared to the scandium analogue with a center at 427 nm. This is consistent with a

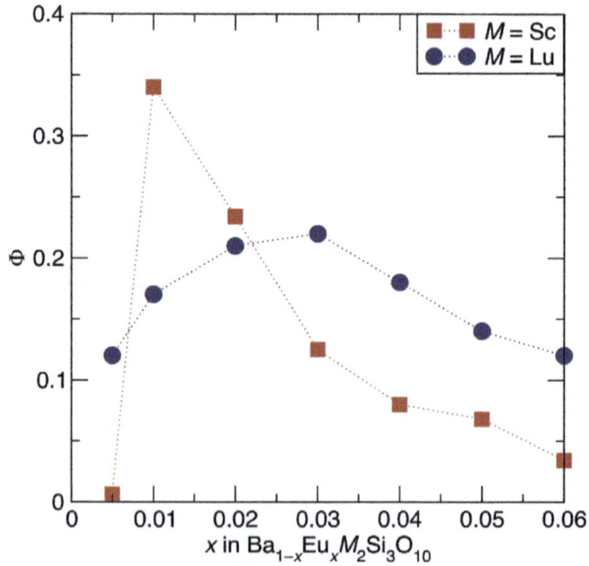

Fig. 2 The Eu^{2+} concentration (x) that yields the highest photoluminescent quantum yield (Φ) for $Ba_{1-x}Eu_xM_2Si_3O_{10}$ (M = Sc, Lu), using a $\lambda_{ex} = 365$ nm. Additional measurements with $\lambda_{ex} = 330$ nm produced comparable results.

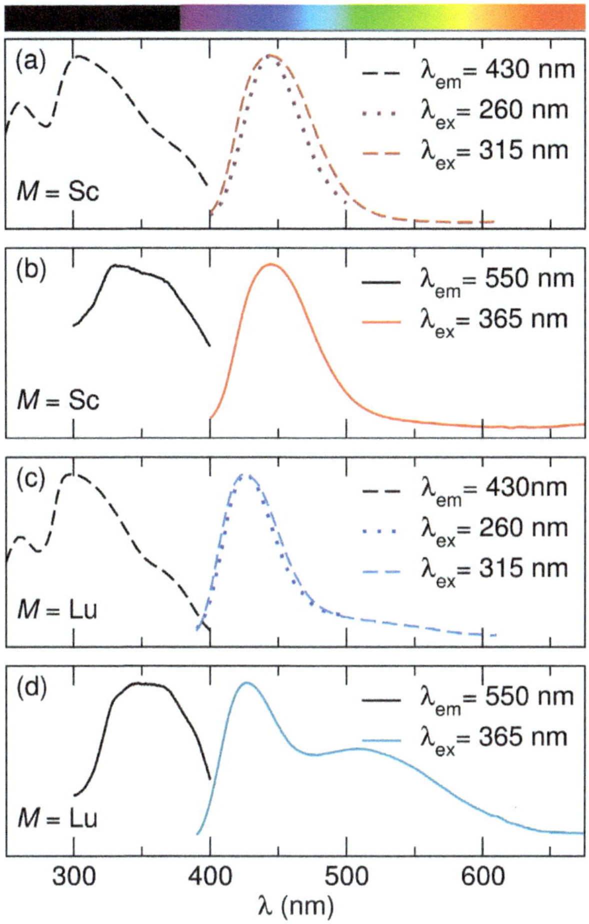

Fig. 3 The florescence spectra for (a and b) $Ba_{0.99}Eu_{0.01}Sc_2Si_3O_{10}$ and (c and d) $Ba_{0.97}Eu_{0.03}Lu_2Si_3O_{10}$ using a variety of excitation wavelengths, including $\lambda_{ex} = 260$ nm, 315 nm, and 365 nm, with emission monitored at $\lambda_{em} = 430$ nm and 550 nm.

decrease in the crystal field splitting due to the larger unit cell and polyhedral volume (see Table 1). In both compounds, there is a small Stokes shift of ≈50 nm due to the substitution of Eu^{2+} into the slightly large Ba^{2+} site. This gives rise to a larger degree of "self-absorption", which is one of the potential causes of the lower than expected Φ.

Surprisingly, when $Ba_{0.97}Eu_{0.03}Lu_2Si_3O_{10}$ is excited at 365 nm the emission spectrum is remarkably different, with the addition of a second, broader emission peak at a longer wavelength. The appearance of this second emission peak requires the addition of a second Gaussian function centered at 527 nm to fit the data. Cooling $Ba_{0.99}Eu_{0.01}Sc_2Si_3O_{10}$ to 77 K and exciting at 365 nm also yields an additional longer wavelength peak, as shown in Fig. 4a. The enhanced emission at the longer wavelength (at 77 K) requires two Gaussian functions to fully describe the emission data. Measuring the emission of $Ba_{0.97}Eu_{0.03}Lu_2Si_3O_{10}$ at 77 K (Fig. 4b) further enhances the magnitude of the two emission peaks, confirming the need for two Gaussian functions to describe the optical data.

The two emission peaks in these compounds are rather surprising considering that the crystal structure contains a single substitution site (Ba1) for the luminescent center. Thus, the longer wavelength emission must arise from a local distortion in the crystal structure that allows additional Eu^{2+} occupation. Because this emission is dramatically red-shifted compared to the main emission peak ($\approx$120 nm) and is extremely broad (FWHM $\approx$ 100 nm at 77 K), it is most likely to arise from the presence of a second, highly-distorted Eu^{2+} site. In this crystal structure, the potential second substitution site is the MO_6 dimer. Because the ionic radius of Eu^{2+} is much larger (1.17 Å; 6-coordinate) than Sc^{3+} (0.745 Å; 6-coordinate) or Lu^{2+} (0.861 Å; 6-coordinate) the substitution at this crystallographic site will require a large local distortion to accommodate the change. The local symmetry of the M-atom site is a general (Wyckoff 4f) position, allowing this site to readily distort without constraint. Nevertheless, the concentration of Eu^{2+} in the M site will be small, but it is clearly consequential to the optical properties. In fact, the presence of two luminescent sites increases the probability of charge-transfer between excited states, which is known to quench luminescence.

Understanding the optical properties of a phosphor as a function of temperature is critical when considering potential device integration. Two primary requirements for a new phosphor are a stable emission intensity (*i.e.*, constant Φ) and robust emission color across a range of temperatures. As shown in Fig. 5a, the emission intensity of $Ba_{0.99}Eu_{0.01}Sc_2Si_3O_{10}$ decreases by 25% from room temperature to 503 K. This decrease in intensity is comparable with that of other thermally robust phosphors such as YAG : Ce^{3+} (ref. 4), which decreases by a similar amount over the same temperature range. The related phosphor,

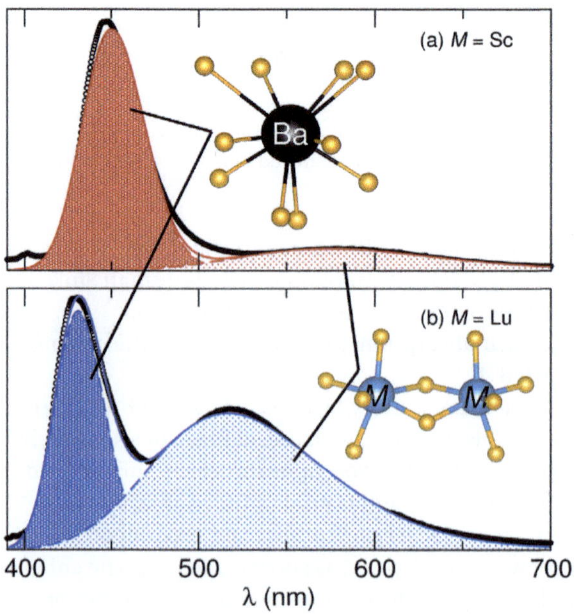

Fig. 4 The emission spectra (λ_{ex} = 365 nm) at 77 K for (a) $Ba_{0.99}Eu_{0.01}Sc_2Si_3O_{10}$ and (b) $Ba_{0.97}Eu_{0.03}Lu_2Si_3O_{10}$ can be fitted with two Gaussian curves, indicating the presence of two independent luminescent centers.

$Ba_9Y_2Si_6O_{24} : Ce^{3+}$, also shows a decrease of around 20% at 503 K.[28] Both of these compounds contain highly connected polyhedra that give an exceptional quantum yield and certainly impact the high temperature properties of these phosphors. However, the emission intensity of $Ba_{0.97}Eu_{0.03}Lu_2Si_3O_{10}$, Fig. 5b, is quite different, with a decrease of nearly 75% of its emission intensity at high temperature.

The effect of Eu^{2+} in the two crystallographic sites becomes more apparent as a function of temperature. As the temperature decreases, the emission from both compounds red-shifts due to crystal field effects. When M = Sc, the main emission peak sharpens at low temperature while the intensity of the second emission peak at longer wavelength ($\approx$ 575 nm) increases slightly below 200 K. The optical properties of M = Lu changes considerably below room temperature. The position and intensity of the main emission peak is fairly stable below 400 K; however, the intensity of the second emission peak increases dramatically below 325 K, as observed in Fig. 5d. The unique appearance of the second emission peak as a function of temperature arises from the shutting down of quenching due to charge transfer. At low temperature, charge transfer over the short distances separating the octahedral dimer sites (3.531(1) Å for Sc^{3+} and 3.635(1) Å for Lu^{3+}) is limited, allowing emission from these substitution sites to emerge. Therefore, the substitution of Eu^{3+} into the Lu^{3+} sites needs to be inhibited to maintain emission intensity as a function of temperature.

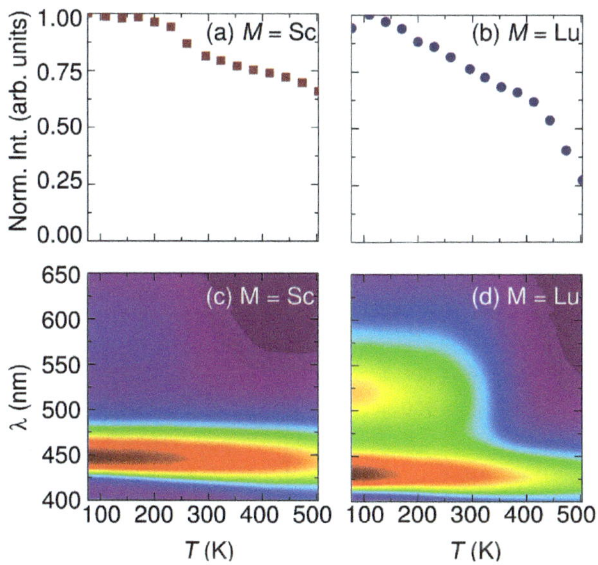

Fig. 5 The relative integrated emission intensity for (a) $Ba_{0.99}Eu_{0.01}Sc_2Si_3O_{10}$ shows a decrease of only 25% of the room temperature intensity at 503 K compared to (b) $Ba_{0.97}Eu_{0.03}Lu_2Si_3O_{10}$ which decreases by 75% over the same range. (c) A minimal shift in $\lambda_{em,max}$ is observed in $Ba_{0.99}Eu_{0.01}Sc_2Si_3O_{10}$ along with an enhancement of the longer wavelength peak below 200 K. (d) A small red-shift in $\lambda_{em,max}$ is observed in $Ba_{0.97}Eu_{0.03}Lu_2Si_3O_{10}$ along with a large increase in the intensity of the peak at 510 nm below 300 K.

The temperature dependent photoluminescence lifetimes provide additional information about the optical properties of these materials, as shown in Fig. 6. Under 365 nm excitation with emission monitored at 433 and 550 nm the luminescence decay requires a bi-exponential function for a correct fit at $\lambda_{em} = 433$ nm and $\lambda_{em} = 550$ nm. Two separate lifetimes, a slow component ($\approx$500 ns) (Fig. 6) and a fast component (<25 ns), compose the signal. The silicone epoxy used to encapsulate the samples was checked for photoluminescence due to the UV excitation wavelength employed. Here, only a fast, mono-exponential photoluminescent lifetime of <2 ns was present on excitation at 365 nm. Interestingly, a sample prepared without rare-earth substitution, *e.g.*, $BaLu_2Si_3O_{10}$, showed luminescence ($\lambda_{ex} = 365$ nm) that required a bi-exponential decay to fit the data. In the unsubstituted compound, the lifetimes were $\approx$2 and 5 ns, respectively. The faster ($\approx$2 ns) component in the unsubstituted compound is certainly from the silicone epoxy. The 5 ns component in these experiments is most likely present due to defect states present in the crystal structure, giving rise to a different recombination rate, which is known to arise in silicates.[33,34] The existence of defect states identified in the host structure from these measurements also

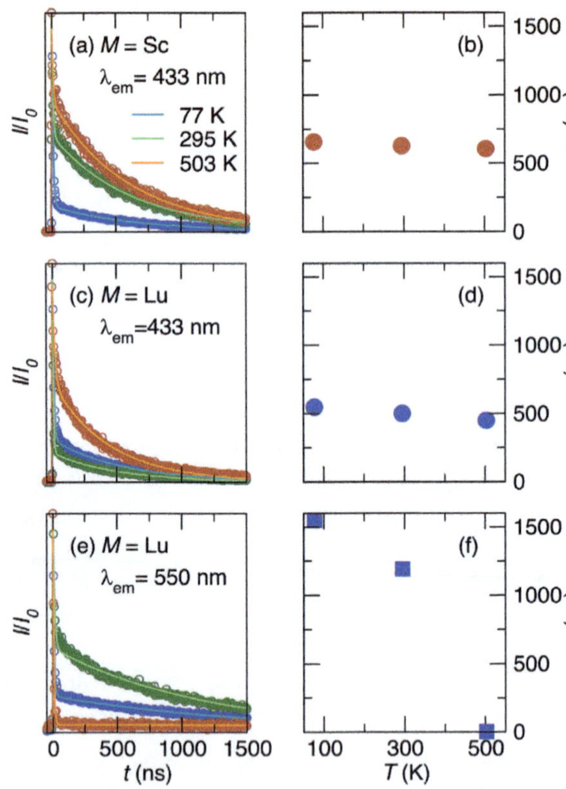

Fig. 6 The luminescence decay as fitted to a bi-exponential function for (a) $Ba_{0.99}Eu_{0.01}Sc_2Si_3O_{10}$ with the extracted lifetimes shown in (b) while the data in (c) are for $Ba_{0.97}Eu_{0.03}Lu_2Si_3O_{10}$ with the corresponding lifetime in (d) when excited at 365 nm and monitored at 433 nm. The lifetime was also monitored at 550 nm for $Ba_{0.97}Eu_{0.03}Lu_2Si_3O_{10}$ as shown in (e) with the extracted lifetimes in (f).

provide a quenching mechanism, further limiting Φ. Focusing on the slower component (τ_2), the lifetime measurements from the fitted data are plotted on the right side of Fig. 6. The slower component, on the time scale of 500 ns to 1000 ns, is in line with previous lifetime measurements of Eu^{2+}. This further confirms that the signal arises from the allowed 5d to 4f transition. Increasing the temperature from 77 K to 503 K accelerates the slower decay time nearly linearly due to the increase in the rate of recombination at elevated temperatures.

4 Conclusions

The interest in photoluminescence from the $Ba_{1-x}Eu_xM_2Si_3O_{10}$ (M = Sc, Lu) phosphor was derived based on its highly connected SiO_4–MO_6 polyhedral network and reasonably high Θ_D. The compounds can be prepared from stoichiometric ratios of starting materials using high temperature ceramic methods. Characterization of these materials using synchrotron X-ray powder diffraction indicates that the materials are nearly pure phase, while optical measurements show that Eu^{2+} substitutes into the Ba^{2+} site as expected, and also surprisingly into the M^{3+} site. Although the atomic radii of Eu^{2+} and M^{3+} are different, the low symmetry of the structure allows it to adapt and incorporate the M atom. The presence of this second site is detrimental to the Φ due to charge-transfer quenching. The temperature dependent photoluminescence highlights the thermally robust emission that arises in highly connected compounds. In fact, the emission intensity of $Ba_{0.99}Eu_{0.01}Sc_2Si_3O_{10}$ drops by only 25% of the room temperature value, which is comparable to industry standard phosphors. The emission intensity of $Ba_{0.97}Eu_{0.03}Lu_2Si_3O_{10}$ decreases by 75%, also because of quenching associated with the second Eu^{2+} site. The temperature dependent lifetime measurements require a bi-exponential function to fit the luminescence decay with a slow relaxation process between 450 and 650 ns corresponding to Eu^{2+} radiative relaxation. From the optical properties identified in these compounds, it is clear that even in highly connected structures, high symmetry is also required to prevent structural distortions that can limit the efficiency of new phosphors.

Acknowledgements

Fellowship support for KAD from the ConvEne IGERT Program (NSF-DGE 0801627) is gratefully acknowledged. KH and CKHB were supported by the MRL-RISE internship program. The research reported here made use of MRL Central Facilities, supported by the MRSEC Program of the NSF under Award no. DMR 1121053. Use of the Advanced Photon Source at Argonne National Laboratory was supported by the U. S. Department of Energy, Office of Science, Office 27 of Basic Energy Sciences, under Contract no. DE-AC02-06CH11357.

References

1 S. Pimputkar, J. S. Speck, S. P. DenBaars and S. Nakamura, *Nat. Photonics*, 2009, **3**, 180–182.
2 H. A. Höppe, *Angew. Chem., Int. Ed.*, 2009, **48**, 3572–3582.

3 N. C. George, K. A. Denault and R. Seshadri, *Annu. Rev. Mater. Res.*, 2013, **43**, 481–501.

4 N. C. George, A. J. Pell, G. Dantelle, K. Page, A. Llobet, M. Balasubramanian, G. Pintacuda, B. F. Chmelka and R. Seshadri, *Chem. Mater.*, 2013, **25**, 3979–3995.

5 N. C. George, A. Birkel, J. Brgoch, B.-C. Hong, A. A. Mikhailovsky, K. Page, A. Llobet and R. Seshadri, *Inorg. Chem.*, 2013, **52**, 13730–13741.

6 J. Brgoch, S. P. DenBaars and R. Seshadri, *J. Phys. Chem. C*, 2013, **117**, 17955–17959.

7 A. Pozharskii, A. Soldatenkov and A. Katritzky, *Heterocycles in Life and Society: An Introduction to Heterocyclic Chemistry, Biochemistry and Applications*, Wiley, 2011.

8 J. D. Furman, B. C. Melot, S. J. Teat, A. A. Mikhailovsky and A. K. Cheetham, *Phys. Chem. Chem. Phys.*, 2011, **13**, 7622–7629.

9 K. A. Denault, J. Brgoch, M. W. Gaultois, A. Mikhailosvsky, R. Petry, H. Winkler, S. P. DenBaars and R. Seshadri, *Chem. Mater.*, 2014, **26**, 2275–2282.

10 K. A. Denault, Z. Cheng, J. Brgoch, S. P. DenBaars and R. Seshadri, *J. Mater. Chem. C*, 2013, **1**, 7339–7345.

11 J. Brgoch, C. K. Borg, K. A. Denault, J. R. Douglas, T. A. Strom, S. P. DenBaars and R. Seshadri, *Solid State Sci.*, 2013, **26**, 115–120.

12 J. Brgoch, S. D. Kloss, K. A. Denault and R. Seshadri, *Z. Anorg. Allg. Chem.*, 2014, **640**, 1182–1189.

13 U. Kolitsch, M. Wierzbicka and E. Tillmanns, *Acta Crystallogr., Sect. C: Cryst. Struct. Commun.*, 2006, **62**, i97–i99.

14 M. Wierzbicka-Wieczorek, U. Kolitsch and E. Tillmanns, *Eur. J. Mineral.*, 2010, **22**, 245–258.

15 W.-R. Liu, C. C. Lin, Y.-C. Chiu, Y.-T. Yeh, S.-M. Jang, R.-S. Liu and B.-M. Cheng, *Opt. Express*, 2009, **17**, 18103–18109.

16 Z. Xia, Y. Liang, D. Yu, M. Zhang, W. Huang, M. Tong, J. Wu and J. Zhao, *Opt. Laser Technol.*, 2014, **56**, 387–392.

17 J. Wang, B. H. Toby, P. L. Lee, L. Ribaud, S. M. Antao, C. Kurtz, M. Ramanathan, R. B. Von Dreele and M. A. Beno, *Rev. Sci. Instrum.*, 2008, **79**, 085105–085107.

18 A. C. Larson and R. B. V. Dreele, *General Structure Analysis System (GSAS)*, Los Alamos National Laboratory Report LAUR 86-748, 2000.

19 B. H. Toby, *J. Appl. Crystallogr.*, 2001, **34**, 210–213.

20 J. C. de Mello, H. F. Wittmann and R. H. Friend, *Adv. Mater.*, 1997, **9**, 230–232.

21 W. Becker, *Advanced Time-Correlated Single Photon Counting Techniques*, Springer, Berlin, Heidelberg, New York, 2005, vol. 81.

22 G. Kresse and J. Furthmuller, *Phys. Rev. B: Condens. Matter Mater. Phys.*, 1993, **47**, 558–561.

23 G. Kresse and J. Furthmuller, *Phys. Rev. B: Condens. Matter Mater. Phys.*, 1996, **54**, 11169–11186.

24 P. Blöchl, *Phys. Rev. B: Condens. Matter Mater. Phys.*, 1994, **50**, 17953–17979.

25 G. Kresse and D. Joubert, *Phys. Rev. B: Condens. Matter Mater. Phys.*, 1999, **59**, 1758–1775.

26 Y. Le Page and P. Saxe, *Phys. Rev. B: Condens. Matter Mater. Phys.*, 2002, **65**, 104104.

27 J. Perdew, K. Burke and M. Ernzerhof, *Phys. Rev. Lett.*, 1996, **77**, 3865–3868.

28 J. Brgoch, C. K. H. Borg, K. A. Denault, A. A. Mikhailovsky, S. P. DenBaars and R. Seshadri, *Inorg. Chem.*, 2013, **52**, 8010–8016.
29 J. Brgoch, C. K. H. Borg, K. A. Denault, S. P. DenBaars and R. Seshadri, *Solid State Sci.*, 2013, **18**, 149–154.
30 R. Shannon, *Acta Crystallogr., Sect. A: Cryst. Phys., Diffr., Theor. Gen. Crystallogr.*, 1976, **32**, 751–767.
31 K. Robinson, G. V. Gibbs and P. H. Ribbe, *Science*, 1971, **172**, 567–570.
32 W. H. Baur, *Acta Crystallogr., Sect. B: Struct. Crystallogr. Cryst. Chem.*, 1974, **30**, 1195–1215.
33 G.-L. Davies, J. E. McCarthy, A. Rakovich and Y. K. Gun'ko, *J. Mater. Chem.*, 2012, **22**, 7358–7365.
34 Z. Bai, M. Fujii, K. Imakita and S. Hayashi, *J. Lumin.*, 2014, **145**, 288–291.

Faraday Discussions

PAPER

Bridging silicon nanoparticles and thermoelectrics: phenylacetylene functionalization†

Shane P. Ashby,[a] Jason A. Thomas,[a] Jorge García-Cañadas,[b] Gao Min,[b] Jack Corps,[c] Anthony V. Powell,[c] Hualong Xu,[d] Wei Shen[d] and Yimin Chao*[a]

Received 14th May 2014, Accepted 26th June 2014
DOI: 10.1039/c4fd00109e

Silicon is a promising alternative to current thermoelectric materials (Bi_2Te_3). Silicon nanoparticle based materials show especially low thermal conductivities due to their high number of interfaces, which increases the observed phonon scattering. The major obstacle with these materials is maintaining high electrical conductivity. Surface functionalization with phenylacetylene shows an electrical conductivity of 18.1 S m^{-1} and Seebeck coefficient of 3228.8 μV K^{-1} as well as maintaining a thermal conductivity of 0.1 W K^{-1} m^{-1}. This gives a ZT of 0.6 at 300 K which is significant for a bulk silicon based material and is similar to that of other thermoelectric materials such as Mg_2Si, PbTe and SiGe alloys.

Introduction

Over the past 20 years there has been increasing pressure for the development of technologies which take advantage of renewable energy sources. This is a result of the environmental impacts and future availability of fossil fuels.[1] There has also been an emphasis on the recycling of waste energy (energy scavenging) to improve the efficiency of many industrial and commercial processes.[2]

Thermoelectric materials are able to take advantage of wasted or unutilized heat sources, such as in furnaces,[3] car exhausts,[4] and solar cells.[5] As a result thermoelectric materials have become an area of great interest.[5d,6] These materials are able to convert a temperature gradient into electrical power, and *vice versa* without mechanical intervention. The power output that current commercial modules produce is however, modest,[7] but the power generated from these

[a]*School of Chemistry, University of East Anglia, Norwich NR4 7TJ, UK. E-mail: y.chao@uea.ac.uk; Fax: +44-1603592003; Tel: +44 1603593146*

[b]*School of Engineering, Cardiff University, Cardiff CF24 3AA, UK*

[c]*Institute of Chemical Sciences, Heriot Watt University, Edinburgh EH14 4AS, UK*

[d]*Advance Materials Laboratory, Department of Chemistry, Fudan University, Shanghai 200433, P. R. China*

† Electronic supplementary information (ESI) available: EDX spectrum. See DOI: 10.1039/c4fd00109e

devices can be used elsewhere for low power applications, *e.g.*, powering sensors or safety feedback loops.

The thermoelectric performance of a material can be expressed in terms of a figure of merit (ZT), given by the following expression:[6b]

$$ZT = \frac{\sigma S^2}{k} T \quad (1)$$

where σ is the electrical conductivity, S is the Seebeck coefficient, k is the thermal conductivity and T is temperature. Commercial thermoelectric materials generally have a ZT value of around 1.

The most common thermoelectric materials used today are based on bismuth telluride.[6a] ZT values of 2.4 at room temperature have been recorded for thin films, using bismuth telluride alloys, Bi_2Te_3/Sb_2Te_3.[8] The major drawback with these materials is that the tellurium required to produce them is toxic and also expensive, due to its low abundance. As a consequence there has been a lot of focus on producing thermoelectric materials from alternative sources such as silicon,[9] conductive polymers,[10] and magnesium silicide.[11]

Silicon is a key material in the semiconductor electronics industry and as a result there is a well-established infrastructure based on these materials.[12] Bulk silicon itself has a relatively high Seebeck coefficient ($-86\ \mu V\ K^{-1}$) and high electrical conductivity when doping is optimized ($3030.3\ S\ cm^{-1}$), which results in a very high power factor.[13] However, it also displays a very high thermal conductivity of $87.3\ W\ m^{-1}\ K^{-1}$ which results in a very low ZT of 0.008.[13]

Silicon nanostructures show much lower thermal conductivities as a result of more efficient boundary scattering of phonons, which can lead to a marked improvement to the figure of merit.[14] For silicon nanowires, alteration of surface roughness, diameter and doping concentrations of nanowires has recently resulted in achieving a high ZT (≈ 1).[9b,c,15] Although it is worth noting that this was obtained from a single narrow nanowire, it is not yet demonstrated if the value can be translated into bulk materials.

Silicon nanowires show great potential as thermoelectric materials but preparation methods typically use harsh conditions or toxic and expensive reagents, such as trisilane,[16] hydrogen fluoride and silver nitrate.[17] It is possible to avoid such issues in the synthesis of silicon nanoparticles (SiNPs).[18]

Highly porous films or pellets of SiNPs show exceptionally low thermal conductivity.[10] These films also display low electrical conductivity, due to the difficulty of transporting electrons across the junctions between SiNPs.[19] Hydrogen terminated SiNPs require high temperature compaction or spark plasma sintering (SPS) to produce a conductive pellet. As a result crystal growth can be observed, significantly compromising the initial thermal conductivity advantages.[13,20]

Introducing suitable functionalization to the nanoscale silicon has been shown to be a method of maintaining the conductivity at the surface,[21] and shows an improvement of the conductivity between two materials, acting as a conductive bridge.[22] This study shows Si–molecule–metal junctions to be more conductive than a Si–H–Hg junction, which is attributed to the alignment of the ligands on the surface. Moreover, the attachment of these molecules gives an increased density of surface states which have energy positions deep into the band gap, which act as slow traps that quench the surface photoconductivity. The use of a

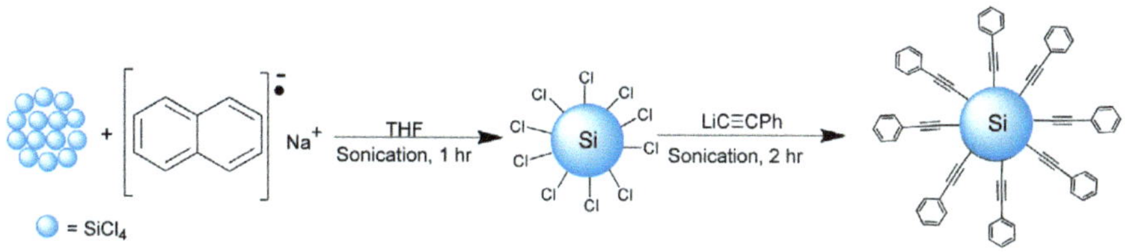

Scheme 1 Synthesis of phenyl acetylene capped SiNPs.

conjugated capping layer, such as phenylacetylene, could give improved conductivity of pellet pressed at ambient temperature compared to that of a silicon based material of equivalent density and porosity.

There are issues with the attachment of such groups to a silicon surface, since the typical hydrosilylation method for SiNPs functionalization does not give desirable high surface coverage.[23] This is due to the ability of such surface agents, once attached to quench the reactive species.[24] However, an alternative method is the reaction of a suitable nucleophile with chloride terminated SiNPs. A well-documented synthesis for chloride terminated SiNPs is the reduction of $SiCl_4$ using a suitable hydride free reducing agent such as sodium naphthalide or sodium silicide.[25] Suitable nucleophiles which are subsequently introduced, include alcohols and organometallic reagents.[26]

In the following report phenylacetylene capped SiNPs are produced using a micelle reduction method.[25,27] This entails the reduction of micelles of $SiCl_4$ using sodium naphthalide to produce chlorine terminated SiNPs, followed by alkylation (functionalization) using lithium phenylacetylide as shown in Scheme 1. The resulting SiNPs were characterized using a combination of Fourier transform infrared spectroscopy (FTIR), ^{1}H-NMR, ^{13}C-NMR, transmission electron microscopy (TEM), X-ray photoelectron spectroscopy (XPS), energy dispersive X-ray spectroscopy (EDX), thermogravimetric analysis (TGA) and differential scanning calorimetry (DSC). The thermoelectric properties of a cold pressed pellet of phenylacetylene capped SiNPs were measured at room temperature. For this material a *ZT* of 0.6 was obtained, which is a high value for a bulk SiNPs pellet[13,20] and is comparable to many thermoelectric materials.[6b,c,28]

Experimental

Materials

Silicon tetrachloride (Sigma, 99%), lithium phenylacetylide (Sigma, 1 M solution in tetrahydrofuran), sodium (Acros, 99.8%), naphthalene (Acros, 99%), tetrahydrofuran (Fisher, 99.5%, dried over sodium/benzophenone and freshly distilled) and toluene (Fisher, 99%).

Synthesis of phenylacetylene capped SiNPs

Silicon tetrachloride was dispersed in tetrahydrofuran by sonication for 15 min and a solution of freshly prepared sodium naphthalide (70 mL) was added. This was prepared by ultrasonication of sodium metal (0.7 g) and naphthalene (2.9 g) for 3 h in dry tetrahydrofuran (70 mL). After the addition of sodium naphthalide solution the mixture was placed in an ultrasonic bath for 1 h to give a dark brown suspension. To this a solution of lithium phenylacetylide (5 mL, 1 M solution in tetrahydrofuran) was added followed by further ultrasonication for 2 h. All precipitate was allowed to settle, the clear liquid was decanted off and all solvent was removed *in vacuo*. The resulting solid was then heated to 373 K under reduced pressure to remove the residual naphthalene. The resulting solid was redissolved in toluene, washed with water to remove any remaining lithium chloride; and the organic layer was dried to give phenylacetylene capped SiNPs, as an orange crystalline solid.

FTIR

The FTIR spectrum was collected using a Perkin-Elmer Spectrum 100 ATR FTIR spectrometer. The background was corrected by subtracting a spectrum of the clean crystal.

^{1}H-NMR and ^{13}C-NMR

Measurements of sample dissolved in $CDCl_3$ were taken using a Bruker 500 MHz NMR spectrometer. These samples were measured relative to chloroform from the lock solvent ($CDCl_3$).

XPS

XPS analysis was carried out on a K-Alpha XPS instrument (Thermo Scientific, East Grinstead, UK). A monochromatic Al Kα X-ray source (1486.6 eV) was used with a spot size of 400 mm diameter. A pass energy of 200 eV and step size of 0.4 eV was used for survey spectra. A pass energy of 40 eV and step size of 0.1 eV was used for high resolution spectra.

TEM

TEM micrographs were taken using a JEOL JEM2000 transmission electron microscope. TEM samples were prepared by drop casting a solution of phenylacetylene capped SiNPs in chloroform onto a 200 mesh copper grid with a holey carbon film. The solvent was evaporated and TEM micrographs were taken at different spots of the grid. The mean diameter and standard deviation were acquired by the measurement of 100 nanoparticles manually from images of different spots on the grid.

TGA and DSC

TGA and DSC analysis was performed using a METTLER-TOLEDO TGA-1. The solid sample (approx. 5 mg) was placed in a 40 μL aluminum pan and measurements were taken during the heating of the sample from 300 to 773 K. The background measurements were taken while heating the empty pan over the same temperature range.

Preparation of pellet and measurement of thermoelectric properties

A dry powder of the sample was pressed uniaxially using a hydraulic press and die into a 13 mm diameter pellet of thickness 1–2 mm by applying the maximum pressure of 800 MPa for 3 min. The pressing was carried out at room temperature.

The Seebeck coefficient was measured at room temperature using a hot probe apparatus.[29] A four probe DC current equipment was used to measure the room temperature electrical resistivity. A current $I = 10$ mA was applied between two probes and the voltage difference V generated at the other two probes was measured. The resistivity R was calculated using

$$R = \frac{V}{I} \times F \times C \times d \tag{2}$$

where F and C are geometrical correction factors and d the pellet thickness. In both measurements the data was taken at different regions of the pellet. The resistivity reported here is the average of two measurements with applied electrical current in two opposite directions. The contacts are spring loaded pressure contact. The influence of the parasitic capacitance is ignored here because the measurement was carried out under DC conditions as thermoelectric devices operate in DC mode.

Thermal diffusivity measurements were performed at 373 K and 423 K using an Anter Flashline 3000 instrument. Measurements were made on pellets of phenylacetylene capped silicon nanoparticles. This instrument determines both the thermal diffusivity (α) and the heat capacity (C_p) of the sample. The thermal conductivity (κ) is calculated from the relationship:

$$\kappa = \alpha C_p \rho \tag{3}$$

where ρ is the sample density. For the determination of the heat capacity, side-by-side testing of a reference material, Pyroceram™ 9606, of known heat capacity, was carried out. The procedure used for the determination of the heat capacity has been described in detail.[30] Efforts to extend the measurements to higher temperatures were unsuccessful as the sample pellet deformed.

Results and discussion

The FTIR spectrum (Fig. 1) shows the peaks expected for the phenylacetylene functionalization of SiNPs. The spectrum shows peaks at 3051 cm^{-1}, representative of aromatic C–H, and at both 1596 and 1487 cm^{-1} which is representative of aromatic C–C bonds. Also clearly visible is the sharp peak at 2161 cm^{-1}, which is characteristic of the C≡C, which, when paired with the absence of a sharp peak at 3300 cm^{-1}, suggests that the terminal side of the alkyne is attached to silicon. This is also supported by the Si–C≡C peaks at 1441 and 1221 cm^{-1}. The Si–O peak between 1100 and 1000 cm^{-1} is observable but is broad and weak when compared with the sharp peaks at 1068 and 1026 cm^{-1} which are also characteristic of phenylacetylene.

Further evidence of phenylacetylene capping can also be observed in the ^{1}H-NMR (Fig. 2a) and ^{13}C-NMR (Fig. 2b) spectra. ^{1}H-NMR shows two multiplet peaks at 7.53 and 7.30 ppm with an integration of 2 : 3, respectively, characteristic of the phenyl protons of phenylacetylene. Also, there is no peak for a proton on the alkyne at 3 ppm. These features are evidence of the molecule being bound to the SiNPs. The ^{13}C-NMR shows the four peaks for the phenyl carbons at 132, 129, 128 and 121 ppm and 2 peaks for the alkyne at 106 and 86 ppm, characteristic of the carbon closest to the phenyl and the silicon respectively.

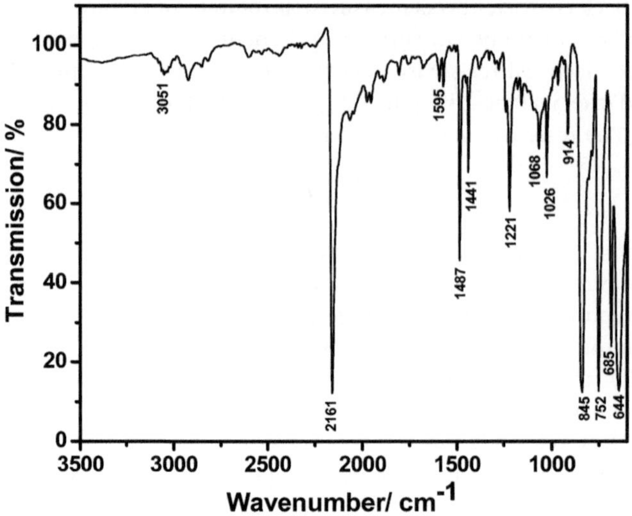

Fig. 1 FTIR spectrum of phenylacetylene capped SiNPs.

The TEM image shows that spherical SiNPs have been synthesized (Fig. 3). The SiNPs show a wide distribution of particle diameters (inset Fig. 3). The mean diameter of the nanoparticles produced is 6 nm with a standard deviation of 1 nm, obtained by analyzing 100 particles from different regions of the grid. The high standard deviation and wide size distribution can be explained by the lack of controlling influence on the micelle size, *i.e.*, lack of use of surfactants.[18a]

The SiNPs are rich in both silicon (inset Fig. 4) and carbon. This can be observed from the XPS survey spectrum of the SiNPs (Fig. 4). Moderate levels of oxygen can also been seen in Fig. 4. However, lower levels of oxygen were observed by EDX (ESI†). This increased oxygen level observed in XPS is typical of X-ray induced oxidation of SiNPs.[31] Additionally the XPS does not show any signals from elemental impurities of potential byproducts: sodium, lithium and chlorine.

The DSC trace (Fig. 5) begins at a flat level of heat flow before a negative peak can be observed at 445 K representing a solid–liquid phase transition. Meanwhile the weight on the TGA trace (Fig. 5) remains close to 100%. After this feature the heat flow into the sample begins to decrease relatively rapidly, which suggests that the sample is beginning to undergo degradation. The DSC trace reaches a maximum at 593 K, where there is no net flow of heat into or out of the sample showing completion of the initial degradation. After 593 K the heat flow into the sample increases again, the TGA trace shows that the weight continues to decrease; becoming more rapid above 700 K. This is likely to be due to continued degradation.

The DSC and TGA traces are significant in that they show to what temperature a material produced from phenylacetylene capped SiNPs can be used or tested. The observation of a solid to liquid phase transition at 448 K suggests that the material could only be used up to approximately 420 K, before the melting process begins warping the material. Also it should be noted that this limits the temperature at which the material can be processed as significant softening occurs at moderately low temperatures. Thus, the material would be suitable for

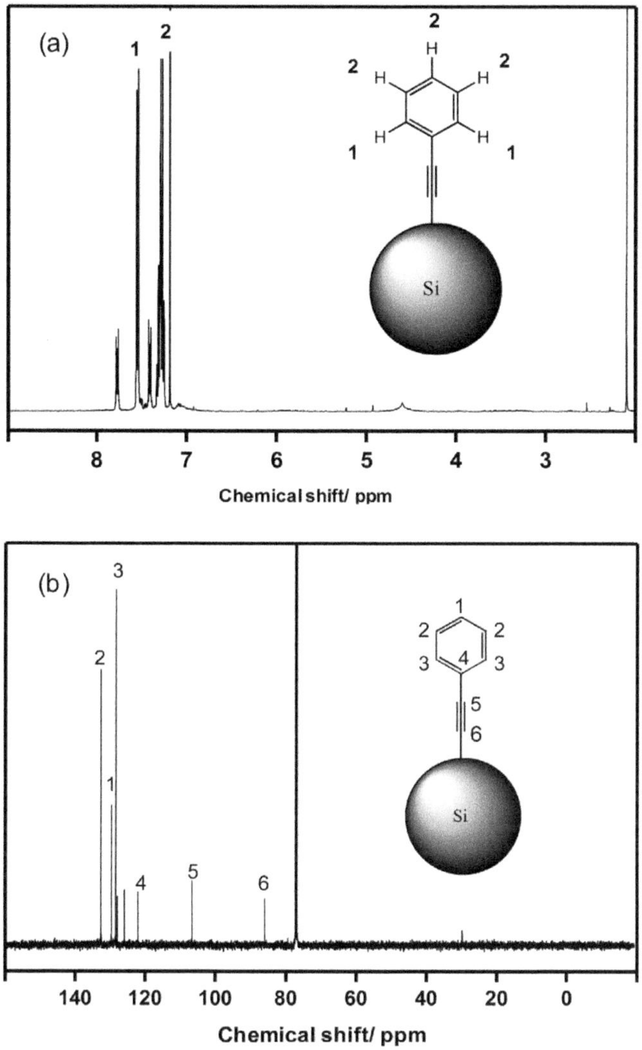

Fig. 2 NMR spectra of phenylacetylene capped SiNPs: (a) ^{1}H-NMR; (b) ^{13}C-NMR.

low temperature thermoelectric applications only. Examples include: energy scavenging from hot water/cooling water outflow from a plant or thermal spring, or the conversion of heat from light harvesting.

The thermoelectric measurements are given in Table 1. The measurements obtained at ambient temperature show that the material produced has an exceptionally high Seebeck coefficient and low thermal conductivity. These two values offset the low electrical conductivity in the equation for the figure of merit (ZT). The thermal conductivity was measured at 373 K and 423 K, and the values were 0.15 and 0.19 W m^{-1} K^{-1}, respectively. The listed values in Table 1 are extrapolated values expressed to one decimal place. This gives an ambient temperature ZT of 0.6. This value is comparative to other materials under investigation in thermoelectrics.[9b,c]

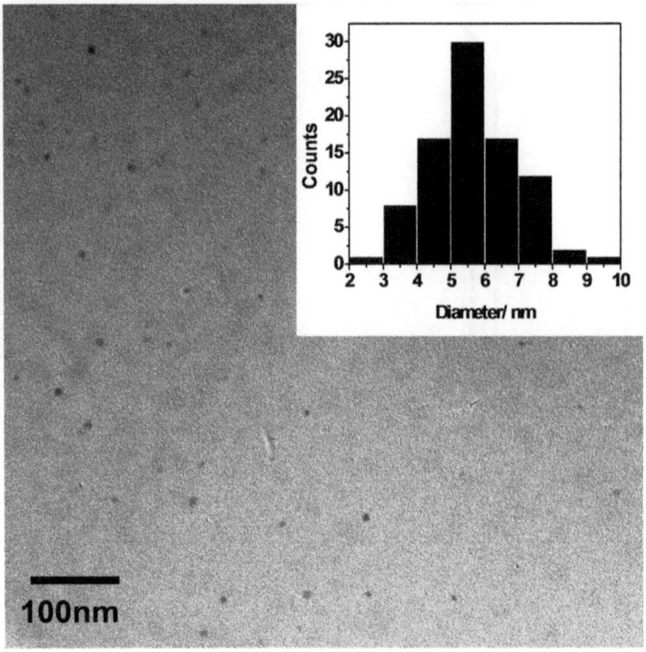

Fig. 3 TEM image and size distribution of phenylacetylene capped silicon nanoparticles. Inset, size distribution of silicon nanoparticles.

The low thermal conductivity values may be in part due to the low density. Because the samples were simply cold pressed, the density might be well below the theoretical (crystallographic) value. The effect of densification on k is of the form below:[32]

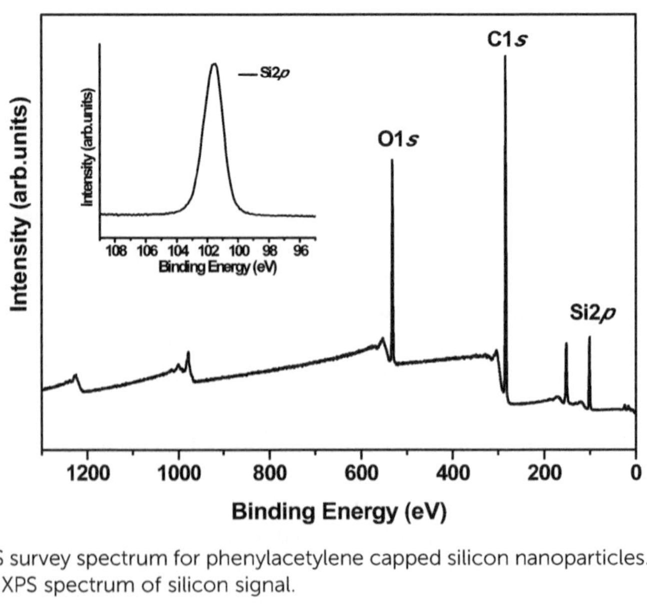

Fig. 4 XPS survey spectrum for phenylacetylene capped silicon nanoparticles. Inset, high resolution XPS spectrum of silicon signal.

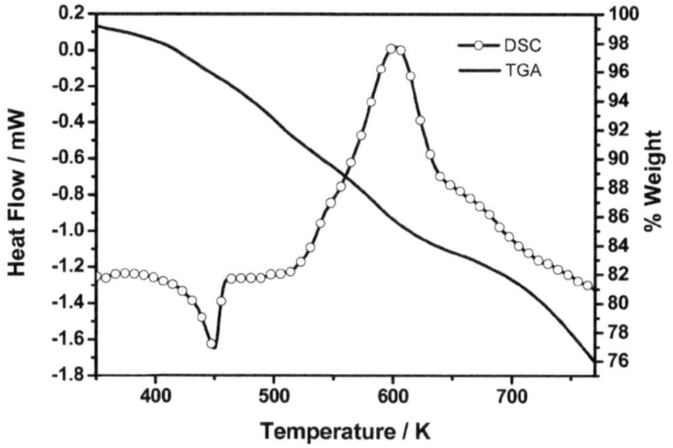

Fig. 5 TGA and DSC analysis of phenylacetylene functionalised silicon nanoparticles.

Table 1 Thermoelectric measurements of a pellet of phenylacetylene capped silicon nanoparticles at room temperature

ρ [kg m^{-3}]	σ [S m^{-1}]	S [μV K^{-1}]	κ [W m^{-1} K^{-1}]	ZT
1022.0	18.1	3228.8	0.1	0.6

$$\frac{k_{\text{porous}}}{k_{\text{dense}}} = 1 - \frac{4}{3}\varphi \quad (4)$$

where φ is the porosity defined as

$$\varphi = 1 - \frac{\rho_{\text{porous}}}{\rho_{\text{dense}}} \quad (5)$$

Unfortunately, given the presence of the organic capping agents, the calculation of the theoretical density is less/not obvious. If the organic component is neglected, then the density is expected to be that of silicon, which is 2329 kg m^{-3}. The measured density is approximately 1022 kg m^{-3}. This suggests, compared to pure silicon, that the sample is approximately 44% dense. Of course this is a gross approximation as it ignores the organic component completely, but it does suggest that the sample is quite porous and that κ is well below that of the dense material.

Relative to other silicon materials of similar porosity, the electrical conductivity of a pellet of phenylacetylene SiNPs is significantly large.[19b,c,33] However, when compared with materials typically involved in thermoelectric applications the electrical conductivity is low.[5c,6b] This conductivity could however be improved and optimized through the introduction of charge carriers, which can be achieved by doping the SiNPs and/or doping the ligands. Both of these should be understood in order to optimize this material.

The magnitude of the Seebeck coefficient is much higher than that which is documented for other silicon based materials.[13,34] This can be attributed to the low carrier concentration discussed above. If the charge carrier concentration is optimized the Seebeck coefficient would be expected to be reduced by a factor of 10 to 100, based on previous work regarding silicon based materials.[13,34] Optimization of doping would mean that a compromise between the magnitude of the electrical conductivity and Seebeck coefficient will need to be made, which in practical terms would make for a more useful material but may alter the *ZT* positively or negatively.

Conclusion

Surface analysis of phenylacetylene functionalized SiNPs shows a surface coverage considerably higher than that previously documented for capping with similar molecules using hydrosilylation based methods. The thermal stability shows that the material produced, when cold pressed undergoes a phase transition at 448 K, so would only be suitable to be used as a thermoelectric material up to 420 K. The thermoelectric performance at ambient temperature is comparable to other silicon based thermoelectric materials, such as silicon nanowires, where a $ZT \approx 1$ was obtained from a single narrow nanowire. Although it is not yet demonstrated if the *ZT* value can be translated into bulk materials. On the other hand, phenylacetylene capped silicon nanoparticles have been readily compacted into bulk materials with a respectable *ZT* of 0.6 prior to optimization of the electrical conductivity by doping. This is a high *ZT* value compared to other silicon based bulk materials and is comparable to other better established thermoelectric materials such as Mg_2Si and SiGe alloys. This result shows a significant step toward silicon based thermoelectrics.

Acknowledgements

SPA is grateful to an Industry CASE studentship sponsored by the UK Engineering and Physical Science Research Council (ESPRC), and European Thermodynamics Ltd. X-ray photoelectron spectra were obtained at the National EPSRC XPS User's Service (NEXUS) at Newcastle University, an EPSRC Mid-Range Facility.

Notes and references

1 (*a*) S. H. Schneider, The Greenhouse Effect: Science and Policy, *Science*, 1989, **243**(4892), 771–781; (*b*) S. Solomon, G.-K. Plattner, R. Knutti and P. Friedlingstein, Irreversible climate change due to carbon dioxide emissions, *Proc. Natl. Acad. Sci. U. S. A.*, 2009, **106**(6), 1704–1709.

2 Y. Qin, X. Wang and Z. L. Wang, Microfibre–nanowire hybrid structure for energy scavenging, *Nature*, 2008, **451**(7180), 809–U5.

3 T. Kajikawa, Approach to the Practical Use of Thermoelectric Power Generation, *J. Electron. Mater.*, 2009, **38**(7), 1083–1088.

4 J. Yang and F. Stabler, Automotive Applications of Thermoelectric Materials, *J. Electron. Mater.*, 2009, **38**(7), 1245–1251.

5 (*a*) D. Kraemer, B. Poudel, H.-P. Feng, J. C. Caylor, B. Yu, X. Yan, Y. Ma, X. Wang, D. Wang, A. Muto, K. McEnaney, M. Chiesa, Z. Ren and G. Chen, High-performance flat-panel solar thermoelectric generators with high thermal concentration, *Nat. Mater.*, 2011, **10**(7), 532–538; (*b*)

C. Lertsatitthanakorn, M. Rungsiyopas, A. Therdyothin and S. Soponronnarit, Performance Study of a Double-Pass Thermoelectric Solar Air Collector with Flat-Plate Reflectors, *J. Electron. Mater.*, 2012, **41**(6), 999–1003; (*c*) A. J. Minnich, M. S. Dresselhaus, Z. F. Ren and G. Chen, Bulk nanostructured thermoelectric materials: current research and future prospects, *Energy Environ. Sci.*, 2009, **2**(5), 466–479; (*d*) T. M. Tritt and M. A. Subramanian, Thermoelectric materials, phenomena, and applications: A bird's eye view, *MRS Bull.*, 2006, **31**(3), 188–194.

6 (*a*) B. Poudel, Q. Hao, Y. Ma, Y. Lan, A. Minnich, B. Yu, X. Yan, D. Wang, A. Muto, D. Vashaee, X. Chen, J. Liu, M. S. Dresselhaus, G. Chen and Z. Ren, High-Thermoelectric Performance of Nanostructured Bismuth Antimony Telluride Bulk Alloys, *Science*, 2008, **320**(5876), 634–638; (*b*) M. G. Kanatzidis, Nanostructured Thermoelectrics: The New Paradigm?, *Chem. Mater.*, 2009, **22**(3), 648–659; (*c*) M. S. Dresselhaus, G. Chen, M. Y. Tang, R. G. Yang, H. Lee, D. Z. Wang, Z. F. Ren, J. P. Fleurial and P. Gogna, New directions for low-dimensional thermoelectric materials, *Adv. Mater.*, 2007, **19**(8), 1043–1053.

7 R. Funahashi, S. Urata, K. Mizuno, T. Kouuchi and M. Mikami, $Ca_{2.7}Bi_{0.3}Co_4O_9/La_{0.9}Bi_{0.1}NiO_3$ thermoelectric devices with high output power density, *Appl. Phys. Lett.*, 2004, **85**(6), 1036–1038.

8 R. Venkatasubramanian, E. Siivola, T. Colpitts and B. O'Quinn, Thin-film thermoelectric devices with high room-temperature figures of merit, *Nature*, 2001, **413**(6856), 597–602.

9 (*a*) J. Tang, H.-T. Wang, D. H. Lee, M. Fardy, Z. Huo, T. P. Russell and P. Yang, Holey Silicon as an Efficient Thermoelectric Material, *Nano Lett.*, 2010, **10**(10), 4279–4283; (*b*) A. I. Boukai, Y. Bunimovich, J. Tahir-Kheli, J. K. Yu, W. A. Goddard and J. R. Heath, Silicon nanowires as efficient thermoelectric materials, *Nature*, 2008, **451**, 168–171; (*c*) A. I. Hochbaum, R. K. Chen, R. D. Delgado, W. J. Liang, E. C. Garnett, M. Najarian, A. Majumdar and P. D. Yang, Enhanced thermoelectric performance of rough silicon nanowires, *Nature*, 2008, **451**, 163–167.

10 O. Bubnova and X. Crispin, Towards polymer-based organic thermoelectric generators, *Energy Environ. Sci.*, 2012, **5**(11), 9345–9362.

11 V. K. Zaitsev, M. I. Fedorov, E. A. Gurieva, I. S. Eremin, P. P. Konstantinov, A. Y. Samunin and M. V. Vedernikov, Highly effective $Mg_2Si_{1-x}Sn_x$ thermoelectrics, *Phys. Rev. B: Condens. Matter Mater. Phys.*, 2006, **74**(4), 045207.

12 Z. Kang, Y. Liu and S.-T. Lee, Small-sized silicon nanoparticles: new nanolights and nanocatalysts, *Nanoscale*, 2011, **3**(3), 777–791.

13 S. K. Bux, R. G. Blair, P. K. Gogna, H. Lee, G. Chen, M. S. Dresselhaus, R. B. Kaner and J.-P. Fleurial, Nanostructured Bulk Silicon as an Effective Thermoelectric Material, *Adv. Funct. Mater.*, 2009, **19**(15), 2445–2452.

14 J. W. Sharp, S. J. Poon and H. J. Goldsmid, Boundary Scattering and the Thermoelectric Figure of Merit, *Phys. Status Solidi A*, 2001, **187**(2), 507–516.

15 N. P. Dasgupta, J. Sun, C. Liu, S. Brittman, S. C. Andrews, J. Lim, H. Gao, R. Yan and P. Yang, 25th Anniversary Article: Semiconductor Nanowires – Synthesis, Characterization, and Applications, *Adv. Mater.*, 2014, **26**(14), 2137–2184.

16 (*a*) A. T. Heitsch, V. A. Akhavan and B. A. Korgel, Rapid SFLS Synthesis of Si Nanowires Using Trisilane with In situ Alkyl-Amine Passivation, *Chem. Mater.*, 2011, **23**(11), 2697–2699; (*b*) H. Chen, J. Xu, P.-c. Chen, X. Fang, J. Qiu, Y. Fu and C. Zhou, Bulk Synthesis of Crystalline and Crystalline Core/

Amorphous Shell Silicon Nanowires and Their Application for Energy Storage, *ACS Nano*, 2011, **5**(10), 8383–8390.

17 K. Peng, Y. Xu, Y. Wu, Y. Yan, S.-T. Lee and J. Zhu, Aligned Single-Crystalline Si Nanowire Arrays for Photovoltaic Applications, *Small*, 2005, **1**(11), 1062–1067.

18 (*a*) J. Wang, S. Sun, F. Peng, L. Cao and L. Sun, Efficient one-pot synthesis of highly photoluminescent alkyl-functionalised silicon nanocrystals, *Chem. Commun.*, 2011, **47**(17), 4941; (*b*) J. P. Wilcoxon and G. A. Samara, Tailorable, visible light emission from silicon nanocrystals, *Appl. Phys. Lett.*, 1999, **74**(21), 3164–3166.

19 (*a*) Z. H. Khan, Electrical and optical properties of thin film of amorphous silicon nanoparticles, *Appl. Surf. Sci.*, 2009, **255**(21), 8874–8878; (*b*) V. M. Aroutiounian and M. Z. Ghulinyan, Electrical conductivity mechanisms in porous silicon, *Phys. Status Solidi A*, 2003, **197**(2), 462–466; (*c*) M. Bouaïcha, M. Khardani and B. Bessaïs, Evaluation of the electrical conductivity of nano-porous silicon from photoluminescence and particle size distribution, *Mater. Sci. Eng., B*, 2008, **147**(2–3), 235–238; (*d*) P. V. More, S. Jeong, J.-Y. Lee, Y.-H. Seo, B.-H. Ryu and Y. Choi, Printable electronics-compatible silicon nanoparticles prepared by the facile decomposition of SiS_2 and their application in a back-to-back Schottky diode, *J. Mater. Chem.*, 2012, **22**(44), 23553–23560.

20 G. H. Zhu, H. Lee, Y. C. Lan, X. W. Wang, G. Joshi, D. Z. Wang, J. Yang, D. Vashaee, H. Guilbert, A. Pillitteri, M. S. Dresselhaus, G. Chen and Z. F. Ren, Increased Phonon Scattering by Nanograins and Point Defects in Nanostructured Silicon with a Low Concentration of Germanium, *Phys. Rev. Lett.*, 2009, **102**(19), 196803.

21 N. Saito, K. Hayashi, H. Sugimura and O. Takai, Microstructured π-Conjugated Organic Monolayer Covalently Attached to Silicon, *Langmuir*, 2003, **19**(26), 10632–10634.

22 R. K. Hiremath, B. G. Mulimani, M. K. Rabinal and I. M. Khazi, Electrical characterization of a phenylacetylene-modified silicon surface *via* mercury probe, *J. Phys.: Condens. Matter*, 2007, **19**(44), 446003.

23 J. A. Kelly, A. M. Shukaliak, M. D. Fleischauer and J. G. C. Veinot, Size-dependent reactivity in hydrosilylation of silicon nanocrystals, *J. Am. Chem. Soc.*, 2011, **133**(46), 19015–19016.

24 (*a*) M. P. Stewart and J. M. Buriak, Exciton-Mediated Hydrosilylation on Photoluminescent Nanocrystalline Silicon, *J. Am. Chem. Soc.*, 2001, **123**(32), 7821–7830; (*b*) J. A. Kelly and J. G. C. Veinot, An Investigation into Near-UV Hydrosilylation of Freestanding Silicon Nanocrystals, *ACS Nano*, 2010, **4**(8), 4645–4656.

25 (*a*) R. K. Baldwin, K. A. Pettigrew, J. C. Garno, P. P. Power, G.-y. Liu and S. M. Kauzlarich, Room Temperature Solution Synthesis of Alkyl-Capped Tetrahedral Shaped Silicon Nanocrystals, *J. Am. Chem. Soc.*, 2002, **124**(7), 1150–1151; (*b*) K. A. Pettigrew, Q. Liu, P. P. Power and S. M. Kauzlarich, Solution synthesis of alkyl- and alkyl/alkoxy-capped silicon nanoparticles *via* oxidation of Mg_2Si, *Chem. Mater.*, 2003, **15**(21), 4005–4011.

26 (*a*) J. G. C. Veinot, Synthesis, surface functionalization, and properties of freestanding silicon nanocrystals, *Chem. Commun.*, 2006, 4160–4168; (*b*) J. M. Buriak, Organometallic chemistry on silicon and germanium surfaces, *Chem. Rev.*, 2002, **102**(5), 1271–1308.

27 S. Ashby, J. Thomas, P. Coxon, M. Bilton, R. Brydson, T. Pennycook and Y. Chao, The effect of alkyl chain length on the level of capping of silicon nanoparticles produced by a one-pot synthesis route based on the chemical reduction of micelle, *J. Nanopart. Res.*, 2013, **15**(2), 1425.

28 V. K. Zaitsev, M. I. Fedorov, E. A. Gurieva, I. S. Eremin, P. P. Konstantinov, A. Y. Samunin and M. V. Vedernikov, Highly effective $Mg_2Si_{1-x}Sn_x$ thermoelectrics, *Phys. Rev. B: Condens. Matter Mater. Phys.*, 2006, **74**(4), 045207.

29 G. Golan, A. Axelevitch, B. Gorenstein and V. Manevych, Hot-probe method for evaluation of impurities concentration in semiconductors, *Microelectron. J.*, 2006, **37**(9), 910–915.

30 (*a*) P. S. Gaal; S. P. Apostolescu, *US Pat.* 6375349, 2002; (*b*) M.-A. Thermitus and P. S. Gaal, in *Proceedings of the 24th International Thermal Conductivity Conference*, 24th International Thermal Conductivity Conference, Lancaster, ed. C. Uher and D. Morelli, Technomic Publishing Co., Lancaster, 1997, p. 340.

31 (*a*) Y. Chao, Q. Wang, A. Pietzsch, F. Hennies and H. Ni, Soft X-ray induced oxidation on acrylic acid grafted luminescent silicon quantum dots in ultrahigh vacuum, *Phys. Status Solidi A*, 2011, **208**(10), 2424–2429; (*b*) P. R. Coxon, S. P. Ashby, M. D. Frogley and Y. Chao, Thermal evaporation and X-ray photostability of dodecyl-passivated silicon nanoparticles, *J. Phys. D: Appl. Phys.*, 2012, **45**(35), 355303.

32 (*a*) P. G. Klemens, Thermal conductivity of inhomogeneous media, *High Temp. – High Pressures*, 1991, **23**, 241–248; (*b*) K. W. Schlichting, N. P. Padture and P. G. Klemens, Thermal conductivity of dense and porous yttria-stabilized zirconia, *J. Mater. Sci.*, 2001, **36**(12), 3003–3010.

33 A. I. Hochbaum, R. K. Chen, R. D. Delgado, W. J. Liang, E. C. Garnett, M. Najarian, A. Majumdar and P. D. Yang, Enhanced thermoelectric performance of rough silicon nanowires, *Nature*, 2008, **451**(7175), 163–U5.

34 A. I. Boukai, Y. Bunimovich, J. Tahir-Kheli, J. K. Yu, W. A. Goddard and J. R. Heath, Silicon nanowires as efficient thermoelectric materials, *Nature*, 2008, **451**(7175), 168–171.

Faraday Discussions

PAPER

Enhanced oxygen evolution activity by NiO_x and $Ni(OH)_2$ nanoparticles

Lucas-Alexandre Stern and Xile Hu*

Received 2nd June 2014, Accepted 5th August 2014
DOI: 10.1039/c4fd00120f

Small nanoparticles of nickel oxide and nickel hydroxide were investigated for the oxygen evolution reaction (OER) in alkaline solutions. These nanoparticles showed excellent catalytic activity, giving a current density of 10 mA cm^{-2} at overpotentials (η) of 300 and 330 mV for $Ni(OH)_2$ and NiO_x, respectively. The catalytic activity was comparable to that of the best mixed metal oxide catalysts under the same conditions. The activity was significantly higher than that of electrochemically formed $Ni(OH)_2$ and NiO_x films, demonstrating the benefit of nanostructuring. Pretreatment of the working electrode was important for the activity and stability of the catalyst layer on the electrode.

1 Introduction

The interest in renewable energy technologies based on solar and wind has grown considerably in the past decades. However, solar and wind energies are intermittent and require efficient storage methods for their wide implementation.[1,2] A possible large-scale storage method is the conversion of the electrical energy from solar and wind into the chemical energy stored in H_2 *via* the electrochemical water splitting reaction. The water splitting reaction consists of two half reactions, namely, the hydrogen evolution reaction (HER) and the oxygen evolution reaction (OER). The OER is the more challenging reaction of the two half reactions as it is a 4 electron and 4 proton process. The overpotential loss due to the sluggish OER constitutes the larger component of the overall overpotential loss for water electrolysis.[3] The development of efficient and cost-effective OER electrocatalysts is therefore an important task in renewable energy research.

A large number of binary metal oxides and ternary mixed metal oxides have been identified as good OER catalysts in alkaline conditions. Among them, nickel oxide and nickel hydroxide are the archetypical catalysts.[4] They have been extensively studied and are employed in alkaline electrolyzers.[5] Furthermore, many of the more advanced mixed oxide catalysts are essentially metal promoted nickel hydroxides. A recent benchmark study showed that NiO_x would require an

Laboratory of Inorganic Synthesis and Catalysis, Institute of Chemical Sciences and Engineering, Ecole Polytechnique Fédérale de Lausanne (EPFL), EPFL-ISIC-LSCI, BCH 3305, Lausanne, CH 1015, Switzerland. E-mail: xile.hu@epfl.ch; Fax: +41 216939305; Tel: +41 216939781

overpotential of more than 420 mV to reach 10 mA cm^{-2} in 1 M NaOH, while other studies have reported higher activity for NiO_x.[6]

As part of our efforts to develop OER catalysts for sunlight-driven water splitting, we decided to reinvestigate the catalytic activity of NiO_x and $Ni(OH)_2$. In particular, we sought to improve their activity using nanostructuring, as previous studies focused either on electrochemically deposited amorphous films or ill-characterized particles. Herein, we show that ultrafine nanoparticles of NiO_x and $Ni(OH)_2$ indeed have significantly higher OER activity than previously reported for bulk samples.

2 Experimental

2.1 Materials

The commercially available materials were used as received. Nickel oxide nanoparticles (NiO 99.8% trace metal basis, <50 nm, Aldrich), nickel chloride ($NiCl_2\cdot 6H_2O$, ReagentPlus®, Aldrich), nickel nitrate ($Ni(NO_3)_2\cdot 6H_2O$ purum p.a. ≥97%, Aldrich), nickel foil (thickness 0.5 mm, 50 × 50 mm, 11.2 g, 99.98% trace metal basis, Aldrich), hydrazine monohydrate ($N_2H_4\cdot H_2O$, reagent grade, 98%, Aldrich), sodium hydroxide (NaOH purum p.a., Reactolab), sulfuric acid (H_2SO_4 95–97%, Merck Millipore), ethanol (absolute alcohol without additive ≥99.8%, Aldrich), Nafion® (117 solution, Aldrich), 2-propanol (HPLC grade, ≥99.8%). The water used was first purified using a Millipore Milli-Q® Integral water purification system (18.2 MΩ cm resistivity).

2.2 Synthesis of ultrafine nickel hydroxide nanoparticles

The synthesis was performed following a previous report.[7] 1.19 g of nickel chloride was dissolved in 50 mL of distilled water in a 250 mL volumetric flask. 200 mL of absolute ethanol was added to form a green solution, which was transferred into a 1 L flask. A second solution was prepared as follows. First, 0.4 g of sodium hydroxide was dissolved in 100 mL of distilled water, then 1.5 mL of hydrazine monohydrate and 100 mL of deionized water were subsequently added. The second solution was added to the first one using a syringe pump at a rate of 0.2 mL per minute and the mixture was stirred. The colour of the resulting solution changed from green to cloudy green to cloudy blue to grey blue as the pH of the solution mixture increased. The addition of the second solution was stopped when the solution mixture reached a pH of ~10. A nanoparticle precipitate was collected from the reaction mixture by centrifugation. The nanoparticles were washed with distilled water and absolute ethanol and then freeze dried.

2.3 Physical characterization

X-ray diffraction (XRD) and grazing incidence X-ray diffraction (GIXRD) measurements were carried out on a X'Pert Philips diffractometer in Bragg–Brentano geometry with Cu $K_{\alpha 1}$ radiation ($\lambda = 0.1540$ nm) and a fast Si-PIN multistrip detector. The tube source was operated at 45 kV and 40 mA. The Scherrer equation was used to calculate the average crystallite size for the powders. The diffraction patterns were analysed and compared to references in the international center of diffraction data (ICDD). To enhance the signals from the samples and suppress the signals from the transparent conductive oxide (TCO) substrates, the X-ray incident angle to the sample surface was fixed at $\alpha \sim 0.1°$. This angle is below the NiO_x critical angle ($\alpha_c = 0.35°$). By the evanescent wave effect, the X-ray penetration depth is reduced to 2 nm.[8] Raman spectra were obtained on a LabRAM HR (HORIBA Jobin Yvon) instrument using a laser excitation of 633 nm. Scanning electron microscopy (SEM) images were taken on a FEI XLF-30 microscope. Transmission electron microscopy (TEM) images were taken on a Philips FEI CM12 instrument with a LaB_6 source operated at a 120 kV accelerating voltage. High resolution TEM (HRTEM) images were taken on a FEI Tecnai Osiris instrument equipped with an 11 Megapixel Gatan Orius CCD camera. Specimens were prepared by ultrasonic dispersion of the samples in ethanol. The slurries were mixed with a micropipette for several suction-release cycles to ensure representative and reproducible TEM samples. A few drops of the mixed suspensions were deposited onto the carbon-coated grids.

2.4 Electrochemical measurements

Electrochemical measurements were recorded using a Gamry Instruments Reference 3000™ potentiostat. A traditional three-electrode configuration was used. For polarization and electrolysis measurements, a platinum wire was used as the auxiliary electrode and a double-junction Ag/AgCl (KCl saturated) electrode was used as the reference electrode. Both the counter and reference electrodes were rinsed with distilled water and dried with compressed air prior to the measurements. Potentials were referenced relative to a reversible hydrogen electrode (RHE) by adding a value of $(0.2 + 0.059 \times \text{pH})$ V. The ohmic drop was corrected using the current interrupt method. A total electrolyte volume of ~50 mL was used to fill the glass cell. For experiments requiring strict Fe-free conditions, several precautions were taken: the electrolyte was purified by removing iron impurities, the glass cell was replaced with a polypropylene beaker, and all electrochemical cell components were cleaned with 1 M H_2SO_4 and rinsed with distilled water prior to the measurements. All of the potentials were converted and referenced relative to the RHE unless stated otherwise. The basic electrolyte, 1 M KOH, which was used throughout all of the electrochemical experiments, was prepared from a standard solution (Titrisol®, Merck Millipore) using distilled water. During the electrochemical experiments, the electrolyte was agitated using a magnetic stirrer rotating at 300 rpm. For all of the electrochemical experiments, a glassy carbon electrode (~0.071 cm^2) was used as the working electrode, except for the bulk NiO_x sample, where a bare nickel electrode (~0.031 cm^2) was used as the support and the precursor for the NiO_x catalyst. For the physical characterization of electrochemically treated samples, fluorine-doped tin oxide (FTO) was used as a substrate, except for the bulk NiO_x sample, where a bare nickel foil was used as the support and the precursor for NiO_x. The nickel foil was dipped in 1 M KOH at 25 °C and a constant potential of 1 V *vs.* RHE was applied for 3 hours. The cyclic voltammetry (CV) experiments were conducted in 1 M KOH at 25 °C using a scan rate of 5 mV s^{-1} across a potential window of 1.2–1.65 V *vs.* RHE. For the stability measurements, chronopotentiometric experiments were performed in 1 M KOH at 25 °C, applying a fixed current density of 10 mA cm^{-2} for 15 hours. Prior to the galvanostatic measurements, the materials were cycled 20 times using CV to fully activate the catalysts. The nanomaterials deposited on FTO were subjected to different potentiostatic experiments in 1 M KOH at 25 °C: either 1.3 V *vs.* RHE for one hour, or 1.5 V *vs.* RHE overnight. The roughness factor was determined from the electrochemically active surface area (ECSA) and the geometric area of the working electrode. The ECSAs of the different materials were calculated from cyclic voltammograms at different sampling rates (0.02, 0.04, 0.06, 0.08, 0.1 and 0.12 V s^{-1}) in 1 M KOH electrolyte at 25 °C at a potential window of 0.9 to 1.1 V *vs.* RHE where no faradaic processes occur.

2.5 Pretreatment of the working electrode

Prior to the loading of the catalyst, the working electrode was pretreated to achieve better performance. First, the electrode was manually polished using alumina powder (CH Instruments, Inc.) with decreasing grain sizes (typically 1, 0.3 and 0.05 μm) on a 73 mm diameter nylon polishing pad (MasterTex, Buehler). Between each polishing step, the electrode was rinsed with deionized water and ultrasonicated in distilled water for 10 seconds. Then, the electrode was dried using compressed air, ultrasonicated in absolute ethanol for 10 seconds, and dried again with compressed air. This polishing process resulted in a shiny mirror finish. Subsequently, the polished electrode was pretreated electrochemically. For this purpose, the bare working electrode underwent 20 CV cycles at a potential window of 0.5 to 1.9 V *vs.* RHE at a scan rate of 0.1 V s^{-1} in 1 M KOH at 25 °C. Finally, the pretreated electrode was rinsed with absolute ethanol and dried with compressed air.

2.6 Electrode preparation

The catalysts were loaded onto a pretreated working electrode *via* drop-casting 10 μL of catalyst ink, equivalent to a loading of 142 μg cm^{-2}. The catalyst suspension was a 1 mL solution consisting of 1 mg of catalyst, 4 μL of Nafion®, 200 μL of 2-propanol, and 796 μL of distilled water. The slurries were ultrasonicated for 5 hours and mixed with a micropipette for several suction–release cycles prior to deposition to ensure representative and reproducible catalyst samples. The temperature of the ultrasonic bath was kept below 45 °C at all times to avoid any undesired heating effects. Once the catalyst was deposited, the electrode was dried in an oven at 50 °C for 10 minutes. For the NiO_x bulk sample (film), a bare pretreated nickel working electrode was used for the measurements. The bulk $Ni(OH)_2$ sample (film) was electrochemically deposited onto the glassy carbon surface following a literature procedure.[9] To be specific, the pretreated electrode was rinsed with absolute ethanol, dried with compressed air, and dipped in a 0.1 M nickel nitrate aqueous solution. Then, a cathodic current density of 16 mA cm^{-2} was applied for 10 seconds. On FTO, the aforementioned process produces only a thin film of material. Thicker films on FTO were produced at a constant potential of −1 V *vs.* RHE for 30 minutes in 0.1 M $Ni(NO_3)_2$. The thicker films were used for the physical characterization of the electrodeposited bulk $Ni(OH)_2$ film.

2.7 Electrolyte purification

The removal of Fe impurities in the commercial 1 M KOH electrolyte was conducted according to a previous report.[10] 2 g of $Ni(NO_3)_2 \cdot 6H_2O$ was added to a 50 mL polypropylene centrifuge tube cleaned with H_2SO_4. This solid was dissolved in 4 mL distilled water. 20 mL of 1 M KOH was then added to precipitate

$Ni(OH)_2$. The mixture was shaken vigorously, centrifuged and the supernatant was decanted. The $Ni(OH)_2$ was subjected to three washing cycles. In each cycle, 20 mL of distilled water and 2 mL of 1 M KOH were added to the centrifuge tube containing $Ni(OH)_2$, which was then dispersed *via* shaking and ultrasonication; the solution mixture was next centrifuged and the supernatant was decanted. This procedure was repeated twice more in order to give high-purity $Ni(OH)_2$. Finally, 50 mL of 1 M KOH was added to this tube containing cleaned $Ni(OH)_2$. The mixture was dispersed by ultrasonication and agitated for 10 minutes. The suspension was allowed to rest for 3 hours. The mixture was then centrifuged and the Fe-depleted KOH solution was decanted and stored in a polypropylene bottle cleaned with H_2SO_4.

3 Results and discussion

3.1 Physical characterization

XRD measurements were conducted on both NiO_x and $Ni(OH)_2$ nanoparticles (Fig. 1a and b). The average crystallite size was calculated using Scherrer's equation:

$$d = \frac{K\lambda}{\beta \cos(\theta)} \tag{1}$$

where K is the shape factor, λ is the X-ray wavelength, β is the full width at half maximum, θ is the Bragg angle and d is the crystallite size. The diffraction pattern of $Ni(OH)_2$ corresponds to α-$Ni(OH)_2$ (red sticks, Fig. 1a). The average crystallite size of $Ni(OH)_2$ is 2.3 nm. The small crystallite size caused a significant broadening of the diffraction peaks. The position of the (006) peak reveals the length of the *c*-axis. The $Ni(OH)_2$ nanoparticles have a larger *d*-spacing along the *c* direction than the standard sample in the database.[11] According to Bode's diagram, the two allotropic forms of $Ni(OH)_2$ (*i.e.*, α-$Ni(OH)_2$ and β-$Ni(OH)_2$) can be oxidized to two distinctive NiOOH forms (*i.e.*, γ-NiOOH and β-NiOOH, respectively).[12] β-NiOOH has a smaller intersheet spacing than γ-NiOOH and α-$Ni(OH)_2$. This elongation is therefore attributed to the partial oxidation of α-$Ni(OH)_2$ to γ-NiOOH.[10] The diffraction pattern of the NiO_x nanoparticles contains characteristic peaks of nickel oxide (purple sticks, Fig. 1b). The average particle size was 6 nm. GIXRD patterns for the bulk $Ni(OH)_2$ film (Fig. 1c) and bulk NiO_x film (Fig. 1d) were also acquired. The broad patterns are due to the amorphous nature of the materials. The presence of different nickel oxides was identified by comparison to standard patterns in the database. Notably, NiOOH was present on both $Ni(OH)_2$ and NiO_x, indicating partial air oxidation.

SEM images of the bulk $Ni(OH)_2$ and bulk NiO_x films are shown in Fig. 2a and d. The bulk $Ni(OH)_2$ film has a surface layer made of dense nanoparticles (Fig. 2a). When the top layer was peeled off, a nanoporous structure was observed (Fig. 2b). The assembly of dense nanoparticles on top of nanostructured layers was also observed from a cross section image of the bulk $Ni(OH)_2$ film (Fig. 2c). The SEM image of the bulk NiO_x film shows a layer of particles aggregated on the nickel foil surface (Fig. 2d). TEM images of the $Ni(OH)_2$ and NiO_x nanoparticles are shown in Fig. 3a and d. The particle sizes observed by TEM are in agreement with the crystallite sizes calculated from the XRD patterns. The $Ni(OH)_2$ particles were well distributed on the carbon grid and only a few aggregates were visible. The selected

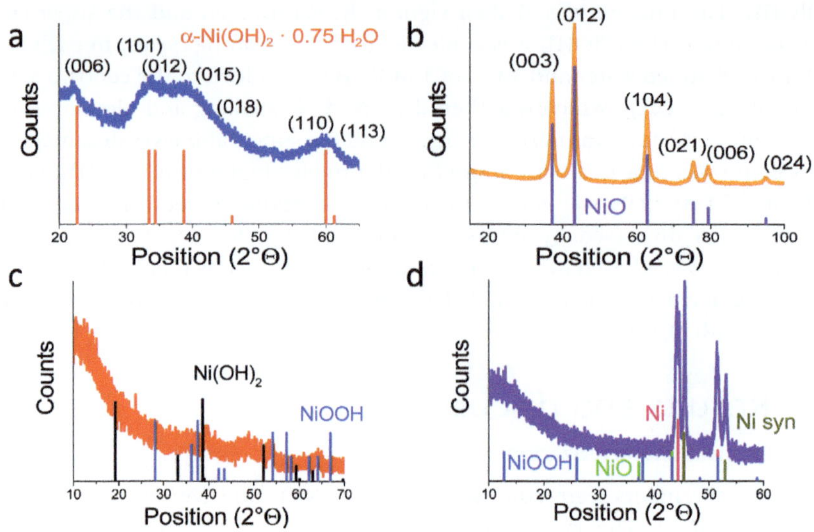

Fig. 1 (a) Powder X-ray diffraction pattern of the $Ni(OH)_2$ nanoparticles. (b) Powder X-ray diffraction pattern of the NiO_x nanoparticles. (c) Grazing incidence X-ray diffraction pattern of the bulk $Ni(OH)_2$ film on FTO. (d) Grazing incidence X-ray diffraction pattern of the bulk NiO_x film on Ni foil. The stick patterns correspond to the diffraction peaks of standard samples in the ICDD database, *i.e.*, Ni (ICDD 00-001-1258), Ni syn (ICDD 04-013-4763), NiO (ICDD 04-006-1773 and 01-078-4359), $Ni(OH)_2$ (ICDD 04-012-5845), α-$Ni(OH)_2 \cdot 0.75H_2O$ (ICDD 00-038-0715) and NiOOH (ICDD 00-006-0075 and 00-027-0956). The low signal to noise ratio of the GIXRD line patterns is attributed to the amorphous nature of the materials.

area electron diffraction (SAED) patterns of the $Ni(OH)_2$ and NiO_x nanoparticles (Fig. 3b and e) are indicative of polycrystalline nanoparticles. Only faint rings were observed for the $Ni(OH)_2$ nanoparticles, likely due to their small size and the random orientation of many crystals. Nevertheless, the diffraction pattern could be indexed using the proper crystal lattices. The larger NiO_x nanoparticles gave a diffraction pattern with more resolved spots (Fig. 3e). The HRTEM image of the $Ni(OH)_2$ nanoparticles shows a homogenous morphology (Fig. 3c). The observation of lattice fringes on the surface suggests that the particles are entirely crystalline and the surface is not amorphous. As prolonged exposure to the intense HRTEM electron beam led to recrystallization of the sample, short exposure HRTEM images (Fig. 3c) were taken at the expense of resolution.

3.2 Electrochemical characterization

3.2.1 Electrochemical surface area and roughness factor. The electrochemical surface areas (ECSAs) of the catalysts were determined from their double-layer capacitance. The capacitance was determined by CV.[13] The area averaged double-layer capacitance (C_{Dl}) can be calculated using the relationship:

$$j = \nu C_{Dl} \tag{2}$$

where j is the capacitive current density and ν is the scan rate. The normalised capacitance (C_{Dl}) can be obtained from the linear dependence of the capacitive

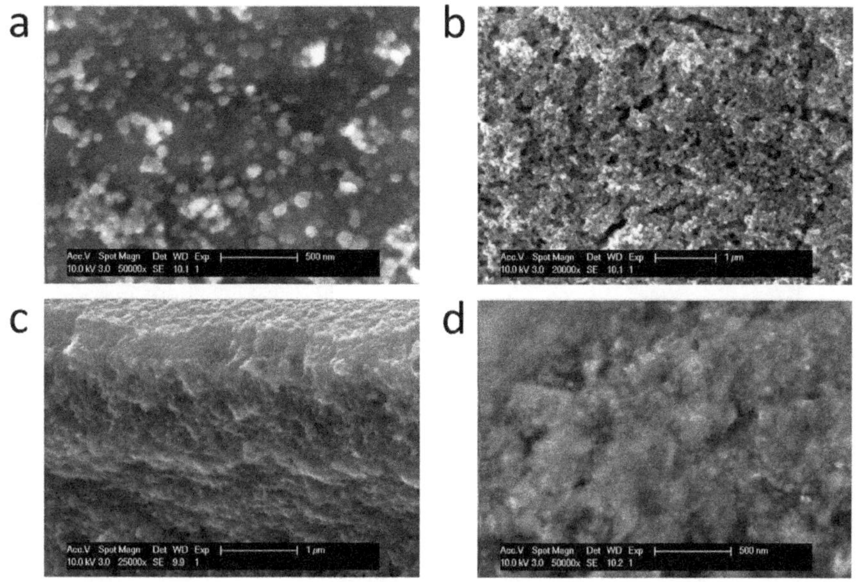

Fig. 2 SEM images of: (a) the as-prepared bulk $Ni(OH)_2$ film; (b) the $Ni(OH)_2$ film after the top layer was peeled off; (c) the cross section of the $Ni(OH)_2$ film; and (d) the bulk NiO_x film on a Ni foil.

current density (j) *vs.* the scan rate, where the slope of Δj *vs.* ν corresponds to twice C_{DI}.[14] The roughness factor (R) is then given as the ratio of the ECSA to the geometric area of the electrode (A), according to:

$$R = \frac{\mathrm{ECSA}}{A} = \frac{C_{\mathrm{DI}}}{C_{\mathrm{s}}} \tag{3}$$

where C_s is the specific capacitance of a standard, flat sample. The theoretical values of the specific capacitance are unavailable for NiO_x and $Ni(OH)_2$. However, for relative comparison, a general C_s value of 25 μF cm^{-2}, corresponding to the specific capacitance of a nickel electrode in 1 M NaOH was applied.[6] Fig. 4 displays the capacitive current densities and the corresponding linear plot for the determination of R for the bulk $Ni(OH)_2$ sample.

The roughness factors for the bulk $Ni(OH)_2$ and nanoparticles of $Ni(OH)_2$ were 12 and 16, respectively. The roughness factors for the bulk NiO_x and nanoparticles of NiO_x were 63 and 30 (Table 1).

3.2.2 **Pretreatment of the working electrode and its effect on the electrochemical activity.** Pretreatment of bare glassy carbon electrodes prior to electrochemical measurements is known to have important consequences on the outcome of the measurements. Many pretreatment procedures have been reported in the literature.[15] Sometimes electrochemical pretreatment of the bare working electrode is essential to obtain high reproducibility of data.[16] In this study, before the deposition of the catalysts, the bare working electrode was subjected to 20 CV cycles in 1 M KOH at 0.1 V s^{-1} from 0.5 to 1.9 V *vs.* RHE. This treatment was expected to change the surface properties of the electrode by: (1) forming oxygenated functional groups at the surface; (2) making the surface more

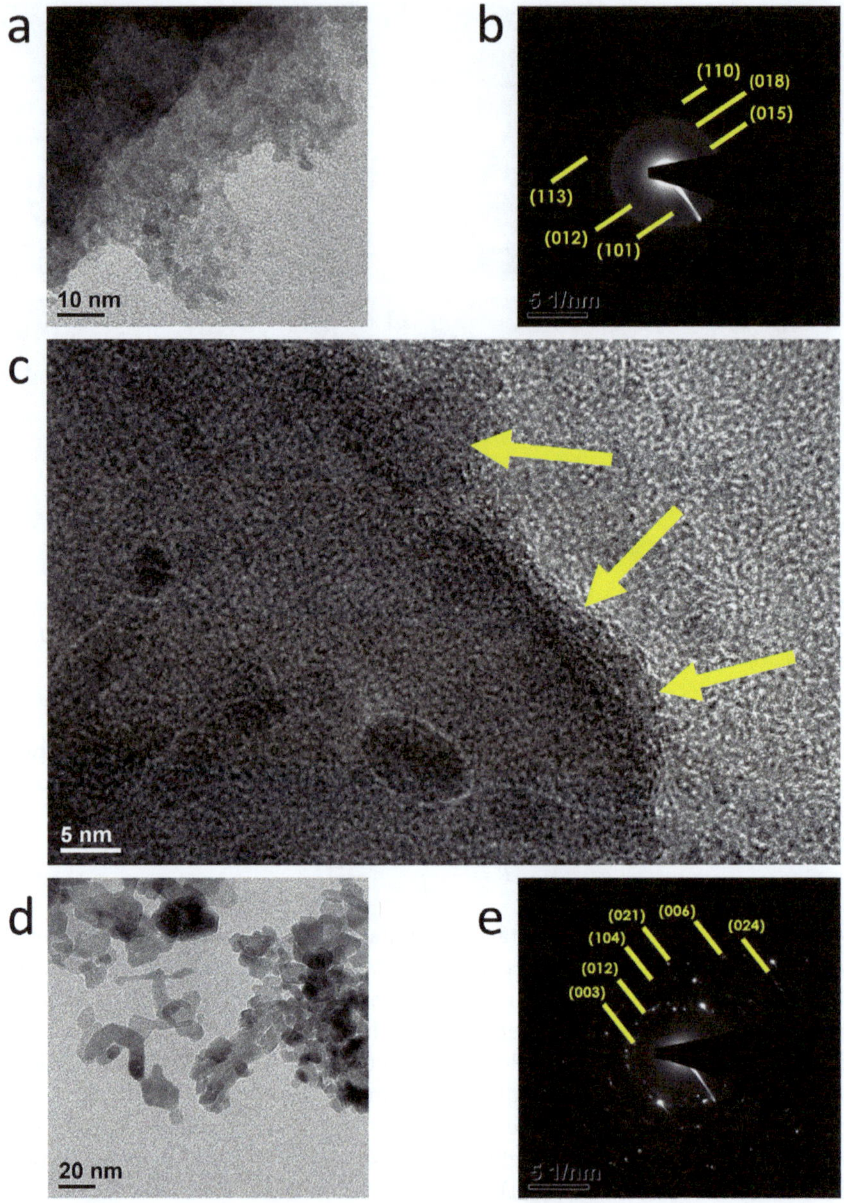

Fig. 3 (a) TEM image of the $Ni(OH)_2$ nanoparticles; (b) SAED patterns of the $Ni(OH)_2$ nanoparticles; (c) HRTEM image of the $Ni(OH)_2$ nanoparticles; selected regions where lattice registry is observed on the surface of the nanoparticles are indicated (yellow arrows); (d) TEM image of the NiO_x nanoparticles; and (e) SAED patterns of the NiO_x nanoparticles.

hydrophilic; and (3) wetting the surface.[17] Fig. 5 shows the effects of the pretreatment on the catalytic performance of both the bulk NiO_x film and fine NiO_x nanoparticles. The activity and stability of the bulk NiO_x film were improved by pretreatment (Fig. 5a), illustrated by the following: (1) the catalytic current was

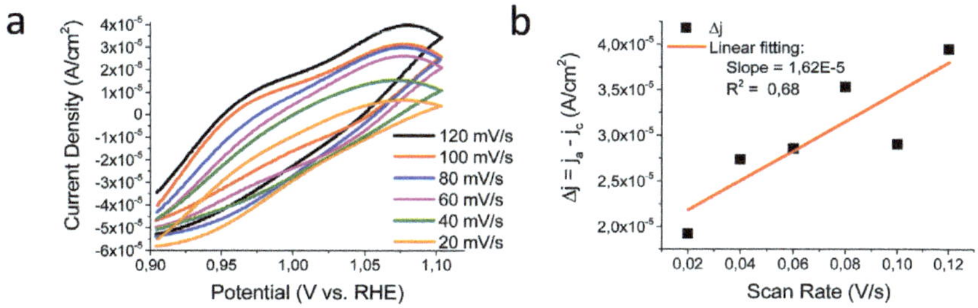

Fig. 4 (a) Scan rate dependent cyclic voltammograms for the bulk $Ni(OH)_2$ films at a potential window where no faradaic processes occur in 1 M KOH. (b) Linear plot of capacitive current density *vs.* scan rate. The value of the capacitive current is given as the difference between the anodic (j_a) and cathodic (j_c) current density. The calculated slope corresponds to twice the averaged double-layer capacitance.

higher in the first CV scan for the pretreated electrode; and (2) the maximum catalytic current was also slightly higher for the pretreated electrode. Similar observations were made for the NiO_x nanoparticles (Fig. 5b). Notably, the catalytic activity of the nanostructured NiO_x on an untreated electrode decayed rapidly, whereas the activity of the same catalyst on a pretreated electrode remained stable. For a pretreated electrode, high activity was obtained after several scanning cycles. On the other hand, for an untreated electrode, a large number of electrochemical cycles was required to reach the optimal activity, which was lower than that of the pretreated electrode. The origins of the increased activity and stability of the pretreated electrodes might be multi-fold. A more hydrophilic surface is advantageous for the deposition of catalysts, especially nanoparticles suspended in aqueous or alcoholic solutions. A higher wetting property facilitates the diffusion of catalysts into the pores of the electrode. The formation of oxygenated functional groups might lead to a stronger interaction between the catalyst and the electrode surface, thereby increasing the catalyst stability. Moreover, the oxygenated functional groups on the surface might transform a precatalyst into its activated form. It was shown that the predominant phase for nickel oxides under alkaline conditions is NiOOH.[18] NiO_x normally needs long and tedious conditioning to be converted to the active $Ni(OH)_2$.[4,18] In the current case, after a short pretreatment of the electrode (10 min), the long conditioning process for NiO_x was no longer necessary. This was probably because the hydroxide functional groups on the electrode surface accelerated the conversion of NiO_x to $Ni(OH)_2$.

Table 1 OER activity and roughness factors of the catalysts

Sample	$\eta_{10\ \mathrm{mA\ cm^{-2}}}$ [mV]	$\eta_{20\ \mathrm{mA\ cm^{-2}}}$ [mV]	Tafel slope [mV dec^{-1}]	Roughness factor [—]
$Ni(OH)_2$ nanoparticles	299	313	53	16
$Ni(OH)_2$ bulk	331	347	47	12
NiO_x nanoparticles	331	347	54	30
NiO_x bulk	327	342	50	63

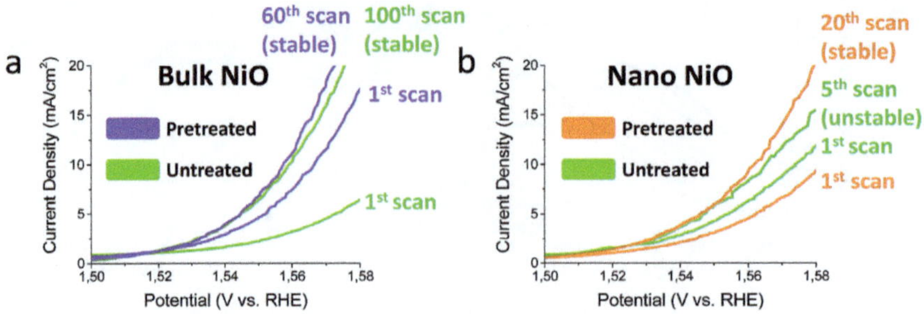

Fig. 5 Cyclic voltammograms in 1 M KOH at a scan rate of 5 mV s^{-1} for the pretreated and untreated working electrodes for (a) the bulk NiO_x films and (b) the NiO_x nanoparticles. For clarity, only the first and highest cathodic curves are shown.

Given the benefits of the pretreatment of the electrodes on the activity and stability of NiO_x, the same pretreatment was conducted before the activity of $Ni(OH)_2$ was measured, even though the effects of the pretreatment were not studied in detail for this catalyst.

3.2.3 Oxygen evolution activity. Fig. 6 shows the cyclic voltammograms for the bulk $Ni(OH)_2$ and NiO_x catalysts in 1 M KOH. The bulk $Ni(OH)_2$ catalyst was prepared by cathodic electrodeposition using a nickel nitrate solution. The bulk NiO_x catalyst was formed *in situ* during the CV measurements using a pretreated nickel electrode. Both catalysts exhibited a small redox wave attributed to a Ni(II)/Ni(III) redox process at potentials lower than the oxygen evolution potentials. The current densities at >1.5 V *vs.* RHE are proportional to the oxygen evolution activity. For the bulk $Ni(OH)_2$ catalyst, consecutive CV scans increased the activity slightly, and the maximum activity was reached after 5 CV cycles. We propose that the as-deposited $Ni(OH)_2$ was quickly converted to NiOOH, the active phase of this catalyst in the OER.[9,18] Previously it was reported that an overpotential of 420 mV was required for this catalyst to reach 10 mA cm^{-2}.[5,6,9,18–20] Higher activity has sometimes been reported in the literature, but it has normally required special experimental conditions, such as a high catalyst loading (>1 mg cm^{-2}),[21] strong alkaline conditions (6 M KOH),[22] or prolonged aging.[23] However, in the present case, high OER activity was obtained after only several CV scans. To reach 10 mA

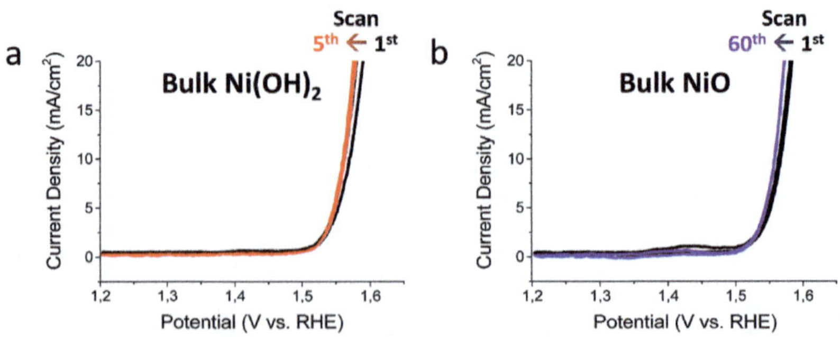

Fig. 6 Cyclic voltammograms for (a) the electrodeposited $Ni(OH)_2$ and (b) the bulk NiO_x in 1 M KOH at a sweeping rate of 5 mV s^{-1}.

cm^{-2} and 20 mA cm^{-2} required overpotentials of 330 mV and 350 mV, respectively. The Tafel slope was 47 mV dec^{-1}. The higher activity of the bulk $Ni(OH)_2$ reported here might be due to the pretreatment of the electrode as described above.

NiO_x was previously reported to require an overpotential of 420 mV for a current density of 10 mA cm^{-2}.[4,24–26] In the present case, overpotentials of 330 mV and 340 mV were necessary to drive current densities of 10 mA cm^{-2} and 20 mA cm^{-2}, respectively. Hours of catalyst conditioning which were previously required were no longer necessary.[2,18,25,26] A Tafel slope of 50 mV dec^{-1} was observed.

Fig. 7 shows the OER activity of the nanostructured $Ni(OH)_2$ and NiO_x particles in 1 M KOH. Recently, Yan *et al.* prepared nanostructured nickel hydroxide particles with different allotropic forms and observed that α-$Ni(OH)_2$ was catalytically more active than β-$Ni(OH)_2$; for α-$Ni(OH)_2$, an overpotential of 331 mV was required for 10 mA cm^{-2}.[10,11] This activity might be considered as the previous benchmark for $Ni(OH)_2$ nanoparticles. Fig. 7a shows that the fine α-$Ni(OH)_2$ nanoparticles synthesized in this study outperformed those in the previous reference. The overpotentials were 300 mV for 10 mA cm^{-2} and 310 mV for 20 mA cm^{-2}. The Tafel slope was 53 mV dec^{-1}. A rapid activation process was observed, where the catalytic activity increased during consecutive CV scans and reached the maximum efficiency at the 30th cycle. During this process, the Ni(II)/Ni(III) peak was shifted to lower potentials. The anodic peak at 1.39 V *vs.* RHE corresponds to the oxidation of α-$Ni(OH)_2$ to γ-NiOOH, the

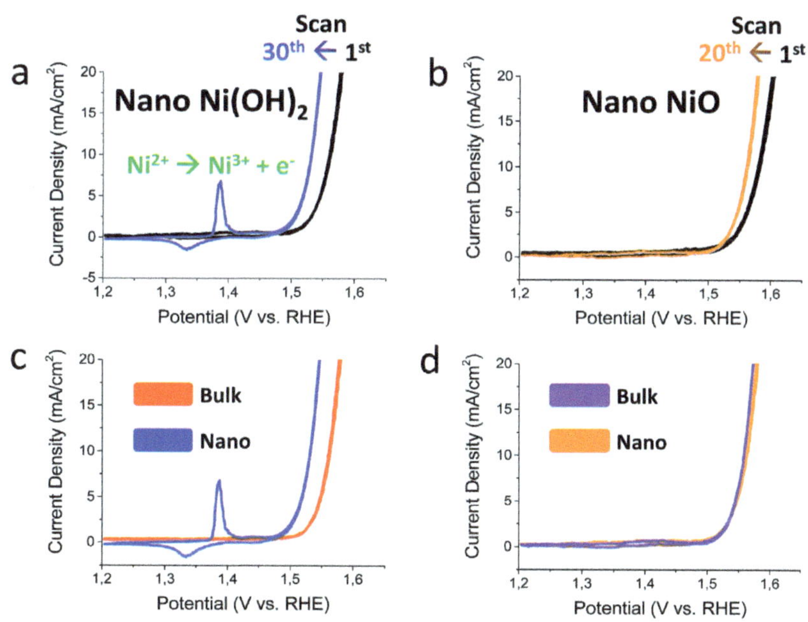

Fig. 7 Cyclic voltammograms for (a) the $Ni(OH)_2$ nanoparticles; (b) the NiO_x nanoparticles; (c) comparison of the OER activity of the bulk $Ni(OH)_2$ and nanostructured $Ni(OH)_2$; and (d) comparison of the stabilized OER activity of the bulk NiO_x and nanostructured NiO_x. The scans were performed in 1 M KOH at a sweeping rate of 5 mV s^{-1}.

active phase of the catalyst. The activity of this $Ni(OH)_2$ nanocatalyst was comparable to that of the best OER catalyst in alkaline media.[11]

The NiO_x-type catalyst has been previously reported to give 10 mA cm^{-2} at $\eta \approx$ 420 mV.[24,27] Fig. 7b shows the cyclic voltammograms for the NiO_x nanoparticles. A stable OER activity was obtained after 20 cyclic scans. The overpotentials to drive 10 mA cm^{-2} and 20 mA cm^{-2} were 330 mV and 350 mV, respectively. The Tafel slope was 54 mV dec^{-1}. Thus, the NiO_x nanoparticles presented here are more active than those which have been previously reported.

Fig. 7c compares the OER activity of the bulk $Ni(OH)_2$ and nanostructured $Ni(OH)_2$. The nanoparticles are significantly more active than the bulk film. The two catalysts have similar surface roughness factors (Table 1). Thus, the difference in activity could not be explained by a difference in the surface area. We propose two factors to explain this difference: (1) electrodeposition to make $Ni(OH)_2$ will result in a mixture of both α- and β-$Ni(OH)_2$, the latter being less active;[28] and (2) nanoparticles have a smaller size and a higher number of facets and edges for the same surface area. If OER activity occurs only at certain facets and edges of $Ni(OH)_2$, then a higher activity is expected for the fine nanoparticles.

Fig. 7d compares the OER activity of the bulk and nanostructured NiO_x catalysts. The two catalysts exhibit no significant difference in activity. The NiO_x bulk film is rougher than the NiO_x nanoparticles (Table 1); on the other hand, the nanoparticles have a smaller size. The compromise between the two factors results in a similar activity.

To further probe the nature of the catalytic active phase, the $Ni(OH)_2$ and NiO_x nanoparticles were deposited on FTO and subjected to Raman and GIXRD measurements after the samples were subjected to electrolysis at two potentials, 1.3 V and 1.5 V *vs.* RHE. At 1.3 V *vs.* RHE, the Ni(II) species is not oxidized to Ni(III), while at 1.5 V, oxidation should occur readily. The Raman spectra of the $Ni(OH)_2$ nanoparticles (Fig. 8a) and NiO_x nanoparticles (Fig. 8b) pre-conditioned at 1.3 V show no significant peaks due to either α-$Ni(OH)_2$ or β-$Ni(OH)_2$. This is likely due to the low Raman scattering cross section for these species, as also observed by Yeo and Bell.[26] When these nanoparticles were subjected to electrolysis at 1.5 V, the spectra changed substantially. The intensity of the Raman signals increases by several fold for both nanomaterials (Fig. 8). The main absorption feature was shifted towards a lower wavenumber. This Raman shift is consistent with the formation of NiOOH.[26] However, due to the small size of the nanoparticles, severe

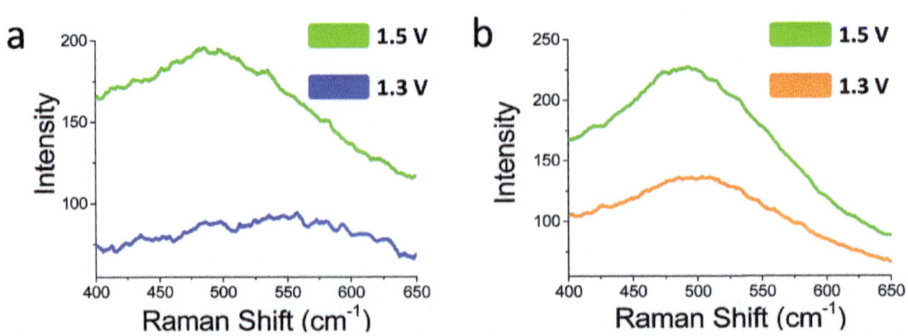

Fig. 8 (a) Raman spectra of the $Ni(OH)_2$ nanoparticles electrolyzed at 1.3 V and 1.5 V *vs.* RHE. (b) Raman spectra of the NiO_x nanoparticles electrolyzed at 1.3 V and 1.5 V *vs.* RHE.

line broadening was observed in the Raman spectra, so further analysis of the phases is impossible.

The formation of NiOOH at oxidative potentials was confirmed by GIXRD. For the $Ni(OH)_2$ nanoparticles preconditioned at 1.3 V, GIXRD indicates only different $Ni(OH)_2$ phases (Fig. 9a). After oxidation at 1.5 V, the conversion of these $Ni(OH)_2$ species to different NiOOH species was indicated by GIXRD (Fig. 9b). For the NiO_x nanoparticles, NiOOH was already present after preconditioning at 1.3 V (Fig. 9c). After oxidation at 1.5 V, the ratio of NiOOH increased largely (Fig. 9d). The formation of NiOOH at potentials just below the onset of oxygen evolution was further confirmed by the electrochromic properties of the materials. The reduced and oxidized states of the $Ni(OH)_2$/NiOOH redox couple exhibit distinct visible colours.[29] Photographs of $Ni(OH)_2$ deposited on FTO are shown in Fig. 10. The material was green at 1.3 V (Fig. 10a), but turned bronze/black upon oxidation at 1.5 V, which is characteristic for NiOOH (Fig 10b).[30]

As mentioned above, the catalytically active phase for the NiO_x and $Ni(OH)_2$ catalysts is nickel oxyhydroxide (NiOOH), which has a layered double hydroxide (LDH) structure.[10,18] The large intersheet spacing in the LDH structure allows the movement of electrons and hydroxide groups between the layers. Consequently, catalysis is not limited to the surface. However, the activity of the NiOOH-based catalysts in the OER might require an additional metal impurity such as Fe.[9,18] Recently, Boettcher *et al.* reported that a part per billion level of iron impurity was sufficient to improve the catalytic activity of $Ni(OH)_2$.[10] It is likely that the system

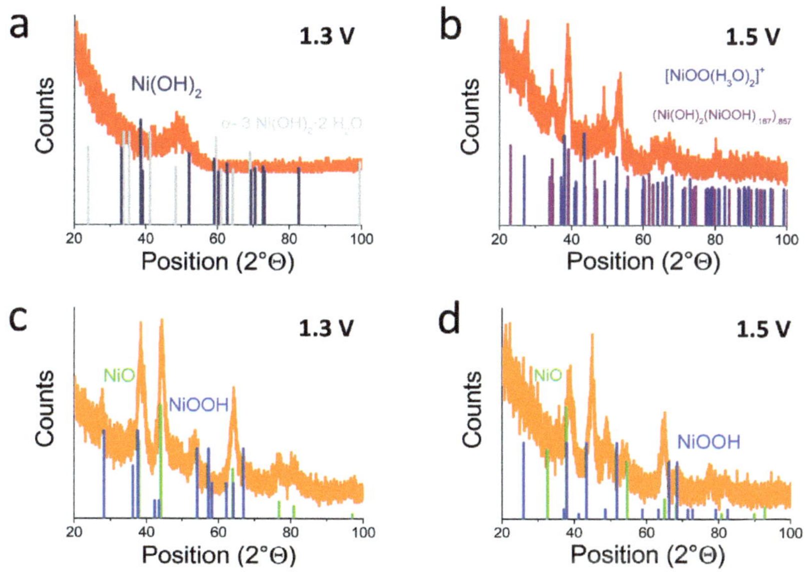

Fig. 9 Grazing incidence X-ray diffraction patterns of (a and b) the $Ni(OH)_2$ nanoparticles and (c and d) the NiO_x nanoparticles. The samples were pretreated electrochemically prior to measurements at specific voltage values (*vs.* RHE). The stick patterns correspond to the standard diffraction peaks of standard samples in the ICDD database, *i.e.*, NiO (ICDD 04-007-8202 and 04-006-6545), α-3 $Ni(OH)_2 \cdot 2H_2O$ (ICDD 00-022-0444), $Ni(OH)_2$ (ICDD 01-089-7111 and 00-014-0117), NiOOH (ICDD 00-027-0956 and 00-006-0075) and $[NiOO(H_3O)_2]^+$ (ICDD 01-073-2408).

Fig. 10 Photographs of the $Ni(OH)_2$ nanoparticles deposited on fluorine-doped tin oxide (FTO), illustrating the electrochromism of the material at different redox states. (a) The electrode preconditioned at 1.3 V. The transparent/light green colour is characteristic for $Ni(OH)_2$. (b) The electrode electrolyzed at 1.5 V; the dark colour corresponds to NiOOH. The red arrow indicates the part of the $Ni(OH)_2$ film that was not submerged in the electrolyte, which kept the green colour of $Ni(OH)_2$.

presented here contains iron impurities, which are ubiquitously present in commercial grade KOH. To confirm this hypothesis, the trace amount of Fe ions in the commercial KOH electrolyte was removed.[10] Fig. 11 compares the OER performance of the $Ni(OH)_2$ nanoparticles in a commercial KOH solution and a Fe-depleted KOH solution. Removal of adventitious iron resulted in a significant decrease of the catalytic activity and the stability of NiOOH. Thus, the Fe impurities were essential for the high activity of the catalysts presented in this work. The active phase, NiOOH, is transformed into $Ni_xFe_{1-x}OOH$ while maintaining its

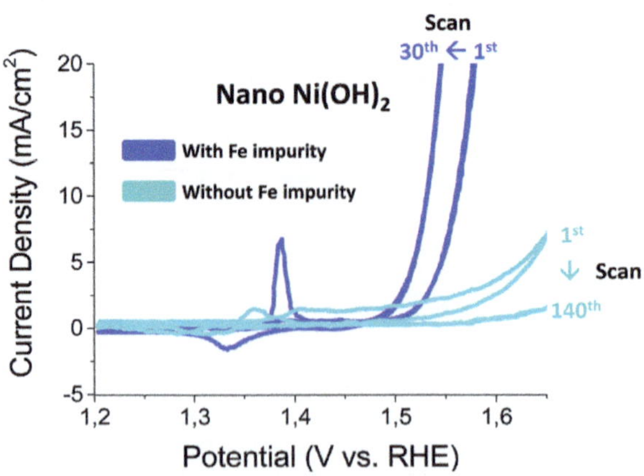

Fig. 11 Cyclic voltammograms collected for the $Ni(OH)_2$ nanoparticles with a scan rate of 5 mV s^{-1}. The difference in activity changes according to the electrolyte used: commercial grade 1 M KOH (blue curves) and purified and iron-free 1 M KOH (cyan curves).

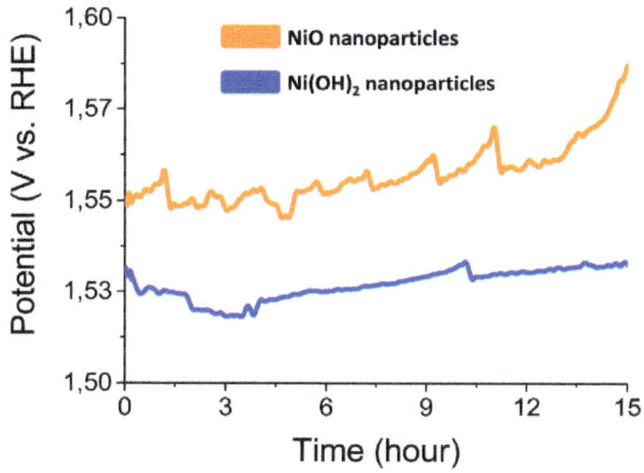

Fig. 12 Chronopotentiometric measurements for the $Ni(OH)_2$ (blue curve) and NiO_x (orange curve) nanoparticles over 15 hours at a constant current density of 10 mA cm^{-2} in 1 M KOH.

LDH structure. It was reported that $Ni_{0.9}Fe_{0.1}OOH$ was a benchmark for the OER in alkaline media, giving 10 mA cm^{-2} at $\eta = 287$ mV.[18] The simple $Ni(OH)_2$ nanoparticles in combination with the adventitious Fe ions already shows a similar activity. The role of Fe is not yet fully understood. It was shown that bimetallic $NiFeO_x$ was a better OER catalyst than FeO_x and NiO_x alone, thus, there is a synergetic effect between the two metals.[31] Fe-doping might enhance the conductivity of the catalyst, and/or induce partial charge transfer to activate the Ni centers.[10] The addition of Fe might promote and/or stabilize the formation of NiOOH with a LDH type of structure, which is beneficial for oxygen evolution.[32,33]

3.2.4 Electrochemical stability. Chronopotentiometric experiments were performed to test the stability of the NiO_x and $Ni(OH)_2$ nanoparticles. A constant current density of 10 mA cm^{-2} was applied and the evolution of the overpotential was recorded over 15 hours. Fig. 12 shows the results. The $Ni(OH)_2$ nanoparticles exhibited higher stability than NiO_x. After 15 h, the overpotential had increased by 5 mV for $Ni(OH)_2$, and 40 mV for NiO_x. Overall, both catalysts have rather high stability.

4 Conclusion

In summary, fine nanoparticles of NiO_x and $Ni(OH)_2$ have been prepared, characterized, and tested for the electrochemical oxygen evolution reaction in an alkaline medium. The activity was compared to that of bulk NiO_x and $Ni(OH)_2$ samples. The $Ni(OH)_2$ nanoparticles are more active than bulk $Ni(OH)_2$, while the NiO_x nanoparticles have a similar activity to bulk NiO_x. The higher activity of the $Ni(OH)_2$ nanoparticles is attributed to their ultrafine size, which results in a higher number of active facets and edges. The simple $Ni(OH)_2$ nanocatalyst gives 10 mA cm^{-2} at an overpotential of only 300 mV, and is one of the best catalysts reported to date. It was also found that pretreatment of the electrode has an important influence on the activity and stability of these Ni-based electrocatalysts.

Acknowledgements

This work was supported by a starting grant from the European Research Council under the European Community's Seventh Framework Programme (FP7 2007–2013)/ERC Grant Agreement 257096.

References

1 M. W. Kanan and D. G. Nocera, *Science*, 2008, **321**, 1072–1075.

2 L. Trotochaud, J. K. Ranney, K. N. Williams and S. W. Boettcher, *J. Am. Chem. Soc.*, 2012, **134**, 17253–17261.

3 J. Suntivich, K. J. May, H. A. Gasteiger, J. B. Goodenough and Y. Shao-Horn, *Science*, 2011, **334**, 1383–1385.

4 Y. Miao, L. Ouyang, S. Zhou, L. Xu, Z. Yang, M. Xiao and R. Ouyang, *Biosens. Bioelectron.*, 2014, **53**, 428–439.

5 M. Merrill, M. Worsley, A. Wittstock, J. Biener and M. Stadermann, *J. Electroanal. Chem.*, 2014, **717–718**, 177–188.

6 C. C. L. McCrory, S. Jung, J. C. Peters and T. F. Jaramillo, *J. Am. Chem. Soc.*, 2013, **135**, 16977–16987.

7 J. Tientong, S. Garcia, C. R. Thurber and T. D. Golden, *J. Nanotechnol.*, 2014, **2014**, e193162.

8 M. D. Irwin, J. D. Servaites, D. B. Buchholz, B. J. Leever, J. Liu, J. D. Emery, M. Zhang, J.-H. Song, M. F. Durstock, A. J. Freeman, M. J. Bedzyk, M. C. Hersam, R. P. H. Chang, M. A. Ratner and T. J. Marks, *Chem. Mater.*, 2011, **23**, 2218–2226.

9 D. A. Corrigan and R. M. Bendert, *J. Electrochem. Soc.*, 1989, **136**, 723–728.

10 L. Trotochaud, S. L. Young, J. K. Ranney and S. W. Boettcher, *J. Am. Chem. Soc.*, 2014, **136**, 6744–6753.

11 M. Gao, W. Sheng, Z. Zhuang, Q. Fang, S. Gu, J. Jiang and Y. Yan, *J. Am. Chem. Soc.*, 2014, **136**, 7077–7084.

12 M. Wehrens-Dijksma and P. H. L. Notten, *Electrochim. Acta*, 2006, **51**, 3609–3621.

13 S. Trasatti and O. A. Petrii, *Pure Appl. Chem.*, 1991, **63**, 711–734.

14 D. Merki, H. Vrubel, L. Rovelli, S. Fierro and X. Hu, *Chem. Sci.*, 2012, **3**, 2515–2525.

15 M. Gross and J. Jordan, *Pure Appl. Chem.*, 1984, **56**, 1095–1129.

16 R. C. Engstrom, *Anal. Chem.*, 1982, **54**, 2310–2314.

17 T. Nagaoka and T. Yoshino, *Anal. Chem.*, 1986, **58**, 1037–1042.

18 L. Trotochaud and S. W. Boettcher, *Scr. Mater.*, 2014, **74**, 25–32.

19 C. He, X. Wu and Z. He, *J. Phys. Chem. C*, 2014, **118**, 4578–4584.

20 P. W. T. Lu and S. Srinivasan, *J. Electrochem. Soc.*, 1978, **125**, 1416–1422.

21 D. E. Hall, *J. Electrochem. Soc.*, 1983, **130**, 317–321.

22 Y. Li, J. Yao, Y. Zhu, Z. Zou and H. Wang, *J. Power Sources*, 2012, **203**, 177–183.

23 C. Bocca, A. Barbucci and G. Cerisola, *Int. J. Hydrogen Energy*, 1998, **23**, 247–252.

24 M. P. Yeager, D. Su, N. S. Marinković and X. Teng, *J. Electrochem. Soc.*, 2012, **159**, A1598–A1603.

25 M. F. Kibria and M. S. Mridha, *Int. J. Hydrogen Energy*, 1996, **21**, 179–182.

26 B. S. Yeo and A. T. Bell, *J. Phys. Chem. C*, 2012, **116**, 8394–8400.

27 F. Chekin, H. Tahermansouri and M. R. Besharat, *J. Solid State Electrochem.*, 2014, **18**, 747–753.
28 R. S. Jayashree and P. V. Kamath, *J. Appl. Electrochem.*, 1999, **29**, 449–454.
29 R. M. Bendert and D. A. Corrigan, *J. Electrochem. Soc.*, 1989, **136**, 1369–1374.
30 M. Fantini and A. Gorenstein, *Sol. Energy Mater.*, 1987, **16**, 487–500.
31 D. A. Corrigan, *J. Electrochem. Soc.*, 1987, **134**, 377–384.
32 F. Song and X. Hu, *Nat. Commun.*, 2014, **5**, 4477, DOI: 10.1038/ncomms5477.
33 M. Gong, Y. Li, H. Wang, Y. Liang, J. Z. Wu, J. Zhou, J. Wang, T. Regier, F. Wei and H. Dai, *J. Am. Chem. Soc.*, 2013, **135**, 8452–8455.

Faraday Discussions

PAPER

Mechanisms for CO oxidation on Fe(III)–OH–Pt interface: a DFT study

Yun Zhao, Guangxu Chen, Nanfeng Zheng* and Gang Fu*

Received 1st July 2014, Accepted 10th July 2014
DOI: 10.1039/c4fd00144c

The full catalytic cycle that involves the oxidation of two CO molecules is investigated here by using periodic density functional calculations. To simulate the nature of $Fe(OH)_x$/Pt nanoparticles, three possible structural models, *i.e.*, $Fe(OH)_x$/Pt(111), $Fe(OH)_x$/Pt(332) and $Fe(OH)_x$/Pt(322), are built. We demonstrate that Fe(III)–OH–Pt stepped sites readily react with CO adsorbed nearby to directly yield CO_2 and simultaneously produce coordinatively unsaturated iron sites for O_2 activation. By contrast, the created interfacial vacancy on $Fe(OH)_x$/Pt(111) prefers to adsorb CO rather than O_2, thus inhabiting the catalytic cycles of CO oxidation. We suggest that such structure sensitivity can be understood in terms of the bond strengths of Fe(III)–OH.

Introduction

Catalytic oxidation of CO to CO_2 is believed to be one of the best-known heterogeneous processes.[1–11] On the one hand, CO oxidation has been regarded as a prototypical reaction which can provide fundamental insight into the structure of the active sites at solid surfaces or interfaces; on the other hand, such a reaction bears considerable practical value in CO removal from car exhausts and the purification of hydrogen fuel in the proton exchange membrane fuel cell (PEMFC) anodes.[12] Developing an efficient strategy to achieve high performance of CO oxidation under ambient conditions is highly desirable. As CO can chemisorb on the noble metal (NM) while reducible transition metal oxides (TMO) usually have relatively high oxygen affinity, it is expected that combining both of them to create a TMO/NM interface could work in synergy to boost the CO oxidation. Based on this concept, numerous catalysts have been successfully developed, such as Au/TiO_2,[1,13] Rh/CeO_2,[2] Pt/FeO_x,[3–8,11] Cu/NiO[9] and Cu/TiO_2(CeO_2)[10,14,15] systems. Among them, the Pt/FeO_x system has been extensively studied experimentally and theoretically, owing to its high activity at low temperatures and high resistance to oxidation.[3–5] It is generally accepted that CO oxidation occurs through the Langmuir–Hinshelwood (L–H) mechanism including multiple elementary steps, such

State Key Laboratory for Physical Chemistry of Solid Surfaces, Collaborative Innovation Center of Chemistry for Energy Materials, College of Chemistry and Chemical Engineering, Xiamen University, Xiamen 361005, China. E-mail: nfzheng@xmu.edu.cn; gfu@xmu.edu.cn

as CO and O_2 adsorption, O–O bond dissociation, and C–O bond formation. However, not all aspects of the reaction have been explored in sufficient depth, and several fundamental questions are still under heavy debate: (i) What is the nature of the Pt–FeO_x interface? (ii) What are the roles that water plays in CO oxidation? (iii) What kinds of intermediates are involved?

As previously reported, both bilayer coordinatively unsaturated O–Fe/Pt and trilayer coordinatively saturated (H)O–Fe–O/Pt interface have been proposed as active sites for CO oxidation. Based on surface science experiments and density functional theory (DFT) calculations, Fu *et al.* suggested that coordinately unsaturated Fe(II) (CUF) sites confined at the Pt–FeOx interface, *i.e.*, O–Fe/Pt, show high activity in CO oxidation under PEMFC conditions.[3] DFT calculations revealed that such high reactivity might originate from the preferential adsorption of oxygen over CO on the CUF sites. It should be pointed out that low-valence ferrous is not stable in oxidative environments. For instance, Ringleb *et al.* showed that FeO(111)/Pt(111) would undergo strong reconstruction under 10–100 mbar partial pressure of oxygen, leading to the formation of trilayer coordinatively saturated O–Fe–O/Pt film, and the latter readily converts into a hydroxyl terminated trilayer with a HO–Fe–O/Pt structure when in contact with water.[8,16] XPS results indicated that Fe present in such a trilayer film is in the +3 oxidation state.[4,8,16] Experimentally, (H)O–Fe–O/Pt was found to be stable in air, but almost inactive towards CO oxidation at room temperature.[4,8]

It has been well documented that water would play a versatile role in CO oxidation. On the one hand, water is generally a promoter for CO oxidation on NM surfaces.[17–20] Gong *et al.* performed DFT calculations to demonstrate that in the presence of H_2O, CO oxidation on Pt(111) began with the reaction of CO and OH, and then the resulting COOH further interacts with another OH to give CO_2 and H_2O.[18] By means of inelastic neutron scattering spectroscopy, Parker further confirmed that two OH groups were required for low temperature CO oxidation.[19] On the other hand, a small amount of water would inhibit CO oxidation over TMO, such as CuO, MnO_x and Co_3O_4.[21–24] Wang *et al.* proposed that the influence of water was different on NM and TMO because there existed significant difference in energy barrier for movement of OH on metal and metal oxide.[24] Obviously, the TMO/NM system is expected to bring more complex chemistry. On the interface of FeO/Pt(111), the reaction of CO with OH was identified experimentally by Huang *et al.*[16] Recent DFT calculations by Li *et al.* suggested that water not only accelerated the reaction, but also promoted interfacial interactions between FeO and Pt(111).[11] Another related topic of interest is water–gas shift (WGS) reaction ($CO + H_2O \rightarrow CO_2 + H_2$), in which TMO/NM catalysts were usually employed.[10,13–15] Most evidence supports that the WGS reaction occurred by an associative mechanism where CO reacted with OH derived from water dissociation, leading to the formation of a COOH intermediate, which further underwent dehydrogenation to generate CO_2.[10,13–15] These results indicate that the OH located at the interfaces between TMO and NM would play a crucial role in CO oxidation.

Very recently, we have successfully fabricated a series of sub-5 nm iron hydroxide-platinum hybrid nanoparticles to demonstrate a smart, collaborative Fe(III)–OH–Pt interface that is highly efficient for CO oxidation.[25] On the basis of the molecular-level characterizations (*i.e.*, aberration-corrected STEM, EXAFS, XANES, XPS, *etc.*), the iron cations were proven to be present as +3, which was surrounded mainly by OH groups. Our DFT calculations and isotope labelling experiments revealed that the OH groups at the Fe(III)–OH–Pt interfaces readily reacted with CO adsorbed nearby to directly yield CO_2 and simultaneously generated coordinatively unsaturated iron sites for O_2 activation. In this contribution, we present a comprehensive study on the fully catalytic cycle involving the oxidation of two CO molecules by using different structural models. The structure sensitivity of different interfaces will be highlighted as well.

Computational methods and model

Spin polarization calculations are carried out by using the Vienna *ab initio* simulation package (VASP).[26,27] The valence electrons are described by plane wave basis sets with a cutoff energy of 400 eV, and the core electrons are replaced by the projector augmented wave pseudopotentials.[28,29] The *k*-points sampling for all structural models is generated following the Monkhorst–Pack procedure with a $2 \times 3 \times 1$ mesh.

Here, we demonstrate how to build a sophisticated catalyst model that incorporates most of the physical/chemical properties characterized by experiments. High resolution-STEM showed that the Pt core in $Fe(OH)_x$/Pt nanoparticles is mainly enclosed with (111) and (100) facets. It was reported previously that iron oxide films, such as FeO, Fe_2O_3 and Fe_3O_4, are able to spread over Pt(111) surface. Thus, we expect that $Fe(OH)_x$ could also deposit onto the Pt(111). Besides the flat (111) surface, two kinds of stepped surfaces, *i.e.*, Pt(332) [$n(111) \times (111)$] and Pt(322) [$n(111) \times (100)$], are also taken into account to simulate the edge sites prevalent in nanoparticles. Here, the $Fe_6O_{18}H_{15}$ cluster is chosen as a model for $Fe(OH)_x$, which is constructed by cutting the Fe_6O_{18} moiety from goethite FeOOH crystal and then saturating most of the outer O atoms with H atoms, giving rise to $Fe_6O_{18}H_{15}$. Such a cluster can be placed on the (111) terrace of the flat or stepped surfaces, resulting in three possible structural models, *i.e.*, $Fe(OH)_x$/Pt(111), $Fe(OH)_x$/Pt(332) and $Fe(OH)_x$/Pt(322), as illustrated in Fig. 1. DFT calculations show that $Fe_6O_{18}H_{15}$ can strongly anchor on the surfaces with a binding energy of 1.05, 1.09 and 1.15 eV/Fe atom for Pt(111), Pt(332) and Pt(322), respectively. Bond valence sum (BVS) calculations show that the average valence of Fe atoms is in the range varied from 2.883 to 2.966, indicating that they are ferric species.[30,31]

Deep inspection shows that such models do reflect the structure features obtained from experimental studies:[25] (1) structural parameters (bond distances, coordination number) of the optimized model structures can fit those from EXAFS experiments; (2) both Bader charge analysis and valence bond simulations shows that Fe is in the +3 valence state, consistent with the XANES and XPS

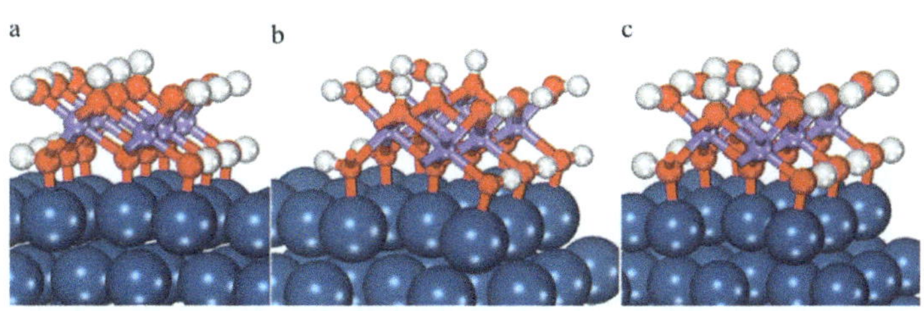

Fig. 1 The possible structural models for Fe(III)–OH–Pt interface: (a) $Fe(OH)_x$/Pt(111), (b) $Fe(OH)_x$/Pt(332) and (c) $Fe(OH)_x$/Pt(322).

Table 1 Valence and structural parameters of $Fe(OH)_x/Pt$ obtained by DFT calculations and experiments

Surfaces	Metal	Valence	M–O (Å)	M–M (Å)
Pt(111)	Pt	+0.192	2.126	2.775
	Fe	+2.966	2.041	3.014
Pt(332)	Pt	+0.203	2.135	2.775
	Fe	+2.905	2.045	3.028
Pt(322)	Pt	+0.181	2.113	2.777
	Fe	+2.883	2.047	3.019
Pt nano (expt)[25]	Pt	δ+	1.99 ± 0.01	2.761 ± 0.001
	Fe	+3	1.97 ± 0.02	3.01 ± 0.02

experiments; and (3) OH/O ratio in our model is equal to 5, close to that deduced from XPS (~4.2); (4) there are still open Pt sites on the surface, consistent with the results of low energy ion scattering spectroscopy.

Exchange and correlation functionals are treated within the Perdew–Burke–Ernzerhof generalized gradient approximation (GGA).[32] Since GGA was reported to be insufficient to describe the Coulombic repulsion between the Fe 3d electrons, here we adopt the GGA+U approximation with U_{eff} of 3.0 eV for Fe atoms.[3,5,33] For spin polarization calculations, the arrangement of spins on different Fe atoms remarkably influences the energies. In the structural models described above, there are two columns of Fe atoms, and three Fe atoms in each column. Our calculations show that antiferromagnetic (AFM) structures are more favourable than ferromagnetic (FM) ones, and the row-wise (RW) arrangement is more favourable than the nearest neighbour one. Thus, all energies reported here use the RW-AFM configuration for Fe atoms.

During structural optimization, the bottom three layers of the slab are fixed at the bulk truncated position, while the top two layers and the adsorbed molecules are fully relaxed. The minimum energy reaction pathways are calculated using the nudged elastic band method. The final transition state (TS) structures are refined by using a quasi-Newton algorithm until the Hellman–Feynman forces on each ion were lower than 0.03 eV $Å^{-1}$. Moreover, frequency calculations are performed to confirm the TS which have only one imaginary frequency. The adsorption energies (ΔE_{ads}) are calculated by using eqn (1):

$$\Delta E_{ads} = E_{ad/sub} - E_{ad} - E_{sub} \tag{1}$$

in which $E_{ad/sub}$, E_{ad}, and E_{sub} are the total energies of the optimized adsorbate/substrate system, the adsorbate in the gas phase, and the clean substrate, respectively.

Results and discussion

Adsorption of CO and O_2 on $Fe(OH)_x/Pt$ surfaces

To boost the low temperature CO oxidation, the key is to depress the CO adsorption and promote the adsorption of O_2. Here, we compare the adsorption energies of CO and O_2 on naked and $Fe(OH)_x$ deposited Pt surfaces, as listed in

Table 2. It was reported previously that the DFT/GGA method would underestimate site preferences for CO adsorption on low-coordination sites on the Pt(111) system, *i.e.*, CO/Pt(111) puzzle.[34,35] Computationally, the differences in energy between the top and hollow site are usually less than 0.3 eV, thus having a minor effect on the following discussion. We show that CO prefers to adsorb on the fcc hollow site of Pt(111) with an energy of −1.87 eV, consistent with previous works of −1.74 (ref. 36) and −1.85 eV,[37] but larger than those of the top site on Pt(111) (−1.64 eV). The Pt(332) and Pt(322) surfaces have five-atom and four-atom wide (111) terraces, respectively, and each unit of both of them contains one column of stepped atoms. Our calculations show that for the stepped surfaces, CO adsorbs preferentially on the edge top. The calculated adsorption energies are −1.96 and −2.05 eV, which are slightly larger than that of CO adsorption on the (111) plane. For O_2 adsorption on the Pt(111), the most stable adsorption occurs at the bridge site with energy of −0.67 eV, in good agreement with those in the previous calculations (−0.63 to 0.71 eV).[11,38] Bader charge analysis shows that adsorbed O_2 is negatively charged by −0.43 *e*, clearly indicating that an electronic rearrangement occurs after adsorption. We also consider the adsorption of O_2 on the edge sites. As shown in Table 2, the calculated adsorption energies for O_2 on the step of Pt(332) and Pt(322) are −0.53 and −0.62 eV, respectively, suggesting that molecular O_2 tends to adsorb on the flat surface. All these indicate that O_2 adsorption would be significantly poisoned by CO adsorption over metallic Pt surfaces since the latter is −1.2 to −1.4 eV more favourable in energy than the former, thus suppressing the low temperature CO oxidation. Indeed, our experiments have also demonstrated that without $Fe(OH)_x$ deposition, pristine Pt nanoparticles exhibited 100% CO conversion only when the temperature is higher than 373 K.

Bader charge analysis (Table 1) shows that the Pt atoms at the $Fe(OH)_x$/Pt interfaces are positively charged by +0.18 to +0.20e, indicating significant charge transfers from Pt 5d orbital to $Fe(OH)_x$. More vacant d orbitals in Pt result in less electron donation to the $2\pi^*$ orbitals of CO, thus weakening the CO adsorption. Actually, calculations on Fe(III)–OH–Pt interfaces yield a significant decrease of adsorption energy for CO (0.5–1.0 eV). Meanwhile, we find that O_2 adsorption is also inhibited at the interfaces with energy decreases of 0.2–0.3 eV. Thus, the differences between CO and O_2 adsorption are still very large, varying from 0.7 to

Table 2 Calculated adsorption energies for CO and O_2 on naked and $Fe(OH)_x$ deposited Pt surfaces (eV)

Surface	CO	O_2
Pt(111)	−1.87, −1.74 (ref. 36), −1.85 (ref. 37), −1.64 (ref. 11)	−0.67, −0.71 (ref. 11), −0.63 (ref. 38)
$Fe(OH)_x$/Pt(111)	−1.43	−0.70
Pt(332)	−1.96, −2.41[a,39], −2.43[b,39]	−0.53
$Fe(OH)_x$/Pt(332)	−1.01	−0.34
Pt(322)	−2.05, −2.41[a,39], −2.43[b,39]	−0.62
$Fe(OH)_x$/Pt(322)	−1.43	−0.31

[a] CO adsorption on Pt(211). [b] CO adsorption on Pt(311).

1.1 eV, and the situation of preferential adsorption of CO to O_2 could not be changed even when the Fe(III)–OH–Pt interfaces have been formed.

The first CO oxidation on $Fe(OH)_x$/Pt interfaces

In WGS reaction, CO oxidation begins with the reaction of CO and OH group to form COOH species.[10,13–15] This process can be viewed as CO insertion, which is prevalent in organometallic chemistry. In our case, hydroxyl species are readily available at the interface so that the CO oxidation is also expected to be triggered by the coupling reaction between the adsorbed CO and interfacial OH. Fig. 2 displays the calculated reaction profiles of the first CO oxidation process over the $Fe(OH)_x$/Pt interfaces and the TS structures are summarized in Fig. 3. Experimentally, the exposed Pt atoms could be saturated by CO even at room temperature. Here, the state in which all open Pt sites except one are occupied by CO is defined as the zero energy reference (i). Interestingly, the adsorption energies for the last CO over different Pt surfaces are calculated to be exothermic by −0.32 to −1.05 eV, much lower than those of low coverage cases. Next, CO can interact with an OH group in the vicinity, leading to the formation of adsorbed CO_2 (iii) directly or *via* COOH intermediate (ii). Our calculations show that COOH species can only be stabilized on Pt(322). This step *via* TS-1 is calculated to be exothermic by 0.16 eV and exhibits a barrier of 0.69 eV. Such a barrier is higher than those on FeO/Pt(111) (0.33 eV),[11] CeO_x/Cu(111) (0.41 eV),[15] but lower that on TiO_2/Au(111) (~1.1 eV).[13] After transferring the H to the neighbour OH group, COOH (iii) can convert into adsorbed CO_2 species (iv) by overcoming a small barrier of 0.20 eV (TS-2). On $Fe(OH)_x$/Pt(111) and $Fe(OH)_x$/Pt(332) interfaces, COOH is not stable since H transfer from COOH to nearby OH occurs spontaneously, and adsorbed CO_2 can be directly formed with the assistance of two OH groups, *i.e.*, TS-1′. The calculated barriers are 0.48 eV and 0.69 eV for Pt(111) and Pt(332), respectively, with respect to CO adsorption (ii). This indicates that the OH present in flat surfaces, *i.e.*, Pt(111) are more active in CO oxidation than those on stepped ones, such as Pt(332) and Pt(322). Subsequently, release of CO_2 takes place to yield a vacancy on the interface, *i.e.*, Fe–□–Pt, with a small activation barrier of 0.16–0.32 eV.

The proposed reaction mechanism can nicely account for the experimental observations. Bader analysis shows that the adsorbed CO_2 (iv) can be viewed as

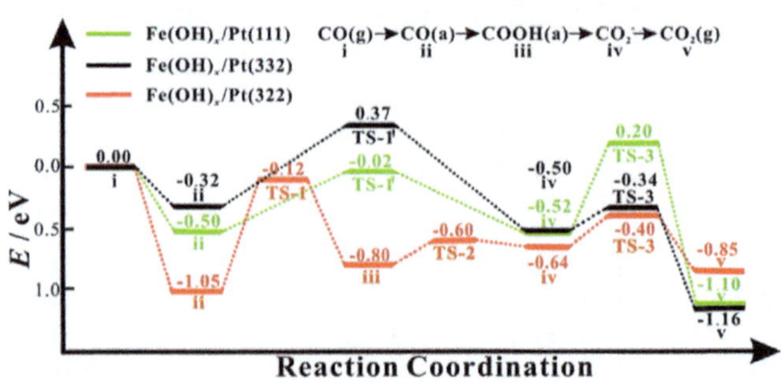

Fig. 2 The proposed reaction mechanisms for the first CO oxidation on the $Fe(OH)_x$/Pt(111), $Fe(OH)_x$/Pt(332) and $Fe(OH)_x$/Pt(322) catalysts.

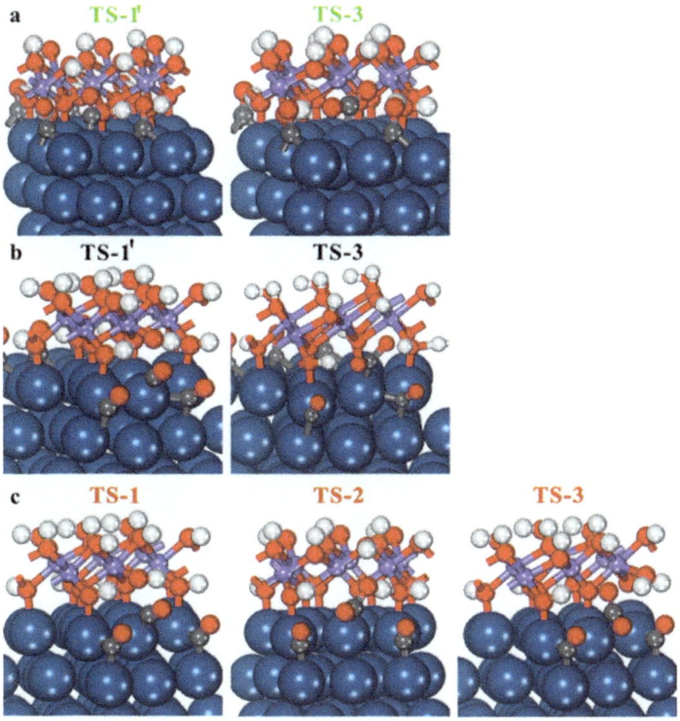

Fig. 3 Optimized TS structures for the first CO oxidation over (a) $Fe(OH)_x$/Pt(111); (b) $Fe(OH)_x$/Pt(332); (c) $Fe(OH)_x$/Pt(322).

$CO_2^{\delta-}$ which carries a significant negative charge of 0.55–0.87e. Experimentally, such species have been identified over both CeO_x/Cu(111) and CeO_x/Au(111) under WGS conditions.[13,15] In addition, we predict that the apparent energy barrier is in the range of 0.00–0.37 eV for CO oxidation by Fe(III)–OH–Pt interface, in good agreement with the experimental value (0.16 eV).[25] Isotope labelling experiments revealed that in the absence of O_2, CO is able to react with the interfacial Fe(III)–OH–Pt sites to yield CO_2, further supporting our mechanism.[25] This finding casts some doubt on the well-accepted assumption that the CO oxidation by FeO_x/Pt should start from O_2 activation.[5,11]

Competitive adsorption of CO and O_2 on Fe–□–Pt

The created interfacial vacancy can be competitively adsorbed by CO and O_2, as shown in Fig. 4. Bader charge analysis reveals that the Pt site under as-formed vacancy at the interface is positively charged by +0.15 to +0.19e. Accordingly, the adsorption energies for CO are predicted to be −1.30, −0.87 and −1.60 eV for Pt(111), Pt(332) and Pt(322), respectively, lower than those on naked Pt surfaces. On the other hand, BVS analysis shows that the average valence of Fe atoms reduces to +2.05 from +2.22. Low valence iron is expected to be preferential to adsorb and activate the O_2 molecule. It should be pointed out that upon O_2 adsorption, a proton would spontaneously transfer from the nearby water to make OOH species. For Pt(111), the adsorption energy of O_2 is

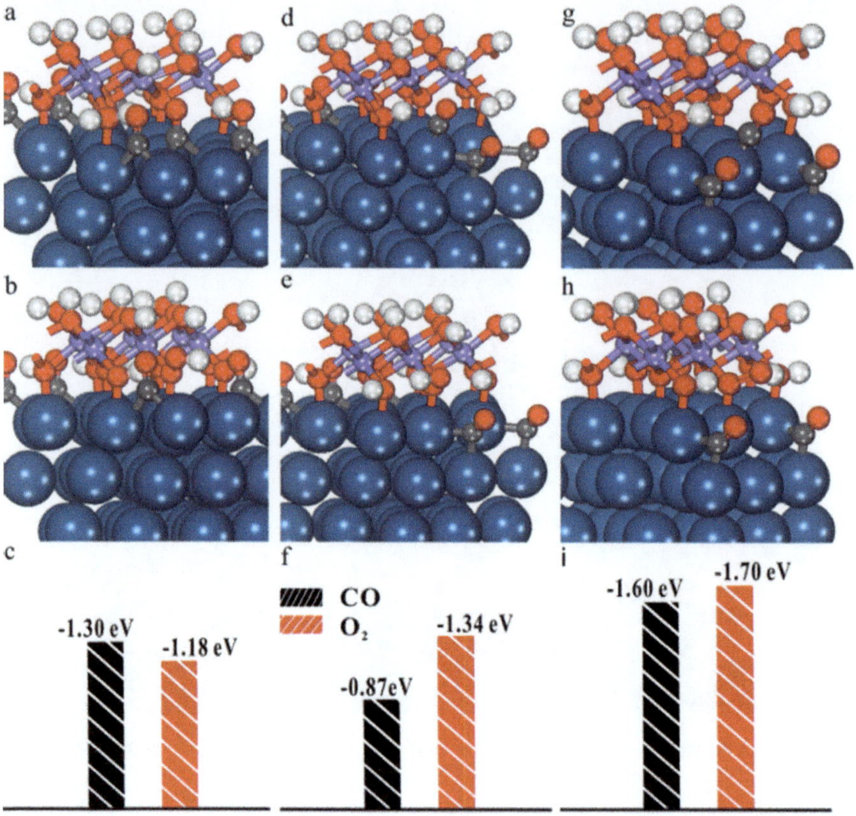

Fig. 4 The adsorption structures of CO and O_2, respectively, on the created interfacial vacancies. (a) and (b) a Pt(111) surface. (d) and (e) a Pt(332) surface. (g) and (h) a Pt(322) surface, (c), (f) and (i). The adsorption energies corresponding to (a)–(b), (d)–(e) and (g)–(h).

exothermic by −1.18 eV, while the stepped surfaces lead to more stable adsorption with energies of −1.34 eV and −1.70 eV for Pt(332) and Pt(322), respectively. Charge and spin analysis show that the O–O groups are negatively charged by −1.20 to −1.24e; meanwhile, the magnetic moments on O atoms are completely quenched. Thus, the enhancement of O_2 adsorption can be attributed to the strong electrostatic attraction between $O_2^{\delta-}$ and Fe cations. From Fig. 4, we can see that the adsorption of O_2 is preferential to CO on stepped sites; whereas CO adsorption is preferential to O_2 (−1.30 *vs.* −1.18 eV) on the terrace. Such a structure sensitivity might originate from the difference in bond strength of Fe(III)–OH. Our calculations show that the strength of Fe(III)–OH decreases in the order −3.30 eV {Pt(111)} > −3.61 eV {Pt(322)} ≈ −3.68 eV {Pt(332)}. A weak Fe(III)–OH bond results in high activity for the reaction of CO and OH, while there is a relatively low adsorption energy for O_2. In this regard, the vacancies on stepped sites can be efficiently healed by oxygen; while Fe(III)–□–Pt(111) site still suffers from CO poisoning. This can explain why trilayer (H)O–Fe–O/Pt(111) structures only exhibit low activity for CO oxidation under room temperature.[4,7,8] In addition, the role that nearby H_2O played should also be highlighted as it could provide H to generate OOH species.

Computationally, the adsorption of H_2O is not so strong that it should be replenished by feed gas, accounting for the fact that the reactivity of CO oxidation is only maintained in the humid air.[25]

Mechanism of the second CO oxidation

Next, we will investigate the second CO oxidation continuing from the interfacial vacancy (*v*), as shown in Fig. 5. The optimized TS structures are illustrated in Fig. 6. For simplicity, here we limit our scope on the Pt(322) surface. By adsorbing O_2 and CO, the interface has been 'saturated' again (vii), with an exothermicity of −1.54 eV. In the following reaction, two possible pathways are considered, namely Path A and Path B. The former involves the coupling between OOH and CO *via* TS-4 with a barrier of 1.38 eV. This step is calculated to be strongly exothermic by −3.82 eV, leading to the formation of CO_2 and the recovery of Fe(III)–OH–Pt interface (i), thus completing the catalytic cycle. On the other hand, the second CO oxidation can also begin with the reaction of CO + OH, *i.e.*, Path B, which is similar to the mechanism for the first CO oxidation. Again, the COOH intermediate cannot be located, and directly transforms into $CO_2^{\delta-}$ species together with adsorbing H_2O_2 (viii) *via* TS-5 with a relatively low barrier of 0.73 eV. This indicates that interfacial OH species is more active than the interfacial OOH towards CO oxidation. Subsequently, $CO_2^{\delta-}$ can overcome a 0.62 eV barrier (TS-6) to liberate a CO_2 and yield a new vacancy (ix). If the vacancy could be occupied by water (x), the HO–OH bond can be readily broken with the assistance of the water. Only a small barrier, 0.34 eV, is required to overcome TS-7, which is followed by an exothermicity of −1.24 eV to the formation of two OH groups, eventually recovering the Fe(III)–OH–Pt interface (i). We suggest that these two pathways are competitive under working conditions because the water adsorption on the vacancy (x) is rather weak (−0.38 eV), downplaying the significance of Path B. Nevertheless, we still anticipate that increasing water pressure might favour Path B over Path A. Another interesting finding is that H_2O_2 species are generated along Path B. This let us recall the well-known Fenton system, *i.e.*, Fe(II) + H_2O_2,[40,41] which is usually used for the oxidation of organic contaminants to carbon dioxide

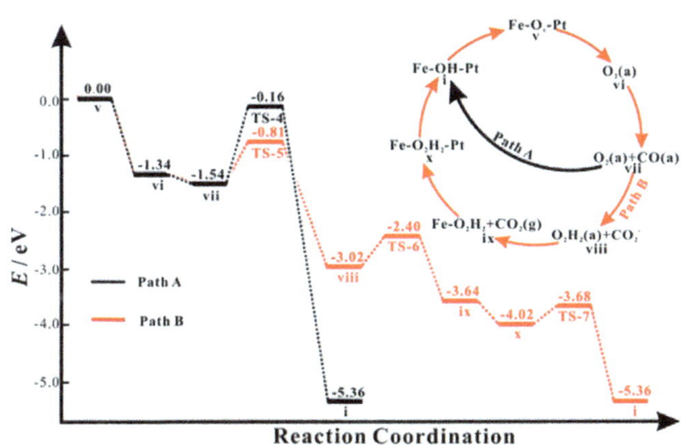

Fig. 5 Possible pathways for the second CO oxidation over Fe(OH)$_x$/Pt(332).

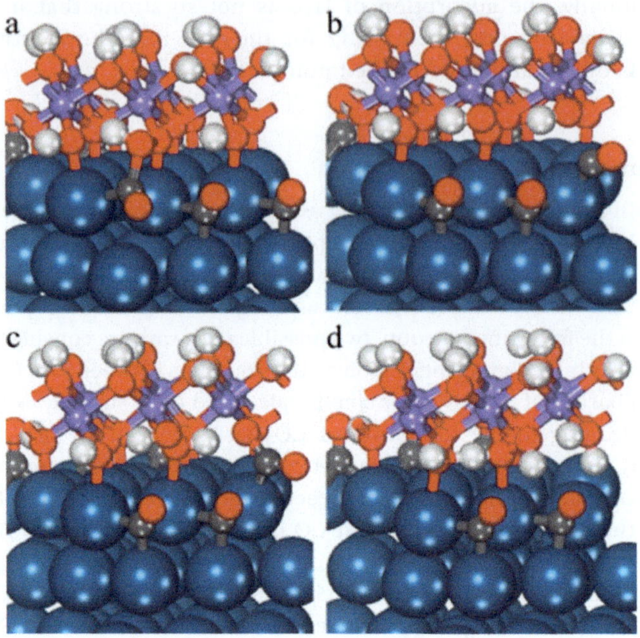

Fig. 6 Optimized TS structures for the second CO oxidation along the energy profiles, (a) TS-4; (b) TS-6; (c) TS-7 and (d) TS-8.

under ambient conditions. We expect that Fe(III)–OH–Pt could also show great promise in the catalytic oxidation of hydrocarbons.

Conclusions

Here we present a theoretical investigation into the full catalytic cycle of CO oxidation by Fe(III)–OH–Pt interface. The main conclusions can be summarized as follows:

(1) Our calculations show that CO oxidation could begin with the coupling between CO and OH with barriers of 0.48–0.93 eV. This indicates that OH present in Fe(III)–OH–Pt is active and liable to oxidizing CO into CO_2.

(2) CO oxidation is found to be a structure sensitive reaction. We show that the performance of different exposing surfaces is dependent on the bond strength of Fe(III)–OH. A weak Fe(III)–OH bond at the $Fe(OH)_x$/Pt(111) interface results in a high activity towards the CO + OH reaction; however, it renders the interfacial vacancy preferentially occupied by CO, inhibiting the activation of O_2. On the other hand, $Fe(OH)_x$ together with the Pt edge site is able to effectively activate both CO and O_2, thus enhancing the low-temperature reactivity of CO oxidation.

(3) In addition, the roles that the adsorbed water plays in adsorption and activation of O_2 have been considered. On one hand, it can stabilize O_2 adsorption by forming OOH species; on the other hand, the O–O bond breaking can be facilitated by the coexistence of water.

(4) We show that during oxygen evolution, H_2O_2 species can be formed such that the Fe(III)–OH–Pt interface might play a crucial role in catalytic oxidation of hydrocarbon under ambient conditions.

We believe that such a detailed mechanistic understanding provides new insights into the roles of the interfacial OH, which will shed light on the design of more efficient TMO/NM catalysts.

Acknowledgements

We thank the financial support from the Ministry of Science and Technology of China (2011CB932403), the National Nature Science Foundation of China (21131005, 20925103, 21373167, 21033006, 21133004, 21333008) and the Fundamental Research Funds of Central Universities. G. Fu thanks Shanghai Supercomputer Center (SSC) for providing computational resources.

Notes and references

1 I. X. Green, W. Tang, M. Neurock and J. T. Yates, *Science*, 2011, **333**, 736.
2 W. Song, A. P. J. Jansen and E. J. M. Hensen, *Faraday Discuss.*, 2013, **162**, 281.
3 Q. Fu, W. X. Li, Y. X. Yao, H. Y. Liu, H. Y. Su, D. Ma, X. K. Gu, L. M. Chen, Z. Wang, H. Zhang, B. Wang and X. H. Bao, *Science*, 2010, **328**, 1141.
4 Y.-N. Sun, L. Giordano, J. Goniakowski, M. Lewandowski, Z.-H. Qin, C. Noguera, S. Shaikhutdinov, G. Pacchioni and H.-J. Freund, *Angew. Chem., Int. Ed.*, 2010, **122**, 4520.
5 B. Qiao, A. Wang, X. Yang, L. F. Allard, Z. Jiang, Y. Cui, J. Liu, J. Li and T. Zhang, *Nat. Chem.*, 2011, **3**, 634.
6 Y. Lei, M. Lewandowski, Y.-N. Sun, Y. Fujimori, Y. Martynova, I. M. N. Groot, R. J. Meyer, L. Giordano, G. Pacchioni, J. Goniakowski, C. Noguera, S. Shaikhutdinov and H.-J. Freund, *ChemCatChem*, 2011, **3**, 671.
7 L. Giordano, M. Lewandowski, I. Groot, Y.-N. Sun, J. Goniakowski, C. Noguera, S. Shaikhutdinov, G. Pacchioni and H.-J. Freund, *J. Phys. Chem. C*, 2010, **114**, 21504.
8 F. Ringleb, Y. Fujimori, H.-F. Wang, H. Ariga, E. Carrasco, M. Sterrer, H.-J. Freund, L. Giordano, G. Pacchioni and J. Goniakowski, *J. Phys. Chem. C*, 2011, **115**, 19328.
9 L.-Y. Gan and Y.-J. Zhao, *J. Phys. Chem. C*, 2012, **116**, 16089.
10 H. Y. Kim and P. Liu, *ChemCatChem*, 2013, **5**, 3673.
11 X.-K. Gu, R. Ouyang, D. Sun, H.-Y. Su and W.-X. Li, *ChemSusChem*, 2012, **5**, 871.
12 G. J. K. Acres, J. C. Frost, G. A. Hards, R. J. Potter, T. R. Ralph, D. Thompsett, G. T. Burstein and G. J. Hutchings, *Catal. Today*, 1997, **38**, 393.
13 J. A. Rodriguez, S. Ma, P. Liu, J. Hrbek, J. Evans and M. Pérez, *Science*, 2007, **318**, 1757.
14 J. A. Rodriguez, J. Graciani, J. Evans, J. B. Park, F. Yang, D. Stacchiola, S. D. Senanayake, S. Ma, M. Pérez, P. Liu, J. F. Sanz and J. Hrbek, *Angew. Chem., Int. Ed.*, 2009, **48**, 8047.
15 K. Mudiyanselage, S. D. Senanayake, L. Feria, S. Kundu, A. E. Baber, J. Graciani, A. B. Vidal, S. Agnoli, J. Evans, R. Chang, S. Axnanda, Z. Liu, J. F. Sanz, P. Liu, J. A. Rodriguez and D. J. Stacchiola, *Angew. Chem., Int. Ed.*, 2013, **52**, 5101.

16 L. Xu, Z. Wu, Y. Zhang, B. Chen, Z. Jiang, Y. Ma and W. Huang, *J. Phys. Chem. C*, 2011, **115**, 14290.
17 J. Bergeld, B. Kasemo and D. V. Chakarov, *Surf. Sci.*, 2001, **495**, L815.
18 X.-Q. Gong, P. Hu and R. Raval, *J. Chem. Phys.*, 2003, **119**, 6324.
19 S. F. Parker, *Chem. Commun.*, 2011, **47**, 1988.
20 R. Caporali, S. Chansai, R. Burch, J. J. Delgado, A. Goguet, C. Hardacre, L. Mantarosie and D. Thompsett, *Appl. Catal., B*, 2014, **147**, 764.
21 G. Y. Wang, W. X. Zhang, Y. C. Cui, H. L. Lian, D. Z. Jiang and T. H. Wu, *Chin. J. Catal.*, 2001, **22**, 408.
22 H. Zou, X. Dong and W. Lin, *Appl. Surf. Sci.*, 2006, **253**, 2893.
23 D. Gamarra and A. Martínez-Arias, *J. Catal.*, 2009, **263**, 189.
24 H.-F. Wang, R. Kavanagh, Y.-L. Guo, Y. Guo, G.-Z. Lu and P. Hu, *Angew. Chem., Int. Ed.*, 2012, **51**, 6657.
25 G. Chen, Y. Zhao, G. Fu, P. N. Duchesne, L. Gu, Y. Zheng, X. Weng, M. Chen, P. Zhang, C.-W. Pao, J.-F. Lee and N. Zheng, *Science*, 2014, **344**, 495.
26 G. Kresse and J. Hafner, *Phys. Rev. B: Condens. Matter Mater. Phys.*, 1993, **48**, 13115.
27 G. Kresse and J. Furthmüller, *Phys. Rev. B: Condens. Matter Mater. Phys.*, 1996, **54**, 11169.
28 P. E. Blöchl, *Phys. Rev. B: Condens. Matter Mater. Phys.*, 1994, **50**, 17953.
29 G. Kresse and D. Joubert, *Phys. Rev. B: Condens. Matter Mater. Phys.*, 1999, **59**, 1758.
30 I. D. Brown and D. Altermatt, *Acta Crystallogr., Sect. B: Struct. Sci.*, 1985, **41**, 244.
31 N. E. Brese and M. O'Keeffe, *Acta Crystallogr., Sect. B: Struct. Sci.*, 1991, **47**, 192.
32 J. P. Perdew, K. Burke and M. Ernzerhof, *Phys. Rev. Lett.*, 1996, **77**, 3865.
33 L. Giordano, G. Pacchioni, J. Goniakowski, N. Nilius, E. D. L. Rienks and H.-J. Freund, *Phys. Rev. B: Condens. Matter Mater. Phys.*, 2007, **76**, 075416.
34 R. A. Olsen, P. H. T. Philipsen and E. J. Baerends, *J. Chem. Phys.*, 2003, **119**, 4522.
35 Y. Yeo, L. Vattuone and D. King, *J. Chem. Phys.*, 1997, **106**, 392.
36 K. Doll, *Surf. Sci.*, 2004, **573**, 464.
37 F. Abild-Pedersen and M. P. Andersson, *Surf. Sci.*, 2007, **601**, 1747.
38 Z. Yang, J. Wang and X. Yu, *Phys. Lett. A*, 2010, **374**, 4713.
39 H. Orita, N. Itoh and Y. Inada, *Surf. Sci.*, 2004, **571**, 161.
40 C. Walling, *Acc. Chem. Res.*, 1975, **8**, 125.
41 L. Chen, J. Ma, X. Li, J. Zhang, J. Fang, Y. Guan and P. Xie, *Environ. Sci. Technol.*, 2011, **45**, 3925.

Faraday Discussions

PAPER

Carbonization of self-assembled nanoporous hemin with a significantly enhanced activity for the oxygen reduction reaction†

Yan Xie,‡[ab] Chizhou Tang,‡[ab] Zhiqiang Hao,[a] Yang Lv,[b] Ruixia Yang,[bc] Xuming Wei,[b] Weiqiao Deng,[bc] Anjie Wang,*[a] Baolian Yi[b] and Yujiang Song*[b]

Received 3rd June 2014, Accepted 9th July 2014

DOI: 10.1039/c4fd00121d

The scarcity and high cost of Pt-based electrocatalysts for the oxygen reduction reaction (ORR) hinder the practical application of proton exchange membrane fuel cells (PEMFCs). It is critical to replace platinum with non-noble metal electrocatalysts (NNMEs). Carbonized metalloporphyrins represent an important class of NNMEs, but most metalloporphyrins are costly and the corresponding NNMEs do not possess a high ORR activity. Herein, we report that the self-assembly of inexpensive hemin leads to porous nanomaterials in water under ambient conditions and subsequent heat-treatment of the unprecedented nanoporous hemin results in a magnetic NNME with a much enhanced ORR activity compared with directly carbonized hemin without self-assembly. The improvement of the ORR activity likely originates from the exposure of more ORR active sites, caused by the surface area increase of the nanoporous hemin after carbonization over that of micro-scale pristine hemin crystals. Moreover, the ORR activity of heat-treated nanoporous hemin is actually comparable to that of commercial Pt/C in alkaline solution. Additionally, the carbonized nanoporous hemin is much better than commercial Pt/C in terms of durability and tolerance to methanol. This study opens up a new avenue to the production of inexpensive metalloporphyrin-based NNMEs with a high ORR performance by using a self-assembly method in combination with traditional pyrolysis.

[a]*School of Chemical Engineering, Dalian University of Technology, Dalian 116024, P.R. China. E-mail: ajwang@dlut.edu.cn; Fax: +86-0411-84986121; Tel: +86-0411-84986121*

[b]*Dalian Institute of Chemical Physics (DICP), Chinese Academy of Sciences, 457 Zhongshan Road, Dalian 116023, P.R. China. E-mail: yjsong@dicp.ac.cn; Fax: +86-0411-84379170; Tel: +86-0411 84379170*

[c]*State Key Laboratory of Molecular Reaction Dynamics, Dalian National Laboratory for Clean Energy, Dalian Institute of Chemical Physics, Chinese Academy of Science, 457 Zhongshan Road, Dalian 116023, P.R. China*

† Electronic supplementary information (ESI) available. See DOI: 10.1039/c4fd00121d

‡ These two authors contributed equally to this work.

1 Introduction

Platinum-based electrocatalysts for the oxygen reduction reaction (ORR) are still predominantly employed in proton exchange membrane fuel cells (PEMFCs).[1–3] However, the scarcity, high cost, poor long-term stability and vulnerability to fuel crossover associated with platinum have impeded widespread commercial application of PEMFCs.[4–6] It is necessary to explore low-cost, efficient and durable alternatives for the replacement of platinum-based electrocatalysts.[7–14] In this regard, great efforts have been contributed to the development of a variety of non-noble metal electrocatalysts (NNMEs) for the ORR, mainly including transition metal macrocycles,[15,16] carbon-supported transition metal oxides,[17–19] doped carbonaceous materials[20–23] and perovskites.[24,25]

In particular, transition metal macrocycles represent a class of promising NNMEs with appreciable electrocatalytic activities in both alkaline and acidic media. In 1964, Jasinski first reported the electrocatalytic performance of cobalt phthalocyanine (CoPc) toward ORR in alkaline solution, which initiated extensive exploration of a variety of metal macrocycles as alternative electrocatalysts for ORR.[26] In order to improve the stability and activity of metal macrocycle-based electrocatalysts, various electrically conductive materials, especially carbon materials, have been used to support NNMEs.[27]

Li *et al.* loaded iron(II) phthalocyanine (FePc) onto KJ300, leading to an improved durability.[28] Dong *et al.* coated single-walled carbon nanotubes with FePc (FePc/SWCNTs), which exhibited a much improved limiting current density.[29] Morozan *et al.* dispersed iron(II) phthalocyanine, cobalt(II) tetra-*tert*-butylphthalocyanine, cobalt(II) 2,3,7,8,12,13,17,18-octaethyl-porphine, and cobalt(II) 5,10,15,20-tetrakis(4-*tert*-butylphenyl)-porphyrin on a high-surface-area support of CNTs (SWCNTs, double-walled CNTs and multi-walled CNTs) for ORR in both alkaline and acidic solutions to improve the ORR activity.[30] Liu *et al.* synthesized an electrocatalyst by functionalizing graphene with iron(II) tetra-aminophthalocyanine (FeTAPc) using an amidation reaction, of which the current at −0.14 V (*vs.* SCE) remained at 83.5% of its initial value after 10 000 s of chronoamperometric testing, whereas a Pt/C electrocatalyst only retained 40.5% of the initial current.[31] However, from the practical application point of view, the stability and activity of metal macrocycles-based electrocatalysts are still low and need to be improved further.[32]

Fortunately, it has been found that carbonization of metal macrocycles is an effective route for the enhancement of their activity and stability toward ORR.[33–38] Dodelet and co-workers adsorbed iron acetate (FeAc) and Cl-Fe(III) tetramethoxyphenyl-porphyrin (Cl-FeTMPP) on various carbon supports and heat-treated the materials at 900 °C in an NH_3-rich atmosphere to increase the N content, thus improving the electrocatalytic activity.[39] Later, Dodelet's group mixed Cl-FeTMPP with commercial carbon black, followed by heat-treatment at 950 °C in NH_3 or Ar. It turned out that NH_3 seemed to be a better pyrolysis atmosphere than Ar.[40] Chang *et al.* loaded cyanocobalamin (vitamin B12) onto carbon black (Vulcan XC-72R) and pyrolyzed it at 500–900 °C in N_2, resulting in a much enhanced activity and durability.[41] Wang *et al.* reported that the heat-treatment of melamine and vitamin B12 on Vulcan XC-72R at 700 °C in N_2 can enhance the electrocatalytic activity.[42] Wu *et al.* used multi-step pyrolysis (600 °C in N_2 and 800 °C in N_2 and

NH_3) to treat FePc and phenolic resin (PhRs) and achieved a high activity and stability.[43] Heat-treatment plays a crucial role in the synthesis of NNMEs, but as-prepared NNMEs cannot satisfy the requirements of practical fuel cells, either. Besides simple pyrolysis, other advanced methods have also been invented to develop metalloporphyrin-based NNMEs for further improvement of the ORR activity and stability.

Dai and co-workers designed and synthesized a class of 2D covalent organic polymers (COPs) with precisely controlled locations of N hetero-atoms and holes using different N-containing ligands, such as triazine, porphyrin and their derivatives *via* a nickel-catalyzed Yamanoto reaction.[44] Loh and co-workers reported a hybrid metal organic framework with enhanced ORR activity by using pyridinium dye-functionalized reduced graphene oxide sheets to link the metalloporphyrin (5,10,15,20-tetrakis(4-carboxyl)-21*H*,23*H*-porphyrin) nodes.[45] Cheon *et al.* prepared ordered porphyrinic carbons with a high ORR activity by impregnation of mesoporous silica templates with metalloporphyrin precursors (FeTMPPCl and CoTMPP).[46] Hijazi *et al.* adsorbed cobalt(II) *meso*-tetraethynylporphyrins on the sidewalls of carbon nanotubes and polymerized the triple bonds *via* Hay-coupling to increase the ORR stability.[47] Recently, we created three types of active sites (oxides, nitrogen dopants and Me–N–C) with a high density on reduced graphene oxide (rGO) to improve the ORR activity and durability in alkaline solution.[48]

In this work, we describe an alternative synthetic strategy to create NNMEs by combing molecular self-assembly (SA) technology with a traditional pyrolysis method. We chose an easily available and low-cost metal macrocycle, hemin, to prove this concept.[49–52] Hemin has frequently been used to synthesize NNMEs. Jiang *et al.* heat-treated the blend of hemin, 5,10,15,20-tetrakis(4-methoxyphenyl)-21*H*,23*H*-porphine cobalt(II) (CoTMPP) and Vulcan XC-72R at 700 °C in N_2 to synthesize NNMEs.[49] Later, the same group pyrolyzed hemin supported on graphene nanoplatelets at 600 °C in N_2 for the preparation of NNMEs.[50] Liang *et al.* supported hemin on Vulcan XC-72R, ball-milled and heat-treated the blend in N_2 at 600 °C to create NNMEs.[51] Xu *et al.* used hemin as an Fe–N-containing precursor to prepare Fe–N-doped mesoporous carbon during a nanocasting process.[52]

Unlike previous hemin-based synthetic methods, we self-assembled hemin into a nanoporous material in water under ambient conditions and heat-treated the nanoporous hemin to create an NNME. Interestingly, the obtained NNME is magnetic and demonstrates an ORR activity superior to that of carbonized pristine hemin and is comparable to that of commercial Pt/C. Additionally, our electrocatalyst is much better than commercial Pt/C in terms of durability and tolerance to methanol.

2 Experimental

Chemicals

Hemin (iron(III) protoporphyrin (IX) chloride, >95%) was purchased from Aladdin Industrial Corporation (Shanghai, China) and used as received without further purification. All aqueous solutions were prepared with ultrapure water (18.2 MΩ·cm at 25 °C) produced from a Millipore water system (Synergy® UV, France).

Synthesis

In a typical synthesis, 60 mL of 10 mM hydrochloric acid is poured into 40 mL of 10 mM sodium hydroxide (aq.) containing 1 mM hemin at room temperature. The addition of acid protonated the carboxylic groups of hemin (Fig. 1a, inset), thus lowering the solubility of hemin and leading to its self-assembly into dark brown precipitates. After 10 min of magnetic stirring, the precipitates were vacuum-filtered, washed with copious amounts of de-ionized water until the pH value of the filtrate reached 7, and dried overnight under ambient conditions. In the following, the samples were heat-treated in a tube furnace at 500–950 °C in an argon atmosphere for 2 h and cooled down to room temperature naturally. The obtained black powder was leached in 0.5 M H_2SO_4 (aq.) at 80 °C for 30 min to remove unstable and inactive species and then thoroughly washed with de-ionized water to a neutral pH value. The final products were collected after overnight drying at 65 °C in an oven. For comparison, the pristine hemin without self-assembly was also treated as described above. The pristine hemin and self-assembled hemin were labelled as P-Hm and SA-Hm, respectively; after heat-treatment at 950 °C and acid leaching, the corresponding samples were marked as P-Hm (950) and SA-Hm (950), respectively.

Characterizations

Thermogravimetric analysis (TGA) was conducted on a Q600 system (TA Instruments, US) using ~5 mg sample with a heating rate of 10 °C min^{-1} in Ar (99.999%) from 25 to 1000 °C. UV-Vis absorption spectra were collected using a Specord S600 spectrophotometer (Analytic Jena, Germany). X-ray diffraction (XRD) spectra were obtained on a X'Pert Pro Super diffractometer (PANalytical, Netherlands) with Cu Kα radiation ($\lambda = 0.154178$ nm) in a 2θ scan mode with a scanning rate of 5 ° min^{-1}. X-ray photoelectron spectra (XPS) were recorded on a ESCALAB 250Xi photoelectron spectrometer (Thermo Scientific, UK). Transmission electron microscopy (TEM) analysis was performed on a Tecnai G2 Spirit (FEI, Czech Republic) operated at 120 keV. High-angle annular dark-field (HAADF) scanning TEM (STEM) and high-resolution TEM (HRTEM) images were taken on a Tecnai G2 F30 S-Twin (FEI, Netherlands) operated at 300 keV. The nitrogen adsorption–desorption measurements were obtained from a QUADRASORB SI instrument (Quantachrome, USA), and the Brunauer–Emmett–Teller (BET)

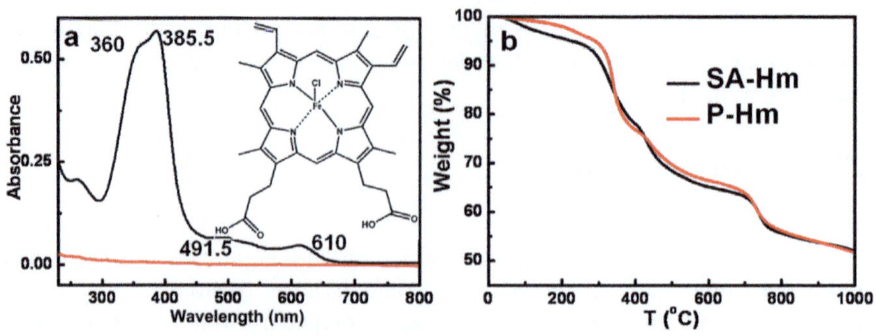

Fig. 1 UV-vis spectrum of hemin and the supernatant after self-assembly (a), TGA curve of P-Hm and SA-Hm (b). Inset to (a) is the molecular structure of hemin.

surface area was calculated by a 6-point measurement over a pressure range of 0.05 to 0.30 P/P_0 at 77.3 K. All electrochemical data in alkaline solution were recorded on a CHI 760D electrochemical workstation (CH Instruments, Shanghai, China). The electrochemical data in acidic solution were collected using an Autolab potentiostat/galvanostat (Echo Chemie BV Model PGSTAT-302N, Netherlands).

Catalyst ink preparation

All the electrochemical data were collected in an electrochemical cell using a Pt mesh as the counter electrode, Hg/HgO or saturated calomel electrode (SCE) as the reference electrode in 0.1 KOH and 0.1 M $HClO_4$, respectively. The electrochemical cell was kept in a water bath at 25 °C for all electrochemical tests. All potentials in acidic solution were referenced to the reversible hydrogen electrode (RHE).

The resultant NNMEs were ground using a pestle and mortar for a better dispersion. Specific amounts of ultrapure water, Nafion solution (5 wt%, Dupont) and ethanol ($V_{water} : V_{ethanol} : V_{Nafion} = 1 : 9 : 0.1$) were mixed with the ground powder and sonicated in a water bath for 5–10 min to form a homogeneous ink (3.9 mg mL^{-1}). Then, 40 μL of the ink was loaded onto a glassy carbon rotating disk electrode (RDE) (5 mm in diameter, 0.19625 cm^2, Pine Instruments, USA), giving an electrocatalyst loading of 0.8 mg cm^{-2}. For comparison, a Johnson Matthey (JM) 20 wt% Pt/C electrocatalyst (1 mg mL^{-1}, 5 μL) was also loaded onto an RDE using the same procedure with a loading of 5 μg_{Pt} cm^{-2}.

Cyclic voltammetry (CV) curves were first carried out on the freshly prepared working electrode by purging with N_2 for 10 min. The electrocatalyst surface was electrochemically cleaned by repeatedly cycling at a positive scan rate of 100 mV s^{-1}. Next, the electrolyte solution was bubbled with a constant flow of O_2 for 10 min before the voltage–current curves were collected. Subsequently, ORR polarization curves were obtained in the O_2-saturated solution at a positive scan rate of 5 mV s^{-1} in 0.1 M KOH solution from −0.8 to 0.2 V (*vs.* Hg/HgO) and 10 mV s^{-1} in 0.1 M $HClO_4$ solution from 0 to 1.2 V (*vs.* RHE). ORR polarization curves of a rotating ring disk electrode (RRDE) (PINE Instrument, USA) were obtained in O_2-saturated 0.1 M KOH aqueous solution ranging from −0.8 to 0.2 V (*vs.* Hg/HgO) with a positive scan rate of 5 mV s^{-1} at 1600 rpm, in which the ring potential was kept constant at 0.2 V (*vs.* Hg/HgO).

For accelerated durability tests (ADTs), the same electrocatalyst-coated glassy carbon RDEs were used as the working electrode. The working electrodes were cycled between −0.4 and 0.2 V (*vs.* Hg/HgO) for a total number of 1500 cycles in O_2-saturated 0.1 M KOH, providing harsh degradation conditions. The scan rate was kept at 100 mV s^{-1} during the process of potential cycling. ORR polarization curves of SA-Hm (950) and 20 wt% Pt/C were collected at a positive scan rate of 5 mV s^{-1} at 1600 rpm in O_2-saturated 0.1 M KOH solution at certain cycles to track the degradation of the electrocatalysts.

As for the methanol tolerance test, the ORR polarization curves of SA-Hm (950) and 20 wt% Pt/C were obtained in O_2-saturated 0.1 M KOH and 0.5 M CH_3OH solution ranging from −0.8 to 0.2 V (*vs.* Hg/HgO) with a positive scanning rate of 5 mV s^{-1} at 1600 rpm.

3 Results and discussion

UV-vis spectrum of hemin in NaOH displays two split Soret bands at 360 and 385.5 nm and Q bands between 500 and 700 nm (Fig. 1a), indicating the formation of a hemin dimer *via* an Fe–O–Fe bridge in alkaline solutions.[53] After the addition of hydrochloric acid, hemin molecules self-assembled into dark brown precipitates with a clear and transparent supernatant. The UV-vis spectrum of the supernatant is featureless (Fig. 1a), verifying that hemin has completely precipitated out during the simple self-assembly process.

It is well known that the heating temperature plays an important role for the ORR activity improvement of the NNMEs. In order to determine which temperature should be utilized to pyrolyze P-Hm and SA-Hm, thermal analysis was conducted. TGA curves (Fig. 1b) demonstrate that the weight loss of SA-Hm is similar to that of P-Hm. The TGA curves of P-Hm and SA-Hm reveal several stages of weight loss. The early weight loss (150–350 °C) can be attributed to the removal of chloride. From 350 to 650 °C, the decomposition of peripheral organic groups like $-CH_3$, $-CH{=}CH_2$ as well as $-CH_2CH_2COOH$ occurred. The weight loss arising from dehydrogenation and decomposition of the hemin backbone happened from 650 to 1000 °C.[54]

According to the thermal analysis results, both P-Hm and SA-Hm were treated at temperatures ranging from 500 to 950 °C. In the case of P-Hm, the half-wave potential ($E_{1/2}$) of the ORR curves shifts positively from 500 to 800 °C, reaches a

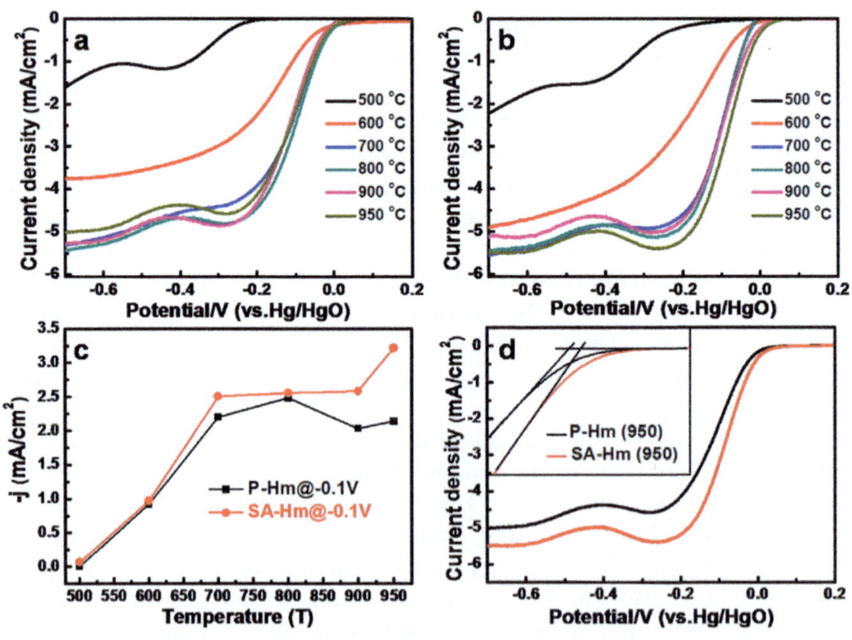

Fig. 2 ORR polarization curves of P-Hm (a) and SA-Hm (b) obtained in the potential range of −0.8 to 0.2 V (*vs.* Hg/HgO) at a scanning rate of 5 mV s^{-1}, 1600 rpm in O_2-saturated 0.1 M KOH. Both P-Hm and SA-Hm were heat-treated at temperatures ranging from 500 to 950 °C. (c) Comparison of current density at −0.1 V (*vs.* Hg/HgO) based on (a) and (b). (d) ORR polarization curve for P-Hm and SA-Hm obtained at 950 °C. Inset is the onset potential comparison.

maximum value at 800 °C, and shifts negatively a little bit from 800 to 950 °C (Fig. 2a). For the case of SA-Hm, the ORR performance increases continuously with the heating temperature from 500 to 950 °C (Fig. 2b). The yield of heat-treated SA-Hm at 950 °C is about 30 wt%. A higher heating temperature may be beneficial for the further enhancement of ORR performance, but was not employed in this study because of the low yield. For further comparison, the current density at −0.1 V (*vs.* Hg/HgO) of the two series of electrocatalysts obtained at different heating temperatures was plotted. It is obvious that the ORR activity of carbonized SA-Hm is always higher than that of carbonized P-Hm at each temperature in terms of the current density (Fig. 2c). Fig. 2d shows the ORR polarization curve of SA-Hm (950) and P-Hm (950), in which the onset potential (E_{onset}) and $E_{1/2}$ of SA-Hm (950) shift positively 11.4 and 24 mV compared with that of P-Hm (950), respectively. This clearly shows that the self-assembly of hemin combined with carbonization is advantageous for the enhancement of ORR performance.

To gain more insight into the electrocatalytic activity of SA-Hm (950), CV curves were recorded in N_2- and O_2-saturated 0.1 M KOH solution at a scan rate of 100 mV s^{-1} (Fig. 3a). Compared with the featureless CV curve in a N_2 atmosphere, the CV curve in an O_2-saturated electrolyte shows a characteristic ORR peak, indicative of a high electrocatalytic activity. ORR polarization curves were collected at different rotating speeds (400–1600 rpm). Fig. 3b demonstrates that

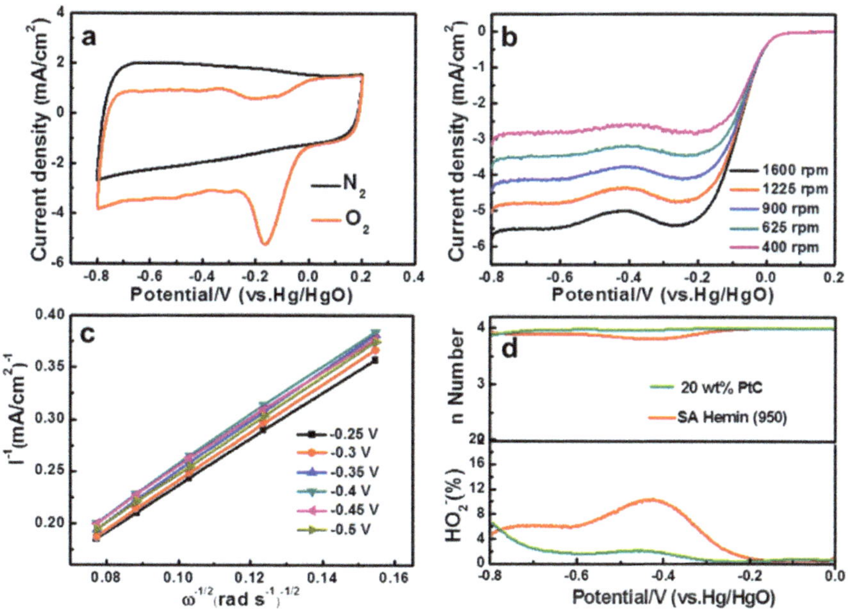

Fig. 3 CV curve of SA-Hm (950) recorded in N_2- and O_2-saturated 0.1 M KOH (aq.) at a scan rate of 100 mV s^{-1} (a); ORR polarization curves of SA-Hm (950) obtained at various rotation rates (400–1600 rpm) with a positive scan rate of 5 mV s^{-1} in O_2-saturated 0.1 M KOH (b); (c) K–L plots at different potentials based on the ORR polarization curves in (b); (d) *n* number and $\%HO_2^-$ of SA-Hm (950) and Pt/C determined at 1600 rpm in O_2-saturated 0.1 M KOH with a positive scan rate of 5 mV s^{-1}.

the diffusion-limited current density of SA-Hm (950) becomes larger with the increasing rotation rates. The linearity of the Koutecky–Levich (K–L) plots suggests first-order reaction kinetics toward the concentration of dissolved O_2,[55] where the observed current and electron transfer number (n) can be expressed as follows:

$$j_{\rm lim} = 0.62nFD^{2/3}\nu^{-1/6}C_0\omega^{1/2} \tag{1}$$

$$\%{\rm HO_2}^- = 200 \times \frac{I_{\rm r}/N}{(I_{\rm r}/N) + I_{\rm d}} \tag{2}$$

$$n = 4 \times \frac{I_{\rm d}}{I_{\rm d} + I_{\rm r}/N} \tag{3}$$

In eqn (1), $j_{\rm lim}$ is the limiting current density; F (96 485 C mol^{-1}) is the Faraday constant; D (1.9×10^{-5} cm^2 s^{-1}) is the diffusion coefficient of O_2 in 0.1 KOH and C_0 (1.2×10^{-6} mol cm^{-3}) is the concentration of O_2 in the electrolyte; ν is the kinetic viscosity of the solution (0.01 cm^2 s^{-1}) and ω is the electrode rotation rate (rad s^{-1}). In eqn (2) and (3), $I_{\rm r}$ and $I_{\rm d}$ are the ring and disk current, respectively. The manufacturer's value of N (the ring collection efficiency) is 37%. The n for ORR was determined to be 3.90 at potentials ranging from −0.25 to −0.50 V (*vs.* Hg/HgO) based on the K–L plots (Fig. 3c), indicating that SA-Hm (950) favours a 4e$^-$ oxygen reduction process. To verify the 4e$^-$ catalytic pathway for the ORR, RRDE measurements were performed to monitor the generation of intermediate peroxide species (${\rm HO_2}^-$) in alkaline solution during the ORR process (Fig. 3d). The evaluated ${\rm HO_2}^-$ yields are less than ~10.3% over the potential range of −0.8 to 0 V (*vs.* Hg/HgO), giving an average n of 3.90. This is consistent well with the average n number based on the K–L plots. This n number is close to that of commercial Pt/C ($n = 4$) (Fig. 3d). This indicates that the ORR process catalyzed by SA-Hm (950) is mainly a 4e$^-$ reaction and mostly leads to the formation of H_2O. The ORR polarization curve of P-Hm (950), SA-Hm (950), and commercial Pt/C measured in both 0.1 M KOH and 0.1 M $HClO_4$ is shown in Fig. 4. This further reveals that the ORR performance of SA-Hm (950) is better than that of P-Hm (950) and close to that of commercial Pt/C in alkaline solution, while there is still a relatively large difference compared with that of Pt/C in an acidic solution.

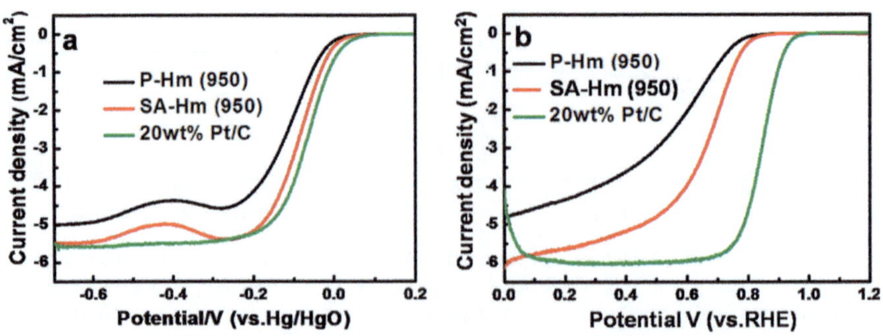

Fig. 4 ORR polarization curve of commercial Pt/C, P-Hm (950), and SA-Hm (950) recorded in 0.1 M KOH (aq.) (a) and 0.1 M $HClO_4$ (aq.) (b) at 1600 rpm. A 5 mV s^{-1} scan rate is used in alkaline solution and 10 mV s^{-1} is used in 0.1 M $HClO_4$ (aq.).

It is necessary to investigate the structural and compositional origin of the high activity of hemin-based electrocatalysts. The P-Hm (950) and SA-Hm (950) powders were subject to nitrogen adsorption–desorption analysis at 77.3 K (Fig. 5a). The N_2 sorption isotherms of P-Hm (950) and SA-Hm (950) exhibit a type IV shape. For each electrocatalyst, there is a relatively large adsorption–desorption hysteresis loop at a relative pressure above 0.4, which is because of the capillary condensation of N_2 in the nanopores. The BET surface area of P-Hm (950) and SA-Hm (950) was determined to as 109.5 and 168.9 $m^3\ g^{-1}$, respectively. The enlarged surface area of SA-Hm (950) should be a key contributor to the improved ORR activity of SA-Hm (950) as compared with P-Hm (950).

Unlike the characteristic XRD pattern of P-Hm crystals with strong and narrow diffraction peaks, the XRD pattern of SA-Hm reveals only two broad peaks (Fig. 5b). This suggests that SA-Hm is disordered and may be much smaller in size than P-Hm. For the heat-treated samples (Fig. 5b), the diffraction peak at 26.0° corresponds to the (002) planes of carbon, whereas the peaks at 33.1, 35.6, 40.8, 49.5, 54.1, 57.4, 62.9 and 77.6° originate from Fe_2O_3 (JCPDS, PDF, File no. 00-024-0072).[56] The small peak at 30.1° likely corresponds to the (220) plane of Fe_3O_4. The Fe_2O_3 and Fe_3O_4 seen should be formed by the oxidation of hemin with a trace amount of oxygen in Ar and/or oxygen species in hemin. The broad peaks at 42.6, 43.0, 43.2 and 44.6° may be indexed to Fe_2N (211), Fe_3O_4 (400), Fe_3N (101) and Fe (110), respectively (JCPDS, PDF, File no. 00-006-0656, 00-001-1111, 00-001-1236 and 00-001-1262).[57,58] Overall, the diffraction peaks of SA-Hm (950) are broader and weaker that those of P-Hm (950). This confirms that SA-Hm (950) has a larger surface area than P-Hm (950), thus contributing to the improvement on the ORR activity.

In addition, XPS analysis shows the presence of C, N, O and Fe in the SA-Hm (950) electrocatalysts (Fig. 6a). The spectrum of Fe2p reveals two peaks, at 723.0 and 711.8 eV, assignable to $Fe2p_{1/2}$ and $Fe2p_{3/2}$ of iron oxides, respectively (Fig. 6b).[59] Iron oxides are well known to be active for the ORR in alkaline solution. The presence of iron oxides accounts for the ORR activity of SA-Hm (950). The N1s spectrum indicates there exist three forms of N, including quaternary N (402.1 eV), pyrrolic N (400.8 eV) and pyridinic N (399.6 eV) as the Fe–N bond[60,61] (Fig. 6c). The N content in SA-Hm (950) with respect to C is about 1%. The N dopant in the

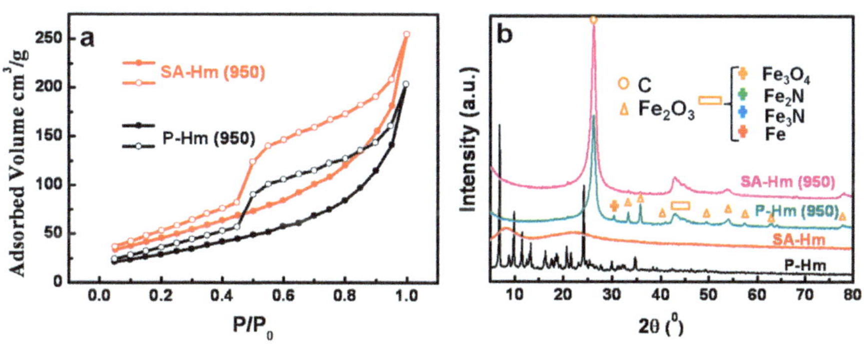

Fig. 5 N_2 adsorption–desorption isotherm of P-Hm (950) and SA-Hm (950) (a); XRD pattern of P-Hm, SA-Hm, P-Hm (950) and SA-Hm (950) (b).

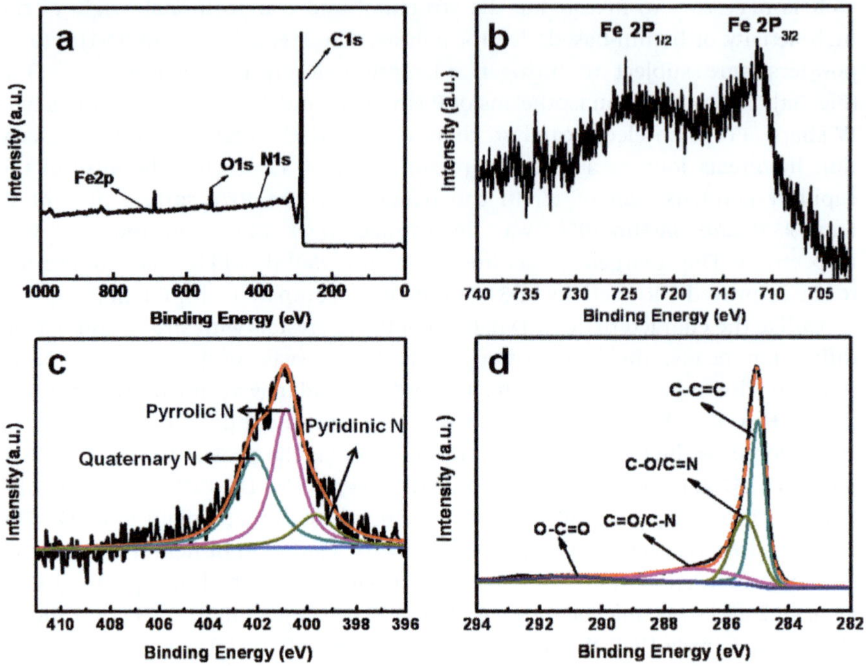

Fig. 6 XPS spectrum of SA-Hm (950) (a); high-resolution XPS spectrum of Fe (b), N (c), and C (d).

carbon matrix could be another contributor to the relatively high ORR activity of SA-Hm (950). Moreover, the C1s spectrum can be peak-fitted into four parts, namely C–C=C, C–O/C=N, C=O/C–N and O–C=O (Fig. 6d). This verifies the presence of the N dopant, which is an important type of active site for the ORR in alkaline solution.

TEM images reveal that pristine hemin exists as macro-scale chunks (Fig. 7a). In contrast, after the self-assembly process hemin molecules are organized into nanowire networks with secondary nanopores (Fig. 7b). After heat-treatment at 950 °C and acid leaching, the resultant P-Hm (950) contains many nanoparticles with an average size of 306 nm, while the average size of the nanoparticles (101 nm) of SA-Hm (950) is much smaller (Fig. 7c, d and Fig. 8). It is necessary to elucidate the composition of the SA-Hm (950). The HAADF STEM image (Fig. 9a) and elemental mapping (Fig. 9b–e) on a selected area show that the bright nanoparticles are iron oxides and the matrix is mainly carbon doped with certain amount of nitrogen. This result agrees with the XRD and XPS analysis. It is worth pointing out that the smaller iron oxide nanoparticles of SA-Hm (950) should lead to a better ORR activity as compared with P-Hm (950) that is composed of larger iron oxide nanoparticles. In addition, HRTEM images of SA-Hm (950) display the characteristic lattice fringes of Fe_2O_3, Fe_3O_4, Fe_3N, and C (Fig. 10a–d), very consistent with the XRD pattern.

In order to evaluate the durability of SA-Hm (950) and commercial Pt/C, ADTs were carried out by potential cycling in the range of −0.4 to 0.2 V (*vs.* Hg/HgO) with a scanning rate of 100 mV s^{-1} in O_2-saturated alkaline solution (Fig. 11a

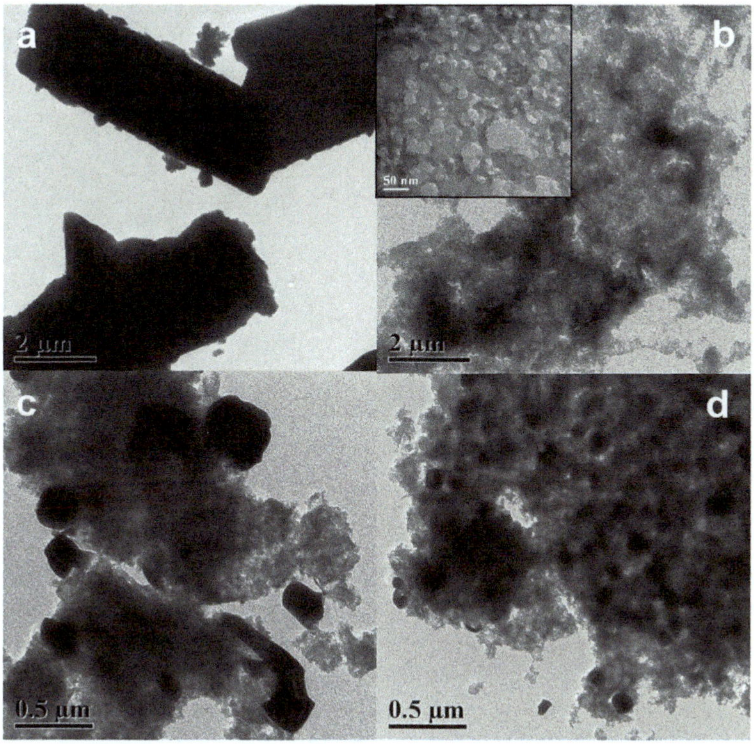

Fig. 7 TEM images of P-Hm (a), SA-Hm (b), P-Hm (950) (c), and SA-Hm (950) (d). Inset is a TEM image of SA-Hm at a high magnification.

and b). The ORR curve of SA-Hm (950) barely shifted during the ADT process (Fig. 11a). In contrast, the ORR curve of commercial Pt/C apparently negatively shifted (Fig. 11b). As shown in Fig. 11c, the current density at −0.1 V (*vs.* Hg/HgO) decreased 8.8% for SA-Hm (950), whereas commercial Pt/C lost 52% of its initial current density after 1500 cycles. This clearly shows that SA-Hm (950) possesses a better durability than that of Pt/C. Additionally, methanol tolerance is regarded as

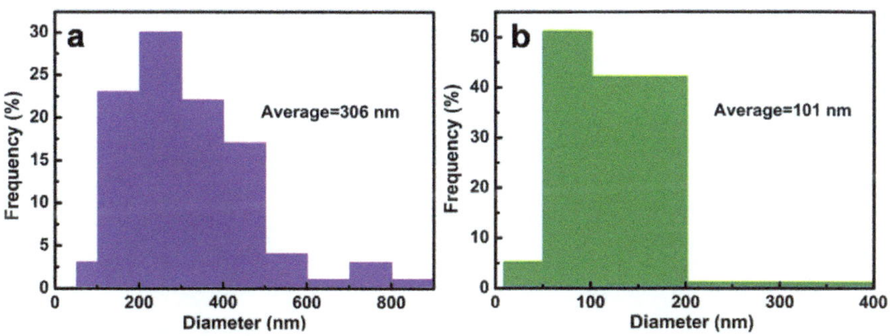

Fig. 8 The size distribution plot of nanoparticles of P-Hm (950) (a) and SA-Hm (950) (b) by measuring 100 randomly selected individual nanostructures.

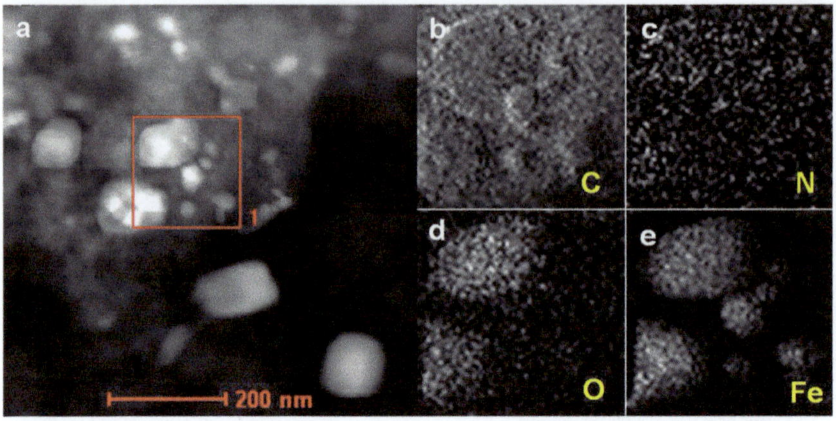

Fig. 9 HAADF STEM image of SA-Hm (950) (a) and elemental mapping of a selected area (b–e).

an important issue for electrocatalysts in direct methanol fuel cells (DMFCs). Thus, the ORR curves of SA-Hm (950) and Pt/C were collected in O_2-saturated 0.1 M KOH (aq.) with 0.5 M methanol at a scanning rate of 5 mV s^{-1} (Fig. 11d). The

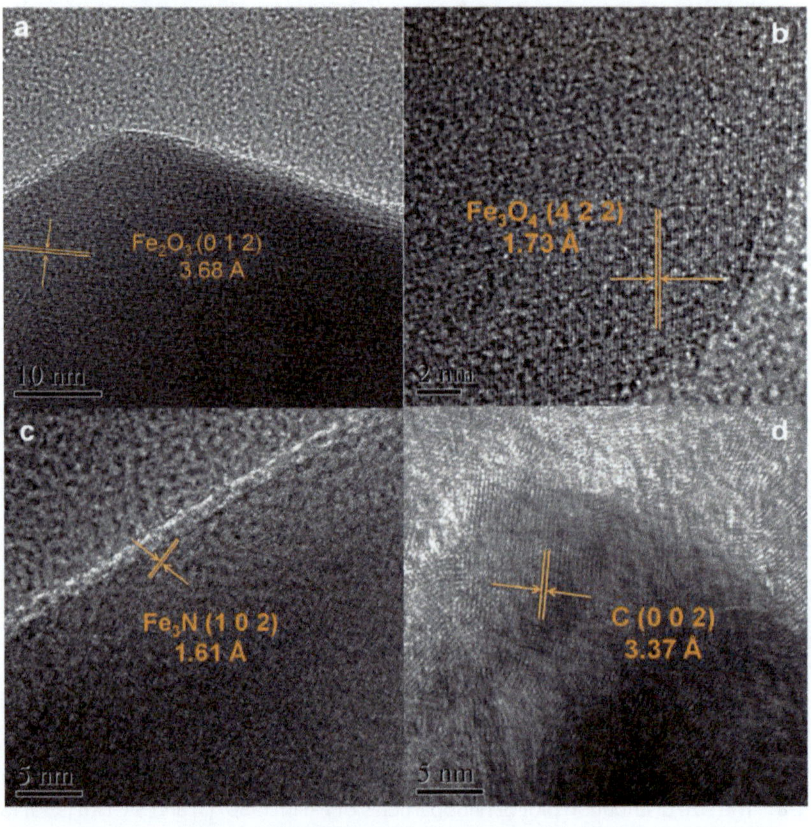

Fig. 10 HRTEM images of SA-Hm (950) of Fe_2O_3 (a); Fe_3O_4 (b); Fe_3N (c); and C (d).

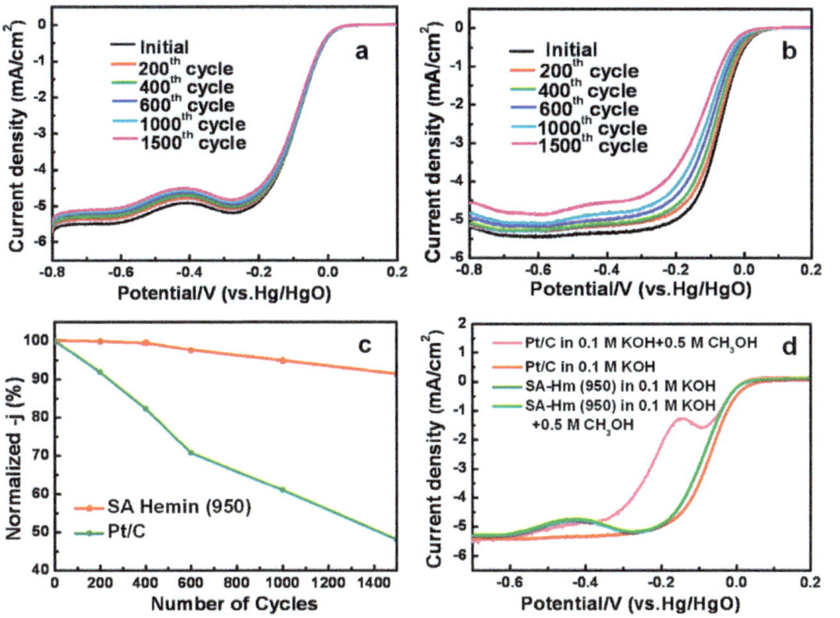

Fig. 11 ORR polarization curves of SA-Hm (950) (a), and commercial 20 wt% Pt/C (b) recorded at certain potential cycles (0, 200, 400, 600, 1000 and 1500) with a positive scanning rate of 5 mV s^{-1} at 1600 rpm in O_2-saturated 0.1 M KOH aqueous solution. The current density degradation of SA-Hm (950) and commercial Pt/C at 0 V (*vs.* Hg/HgO) during the potential cycling process from −0.4 to 0.2 V at a scan rate of 100 mV s^{-1} (c); CV curves of SA-Hm (950) and commercial Pt/C in N_2-saturated 0.1 M KOH (aq.) alone and 0.1 M KOH with 0.5 M methanol in the potential range of −0.8 to 0.2 V at a scan rate of 5 mV s^{-1} (d).

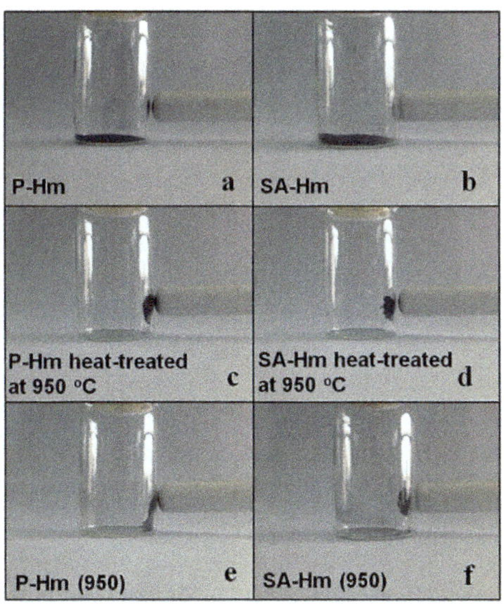

Fig. 12 Magnetic property of the obtained samples: (a) P-Hm, (b) SA-Hm, (c) P-Hm heat-treated at 950 °C, (d) SA-Hm heat-treated at 950 °C, (e) P-Hm (950), and (f) SA-Hm (950).

ORR curve of commercial Pt/C negatively shifted in the presence of 0.5 M methanol, suggesting a serious poisoning effect by methanol. Unlike commercial Pt/C, the ORR curve of SA-Hm (950) remained unchanged after the addition of methanol and demonstrated an excellent tolerance to methanol poisoning. SA-Hm is inactive to methanol oxidation, thus resulting in its tolerance to methanol.

Both P-Hm and SA-Hm show no magnetic properties (Fig. 12a and b). Interestingly, a magnetic property was observed after heat-treatment of P-Hm and SA-Hm owing to the formation of iron oxides (Fig. 12c and d). After leaching with acidic solution, the magnetic property remained, possibly due to encapsulation of iron oxides nanoparticles within the carbon materials (Fig. 12e and f). This magnetic property makes the electrocatalyst suitable for collection and re-utilization.

4 Conclusions

In summary, inexpensive hemin molecules were readily self-assembled into porous nanomaterials in water under ambient conditions. Subsequent heat-treatment of the unprecedented nanoporous materials led to a magnetic NNME with a much improved ORR activity relative to directly carbonized hemin without self-assembly. The enhanced ORR activity possibly originates from the higher surface area of nanoporous hemin after heat-treatment compared with that of micro-scale pristine hemin. Actually, the ORR activity of carbonized nanoporous hemin is comparable to commercial Pt/C in 0.1 M KOH. Furthermore, the heat-treated nanoporous hemin shows a much better performance than commercial Pt/C in terms of stability and tolerance to methanol. The molecular self-assembly of inexpensive metalloporphyrin combined with traditional pyrolysis may open a new avenue for the creation of novel NNMEs with a high electrocatalytic performance.

Acknowledgements

This work was partially supported by the National Key Basic Research Program of China (973 Project 2012CB215500), and National Natural Science Foundation of China (NSFC 21003114, 21103163 and 21373211).

References

1 Z. W. Chen, M. Waje, W. Z. Li and Y. S. Yan, *Angew. Chem., Int. Ed.*, 2007, **46**, 4060–4063.

2 M. K. Debe, *Nature*, 2012, **486**, 43–51.

3 P. P. Wells, Y. Qian, C. R. King, R. J. K. Wiltshire, E. M. Crabb, L. E. Smart, D. Thompsett and A. E. Russell, *Faraday Discuss.*, 2008, **138**, 273–285.

4 R. Borup, J. Meyers, B. Pivovar, Y. S. Kim, R. Mukundan, N. Garland, D. Myers, M. Wilson, F. Garzon, D. Wood, *et al.*, *Chem. Rev.*, 2007, **107**, 3904–3951.

5 R. Service, *Science*, 2002, **296**, 1222–1224.

6 B. Winther-Jensen, O. Winther-Jensen, M. Forsyth and D. R. MacFarlane, *Science*, 2008, **321**, 671–674.

7 E. Proietti, F. Jaouen, M. Lefevre, N. Larouche, J. Tian, J. Herranz and J. P. Dodelet, *Nat. Commun.*, 2011, **2**, 416.

8 M. Lefevre, E. Proietti, F. Jaouen and J. P. Dodelet, *Science*, 2009, **324**, 71–74.
9 G. Wu, K. L. More, C. M. Johnston and P. Zelenay, *Science*, 2011, **332**, 443–447.
10 F. Jaouen, E. Proietti, M. Lefevre, R. Chenitz, J. P. Dodelet, G. Wu, H. T. Chung, C. M. Johnston and P. Zelenay, *Energy Environ. Sci.*, 2011, **4**, 114–130.
11 Z. Chen, D. Higgins, A. Yu, L. Zhang and J. Zhang, *Energy Environ. Sci.*, 2011, **4**, 3167–3192.
12 R. Bashyam and P. Zelenay, *Nature*, 2006, **443**, 63–66.
13 E. Proietti, F. Jaouen, M. Lefèvre, N. Larouche, J. Tian, J. Herranz and J.-P. Dodelet, *Nat. Commun.*, 2011, **2**, 416.
14 S. K. Bikkarolla, P. Cumson, P. Joseph and P. Papakonstantinou, *Faraday Discuss.*, 2014, DOI: 10.1039/c4fd00088a.
15 C. Medard, M. Lefevre, J. P. Dodelet, F. Jaouen and G. Lindbergh, *Electrochim. Acta*, 2006, **51**, 3202–3213.
16 L. B. Wu, Y. Nabae, S. Moriya, K. Matsubayashi, N. M. Islam, S. Kuroki, M. Kakimoto, J. Ozaki and S. Miyata, *Chem. Commun.*, 2010, **46**, 6377–6379.
17 H. T. Chung, J. H. Won and P. Zelenay, *Nat. Commun.*, 2013, **4**, 1922.
18 Y. Liang, Y. Li, H. Wang, J. Zhou, J. Wang, T. Regier and H. Dai, *Nat. Mater.*, 2011, **10**, 780–786.
19 S. J. Guo, S. Zhang, L. H. Wu and S. H. Sun, *Angew. Chem., Int. Ed.*, 2012, **51**, 11770–11773.
20 X. Wang, X. Li, L. Zhang, Y. Yoon, P. K. Weber, H. Wang, J. Guo and H. Dai, *Science*, 2009, **324**, 768–771.
21 S. Wang, L. Zhang, Z. Xia, A. Roy, D. W. Chang, J. B. Baek and L. Dai, *Angew. Chem., Int. Ed.*, 2012, **51**, 4209–4212.
22 L. Qu, Y. Liu, J. B. Baek and L. Dai, *ACS Nano*, 2010, **4**, 1321–1326.
23 Y. Wang, J. Zhang, X. Wang, M. Antonietti and H. Li, *Angew. Chem., Int. Ed.*, 2010, **49**, 3356–3359.
24 D. Meadowcroft, *Nature*, 1970, **226**, 847–848.
25 E. J. Crumlin, E. Mutoro, Z. Liu, M. E. Grass, M. D. Biegalski, Y. L. Lee, D. Morgan, H. M. Christen, H. Bluhm and Y. Shao-Horn, *Energy Environ. Sci.*, 2012, **5**, 6081–6088.
26 R. Jasinski, *Nature*, 1964, **201**, 1212–1213.
27 R. Chen, H. Li, D. Chu and G. Wang, *J. Phys. Chem. C*, 2009, **113**, 20689–20697.
28 W. Li, A. Yu, D. C. Higgins, B. G. Llanos and Z. Chen, *J. Am. Chem. Soc.*, 2010, **132**, 17056–17058.
29 G. Dong, M. Huang and L. Guan, *Phys. Chem. Chem. Phys.*, 2012, **14**, 2557–2559.
30 A. Morozan, S. Campidelli, A. Filoramo, B. Jousselme and S. Palacin, *Carbon*, 2011, **49**, 4839–4847.
31 Y. Liu, Y. Y. Wu, G. J. Lv, T. Pu, X. Q. He and L. L. Cui, *Electrochim. Acta*, 2013, **112**, 269–278.
32 H. Alt, H. Binder and G. Sandsted, *J. Catal.*, 1973, **28**, 8–19.
33 R. Holze, I. Vogel and W. Vielstich, *J. Electroanal. Chem.*, 1986, **210**, 277–286.
34 M. Lefevre, J. P. Dodelet and P. Bertrand, *J. Phys. Chem. B*, 2005, **109**, 16718–16724.
35 M. Tsionsky and O. Lev, *J. Electrochem. Soc.*, 1995, **142**, 2132–2138.
36 E. Claude, T. Addou, J. M. Latour and P. Aldebert, *J. Appl. Electrochem.*, 1998, **28**, 57–64.
37 P. Gouerec and M. Savy, *Electrochim. Acta*, 1999, **44**, 2653–2661.

38 H. Jahnke, M. Schönborn and G. Zimmermann, *Top. Curr. Chem.*, 1976, **61**, 133–181.

39 C. Medard, M. Lefevre, J. Dodelet, F. Jaouen and G. Lindbergh, *Electrochim. Acta*, 2006, **51**, 3202–3213.

40 F. Charreteur, F. Jaouen and J. P. Dodelet, *Electrochim. Acta*, 2009, **54**, 6622–6630.

41 S. T. Chang, C. H. Wang, H. Y. Du, H. C. Hsu, C. M. Kang, C. C. Chen, J. C. Wu, S. C. Yen, W. F. Huang and L. C. Chen, *Energy Environ. Sci.*, 2012, **5**, 5305–5314.

42 C. H. Wang, H. C. Huang, S. T. Chang, Y. C. Lin and M. F. Huang, *RSC Adv.*, 2014, **4**, 4207–4211.

43 L. Wu, Y. Nabae, S. Moriya, K. Matsubayashi, N. M. Islam, S. Kuroki, M.-a. Kakimoto, J. i. Ozaki and S. Miyata, *Chem. Commun.*, 2010, **46**, 6377–6379.

44 Z. Xiang, Y. Xue, D. Cao, L. Huang, J. F. Chen and L. Dai, *Angew. Chem. Int. Ed.*, 2014, **53**, 2433–2437.

45 M. Jahan, Q. Bao and K. P. Loh, *J. Am. Chem. Soc.*, 2012, **134**, 6707–6713.

46 J. Y. Cheon, T. Kim, Y. Choi, H. Y. Jeong, M. G. Kim, Y. J. Sa, J. Kim, Z. Lee, T. H. Yang and K. Kwon, *Sci. Rep.*, 2013, **3**, 2715.

47 I. Hijazi, T. Bourgeteau, R. Cornut, A. Morozan, A. Filoramo, J. Leroy, V. Derycke, B. Jousselme and S. Campidelli, *J. Am. Chem. Soc.*, 2014, **136**, 6348–6354.

48 Y. Xie, H. Li, C. Tang, S. Li, J. Li, Y. Lv, X. Wei and Y. Song, *J. Mater. Chem. A*, 2014, **2**, 1631–1635.

49 R. Jiang and D. Chu, *J. Power Sources*, 2014, **245**, 352–361.

50 R. Jiang, D. T. Tran, J. McClure and D. Chu, *Electrochem. Commun.*, 2012, **19**, 73–76.

51 Z. X. Liang, H. Y. Song and S. J. Liao, *J. Phys. Chem. C*, 2011, **115**, 2604–2610.

52 J. Xu, T. Zhao and L. Zeng, *Int. J. Hydrogen Energy*, 2012, **37**, 15976–15982.

53 B. R. Wood, S. J. Langford, B. M. Cooke, J. Lim, F. K. Glenister, M. Duriska, J. K. Unthank and D. McNaughton, *J. Am. Chem. Soc.*, 2004, **126**, 9233–9239.

54 Q. Wang, Z. Zhou, D. Chen, J. Lin, F. Ke, G. Xu and S. Sun, *Sci. China: Chem.*, 2010, **53**, 2057–2062.

55 A. J. Bard and L. R. Faulkner, *Electrochemical Methods: Fundamentals and Applications*, Wiley, New York, 2nd edn, 2001.

56 M. S. Islam, Y. Kusumoto and M. Abdulla Al Mamun, *Mater. Lett.*, 2012, **66**, 165–167.

57 C. L. Snow, Q. Shi, J. Boerio Goates and B. F. Woodfield, *J. Phys. Chem. C*, 2010, **114**, 21100–21108.

58 Z. Wen, S. Ci, F. Zhang, X. Feng, S. Cui, S. Mao, S. Luo, Z. He and J. Chen, *Adv. Mater.*, 2012, **24**, 1399–1404.

59 Z.-S. Wu, S. Yang, Y. Sun, K. Parvez, X. Feng and K. Müllen, *J. Am. Chem. Soc.*, 2012, **134**, 9082–9085.

60 Z. Xiang, Y. Xue, D. Cao, L. Huang, J. F. Chen and L. Dai, *Angew. Chem., Int. Ed.*, 2014, **53**, 2433–2437.

61 K. Gong, F. Du, Z. Xia, M. Durstock and L. Dai, *Science*, 2009, **323**, 760–764.

Faraday Discussions

PAPER

A comparative study of CO adsorption on tetrahexahedral Pt nanocrystals and interrelated Pt single crystal electrodes by using cyclic voltammetry and *in situ* FTIR spectroscopy

Hai-Xia Liu, Na Tian,* Jin-Yu Ye, Bang-An Lu, Jie Ren, Zhi-Chao Huangfu, Zhi-You Zhou and Shi-Gang Sun*

Received 19th June 2014, Accepted 22nd August 2014
DOI: 10.1039/c4fd00136b

This study focuses on CO adsorption at tetrahexahedral Pt nanocrystals (THH Pt NCs) by using cyclic voltammetry and *in situ* FTIR spectroscopy. Since the electrochemically prepared THH Pt NCs in this study are enclosed by {730} facets which could be considered by a subfacet configuration of 2{210} + {310}, we have also studied CO adsorption on the interrelated Pt(310) and Pt(210) single crystal electrodes as a comparison. Cyclic voltammetry results demonstrated that CO adsorbs dominantly on the (100) sites of THH Pt NCs at low CO coverage ($\theta_{CO} \leq 0.135$), while on both (100) and (110) sites at higher CO coverage. On ordered Pt(310) and Pt(210), *i.e.* they were flame annealed and then cooled in H_2 + Ar, CO adsorption also illustrates relative priority on (100) sites at low CO coverage; while at high CO coverage or on oxygen-disordered Pt(310) and Pt(210) when they were cooled in air after flame annealing, the adsorption of CO presents a weak preference on (100) sites of Pt(310) and even no preference at all on (100) sites of Pt(210). *In situ* FTIR spectroscopic studies illustrated that CO adsorption on THH Pt NCs yields anomalous infrared effects (AIREs), which are depicted by the Fano-like IR feature on a dense distribution (60 μm^{-2}) and the enhancement of abnormal IR absorption on a sparse distribution (22 μm^{-2}) of THH Pt NCs on glassy carbon substrate. Systematic investigation of CO coverage dependence of IR features revealed that, on THH Pt NCs, the IR band center (ν_{COL}) of linearly bonded CO (CO_L) is rapidly shifted to higher wavenumbers along with the increase of CO coverage to 0.184, yielding a fast linear increase rate with a high slope ($d\nu_{COL}/d\theta_{CO}^{IR}$ = 219 cm^{-1}); when $\theta_{CO}^{IR} > 0.184$, the increase of ν_{COL} with θ_{CO} slows down and deviates drastically from linearity. In contrast, the ν_{COL} on the ordered Pt(310) electrode maintains a linear increase with θ_{CO}^{IR} for the whole range of θ_{CO}^{IR} variation, and gives a much smaller increase rate of slope 74.3 cm^{-1}. The significant differences in CO

State Key Lab of Physical Chemistry of Solid Surfaces, Department of Chemistry, College of Chemistry and Chemical Engineering, XiamenUniversity, Xiamen, 361005, China. E-mail: tnsd@xmu.edu.cn; sgsun@xmu.edu.cn

adsorption behavior on THH Pt NCs and on interrelated Pt single crystal planes demonstrated clearly the unique properties of nanoparticles enclosed by high-index facets.

1. Introduction

High-index faceted metal nanocrystals have received great attention due to their high catalytic activity originating from the high density of low-coordinated step atoms on their surface.[1,2] In recent years, we have devoted our effort to the synthesis of high-index faceted metal nanocrystals by developing electrochemical methods, in which the square-wave potential (SWP) method has a unique advantage of synthesizing metal nanocrystals of high surface energy.[3] Our systematic studies led to the successful preparation of a series of Pt-group metal catalysts enclosed with high-index facets, including: (1) the tetrahexahedral metal nanocrystals (THH metal NCs) with $\{hk0\}$ high-index facets, such as THH Pt NCs,[3] THH Pd NCs[4] and THH Rh NCs;[5] (2) the trapezohedral metal nanocrystals (TPH metal NCs) with $\{hkk\}$ high-index facets, such as TPH Pt NCs[6] and TPH Pd NCs;[7] (3) the hexoctahedral Pt nanocrystals (HOH Pt NCs) with $\{hkl\}$ high-index facets;[8] and (4) the Pd- and Pt-based alloy metal nanocrystals with high-index facets, such as THH PdPt NCs,[9] THH PtRh NCs and TPH PtRh NCs.[10] In the SWP method, the parameters that govern the composition, the type of high-index facets and the size of high-index faceted metal nanocrystals consist mainly of metal precursors and their concentration in solution, the SWP (the frequency, the upper and lower limit potentials), and the growth time. Generally, the high-index facets are mainly created by the repetitive adsorption–desorption of oxygen species mediated by the SWP.[3] The growth rate of nanocrystals is controlled mainly by the lower limit potential. As for different metals, the upper and lower limit potentials of the SWP should be adjusted and optimized on the basis of the adsorption–desorption behavior of oxygen species on them. Usually, the upper limit potential for Pt(1.0–1.20 V *vs.* SCE) is significantly higher than that for Pd and Rh (about 0.70 V), because the latter two elements can be oxidized at lower potentials and the formed oxygen species on them are more difficult to be reduced electrochemically than those on Pt.

Although the Miller index that represents the atomic arrangement of the facets of high-index faceted metal nanocrystals can be determined by transmission electron microscopy (TEM), the real surface structure and its evolution under chemical conditions is less studied, though it is most important for understanding the origin of the excellent catalytic properties of high-index faceted metal nanocrystals. The most frequently prepared THH Pt NCs by electrochemically shape-controlled synthesis are bounded with {730} facets.[3,8] It is known that a single crystal plane with a (730) index could be considered as a subfacet configuration of 2(210) + (310), *i.e.* it contains periodically two (210) subfacets and one (310) subfacet. Although the surface structure–catalytic functionality of metal single crystal planes used as model electrocatalysts has been investigated extensively, especially in case of platinum group metals,[11–13] the surface structure–catalytic functionality of metal nanoparticles employed as real catalysts in industrial applications is far from understood. The shape-controlled synthesized metal nanoparticles, such as THH Pt NCs, expose smooth facets with well-defined surface atomic structure as single crystal planes, while holding a nanometer particle size, and particular corner and edge sites. So, the surface structure of the nanoparticle is more complex, which could consequently lead to very different properties of the nanoparticles from the interrelated single crystal planes with the same atomic structure of facets that enclose the nanoparticles.

CO is one of most attractive molecules in electrocatalysis due to the following reasons: (i) CO is the poisonous intermediate involved in oxidation or dissociation of most organic molecule fuels;[14,15] (ii) the blank $2\pi^*$ orbital of CO can accept d-electrons from transition metals to form d–π^* back-donation bond, enhancing consequently the interaction between CO and Pt sites;[16,17] and (iii) CO electro-oxidation is a surface structure sensitive reaction.[18,19] Due to the strong interaction between CO and Pt, CO usually serves as a probe molecule to detect the surface active sites of Pt catalysts.

It has been reported that, for CO adsorption on basal planes of Pt single crystal, the bridge-bonded CO (CO_B) is favorable on Pt(111) and Pt(100) in both electrochemical and UHV conditions at low coverage,[20] while the linearly-bonded CO (CO_L) is dominant on the Pt(110) single crystal electrode.[20,21] Most studies concerning CO adsorbed on Pt high-index planes indicate that CO prefers to adsorb on steps, but CO randomly adsorbs on steps and terraces on the surface of Pt(S) [n(111) × (110)] single crystal planes.[22]

In the current study, we employ CO as probe molecule to investigate the surface sites on THH Pt NCs by cyclic voltammetry (CV) and *in situ* FTIR spectroscopy. The results are compared with interrelated Pt(210) and Pt(310) single crystal planes, in order to understand the unique properties of the THH Pt NCs.

2. Experimental

2.1 Materials and apparatus

H_2PtCl_6 (99.99%, Sigma Aldrich), sulfuric acid (96%, Merck), perchloric acid (super pure) were used as received. All solutions were prepared using super-pure water (18 MΩ cm) generated from a Milli-Q system (Nihon Millipore Ltd.). Carbon monoxide gas (99.99%) was purchased from Xiamen Xinhang Co.

THH Pt NCs were prepared electrochemically on a glassy carbon (GC) electrode ($\Phi = 6$ mm) by using the same procedure described in our previous work.[3] Prior to depositing THH Pt NCs, the GC was polished mechanically by using successively alumina powder of sizes 5, 1, 0.3 μm, and finally cleaned in an ultrasonic bath. The counter electrode and reference electrode are platinum foil and a saturated calomel electrode (SCE), respectively. The electrochemical experiments were carried out in a standard three electrode cell at room temperature (~25 °C). Potential control was achieved with a PAR 263A potentiostat/galvanostat operated through home-developed software.

All solutions were purged by bubbling pure N_2 gas (99.99%) for 15 min before experiments. A nitrogen gas flow was maintained above the solution during measurement to prevent interference of atmospheric oxygen and impurities.

2.2 Preparation of THH Pt nanocrystals

THH Pt NCs were prepared by using a square-wave potential method.[10] Briefly, THH Pt NCs were deposited on GC electrode by applying a potential step from

1.20 V to a reduction potential of −0.30 V, holding for 0.14 s in 1.93 mM H_2PtCl_6 + 0.1 M H_2SO_4 solution. A square-wave potential of frequency 100 Hz with lower potential 0.03 V and upper potential 1.06 V was applied for 20 min to grow the THH Pt NCs. The deposited THH Pt NCs were characterized by cyclic voltammetry in 0.1 M H_2SO_4 solution and then imaged by scanning electron microscopy (SEM, Hitachi S-4800).

2.3 Preparation of Pt single crystal electrodes

Platinum single crystal beads were oriented, cut and polished according to Clavilier's method.[23] The electrode was annealed in a H_2–O_2 flame at about 1200 °C making it clean and ordered, and then cooled down to room temperature in H_2 + Ar flow or air (atmosphere). The flame treated electrode was immediately transferred into an electrochemical cell or *in situ* FTIR spectroelectrochemical cell with its surface protected by super-pure water for further investigation.

2.4 CO adsorption experiments

The THH Pt NCs electrode was placed in a 0.1 M H_2SO_4 solution. In order to obtain a low CO coverage, with the electrode potential held at −0.10 V CO gas was bubbled through the solution. The CO coverage was controlled by adjusting the time of CO bubbling. The CO supply was then turned off and N_2 gas was introduced to purge the solution for 10 min in order to remove completely dissolved CO in solution. Finally, oxidation of the adsorbed CO (CO_{ad}) was investigated by cyclic voltammetry. The electrode potential was scanned starting from the initial potential of −0.10 V down to −0.25 V, and then a complete cyclic voltammogram with an upper limit of 0.75 V was recorded. The scan rate was 50 mV s^{-1}.

The coverage of CO (θ_{CO}) was calculated by the charge of CO_{ad} oxidation to that of one monolayer CO_{ad} oxidation, and was defined as:

$$\theta_{CO} = \frac{Q^{CO}}{Q^{CO}_{1ML-CO}} \quad (1)$$

Where Q^{CO} and Q^{CO}_{1ML-CO} are the charges of CO_{ad} oxidation of the sub monolayer CO_{ad} and one monolayer CO_{ad}, respectively. It is worthwhile noting that this definition stands for the ratio of adsorbed CO *versus* the saturation adsorption of a CO_{ad} monolayer.

The procedure of yielding different coverage of CO adsorbed on Pt single crystal electrodes was similar with that of CO adsorption on the THH Pt NCs. In order to prevent the perturbation of surface atomic structure at high electrode potentials, the cyclic potential range was limited between −0.25 and 0.60 V in cyclic voltammetry studies of Pt single crystal electrodes.

The electrochemical active surface area (ECSA) of the THH Pt NCs electrode was measured according the method reported in the literature[24] through ECSA = Q_H/(210 μC cm^{-2}), where Q_H is the charge of hydrogen desorption by integration of the CV curve recorded in 0.1 H_2SO_4 solution between −0.25 to 0.15 V.

2.5 Electrochemical *in situ* FTIR reflection spectroscopy

In situ FTIR spectroscopy experiments were carried out on a Nexus 870 FTIR spectrometer (Nicolet) equipped with a liquid-nitrogen-cooled MCT-A detector. A thin layer *in situ* FTIR cell with CaF_2 window was utilized.[25] A total of 400 single-beams at a resolution of 2 cm^{-1} were co-added to increase the signal-to-noise ratio. The *in situ* FTIR spectra of various coverage of CO_{ad} was measured using the following procedure: (i) holding the potential at −0.10 V, bubbling a few bubbles of CO; (ii) after removing completely dissolved CO in solution by bubbling N_2, the electrode was pushed to the window and the sample spectra were collected at a series sample potentials E_S; (iii) the reference spectrum was acquired finally at E_R (0.75 V for THH Pt NCs, and 0.60 V for Pt single crystal electrodes), at which potential the adsorbed CO is oxidized completely. The resulting spectra were reported as the relative change in reflectivity, that is,

$$\frac{\Delta R}{R}(E_S) = \frac{R(E_S) - R(E_R)}{R(E_R)} \tag{2}$$

where $R(E_S)$ and $R(E_R)$ are single-beam spectra collected at the sample potential E_S and reference potential E_R, respectively.

3. Results and discussions

3.1 Cyclic voltammetry studies

Fig. 1A shows the SEM image of THH Pt NCs with a distribution density of about 60 μm^{-2}. The high magnification SEM image is shown in the inset (a). The surface facets of the THH Pt NCs prepared under present conditions are mainly {730} facets,[8] which can be denoted as 2{210} + {310} or 3{100} × {110} + 2{100} × {110}. The inset (b) is an atomic model of a THH Pt NC enclosed with {730} facets. The size distribution histogram of THH Pt NCs in Fig. 1B indicates that the mean size is about 54 nm.

Fig. 1C and D illustrate the CV curves of CO_{ad} stripping on THH Pt NCs at various CO_{ad} coverages. From the features of hydrogen adsorption–desorption on a clean THH Pt NCs ($\theta_{CO} = 0$), two potential regions may be discerned. One is the potential region between −0.25 V and −0.106 V for hydrogen desorption on (110) sites, another potential region is from −0.106 V to 0.15 V for hydrogen desorption on (100) sites. We can observe from Fig. 1C that the hydrogen desorption peak at around −0.02 V (corresponding to Pt(100) sites) is blocked faster than that at −0.16 V (corresponding to Pt(110) sites). At low CO coverage ($\theta_{CO} \leq 0.135$), the peak at −0.02 V was obviously blocked, while the peak at −0.16 V almost remains unblocked. This implies that CO adsorbs dominantly on Pt(100) sites at such a low CO_{ad} coverage. When $\theta_{CO} \geq 0.175$, CO adsorbs on both Pt(100) sites and (110) sites, as illustrated by the simultaneous blocking of both hydrogen adsorption peaks.

The voltammogram of the THH Pt NCs ($\theta_{CO} = 0$) presented in Fig. 1C exhibits a strong peak for oxygen adsorption around 0.6 V with its onset potential at 0.46 V. However, for the stripping of CO_{ad} with low coverage, the onset potential of CO_{ad} oxidation is measured to be about 0.27 V, which is negatively shifted by about 0.19 V than the onset potential of Pt–OH formation on the surface of THH Pt NCs. This may be attributed to the adsorption of H_2O on Pt(110) step sites that promoted the oxidation of CO adsorbed on Pt(100) sites. Lebedeva *et al.* have studied in detail CO stripping on a series of Pt(S) [n(111) × (110)] single crystal electrodes in acidic media,[26,27] and demonstrated that the rate constant for CO_{ad} oxidation is a linear function of the step density, suggesting that indeed OH_{ad} formation occurs preferentially on the (110) steps. A previous study[28] of CO_{ad} monolayer oxidation

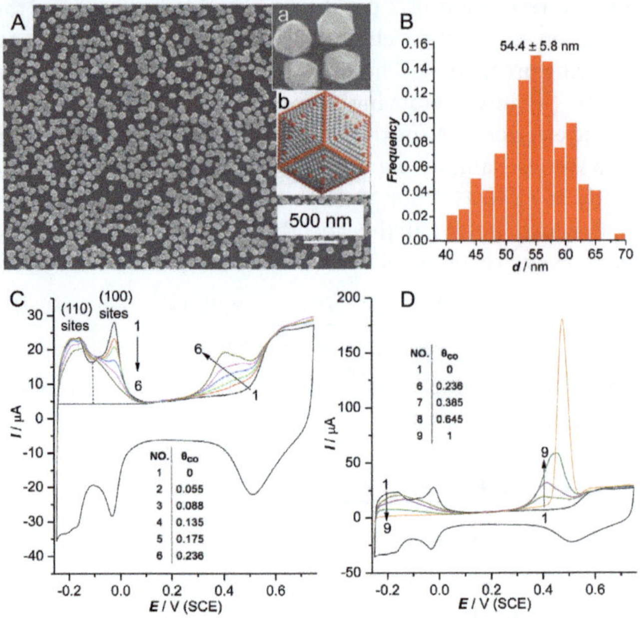

Fig. 1 (A) SEM image of THH Pt NCs. The inset (a) is a high magnification image of the sample, which illustrates clearly the THH shape, and the inset (b) depicts the atomic model of a THH Pt NC with {730} facets. (B) Size distribution histogram for the THH Pt NCs by counting 200 nanoparticles. (C and D) CV curves of CO stripping on THH Pt NCs electrodes with different CO coverage, in 0.1 M H_2SO_4 solution, scan rate 50 mV s^{-1}.

on stepped Pt(s) [(n − 1)(100) × (110)] surfaces indicated that for the short terraces ($n < 7$) the CO_{ad} oxidation peak shifts toward lower potential as the step density increases, and for the long terraces ($n > 7$) the result is reverse. On the THH Pt NCs enclosed by {730} facets, the value of n is small than 7, so the results of the high activity for CO_{ad} oxidation is consistent with the literature results. As illustrated in Fig. 1D, the onset potential and the peak potential are both shifted toward higher potential along with increasing the CO coverage. It can be observed that the oxidation of a monolayer of CO_{ad} gives a small per-oxidation current between 0.32–0.41 V, and a large and sharp current peak at about 0.475 V.

To understand the surface structure effect of the THH Pt NCs, we investigated the behavior of CO adsorption on interrelated Pt single crystal electrodes. Fig. 2 shows the cyclic voltammograms of hydrogen desorption potential region (−0.25–0.15 V) recorded on ordered (cooled in H_2 + Ar) and oxygen-disordered (cooled in air) Pt(210) and Pt(310) single crystal electrodes at different CO_{ad} coverages, respectively. It is known that the surface cooled in air is usually less ordered than that cooled in H_2 + Ar, since the oxygen adsorption can induce surface defects.[29] We observe from Fig. 2 that CO adsorption can suppress the hydrogen desorption current on both (100) and (110) sites for the whole CO_{ad} coverage. This is different from that on THH Pt NCs, which show a dominant preferential adsorption on (100) sites at low CO coverage of $\theta_{CO} \leq 0.135$.

We can use the ratio of the hydrogen desorption charge between −0.25 and −0.106 V over the total hydrogen desorption charge to estimate approximately the proportion of (110) sites, and that between −0.106 and 0.15 V for the proportion

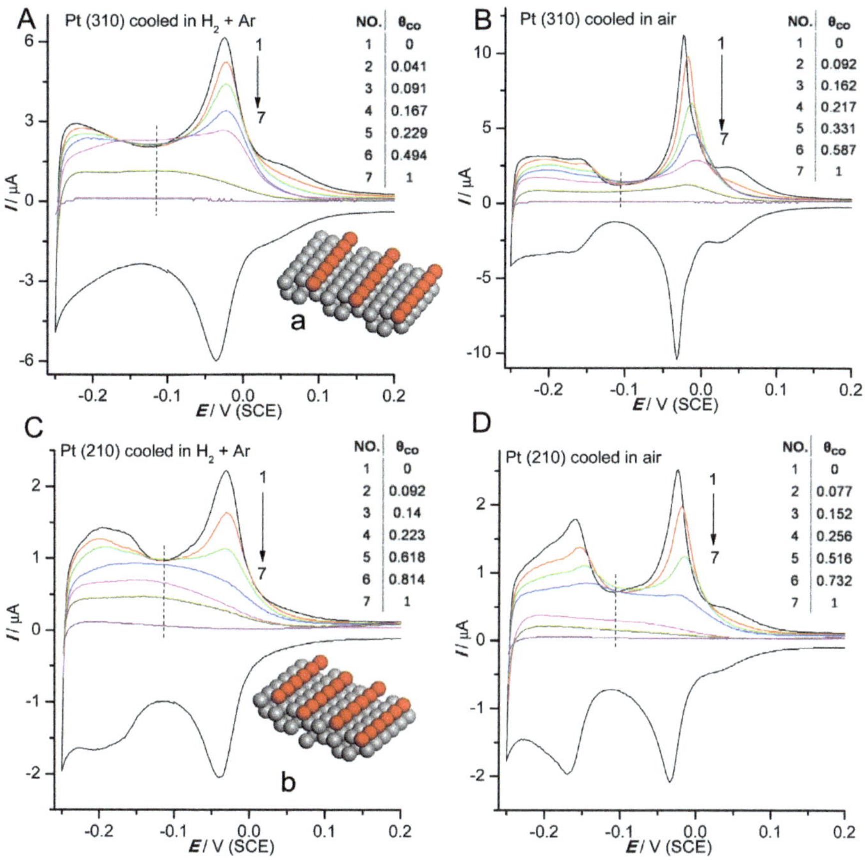

NO.	θ_{CO}
1	0
2	0.041
3	0.091
4	0.167
5	0.229
6	0.494
7	1

NO.	θ_{CO}
1	0
2	0.092
3	0.162
4	0.217
5	0.331
6	0.587
7	1

NO.	θ_{CO}
1	0
2	0.092
3	0.14
4	0.223
5	0.618
6	0.814
7	1

NO.	θ_{CO}
1	0
2	0.077
3	0.152
4	0.256
5	0.516
6	0.732
7	1

Fig. 2 Cyclic voltammograms of the hydrogen adsorption region for Pt(210) and Pt(310) single crystal electrodes at different CO coverage. (A) Pt(310) cooled in H_2 + Ar; (B) Pt(310) cooled in air; (C) Pt(210) cooled in H_2 + Ar; (D) Pt(210) cooled in air. 0.1 M H_2SO_4 solution, scan rate 50 mV s^{-1}. The insets represent the atomic models of Pt(310) (a) and Pt(210) (b), in which the red atoms are the (110) steps and the gray ones are (100) terraces.

of (100) sites on the clean surface of the THH Pt NCs, *i.e.* $\theta_{CO} = 0$. The experimental result indicates a proportion of (110) sites at 51%, which is slightly larger than that from theoretical calculation of a Pt(730) single crystal plane (43%). This may be caused by corner and edge atoms on the THH Pt NCs. For the Pt(210) and Pt(310) cooled in air, the proportions of CO_{ad} blocked (110) sites and (100) sites were measured from the hydrogen adsorption charges in the region between −0.25 and −0.106 V and that between −0.106 and 0.15 V (the separation potential −0.106 corresponding to the lowest current between the two regions) over the total hydrogen adsorption charge; and for Pt(310) and Pt(210) cooled in H_2 + Ar, the proportions of CO_{ad} blocked (110) sites and (100) sites were measured from the hydrogen desorption charges in the region between −0.25 and −0.114 V and that between −0.114 and 0.15 V over the total hydrogen adsorption charge. The results were nearly 1 : 1 ((110) : (100)) for Pt(210) and 1 : 2 for Pt(310), respectively, which are the same as those from theoretical calculation of these two single crystal planes.

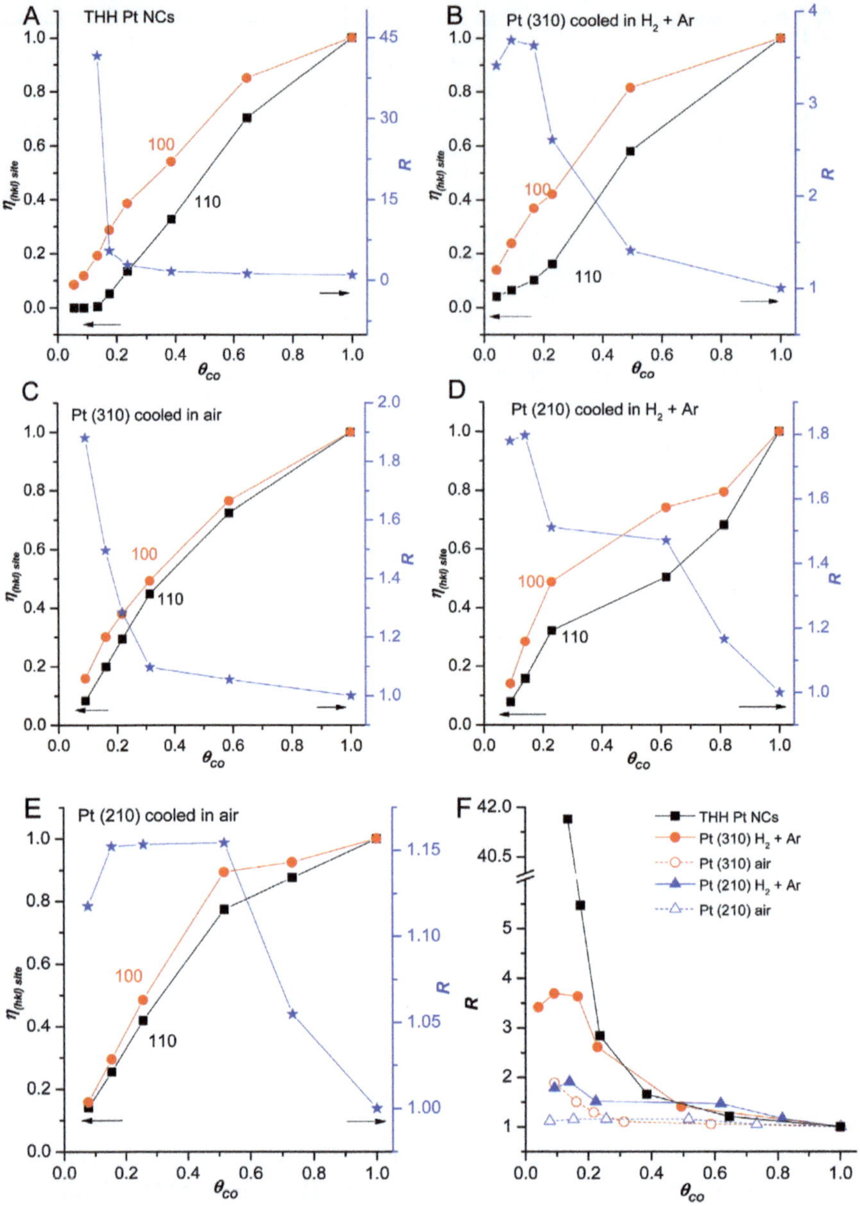

Fig. 3 Coverage dependence of $\eta_{(hkl)\ \mathrm{sites}}$ and R on THH Pt NCs (A), Pt(310) cooled in H_2 + Ar (B), Pt(310) cooled in air (C), Pt(210) cooled in H_2 + Ar (D), and Pt(210) cooled in air (E). Comparison of R measured on all investigated electrodes (F).

In order to analyze the preferential adsorption of CO on the (100) sites clearly, we define the proportion of (100) sites ($\eta_{(100)\ \mathrm{sites}}$) and (110) sites ($\eta_{(110)\ \mathrm{sites}}$) that are blocked by adsorbed CO at various CO coverages on THH Pt NCs and Pt(210), Pt(310) single crystal electrodes. The (110) sites and (100) sites blocked by CO adsorption were obtained by subtracting the hydrogen desorption charge integrated from the CV curves of the corresponding potential regions mentioned above before CO stripping from that after CO stripping. The proportion ($\eta_{(hk0)\text{ sites}}$) of CO-blocked (100) or (110) sites was calculated by CO blocked (100) sites (or (110) sites) over the total (100) sites (or (110) sites) as:

$$\eta_{(hk0)-\text{sites}} = \frac{Q^{\text{H}}_{(1k0)\text{ sites}} - Q^{\text{H}}_{(1k0)\text{ sites}-\text{CO}}}{Q^{\text{H}}_{(1k0)\text{ sites}}} = 1 - \frac{Q^{\text{H}}_{(1k0)\text{ sites}-\text{CO}}}{Q^{\text{H}}_{(1k0)\text{ sites}}} \quad \left(k = 1 \text{ or } 0\right) \tag{3}$$

Where $Q^{\text{H}}_{(1k0)\text{ sites-CO}}$ is the charge of hydrogen desorption on (1*k*0) sites with adsorbed CO, $Q^{\text{H}}_{(1k0)\text{ sites}}$ is the charge of hydrogen desorption on (1*k*0) sites after CO_{ad} stripping without any CO adsorption.

The results in Fig. 3A indicate that, on THH Pt NCs, the $\eta_{(100)\text{ sites}}$ is about 0.084, 0.117 and 0.192 corresponding respectively to θ_{CO} at 0.055, 0.088 and 0.135, and $\eta_{(110)\text{ sites}}$ is always near zero at these low CO coverages, indicating that CO adsorbed predominantly on (100) sites at such a low CO coverage ($\theta_{CO} \leq 0.135$); following the increase of θ_{CO} to 0.175, the $\eta_{(110)\text{ sites}}$ begin increasing; with further increasing CO coverage, both $\eta_{(100)\text{ sites}}$ and $\eta_{(110)\text{ sites}}$ increase rapidly, but $\eta_{(100)\text{ sites}}$ are always slightly bigger than $\eta_{(110)\text{ sites}}$ for all θ_{CO} except $\theta_{CO} = 1$, at which the $\eta_{(100)\text{ sites}}$ equal the $\eta_{(110)\text{ sites}}$. These results suggest that CO adsorption is always preferred on the (100) sites of the THH Pt NCs' surface. The degree of preferential adsorption of CO on (100) sites to that on (110) sites is expressed as the ratio (R) of $\eta_{(100)\text{ sites}}$ over $\eta_{(110)\text{ sites}}$, *i.e.*,

$$R = \frac{\eta_{(100)\text{ sites}}}{\eta_{(110)\text{ sites}}} \tag{4}$$

R stands for the priority of CO adsorption on (100) sites. The larger the value of R, the greater the priority of CO adsorption on (100) sites. The CO coverage dependence of R for all electrodes investigated are compared in Fig. 3. For the THH Pt NCs, R is very large at low θ_{CO}, and is decreased with the increase of θ_{CO}, *e.g.*, 41.6 at $\theta_{CO} = 0.135$ and 5.5 at $\theta_{CO} = 0.175$.

For Pt(310) and Pt(210) single crystal electrodes cooled in H_2 + Ar or in air after flame annealing, as shown in Fig. 3B–E, the values of $\eta_{(100)\text{ sites}}$ are always bigger than those of $\eta_{(110)\text{ sites}}$, and the R occurs in a decreasing trend with the increase of θ_{CO}. The Pt(310) cooled in H_2 + Ar (Fig. 3B) demonstrates the largest difference between $\eta_{(100)\text{ sites}}$ and $\eta_{(110)\text{ sites}}$, and therefore it has the largest R among all the single crystal electrodes studied. For the ordered Pt(310), R is 3.6 at $\theta_{CO} \leq 0.167$, while R is about 1.5 at $\theta_{CO} = 0.162$ for oxygen-defected Pt(310). When $\theta_{CO} \leq 0.14$ the R is about 1.8, measured on the ordered Pt(210), but the R is near to 1 for the oxygen-defected Pt(210), which indicates that there is almost no priority of CO adsorption on (100) sites on the oxygen-defect Pt(210) electrode.

The variations of R against θ_{CO} for all electrodes investigated are compared in Fig. 3F. It is clear that at low CO coverage ($\theta_{CO} < 0.2$) the THH Pt NCs electrode has the largest R value among all the electrodes. The preferential adsorption of CO on (100) sites for all electrodes is decreased in the order of THH Pt NCs > ordered Pt(310) > oxygen-defected Pt(310) ≈ ordered Pt(210) > oxygen-defected Pt(210). It is interesting to see that the R values of ordered Pt(310) and Pt(210) are always larger than those of oxygen-defected Pt(310) and Pt(210), respectively, indicating that a high R value corresponds to a more ordered surface. These results imply that the small facets on the surface of THH Pt NCs are even more ordered than those of Pt(310) and Pt(210) single crystal planes.

3.2 *In situ* FTIR spectroscopic studies

CO adsorption on THH Pt NCs and Pt(310) single crystal electrodes is further studied by using *in situ* FTIR spectroscopy. The CO coverage in the *in situ* FTIR studies was defined as,

$$\theta_{CO}^{IR} = \frac{I_{CO}^{COL} + I_{CO}^{COB}}{I_{1ML-CO}^{COL} + I_{1ML-CO}^{COB}} \tag{5}$$

In which the I_{CO}^{COL} and I_{CO}^{COB} are the integrated intensity of the IR bands of CO_L and CO_B at different coverage, respectively. I_{1ML-CO}^{COL} and I_{1ML-CO}^{COB} are the integrated intensity of the IR bands of CO_L and CO_B for CO saturation adsorption (one monolayer CO).

The *in situ* FTIR spectra of CO adsorption on the THH Pt NCs with a high distribution density of 60 μm^{-2} displays three obvious IR bands, as illustrated in Fig. 4A. One band between 2000 and 2100 cm^{-1} and another one around 1782 cm^{-1} are assigned respectively to linearly bonded CO (CO_L) and bridge coordinated CO (CO_B) species according to the previous study.[30] The positive-going band around 2343 cm^{-1} is attributed to IR absorption of solution CO_2 species produced solely from CO_{ad} oxidation at reference potential (0.75 V). From the definition of *in situ* FTIR spectra by eqn (2), the IR absorption of CO_{ad} at E_S should yield negative-going bands. However, the CO_B band is in the opposite direction to negative-going, and the CO_L band appears in an asymmetric bipolar shape with

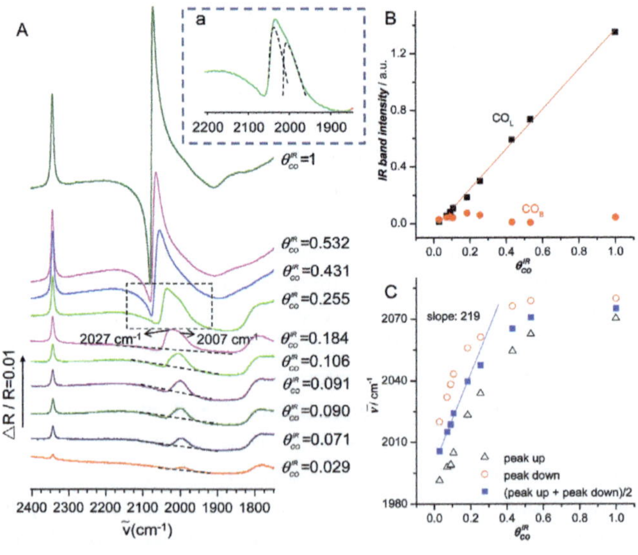

Fig. 4 (A) Coverage dependence of *in situ* FTIR spectra for CO adsorbed on THH Pt NCs of high distribution density (60 μm^{-2}) in 0.1 M $HClO_4$ solution, $E_S = -0.10$ V, $E_R = 0.75$ V (SCE). (B) Coverage dependence of band intensities of CO_L and CO_B on THH Pt NCs in Fig. 4A. (C) The IR band center of CO_L plotted against the coverage of CO (θ_{CO}^{IR}) in Fig. 4A, hollow dark triangles represent the center of the positive peak of CO_L asymmetric bipolar band; the hollow red circles stand for the center of the negative peak of the CO_L asymmetric bipolar band; and the filled blue squares are the average wavenumber of the centers of positive and the negative peaks.

enhanced intensity. These are typical anomalous IR effects (AIREs) of nanomaterials for CO adsorption,[31–34] among which the asymmetric CO_L bipolar IR band has been denoted as a Fano-like IR spectral line shape.[34]

Fig. 4B demonstrates the IR band intensity of CO_L (I_{COL}) and CO_B (I_{COB}) measured in spectra of THH Pt NCs of high distribution density. I_{COL} is weaker than I_{COB} at very low CO coverage ($\theta_{CO}^{IR} = 0.029$), but I_{COL} increases more quickly than I_{COB}, yielding a linear variation with the increase of CO coverage. In contrast, the I_{COB} decreases at $\theta_{CO}^{IR} > 0.184$. Since CO_L can adsorb on both (100) sites or (110) sites and CO_B adsorbed mainly on Pt(100) sites, the amount of CO_B adsorbed on Pt(100) sites is certainly larger than that of CO_L on Pt(110) sites at CO coverage of 0.029. We can thus conclude that CO prefers to adsorb on Pt(100) sites at a very low CO coverage of 0.029.

The IR band center of CO_L shifts to high wavenumbers with the increase of CO coverage, as can be seen in Fig. 4C. With the increase of CO coverage, CO_L band blue-shifts linearly and quickly from 2006 cm^{-1} to 2040 cm^{-1} when $\theta_{CO}^{IR} \leq 0.184$, and produced a big slope of about 219 cm^{-1}. When $\theta_{CO}^{IR} > 0.184$; the increase of the band center of CO_L is deviated from linearity and increased relatively slowly from 2040 cm^{-1} to 2075 cm^{-1} as the CO coverage increases from 0.184 to 1. This demonstrates that the behavior of CO adsorption is different for $\theta_{CO}^{IR} < 0.184$ and $\theta_{CO}^{IR} > 0.184$. From IR spectra (Fig. 4A), the broad positive branch of the asymmetric bipolar CO_L band could be deconvoluted into two peaks, as illustrated in the inset (a) for the spectrum of $\theta_{CO}^{IR} = 0.255$. In the spectrum of $\theta_{CO}^{IR} = 0.184$ we indeed can discern two CO shoulder peaks (2007 and 2027 cm^{-1}) on CO_L bands. According to previous reports[26] the spectral bands for CO_L adsorbed on (100) steps are *ca.* 10 cm^{-1} red-shifted with respect to those for CO on (110) steps at all CO fractional coverages. For the {730} facets on THH Pt NCs, the width of (100) terrace is quite narrow, the (100) sites on THH Pt NCs behave essentially as (100) steps, as evidenced by the hydrogen adsorption–desorption peaks at −0.02 V in the cyclic voltammogram (Fig. 1C). The observed CO shoulder peaks at low and

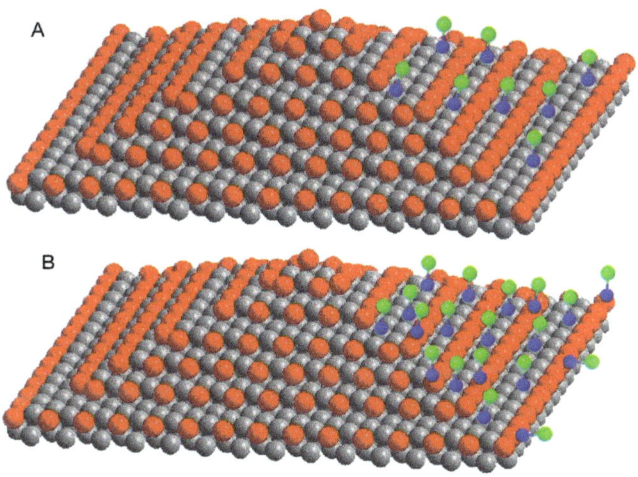

Fig. 5 The models of low coverage (A) and high coverage (B) of CO adsorbed on the surface of THH Pt NCs (only a half square pyramid of the THH Pt NCs is shown. And red: steps; grey: terraces; blue: C, green: O).

high wavenumbers can be assigned to CO_L on (100) sites and (110) sites, respectively. At a CO coverage higher than 0.184, the amount of CO_L on (110) sites increases obviously. These FTIR results are consistent with those obtained from cyclic voltammetric studies.

From the above results, we may propose models for CO adsorption at different coverage. As presented in Fig. 5A, at lower coverage of CO, both CO_L and CO_B adsorbed on (100) sites. Along with increasing the coverage of CO (Fig. 5B), CO_L and CO_B adsorbed on both the (100) and (110) sites, and the amount of CO_L is larger than that of CO_B.

The *in situ* FTIR spectra of CO adsorption on Pt(310) cooled in air after flame treatment with different CO coverages at −0.10 V are displayed in Fig. 6A. The negative-going bands at 2002 and 1798 cm^{-1} can be assigned as CO_L and CO_B, respectively, and the positive-going band around 2343 cm^{-1} is attributed to IR absorption of solution CO_2 species produced by CO oxidation at reference potential (0.60 V). It is evident that the IR features of CO adsorption on a bulk electrode are normal, *i.e.* the direction of the IR bands of adsorbed CO follow the *in situ* FTIR spectra definition of eqn (2).

As presented in Fig. 6B, the IR band intensities of I_{COL} and I_{COB} on Pt(310) single crystal electrode cooled in air are almost linearly increased with the increase of θ_{CO}^{IR}. At low CO coverage ($\theta_{CO}^{IR} \leq 0.117$), I_{COL} is nearly equal to I_{COB}, but I_{COL} increases quickly with the increase of θ_{CO}^{IR}. while I_{COB}increases slowly. Fig. 6C

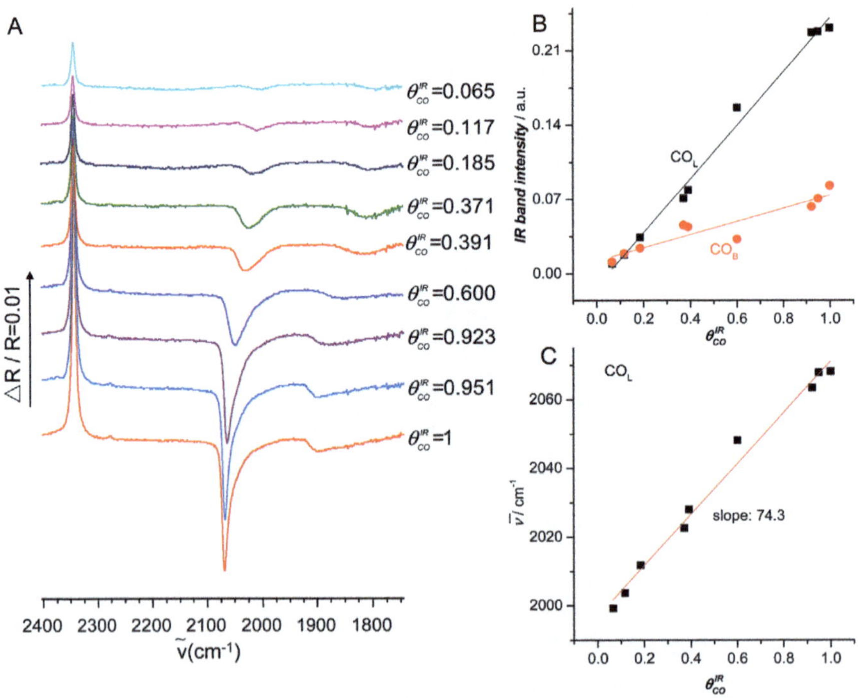

Fig. 6 (A) Coverage dependence of *in situ* FTIR spectra for CO adsorbed on Pt(310) cooled in air after flame treatment, in 0.1 M $HClO_4$ solution, $E_S = -0.10$ V, $E_R = 0.60$ V (SCE). (B) Coverage dependence of the band intensity of CO_L and CO_B in air in (A). (C) IR band center of CO_L plotted against the coverage of CO_L and CO_B.

shows the variation of IR band centers of CO_L and CO_B against the CO coverage in the whole θ_{CO}^{IR} range. Along with increasing the CO coverage, the CO_L band center is linearly blue shifted from 1998 cm^{-1} to 2070 cm^{-1}, producing a slope about 74.3 cm^{-1} that is only about one third of the value obtained on THH Pt NCs. The fact that the IR band centers of both CO_L and CO_B linearly blue shift with the increase of CO coverage in the whole CO coverage indicates that CO_L adsorbs homogeneously on (100) and (110) sites of Pt(310), which is consistent with the results from studies of cyclic voltammetry.

4. Discussions

As shown in Fig. 4, the *in situ* FTIR spectra of CO_{ad} on THH Pt NCs with a distribution density of 60 μm^{-2} demonstrate asymmetric bipolar IR features and the center of the CO_L band increases nonlinearly with the increase of CO coverage. As the anomalous IR effects of nanoparticles depend strongly on aggregation state of the nanomaterials,[31,35] we have also prepared THH Pt NCs with a relatively low distribution density of about 22 μm^{-2}. The mean size of the THH Pt NCs is about 53 nm, that is similar to the THH Pt NCs of high distribution density studied above. The high magnification of the sample and the size distribution histogram are illustrated in Fig. 7A and B, respectively.

Fig. 7C displays the *in situ* FTIR spectra of CO adsorbed on THH Pt NCs with a low distribution density at −0.10 V at different CO coverage. The positive-going bands around 2010 and 1805 cm^{-1} are assigned to CO_L and CO_B, respectively. Interestingly, the CO_L on the THH Pt NCs with low distribution density yields only an abnormally enhanced IR band and loses the "Fano-like" IR feature of the asymmetric bipolar line shape. The positive band around 2343 cm^{-1} is ascribed to IR absorption of solution CO_2 species produced by CO oxidation at a reference potential (0.75 V).

As presented in Fig. 7D, the IR band intensity of I_{COL} and I_{COB} in the spectra of CO adsorbed on THH Pt NCs with low distribution density both increased linearly with the increase of θ_{CO}^{IR}. This is different from the variations of I_{COL} and I_{COB} in the spectra of CO adsorbed on the THH Pt NCs with high distribution density (shown in Fig. 4B). Although the I_{COL} is still much larger than I_{COB} and gives a fast linear increase with the increase of θ_{CO}^{IR}, the I_{COB} maintains a slow and linear increase in the whole θ_{CO}^{IR} range.

The variation of the IR band center of CO_L with the increase of CO coverage is shown in Fig. 7E. We observe a similar variation trend to that presented in Fig. 4C for CO adsorption on THH Pt NCs with a high distribution density. The CO_L band center is blue shifted linearly from 2010 cm^{-1} to 2040 cm^{-1} for $\theta_{CO}^{IR} < 0.181$, still yielding a large slope of about 256 cm^{-1}. When $\theta_{CO}^{IR} > 0.181$, the increase of the CO_L band center deviates from linearity, and becomes slow and nonlinear. The CO_L band center increased from 2040 cm^{-1} to 2070 cm^{-1} along with the increase of θ_{CO}^{IR} from 0.181 to 1. These results demonstrate that the THH Pt NCs have anomalous IR effects for CO adsorption, and the anomalous IR features depend on the distribution density of the THH Pt NCs on GC substrate. Although it is difficult to discern clearly two shoulder peaks on THH Pt NCs with low distribution density in the spectra displays in Fig. 7A, the CO_L bands are obviously broadened at a CO coverage larger than 0.181.

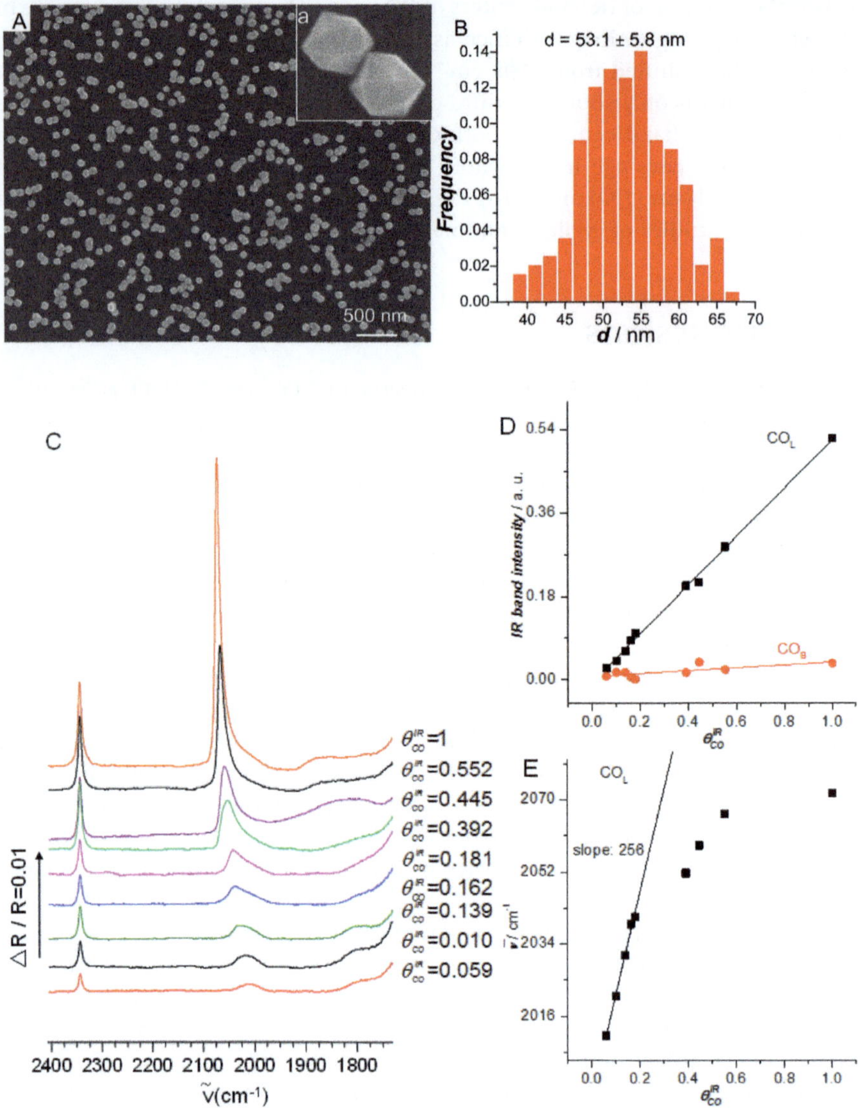

Fig. 7 (A) SEM image of THH Pt NCs with relatively low distribution density of 22 μm^{-2}. The inset (a) is a high magnification image of the sample and illustrates clearly the characteristic THH shape. (B) Size distribution histogram of the THH Pt NCs. (C) Coverage dependence of *in situ* FTIR spectra for CO adsorbed on THH Pt NCs electrode in 0.1 M $HClO_4$ solution, $E_S = -0.10$ V, $E_R = 0.75$ V. (D) Coverage dependence of IR band intensity of CO_L and CO_B on THH Pt NCs in (C). (E) IR band center of CO_L plotted against θ_{CO}^{IR} in (C).

Compared to the linear blue shift of the CO_L band center with the increase of CO coverage on Pt(310) electrode cooled in air, the changing tendency of CO adsorption on THH Pt NCs has two stages. For the first stage of CO adsorption ($\theta_{CO}^{IR} < 0.181$), the slope was 256 cm^{-1} on THH Pt NCs with a low distribution density of 22 μm^{-2} (Fig. 7E) and 219 cm^{-1} on THH Pt NCs with a high distribution density of 60 μm^{-2} (Fig. 4C); both are much larger than that on the Pt(310)

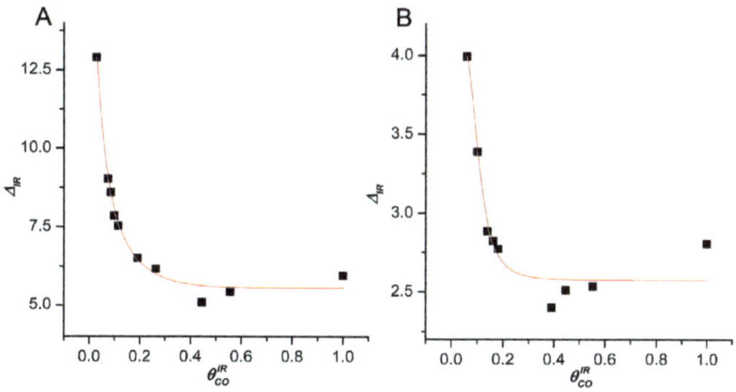

Fig. 8 Coverage dependence of IR enhancement factor (Δ_{IR}) on THH Pt NCs with high distribution density (A) and low distribution density (B).

electrode cooled in air (only 74.3 cm^{-1}, Fig. 6C). As stated above, on THH Pt NCs CO adsorbed dominantly on (100) sites at $\theta_{CO}^{IR} < 0.181$, while CO adsorbed on both (100) and (110) sites on the Pt(310) electrode cooled in air. For a given low CO coverage, the local CO coverage on THH Pt NCs should be larger than that on Pt(310) electrode cooled in air. For CO adsorption on Pt, there is dipole–dipole interaction of neighboring CO, which can decrease the d–π* back donation. This consequently enhances the C≡O bond and weakens the CO adsorption, as well as increasing the frequency of CO stretching vibration.[36] The dipole–dipole interaction of neighboring CO increases with the increase of CO coverage, resulting in the blue shift of the CO band with increasing coverage. The dipole–dipole interaction should be much stronger on THH Pt NCs than on the Pt(310) electrode cooled in air, which may be ascribed to the relatively higher local CO coverage and the particularity of the nanosize, corner and edge sites on the THH Pt NCs.

The frequency of the CO_L band is different for different samples. For a CO coverage of ~0.2, the band centers of CO_L are at 2042 and 2047 cm^{-1} on THH Pt NCs with both high and low distribution density, respectively, while it is only 2012 cm^{-1} on a Pt(310) single crystal cooled in air. These differences can also be explained by the stronger dipole–dipole interaction of neighboring CO on THH Pt NCs. Another factor may be due to the confinement of the facets on the nanoscale and the local environment (corner and edge sites) difference between nanoparticles and single crystal plane.

The preferential adsorption of CO on all electrodes investigated is decreased in the order of: THH Pt NCs > Pt(310) cooled in H_2 + Ar > Pt(310) cooled in

Table 1 The real CO coverage (θ_R) corresponding to $\theta_{CO} = 1$ on all electrodes investigated in Fig. 1 and 2

Electrode	θ_R
THH Pt NCs	0.81
Pt(210) cooled in air	0.73
Pt(210) cooled in H_2 + Ar	0.69
Pt(310) cooled in air	0.67
Pt(310) cooled in H_2 + Ar	0.69

Table 2 The real CO coverage (θ_{CO}^{IR}) for the THH Pt NCs with different distribution density in Fig. 4 and 7

THH Pt NCs of high distribution density (60 μm^{-2})		THH Pt NCs of low distribution density (22 μm^{-2})	
θ_{CO}^{IR}	θ_{R}^{IR}	θ_{CO}^{IR}	θ_{R}^{IR}
0.029	0.012	0.059	0.038
0.071	0.043	0.100	0.076
0.090	0.058	0.139	0.123
0.091	0.064	0.162	0.147
0.106	0.079	0.181	0.168
0.184	0.159	0.391	0.388
0.255	0.234	0.445	0.445
0.431	0.428	0.552	0.552
0.532	0.532	1	1
1	1		

air ≈ Pt(210) cooled in H_2 + Ar > Pt(210) cooled in air. For Pt(210) cooled in air, there was no preferential adsorption on different sites, referring to the results of the Feliu group[22] and Kim *et al.*[37] The Feliu group investigated the behavior of CO adsorption on Pt high-index single crystal planes of (776) and (554) and low-index (111) plane, and demonstrated that CO adsorbed on terraces and steps randomly.[22] They also investigated CO adsorption on well-defined Pt nanoparticles enclosed with low-index facets (including cubes, octahedral tetrahedral nanoparticles, truncated octahedral and tetrahedral nanoparticles with a size of about 4–5 nm) in alkaline conditions,[38] and found that there is no preferential adsorption of CO on different surface sites. The significant priority of CO adsorption on (100) sites on THH Pt NCs manifests an exceptional property of high-index facets at the nanoscale.

The IR features are different for THH Pt NCs with a low distribution density and those with a high distribution density. With the increase of the distribution density of THH Pt NCs, the IR band of CO_L transformed from abnormal

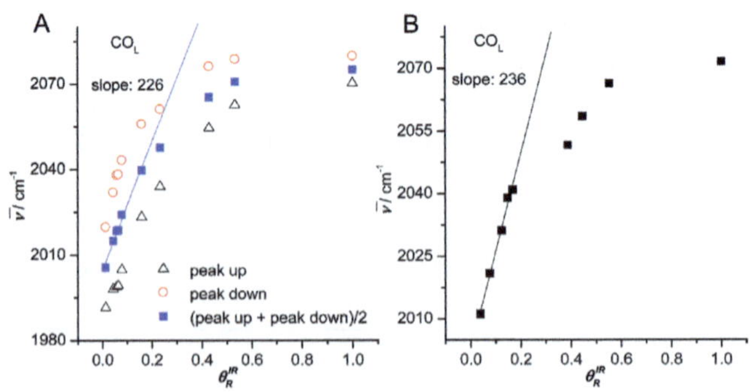

Fig. 9 IR band center of CO_L plotted against θ_R^{IR} for THH Pt NCs with high distribution density (60 μm^{-2}) (A), and low distribution density (22 μm^{-2}) (B). In A, the hollow dark triangles represent the positive peak of the asymmetric bipolar CO_L band; the hollow red circles stand for the negative peak of the asymmetric bipolar CO_L band; and the filled blue squares are the average wavenumbers of the positive and the negative peak.

enhancement IR features (distribution density of 22 μm^{-2}) to Fano-like IR features (distribution density of 60 μm^{-2}). Since the size of THH Pt NCs is similar, this transition should relate to difference in the distribution density of THH Pt NCs. Our previous results on Co nanoparticles[34] demonstrated that with the growth of nanoparticles, the IR band changes from abnormal enhancement IR features to Fano-like IR features, and finally to enhanced IR absorption features. The anomalous IR effects are peculiar to nanostructured materials, which usually relate to the structure, size and distribution density of the nanoparticles.

The IR absorption of CO_L on THH Pt NCs has been significantly enhanced. The IR enhancement factor (Δ_{IR}) is defined as:

$$\Delta_{IR} = \frac{\left[\frac{I_{COL}+I_{COB}}{I_{CO_2}}\right]_{THH}}{\left[\frac{I_{COL}+I_{COB}}{I_{CO_2}}\right]_{Pt(310)}} \tag{6}$$

Fig. 8 displays the coverage dependence of Δ_{IR} for THH Pt NCs with high and low distribution densities. Δ_{IR} decreases hyperbolically against the increase of CO coverage on THH Pt NCs with both high and low distribution density, while the Δ_{IR} of the former is larger than that of the latter one. The highest value of Δ_{IR} for the THH Pt NCs electrode with a high distribution density is about 12.8 at the lowest CO coverage ($\theta_{CO}^{IR} = 0.029$). Δ_{IR} decreases sharply with the increase of CO coverage, drops to about 5.6 at CO coverage of 0.445, and then keeps stable with further CO coverage increasing. However, for the THH Pt NCs with low distribution density, Δ_{IR} is only 4.0 at the lowest CO coverage ($\theta_{CO}^{IR} = 0.059$), and then decreases sharply with the increase of CO coverage. It decreases to about 2.8 at $\theta_{CO}^{IR} = 0.181$, and then reaches a progressively steady state at about 2.6 with the increase of CO coverage. It is evident that the THH Pt NCs of high distribution density have a higher IR enhancement factor towards CO adsorption than those of low distribution density.

It should be noted that, in the present paper, the CO coverage (θ_{CO}) used in cyclic voltammetry studies was calculated by taking the ratio of the oxidation charge of sub monolayer CO_{ad} to that of one monolayer CO_{ad} (*i.e.* the saturation adsorption), which is different from that used in Weaver's work.[39] According to ref. 32, the real CO coverage (θ_R) could be expressed as: $\theta_R = \frac{Q_{CO}}{2Q_H}$. The real CO coverage ($\theta_R$) corresponding to $\theta_{CO} = 1$ on all electrodes investigated in Fig. 1 and 2 were calculated and listed in the Table 1. The θ_R corresponding to $\theta_{CO} = 1$ on THH Pt NCs is around 0.8, and it is about 0.7 for $\theta_{CO} = 1$ on Pt(210) and Pt(310) single crystal electrodes.

For the θ_{CO}^{IR} denoted in the FTIR spectra, it is calculated according to the formula (5). Since at different CO coverage the CO_L band shows different IR enhancement (Fig. 8), the "real" CO coverage (θ_R^{IR}) should be different. The θ_R^{IR} can be calculated by taking into account the influence of the IR enhancement factor, at the same time omitting the contribution of CO_B (since its intensity is quite low), by using the following formula:

$$\theta_R^{IR} = \frac{I_{CO}^{COL}/\Delta_{IR-CO}}{I_{1ML-CO}^{COL}/\Delta_{IR-1ML-CO}} \tag{7}$$

The θ_R^{IR} calculated is listed in Table 2.

Fig. 9 demonstrates the IR band center of CO_L plotted against θ_R^{IR} for THH Pt NCs electrode with high distribution density (60 μm^{-2}) and low distribution density (22 μm^{-2}). The calibration of the CO coverage changes the slope slightly. For THH Pt NCs with a high distribution density (60 μm^{-2}) the slope increases from 219 to 226 cm^{-1}, and it decreases from 256 to 236 cm^{-1} for THH Pt NCs with a low distribution density (22 μm^{-2}).

5. Conclusions

We investigated CO adsorption on THH Pt NCs and interrelated Pt single crystal planes by electrochemical cyclic voltammetry and *in situ* FTIR spectroscopy. The results revealed that CO adsorption on Pt THH NCs exhibits significant priority on (100) sites at low CO coverage. The preferential adsorption of CO on (100) sites decreases in the order of Pt THH NCs > Pt(310) cooled in H_2 + Ar > Pt(310) cooled in air ≈ Pt(210) cooled in H_2 + Ar > Pt(210) cooled in air.

The *in situ* FTIR study demonstrates that, in comparison with Pt(310) singe crystal electrode, the THH Pt NCs exhibit anomalous IR effects (AIREs) towards CO adsorption. The anomalous IR features of CO_{ad} depend on the distribution density of THH Pt NCs, and are transformed from enhancement of abnormal IR absorption (22 μm^{-2}) to a Fano-like IR spectral line shape (60 μm^{-2}) along with increasing distribution density. The coverage-dependent vibration frequency of CO_L on THH Pt NCs demonstrates two segments. The center of the CO_L band is blue shifted quickly and linearly with the increase of CO coverage when $\theta_{CO}^{IR} < 0.18$, yielding a large slope of 219, and then deviates from linearity and approaches finally a steady variation for $\theta_{CO}^{IR} > 0.18$. The variation of the center of the CO_L band on Pt(310) cooled in air is quite different. This manifests a linear relationship for the whole range of θ_{CO}^{IR} and gives a much smaller slope of 74.3. The significant priority of CO adsorption on the (100) sites of THH Pt NCs and the AIREs of the THH Pt CNs demonstrate the unique properties of the THH Pt NCs that are significantly different from the interrelated single crystal planes of Pt(310) and Pt(210). The unique properties are ascribed to the confinement of the Pt high-index planes into nanosized facets, the particular corner and edge sites on the surface of high-index faceted nanoparticles. The results in this paper have enriched the understanding of surface structure effects of Pt nanocrystals enclosed with high-index facets.

Acknowledgements

This study was supported by the National Basic Research Program of China (2012CB215500), NSFC (21222310, 21373175, 21361140374, 21321062, and 21229301), and the foundation for the Author of National Excellent Doctoral Dissertation of China (201126).

References

1 Z.-Y. Zhou, N. Tian, J.-T. Li, I. Broadwell and S.-G. Sun, *Chem. Soc. Rev.*, 2011, **40**, 4167–4185.

2 Z. Quan, Y. Wang and J. Fang, *Acc. Chem. Res.*, 2013, **46**, 191–202.

3 N. Tian, Z. Y. Zhou, S. G. Sun, Y. Ding and Z. L. Wang, *Science*, 2007, **316**, 732–735.

4 N. Tian, Z. Y. Zhou, N. F. Yu, L. Y. Wang and S. G. Sun, *J. Am. Chem. Soc.*, 2010, **132**, 7580–7581.

5 N. F. Yu, N. Tian, Z. Y. Zhou, L. Huang, J. Xiao, Y. H. Wen and S. G. Sun, *Angew. Chem., Int. Ed.*, 2014, **53**, 5097–5101.

6 Y. Y. Li, Y. X. Jiang, M. H. Chen, H. G. Liao, R. Huang, Z. Y. Zhou, N. Tian, S. P. Chen and S. G. Sun, *Chem. Commun.*, 2012, **48**, 9531–9533.

7 N. Tian, Z.-Y. Zhou and S.-G. Sun, *J. Phys. Chem. C*, 2008, **112**, 19801–19817.

8 J. Xiao, S. Liu, N. Tian, Z. Y. Zhou, H. X. Liu, B. B. Xu and S. G. Sun, *J. Am. Chem. Soc.*, 2013, **135**, 18754–18757.

9 Y. J. Deng, N. Tian, Z. Y. Zhou, R. Huang, Z. L. Liu, J. Xiao and S. G. Sun, *Chem. Sci.*, 2012, **3**, 1157–1161.

10 N. Tian, J. Xiao, Z. Y. Zhou, H. X. Liu, Y. J. Deng, L. Huang and B. B. X. a. S. G. Sun, *Faraday Discuss.*, 2013, **162**, 77–89.

11 N. Hoshi, O. Koga, Y. Hori and T. Ogawa, *J. Electroanal. Chem.*, 2006, **587**, 79–85.

12 C. Fan, Y. Fan, C. Zhen, Q. Zheng and S. Sun, *Sci. China, Ser. B: Chem.*, 2007, **50**, 593–598.

13 F. J. Vidal-Iglesias, J. Solla-Gullón, J. M. Campiña, E. Herrero, A. Aldaz and J. M. Feliu, *Electrochim. Acta*, 2009, **54**, 4459–4466.

14 S. G. Sun, *Chapter in Electrocatalysis*, ed. J. Lipkowski and P. N. Ross, Wiley-VCH, Inc., 1998, pp. 243–290.

15 Y. J. Fan, Z. Y. Zhou, C. H. Zhen, S. P. Chen and S. G. Sun, *Electrochem. Commun.*, 2011, **13**, 506–508.

16 F. Illas, F. Mele, D. Curulla, A. Clotet and J. M. Ricart, *Electrochim. Acta*, 1998, **44**, 1213–1220.

17 G. F. Cabeza, N. J. Castellani and P. Legare, *Surf. Rev. Lett.*, 1999, **6**, 369–381.

18 M. J. S. Farias, E. Herrero and J. M. Feliu, *J. Phys. Chem. C*, 2013, **117**, 2903–2913.

19 M. J. S. Farias, C. Busó-Rogero, R. Gisbert, E. Herrero and J. M. Feliu, *J. Phys. Chem. C*, 2014, **118**, 1925–1934.

20 S. C. Chang and M. J. Weaver, *Surf. Sci.*, 1990, **238**, 142–162.

21 S. C. Chang and M. J. Weaver, *J. Phys. Chem.*, 1991, **95**, 5391–5400.

22 M. J. S. Farias, A. A. Tanaka, G. Tremiliosi and J. M. Feliu, *Electrochem. Commun.*, 2011, **13**, 338–341.

23 J. Clavilier, D. Aemand, S. G. Sun and M. Petit, *J. Electroanal. Chem.*, 1986, **205**, 267–277.

24 T. Biegler, D. A. J. Rand and R. Woods, *J. Electroanal. Chem.*, 1971, **29**, 269–277.

25 S. G. Sun and Y. Lin, *Electrochim. Acta*, 1998, **44**, 1153–1162.

26 N. P. Lebedeva, A. Rodes, J. M. Feliu, M. T. M. Koper and R. A. van Santen, *J. Phys. Chem. B*, 2002, **106**, 9863–9872.

27 N. P. Lebedeva, M. T. M. Koper, J. M. Feliu and R. A. van Santen, *J. Phys. Chem. B*, 2002, **106**, 12938–12947.

28 F. J. Vidal-Iglesias, J. Solla-Gullon, J. M. Campina, E. Herrero, A. Aldaz and J. M. Feliu, *Electrochim. Acta*, 2009, **54**, 4459–4466.

29 L. A. Kibler, A. Cuesta, M. Kleinert and D. M. Kolb, *J. Electroanal. Chem.*, 2000, **484**, 73–82.

30 K. Mikita, M. Nakamura and N. Hoshi, *Langmuir*, 2007, **23**, 9092–9097.

31 G. Q. Lu, S. G. Sun, L. R. Cai, S. P. Chen, Z. W. Tian and K. K. Shiu, *Langmuir*, 2000, **16**, 778–786.

32 H. Gong, S. G. Sun, Y. J. Chen and S. P. Chen, *J. Phys. Chem. B*, 2004, **108**, 11575–11584.

33 H. C. Wang, S. G. Sun, J. W. Yan, H. Z. Yang and Z. Y. Zhou, *J. Phys. Chem. B*, 2005, **109**, 4309–4316.

34 Q. S. Chen, S. G. Sun, J. W. Yan, J. T. Li and Z. Y. Zhou, *Langmuir*, 2006, **22**, 10575–10583.

35 Y. X. Jiang, S. G. Sun and N. Ding, *Chem. Phys. Lett.*, 2001, **344**, 463–470.

36 S. C. Chang and M. J. Weaver, *J. Chem. Phys.*, 1990, **92**, 4582–4594.

37 C. S. Kim and C. Korzeniewski, *Anal. Chem.*, 1997, **69**, 2349–2353.

38 M. J. S. Farias, F. J. Vidal-Iglesias, J. Sola-Gullon, E. Herrero and J. M. Feiu, *J. Electroanal. Chem.*, 2014, **716**, 16–22.

39 R. Gómez, J. M. Feliu, A. Aldaz and M. J. Weaver, *Surf. Sci.*, 1998, **410**, 48–61.

Faraday Discussions

DISCUSSIONS

Energy-related catalytic and other materials: general discussion

Joachim Maier, Galen Stucky, Rudolf Holze, Gang Chen, Jiafang Xie, Shi-Gang Sun, Guoxiong Wang, Beien Zhu, Yiren Zhong, Pengfei Liu, Lee Cronin, Gang Fu, Shihe Yang, Mingquan Yu, Clare Grey, Andrew Mount, Wee Shong Chin, Fuping Pan, Zhonghua Li, Zhongqun Tian, Dehui Deng, Nanfeng Zheng, Ram Seshadri, Yujiang Song, Xile Hu and Yimin Chao

DOI: 10.1039/c5fd90003d

Joachim Maier opened the discussion of the paper by Ram Seshadri: The Debye temperature is not well-defined in complex crystals unless referring to extremely low temperatures. To what conditions do the DFT-calculations refer?

Ram Seshadri replied: We use the quasi-harmonic approximation with 0 K DFT for the computation, and this fits to the very low-temperature heat capacity as one of the experimental confirmations.

Lee Cronin asked: Can you use optimisation algorithms to search the state space? Can you use a population based algorithm? Could you make a proxy and minimise the proxies?

Ram Seshadri responded: All of these are in principle possible. The impediments at this time are a lack of expertise on the part of my research group, and also, the lack of experimental databases for some of the machine-learning strategies to be applied to.

Andrew Mount asked: One opportunity to enhance data-driven discovery in general is to include additional functional relevance to applications to inform better discovery. What are the opportunities for incorporating *e.g.* phosphorescence data to inform better phosphor discovery?

Ram Seshadri answered: I completely agree that other information could be added to our data-mining and searching strategies. The issue is the skills required for many of the machine-learning type approaches that one could employ: skills that us materials chemists may not possess. My group is in the process of mining a lot of experimental data such as Stokes' shifts, and perhaps something may emerge.

Clare Grey asked: The nitrides, oxyfluorides and YAGs all show high efficiencies and yet their Debye temperatures are quite different.

What else can you learn about the structural motifs to gain insight into what controls high efficiency? It seems like you are looking for structural motifs that are essentially the opposite of those that people look for in trying to identify materials with negative thermal expansion. You have essentially achieved 100% efficiency. So at one level, this seems like the end of the story! What is left then is the prediction of wavelengths for emission. You made the point that its difficult to predict luminescent properties from DFT methods because of the problems with accurate calculations of band gaps and electronic properties in materials involving 4f electrons. It is, however, straightforward to predict crystal field splittings with relatively simple computational/semi-empirical approaches. Could you include some of these approaches into your screening methods?

Ram Seshadri answered: This is absolutely important. We find a highly connected structure to be indicative of rigidity and good phosphor performance, and in fact, the garnet structure is absolutely great as an example of a highly connected structure. Likewise, the point about materials (and structures) displaying negative thermal expansion, like the canonical material ZrW_2O_8 can be expected, as you say, to be terrible phosphor hosts from the perspective of rigidity. At this stage, we are trying to think about factors such as the crystal field splitting and their computation. Frankly, parametrized approaches using experimental data may work as well and more robustly.

Clare Grey said: Are there other parameters you could pick out to put into the optimisation processes (looking at your slide showing performance of different materials)? The parameters you are looking for are inverse to those that people look for for negative thermal expansion materials.

Ram Seshadri responded: The parameters that we would like to add are absolute band positions, *i.e.* the position of the valence band maximum and the conduction band minimum with respect to the vacuum level. Unfortunately, these are not easy to compute. On the point regarding materials displaying negative thermal expansion (NTE), the point is very well taken. Phosphor hosts are emphatically not like NTE materials.

Gang Fu commented: Can you design some new structures for energy materials on the basis of your calculations?

Ram Seshadri replied: On the basis of what I showed, no; we cannot design a new structure, only look at existing structures but potentially with new compositions.

Galen Stucky commented: You calculated the band gaps to help define the optical performance parameter space. Do you have a way of defining the excited states as part of the performance parameter search space?

Ram Seshadri answered: We use calculations of the band gaps only insofar as we wish to ensure that the f-to-d transition on the activator ion (Eu^{2+} or Ce^{3+}) lies within the optical window of the band gap. Calculating the crystal-field broadening of the d states is not something that we have so far attempted.

Nanfeng Zheng asked: Is it possible to apply the developed strategy to prepare efficient white-light phosphors using rare earth elements? The production of rare earth compounds from rare earth ores has caused serious environmental problems in China.

Ram Seshadri replied: I think the question is whether the strategies could be used in a manner that avoids rare-earth elements, and the answer is very much so. The strategy does not care about the precise nature of elements.

Shihe Yang enquired: It seems to me that structural rigidity is more relevant to organic molecules because they may have many different conformations. Indeed it is well known that rigid organic molecules emit light more efficiently, of course in the relative sense, because of less nonradiative pathways. But when you talk about rigidity of inorganic materials, what are you really referring to? Can you elaborate more on the differences between inorganic and organic light emitting materials with respect to your thesis?

Ram Seshadri replied: This is correct, that rigid molecules make better emitters than "floppy" molecules with many degrees of torsional freedom. The point that we are making with crystal structures is that such "floppiness" may not be readily recognizable simply by looking at the crystal structures, and this is where the (computable) proxy that we employ, the Debye temperature, becomes very useful.

Zhonghua Li remarked: Can you obtain monochromatic light, then enhance the efficiency?

Ram Seshadri responded: I am not sure that I fully understand the question. The goal of white lighting is not to get monochromatic light, but to fill out the spectrum, so monochromatic light is somewhat pointless.

Galen Stucky asked: Do we have perfect white light LEDs? How might the performance be improved for other desired applications?

Ram Seshadri replied: We are really close to perfect white light, and in fact, we have learned some interesting things in recent years. One of these is that a little bit of near-UV in white light is actually a good thing, since so many everyday objects (like white shirts) have had optical brightening agents added to them. The best solid-state white light devices (in terms of color) are made today using near-UV diodes and red/green/blue phosphors, and these have color rendition indices close to 100.

Galen Stucky opened the discussion of the paper by Yimin Chao: Could you give more details on the doping and functionalisation for generating p and n type particles? What are the resultant carrier densities and doping levels?

Yimin Chao responded: The functionalization is a nucleophilic substitution reaction with chloride termination on the particle surface and there is potential for controlling doping on both the particle and the ligand. However in this work we have not actively doped this material and therefore have not measured the carrier densities or doping levels. In the paper we suggest that charged surface states may be responsible for the electrical resistivity observed but it is difficult to be sure. There is scope for both doping the particles and doping the ligands. The ligands can be doped using typical oxidation processes as used by conductive polymers, although realistically they can only be p-doped, as n doped organics tend to be less stable (difficult using phenyl acetylene). On the other hand the particles can be both n and p doped by using a starting material of known carrier concentration. With regards to producing both p and n doped materials, it has been shown with Bi_2Te_3 - PEDOT:PSS nanocomposites that n type materials can be achieved without the complication of n doping the organic component (please see B. Zhang, J. Sun, H. E. Katz, F. Fang and R. L. Opila, *ACS Appl. Mater. Interfaces*, 2010, **2**, 3170–3178).

Galen Stucky said: Does $ZT = 0.6$ because the thermal conductivity is low or because the power factor is high? What is the value of the experimental thermal conductivity?

Yimin Chao responded: ZT is high mainly due to an extremely low thermal conductivity although the power factor is reasonable in the region of 10^{-5} V m^{-1} K^{-1}, which is 10 times lower than that of the best conductive polymer and 100 times lower than bismuth telluride (please see O. Bubnova and X. Crispin, *Energy Environ. Sci.*, 2012, **5**, 9345–9362 for values). The experimental value of thermal conductivity here in the paper is 0.1 W m^{-1} K^{-1}.

Shihe Yang opened the discussion of the paper by Yimin Chao: One question concerns the use of organic molecules for thermoelectric devices. What are the parameters you seek to enhance in the thermoelectic figure of merit? Second, it seems that much work is focused on reducing phonon conductivity. How much room is there in improving the thermal power Q by nanostructure engineering? There had been much discussion in this direction early on, but what is the current situation?

Yimin Chao responded: By the introduction of organics we are aiming to reduce the thermal conductivity. As such, organic chains on their own display very low thermal conductivities. Additionally these types of materials are more versatile with regards to low-temperature processing techniques such as printing. So essentially we have, as workers before us, focused on reducing the phonon conduction even though our approach is different.

Using nanostructuring engineering, it has been possible to match the thermopower of bulk silicon. There were significant advancements in the ZT early on when compared to bulk silicon. These materials show a significant decrease in the lattice themal conductivity. At high temperatures these materials look promising but little improvement using bulk nanostructuring has been made since publications by Bux *et al.* in 2009. With regards to low temperature performance, a more significant advancement using nanostructuring lies with the production of nanoporous ribbons or nanowires that do offer promise for good thermopower and a *ZT* close to 0.4 (Tang *et al.* 2010) and 1 (Hochbaum *et al.* and Boukai *et al.* 2008) at RT respectively. However, in terms of industrial scale production, bulk nanostructuring of silicon is still very attractive, especially if *ZT* can be improved to match SiGe alloys.

Yiren Zhong opened the discussion of the paper by Xile Hu: I have noticed the CV scans of nano-$Ni(OH)_2$ and NiO are quite different; that is the CV of $Ni(OH)_2$ has a pair of peaks, why does this happen?

Xile Hu replied: The peak you refer to is the Ni(II) to Ni(III) peak and its position and shape are different for different catalysts. A more active catalyst tends to have higher intensity for this peak.

Mingquan Yu asked: In your experiment, iron impurities were removed on purpose. As far as I know, iron-doping usually enhance the activity of the catalyst a lot. So, could you please explain the synergetic effect between iron and nickle?

Xile Hu replied: This is still an open question. In the paper we listed two possibilities such as the synergetic effect of Ni-O-Fe units, or doping of Fe leading to electronic structure change. Professor Maier mentioned the possibility of defect formation by Ni vacancy. It will be an important research topic for the near future.

Joachim Maier added: Could the Fe-effect be a formation of point defects such as nickel vacancies (doping effect) which would be extremely basic? *Cf.* R. Merkle and J. Maier, *Top. Catal*, 2006, **38**, 141.

Xile Hu responded: This is a very interesting idea. We will look into it.

Guoxiong Wang enquired: Have you investigated the effect of the amount of iron content on OER activity?

Xile Hu responded: We have done it only partially. We should do more; thank you for the suggestion.

Pengfei Liu asked: According to the CV curve of $Ni(OH)_2$ nanoparticles with Fe impurity, the anodic peak around 1.39V *vs.* RHE corresponds to the phase change $Ni(OH)_2$ to NiOOH. My question is: does the anodic peak always mean that the metal element valence state changes; does the anodic peak have something to do with the OER performance. If do, how does it influence this?

Xile Hu replied: The position and shape of this redox peak are indicative of the phase of the $Ni(OH)_2$ species. A certain phase is more active than the other. Our paper has a discussion of this, and has cited several papers on this topic.

Pengfei Liu spoke: As we know, the HER is easier in acid electrolyte and the OER is easier in basic electrolyte. Some HER catalysts, for example, sulfides, carbides, nitrides, phosphides and some carbon materials are active in acid electrolyte but less active and more unstable in basic electrolyte. Some OER catalysts like perovskites, amorphous metal oxides and carbon materials are also less active and more unstable in acid electrolyte. My question is that, since we should use the same electrolyte for water electrolysis, should we develop active and stable HER catalysts in basic electrolyte or develop the active and stable OER catalysts in acid electrolyte? Which is more important: perhaps both?

Xile Hu answered: You are absolutely right. In general HER is easier in acid and OER is easier in base. Now it is time to combine them. It appears that HER in base is more promising than OER in acid for most non-precious catalysts due to the instability of many oxides in acid. Watch out, we will soon publish a paper on photoelectrochemical HER in base for the first time.

Dehui Deng asked: Why does the performance of the material with Fe impurity increase after 30 scans (looking at a plot of current density *vs.* potential)? Can you run for longer scans?

Xile Hu replied: During the first 30 scans, the catalyst undergoes a transformation to form the active phase, likely layered NiOOH species. During this process, the activity increases. Further scans do not give higher activity, as once the phase is formed, it is stable and its activity is constant.

Gang Fu enquired: Where is the iron: on the Ni(OH)2 or on the carbon? What is the valence of the iron?

Xile Hu replied: Iron should be close to nickel in the catalyst layer. The oxidation state of iron is likely to be Fe(III).

Fuping Pan addressed Xile Hu: Much research is dedicated to develop low-cost yet efficient catalysts to replace noble metals for energy-related reactions, such as ORR and OER. How far can the non-precious catalyst go? Can they replace completely precious metal-based catalysts in practical application?

Xile Hu answered: For ORR and OER under certain conditions, non-precious catalysts can do better than precious metals. For instance, in alkaline conditions, such catalysts exist and they can be further improved. Also even if non-precious catalysts have lower performance under certain conditions, for example, in acidic solutions, as long as the performance per cost is higher, they can be used.

Guoxiong Wang asked: For OER a conductive material is not necessary, why? It is possible to enhance the OER activity by adding some electroactive material?

Xile Hu replied: Actually for OER you also need highly conductive catalysts. Often the catalysts are thin and porous, so you don't notice the problem in preliminary tests. Adding electroactive materials is a good idea.

Jiafang Xie enquired: I'm interested in the role that Fe played in your system for its abnormal enhancement of the peak current in the CV sweep. I noticed that with Fe impurities, the potential of the peak current in CV is different to that of without Fe impurities. As far as I know, this is likely to imply an different reaction in which Fe and/or your catalyst participate. As a similar example, copper quickly loses its ability to catalyze electrochemical carbon dioxide reduction to hydrocarbons when there is trace Fe impurity; the current is even larger, but only H_2 can be detected. The explanation is that Fe atoms occupy the active sites on the surface of the copper and H is highly adsorbed on Fe. So, did you try to detect the product and make a comparison? I think this would be helpful.

Xile Hu responded: Very good suggestion! We will look into it.

Jiafang Xie then asked: I'm very interested in the role that Fe plays in your system. In your paper, you showed a CV of a comparison of with and without the removal of Fe. For a more complete comparison, experiments of adding Fe into the electrolyte or just doping Fe onto your catalyst should be conducted; I wonder if you have done these? And if you have, what happened? There result would help us determine the role of Fe impurities in your system.

Xile Hu replied: Again very good suggestions! We have not done this and we will do it.

Jiafang Xie added: This is a small question about the details of your experiment. In your work, when the final suspension was obtained, it was allowed to rest for 3 h before further centrifugation. I wonder what is the meaning of this step. Is it of special purpose?

Xile Hu answered: This is just to separate the solid from the solution. You can probably do it by a centrifuge, which would be faster.

Shi-Gang Sun opened the discussion of the paper by Gang Fu: In the theoretical modeling, what are the effects of iron oxide and nickel oxide that may change the structure. Also, what is the effect of substrates of different single crystal planes?

Gang Fu responded: Our DFT calculations suggested that the OH groups at Fe_{3+}-OH-Pt react readily with CO adsorbed nearby to yield directly CO_2 and produce simultaneously coordinatively unsaturated Fe sites for O_2 activation. Ni_{2+} plays a key role in stabilizing the interface against dehydration through the strong interaction between $Ni(OH)_x$ and $Fe(OH)_x$. We have explored different Pt planes, including flat (111) surface and stepped (332) and (322) surfaces, and found that CO oxidation is a structure sensitive reaction, and the performance on different exposed surfaces depends on the bond strength of Fe(III)–OH.

Beien Zhu asked: There are two well known mechanisms for CO oxidation. One is the Eley–Rideal mechanism which concerns CO and O_2 forming OCOO; another is the Langmuir–Hinshelwood mechanism which involves the dissociation of O_2. In your calculations, did you consider both mechanisms?

Gang Fu answered: We only considered the Langmuir–Hinshelwood mechanism because there were still open Pt sites which can strongly adsorb CO. In our mechanism, CO oxidation is triggered by CO oxidative coupling with interfacial OH groups at Fe-OH-Pt, simultaneously creating coordinatively unsaturated ferrous sites (Pt–□–Fe) for the following O_2 activation. We also showed that CO can promote O–O bond cleavage *via* an [OC–O–O(H)Fe] transition state.

Beien Zhu asked: In your work, how do you prevent the supported Fe nanoparticles from aggregating?

Gang Fu responded: Despite exhibiting a high catalytic activity in CO oxidation, the $Fe(OH)_x$/Pt catalyst did not work steadily for a long time, mainly due to the easy dehydration of $Fe(OH)_x$ sub-monolayers into three-dimensional iron oxyhydroxide or iron oxide. To prevent the dehydration-induced loss of interfacial Fe(III)-OH-Pt sites, it was undertaken to dope the overgrowth $Fe(OH)_x$ sub-monolayer with Ni^{2+} to form Pt/$FeNi(OH)_x$ hybrid nanoparticles. The resulting catalysts can maintain high catalytic activity for more than 28 h.

Yiren Zhong enquired: Is the $Fe(OH)_x$ on the Pt surface amorphous or crystallized?

Gang Fu answered: $Fe(OH)_x$ was deposited on the sub 5-nm Pt nanoparticles such that there is no long range periodicity in its structure, *i.e.* it is amorphous. However, the local structure of $Fe(OH)_x$ was found to be well-defined because XANES showed that the structure of $Fe(OH)_x$ is similar to goethite FeOOH. Computationally, we constructed $Fe(OH)_x$ models by cutting the Fe_6O_{18} moiety from the goethite FeOOH crystal, and periodic density functional theory was performed.

Yiren Zhong asked: According to your experience, could you give us a comment or a guess on which state of catalyst has more efficient catalytic effect, amorphous or crystal?

Gang Fu replied: The efficiency of catalysts depends critically on how many active sites they contain rather than their crystal states. Fundamentally, researchers preferred to use model systems, such as single crystals or well-defined nano catalysts to explore the catalytic mechanisms. For example, we have demonstrated here that well-defined hybrid nanoparticles are an ideal model system to study interfacial effects. Based on our understanding of the interfacial effects, an alloy-assisted strategy was further developed to produce a more practical Pt-based nanocatalyst. Although the catalyst did not have easily characterized catalytic interfaces, it contained much more abundant Fe-OH–Pt catalytic interfaces and displayed a much higher catalytic performance than the well-defined systems. So, efficient catalysts might not have well-defined structures for characterization. They could be either amorphous or crystalline. But they have to contain abundant catalytically active sites.

Fuping Pan enquired: Can the water absorb and occupy the active sites? What is the accurate role of water?

Gang Fu responded: The adsorption of water on the active sites is relatively weak, which cannot compete with the adsorption of CO and O_2. Therefore, water is unable to block the active sites. However, water is necessary for low-temperature CO oxidation because the Pt/Fe(OH)$_x$ catalyst did not work steadily in dry streams. Based on our DFT calculations, we proposed that the water can promote O_2 activation and recover the interfacial OH.

Zhonghua Li asked: Why did you choose one layer Fe(III)–OH instead of multiple layers?

Gang Fu answered: Experimentally, we used a wet-chemical method to fabricate Fe-OH–Pt interfaces by atomically thick Fe-OH partially covering the surface of monodisperse Pt nanocrystals. In order to match the structure characterized by experiment, we built catalyst models with a single layer but without multiple layers of Fe(OH)$_x$ deposited on different Pt surfaces, such as Pt(111), Pt(332) and Pt(322).

Shihe Yang enquired: Presumably, the charge of the active center should change during the reaction. Does proton-coupled electron transfer have a role to play here? This could lower the activation energy. The fact that water is needed seems to speak in favor of this mechanism.

Gang Fu replied: CO oxidation involves a two-electron redox reaction. In our mechanism, CO oxidation began with the adsorbed CO and interfacial –OH moieties, which can be viewed as a two-electron reductive elimination. During this process, the Fe(OH)$_x$ species serve as oxidants, which will be reduced by two units. For the first CO oxidation, only proton transfers take place. Proton-coupled electron transfer plays a key role in forming the OOH species, in which the electron donor is Fe^{2+}, while the proton donor is adsorbed H_2O. The role the water plays has been emphasized in our paper. On one hand, the water can stabilize O_2 adsorption by forming OOH species; on the other hand, the O–O bond breaking might also be facilitated by the coexistence of water.

Gang Chen asked: How large are your iron nanoparticles? Can you estimate what percentage of atoms at the interface are of interest in your study? If so, you can estimate how much signal in your XANES spectra is from the interfacial Fe atoms.

Gang Fu responded: Experimentally, we fabricated Fe-OH–Pt interfaces by partially covering the surface of mono disperse Pt nanoparticles (which are <5 nm in diameter) with Fe-OH submonolayers through a wet-chemical method. Thus, nearly 100% of Fe atoms are at the interface, and the area of iron hydroxide is less than 25 nm^2. We estimated that about 20–30% Fe atoms are on the boundary to form interfacial Fe-OH–Pt groups. X-Ray absorption near-edge structure (XANES) and extended X-ray absorption fine structure (EXAFS) spectra gave an average result indicating that Fe atoms are located in a distorted octahedral environment and the bond distances of Fe–O and Fe–Fe are 1.97 ± 0.02 Å and 3.01 ± 0.02 Å, respectively. In our theoretical models, all Fe atoms are saturated by O or OH, *i.e.*

in six-fold coordination, and the calculated bond distances are in good agreement with experimental values.

Rudolf Holze opened the discussion of the paper by Yujiang Song: I definitely disagree with your claim that dioxygen-reducing electrodes based on heat-treated metal-macrocycles have not matched so far platinum/carbon-based ones. Already in 1983/1984 the E. Yeager group (CWRU, Cleveland) reported performance data for real gas-fed electrodes based on carbon-supported heat-treated (800 °C, argon) μ-oxo-iron-*meso*-1,5,10,15,20-tetramethoxyphenylporphyrin (not for samples deposited on disk supports *etc.*) showing performance equal or better to/than the best available Pt/C (by Stonehart; see *e.g.* R. Holze, E. Yeager, *Dechema-Monographie* 1985, **98**, 499 & E. Yeager, *Electrochim. Acta*, 1984, **29** 1527, also R. Holze, I. Vogel and W. Vielstich, *J. Electroanal. Chem.*, 1986, **210**, 277). Actually the idea of heat-treatment goes back to Grünig *et al.*, neither to Jasinski – who introduced a molecule from nature into dioxygen activation – nor any later investigator (see: Gabriele Grünig, "Die Dibenzotetraazaannulene und ihre Pyrolyseprodukte als Elektrokatalysatoren für die katodische Sauerstoffreduktion im sauren Medium", Dissertation, Technische Universität Dresden 1983; G. Grünig, S. Gamburtsev, I. Iliev, A. Kaisheva and K. Wiesener, *Sov. Electrochem.*, 1983, **19**, 1408; S. Gamburtsev, G. Grünig, A. Kaisheva, I. Iliev and K. Wiesener, *Elektrokhimiya*, 1984, **20**, 500; G. Grünig, E.-G. Jäger, U. Müller and K. Wiesener, *Z. Phys. Chem., Leipzig*, 1986, **267**, 994; G. Grünig, K. Wiesener, S. Gamburzev, I. Iliev and A. Kaisheiva, *J. Electroanal. Chem.*, 1983, **159**, 155; G. Grünig, K. Wiesener, A. Kaisheva, S. Gamburtsev and I. Iliev, *Elektrokhimiya*, 1983, **19**, 1571; for an overview see also K. Wiesener *et al.*, *Mater. Chem. Phys.*, 1989, **22**, 457). The statement by Alt *et al.* (H. Alt, H. Binder, and G. Sandstede, *J. Catal.*, 1973, **28**, 8) was certainly valid then – it pertained to the macrocycles without the heat-treatment, and it was made a few days before the work by Grünig *et al.* and Yeager *et al.* Indeed metal macrocycles without heat-treatment will most likely remain insufficiently stable except for applications in biofuel cells operating under very mild, almost biological conditions. Your ref. 33 is unfortunately not very helpful (V. S. Bagotzky, Y. B. Vassiliev and O. A. Khazova, *J. Electroanal. Chem.*, 1977, **81**, 229). The publication is entitled: "Generalized scheme of chemisorption, electrooxidation and electroreduction of simple organic compounds on platinum group metals" and does not contain – as far as could be verified – anything pertaining to the subject discussed here. Jahnke *et al.* (H. Jahnke, M. Schönborn and G. Zimmermann, "N_4-Chelate als Katalysatoren in der Brennstoffzelle", *Bosch Technische Berichte*, 1973, **4**, 98; H.G. Jahnke, M.F. Schönborn and G. Zimmermann, "Transition metal chelates as catalysts for fuel cell reactions", *Proceedings on the Symposium "Electrode Materials and Processes for Energy Conversion and Storage"*, The Electrochemical Society, Princeton, NJ, USA, 1974, 303; H. Jahnke, M. Schönborn and G. Zimmermann, "Organic dyestuffs as catalysts for fuel cells", *Top. Curr. Chem.*, 1976, **61**, 133) did not study heat-treated macrocycles (your reference H. Jahnke, M. Schönborn and G. Zimmermann, in *Physical and Chemical Applications of Dyestuffs*, ed. F. P. Schäfer, H. Gerischer, F. Willig, H. Meier, H. Jahnke, M. Schönborn and G. Zimmermann, Springer, Berlin, Heidelberg, 1976, vol. 61, pp. 133–181). I suggest to keep open the question whether your catalyst preparation is indeed better in terms of both stability and activity until measurements with gas-fed electrodes under conditions as close to reality as possible (*i.e.* gas-fed electrodes at fuel cell operation temperatures) have been performed.

The methanol tolerance of these catalysts because of the absence of mixed potential establishment (*i.e.* practically no catalytic activity for methanol *etc.* oxidation) has been reported before (R. Holze, I. Vogel and W. Vielstich, *J. Electroanal. Chem.*, 1986, **210**, 277).

Yujiang Song answered: Thank you for providing us many important references.

1. Heat-treated metalloporphyrin supported on carbon is a class of promising non precious metal (NPM) electrocatalysts. However, the NPM electrocatalysts still cannot replace Pt based electrocatalysts.

2. I agree with you that the heat-treatment idea came from Grünig *et al.* instead of Jasinski.

3. Thanks for helping us find the mistakes in two references, which will be corrected in the revised version of our article.

4. Thank you for your insightful suggestion on the measurement using gas-fed electrodes. In the future, we will conduct the experiments for further confirmation. However, the data collected using a rotating disk electrode are also reliable. In this study, one of our major claims is that self-assembled hemin after heat treatment performs better than pyrolyzed primitive hemin.

5. We agree with you that the methanol tolerance of the NPM electrocatalysts has been reported before. We will add in the reference in the revised version of our article.

Shi-Gang Sun asked: The NNM ORR catalysts are important for fuel cell development. In order for rational design and synthesis best NNM ORR catalysts, understanding the ORR active center is the priority issue. Have the authors identified (or just guessed) the ORR center for their catalysts developed? Do they have references for some recent papers concerning this investigation, such as: Q. Wang *et al.*, *J. Am. Chem. Soc.*, 2014, **136**, 10882–10885; M. S. Thorum *et al.*, *J. Phys. Chem. Lett.*, 2011, **2**, 295–298 *etc.*

Yujiang Song responded: The active sites of non-noble metal electrocatalysts for ORR are currently under debate. However, it is generally accepted that perovskites, transition metal oxides, doped carbon, metallomacrocylces, and so on are active for ORR. We did not mention the two articles in the article. If possible, we will reference them in the revised version of the article.

Fuping Pan enquired: (1) There are large iron particles in catalyst, so does the structure of hemin change after thermal treatment? (2) Between the iron particles and carbon shell, which is more important for the oxygen reduction reaction?

Yujiang Song responded: (1) Hemin molecules do decompose during the process of heat-treatment, forming iron oxide nanoparticles. (2) We have no evidence to conclude which part is more important. It is possible that both components are important.

Gang Chen asked: What is the driving force to make the different shapes of self-assembled nanostructures? Are they formed from the same macromolecules?

Yujiang Song answered: The non-covalent driving force includes hydrogen bonding, electrostatic interaction, π–π stacking, and so on. Different types of metalloporphyrin molecules have been used in the self-assembly process.

Jiafang Xie asked: In the preprint of your paper, your nanomaterial showed a more negative half-wave potential than Pt/C, which means a larger activation barrier that your nanomaterial needed to across., but your nanomaterial showed a excellent activity to ORR, is there any possible explanation to this phenomenon?

Yujiang Song responded: Self-assembled hemin after pyrolysis performed better than heat-treated pristine hemin toward ORR. I am afraid that there might be some misunderstanding here.

Guoxiong Wang opened the discussion of the paper by Shi-Gang Sun: In which potential window is your system stable?

Shi-Gang Sun replied: The electrochemical stability of THH Pt NCs is fairly good. THH Pt NCs in this paper were prepared using a square-wave potential with $E_L = 0.03$ V and $E_U = 1.06$ V (*vs.* SCE), so they should be stable in this potential window. We also carried out electrochemical stability tests by using potential cycling between 0.6–1.1 V (*vs.* RHE). After 1000 potential cycles, the loss in electroactive surface area was less than 3% (Liu *et al.*, *Phys. Chem. Chem. Phys.*, 2012, **14**, 16415–16423).

Guoxiong Wang enquired: Have you tried other electrochemical reactions on the THH metal nanocrystal?

Shi-Gang Sun responded: We have tested the THH metal nanocrystal for the electrocatalytic oxidation of a series of small organic molecules (such as formic acid, ethanol, methanol, and ethylene glycol) on THH NCs of Pt, Pd, Rh as well as their alloy (*Angew. Chem., Int. Ed.* 2014, **53**, 5097–5101; *ACS Catal.*, 2012, **2**, 708–715; *Faraday Discuss.*, 2013, **162**, 77–89; *Chem. Sci.*, 2012, **3**, 1157–1161; *J. Am. Chem. Soc.*, 2011, **133**, 12930–12933; *J. Am. Chem. Soc.*, 2010, **132**, 7580–7581). Recently, we also tested the oxygen reduction reaction (ORR) and found that high-index faceted Pt nanocrystals exhibit higher ORR activity than the commercial Pt catalyst.

Beien Zhu asked: It has been observed in recent research that active gases like CO or O_2 will cause the material surface to restructure under certain pressure. When you increased the CO coverage on the Pt surface, did you consider this possibility or see some surface reconstruction?

Shi-Gang Sun answered: For a clean single-crystal electrode after flame annealing, the surface structures frequently deviate from their ideal structure, *e.g.*, for Pt(100), it changes into a structure of (5×20). After CO adsorption, it reform into (1×1) ideal structure. While oxygen adsorption may induce serious surface restructuring (formation of facets), since oxygen atoms can invade the lattice of the host materials. We do not actually know whether there is surface reconstruction on Pt nanocrystals or not, because it is very challenging to investigate the surface atomic arrangement of nanoparticles, and to monitor their kinetic processes under electrochemical conditions. Although STM can provide atomic resolution images, this method is limited to bulk single crystal surfaces. It is very difficult to obtain atomic-resolved STM images on nanoparticles due to their rough and uneven surfaces.

Zhonghua Li enquired: Where is the active site? On the corner or the edge?

Shi-Gang Sun replied: Usually, low coordination atoms have high reactivity. On normal nanoparticles, both the corner and edge are such sites. On high-index faceted nanocrystals, many step atoms also exist on the facets besides the corner and edge. Besides, the active site is usually not limited to a single atom. Frequently, electrocatalytic reactions need a steric active site that consists of several vicinal surface atoms (*e.g.*, step-terrace atoms) with a synergetic effect (Tian *et al.*, *J. Phys. Chem. C*, 2008, **112**, 19801–19817).

Rudolf Holze asked: The observed change of the band position *vs.* coverage is too large to be explained with Stark-tuning, and it also appears too large for an explanation based on electron-donation/backdonation. Have you considered dipole–dipole interactions which might be particularly effective at very high coverages where the observed band positions deviate strongly from the interpolated line? See *e.g.* W. Tornquist, F. Guilleaume and G. L. Griffin, *Langmuir*, 1987, **3**, 477. and references therein.

Shi-Gang Sun answered: The IR band position ($\nu_{\rm CO}$) of adsorbed CO is a function of both electrode potential (E) and CO coverage (θ):

$$\nu_{\rm CO} = f(\theta, E)$$

$$\mathrm{d}\nu_{\rm CO} = \left(\frac{\partial \nu_{\rm CO}}{\partial \theta}\right)_E \mathrm{d}\theta \left(\frac{\partial \nu_{\rm CO}}{\partial E}\right)_\theta \mathrm{d}E$$

At a fixed coverage, we have determined that the Stark-tuning rate $\left(\frac{\partial \nu_{\rm CO}}{\partial \theta}\right)_\theta$ is about 25 cm^{-1} V^{-1} on Pt(310) and between 21 and 23 cm^{-1} V^{-1} on THH Pt NCs, which are common values for adsorbed CO on Pt. As a consequence, the large slope (>200 cm^{-1}) we measured should be attributed to $\left(\frac{\partial \nu_{\rm CO}}{\partial \theta}\right)_E$ at a fixed potential, which is not a Stark-tuning rate. We suggest that both electron-donation/back–donation and dipole–dipole interactions may contribute to such large slope.

The fact that the $\left(\frac{\partial \nu_{\rm CO}}{\partial \theta}\right)_E$ on THH Pt NCs is much larger than that on bulk single-crystal electrodes could be attributed to preferential CO adsorption on (100) sites on THH Pt NCs at low coverage, as illustrated in the paper. Although the overall CO coverages are the same, the local adsorbed CO density on the (100) terraces of (310) and (210) subfacets of THH Pt NCs is much larger than that on (100) terraces of (310) or (210) Pt single crystal planes, resulting in stronger dipole–dipole interactions that induce a big blue-shift of CO band position.

Mingquan Yu opened a general discussion by addressing Yujiang Song, Xile Hu and Shi-Gang Sun: When we conduct electrochemical tests, catalysts usually exhibit a redox wave. However, the redox wave would get smaller-and-smaller, even seem to disappear with slowing scan rate. So I'd like to ask what happens when the scan rate gets slower?

Yujiang Song replied: I guess you are talking about cyclic voltammetry (CV) curves. When you use a slower scan rate, the current density should get smaller. The current density is proportional to the square root of scan rate.

Xile Hu continued: In electrochemistry, scan rate dependent voltammetry will give information about the influence of mass transport as well as capacitive behavior of the test molecule. You have to analyze it case-by-case.

Shi-Gang Sun concluded: The relationship between redox current and scan rate depends on the types of electrochemical reactions and their reversibility. The electrooxidation reactions of most of small organic molecules are usually kinetic-controlled, but not diffusion-controlled (that is, mass transfer rate is larger than intrinsic electrochemical kinetics). So, as the potential scan rate decreases, the oxidation current will gradually decline and trend to a stable value, but not zero. As for adsorbed species reaction, *e.g.*, hydrogen adsorption on Pt, the current can really trend to zero as scan rate decreases.

Mingquan Yu addressed Yujiang Song, Xile Hu and Shi-Gang Sun: Has it become necessary for us to set standard electrochemical test conditions to compare the activity of different catalysts? As different groups often choose to use different test conditions, it is hard for us to compare the activity, so I presume setting a standard electrochemical test condition is important and urgent.

Yujiang Song replied: It is necessary to benchmark how to evaluate the electrocatalysts toward oxygen reduction reaction. I agree with you that it is critical and urgent to set up a standard for electrocatalyst examination.

Xile Hu followed up these remarks: I agree. In fact, standard testing conditions already exist. It is only because many active researchers in the field do not have a background in electrochemistry that we see papers reporting results under non-standard conditions. This situation will improve soon.

Shi-Gang Sun concurred: I also agree that a standard electrochemical test is very important in order to compare the results reported by different groups worldwide. Unfortunately, such standards have not been established yet due to the diversity of reactants, catalyst materials, and electrolyte. This requires that academic societies (*e.g.*, IUPAC, ISE or ECS) promulgate the relevant standards.

Joachim Maier addressed Gang Fu and Shi-Gang Sun: Do you know why, in catalysis, people do not concentrate more on point defects in the surfaces? Otherwise it is like dealing with catalysis on a water droplet and ignoring H^+ and OH^- ions in the surface layer.

Gang Fu answered: In the catalysis, the catalytic sites should be considered as "defect sites. The "defect" sites change their molecular structures during catalysis.

Shi-Gang Sun added: We agree that point defects are also important in catalysis. For example, Ertl and co-workers observed by STM that the decomposition of NO exclusively occurs on line defects (*i.e.*, edge sites) of Ru(0001) surface (Ertl *et al.*, *Science*, 1996, **273**, 1688–1690). However, it is difficult to control and monitor point defects on surfaces, especially on nanocrystals. So, it is challenging to study the role played by point defects during catalytic reactions at present.

Zhongqun Tian continued the general discussion. The speakers have very nicely shown that catalytic reaction sites can be more well-defined with new synthesis methods, *in situ* characterization and theoretical calculation. What could be the next generation surface for energy chemistry with the synergetic effects, rationally designed paths and higher stability?

Xile Hu spoke first: With increasing understanding of catalysis at a molecular level, researchers will be able to develop next generations of catalysts that are highly active per atom, multifunctional due to synergetic effects, and with high stability once the decomposition pathways are known and inhibited.

Yujiang Song continued: This is a very broad and challenging question! My answer will only focus on non-noble metal electrocatalysts (NNMEs) for the oxygen reduction reaction (ORR). In an alkaline medium, NNMEs are close to replacing platinum based electrocatalysts. If we follow the three rules for the rational design and synthesis of NNMEs (high surface area, high density of ORR active sites, and highly ordered active sites), the next step would be the creation of NNMEs with highly ordered active sites at a high density. In an acidic medium, it is not clear what is the exact nature of the ORR active sites of NNMEs even at the nano-scale. In this scenario, the first task should be the identification of the ORR active sites and then to follow the three rules to develop the next generation of NNMEs for acidic conditions.

Gang Fu added: In our opinion, the next generation surfaces should be multicomponent and multifunctional, which will enable simultaneous activation of multiple reactants having different properties, and conversion of them into the products with high selectivity. We have demonstrated that noble metal (NM)–transition metal hydroxide interface systems have a synergy effect to boost CO oxidation because CO can chemisorb on the NM, while transition metals usually have a relatively high oxygen affinity. It should be pointed out that CO oxidation is just a 2e redox process. For more complicated cases, a smart, sophisticated design is required. Generally speaking, to design a multicomponent catalyst, we should understand the functions of each component, and then combine them in a reasonable way to make the active center, and then prevent the loss of activity during the reaction process. To accomplish these steps, enormous efforts should be made to develop new synthesis methods, new characterization techniques, and new theoretical methods as well.

Shi-Gang Sun concluded: The next generation surface for catalysis should have high-density of active sites and optimal surface atomic arrangement for a certain reaction. This needs to understand the surface structure–catalytic function relationships at the atomic level by a combination of experiment and theory. Another approach is to construct atom–nano–micro multiscale architectures with synergetic effects. In short, we need to engineer rationally multiscale structured catalysts from the atomic to nano and to micro structure, as well as with a local to long-range environment of active sites.

Lee Cronin addressed Yujiang Song, Xile Hu, Gang Fu and Shi-Gang Sun: How do we more intelligently discover novel spaces in chemistry and exploit the interplay between theory and experiment in the understanding, design, and prediction of new catalysts?

Yujiang Song answered: It is necessary to have experimentalists and theorists to communicate more efficiently, thus building up theoretical models that are close to real reaction systems. I tend to claim that only theoretical models that have been proven in real reactions are reliable. Otherwise, theoretical models should be re-constructed and optimized.

Xile Hu noted: There is no smarter method in research than giving intelligent people the opportunity to do what they think is interesting. Novel discoveries will occur along the way. To further exploit the interplay between experiments and theories, we need better communication and to work on the same goal.

Gang Fu continued: Rational design should be based on our understanding of the surface structure and reaction mechanism under experimental conditions. However, both of them are not fully understood due to the compositional and/or structural inhomogeneity of conventional catalysts. Thus, we should develop well-defined systems (*e.g.* nanoparticles), which can decouple complicated factors, and allow the application of advance characterization techniques to established structure–activity relationships. To gain deeper insight, theoretical calculations should be performed to explain the experimental results, and further elucidate the intrinsic factors in catalysis. We expect that fruitful interaction between theory and experiment could promote the development of novel catalysts.

Shi-Gang Sun added: In our opinion, understanding the surface structure–catalytic function relationships of nanoparticles is of crucial importance in the design and predication of new catalysts. This requires the development of nanocrystal-based model nanocatalysts with a well-ordered and tunable surface structure, which could bridge the huge gap between bulk single-crystal model catalysts and practical nanocatalysts in terms of size and structural complexity, and provide catalytic reaction principles more related to practical nanocatalysts. Theoretical methods can provide new insight into catalytic reaction mechanism and surface processes. On basis of this knowledge and by combination of experiment and theory, we may efficiently predict and design new catalysts.

Wee Shong Chin communicated: While Professor Chao has discussed the use of phenylacetylene cross-linked to Si NPs, could he comment on the relative amount and the effect of this on the resultant material properties?

Yimin Chao communicated in reply: In the paper we have not mentioned crosslinking so I assume you mean be coverage and overlap. The relative surface coverage is approximately 70% of the available surface. The effect the coverage has on the thermoelectric properties of the material has not been studied, as the coverage is difficult to control, but at a lower surface coverage we would expect a higher concentration of surface oxide and as a result a lower electrical conductivity. This is a problem with many semiconductor organic nanocomposites. Additionally the level of overlap between the chains would be expected to have a positive effect on the materials' performance, as it would give better electron transfer between chains.

Wee Shung Chin commented: [illegible] has discussed the [illegible] would [illegible] comment on [illegible] and the effect of this on the resultant material properties?

Wenle Chao commented [illegible] In [illegible] paper we [illegible] to measure [illegible] coverage and [illegible]. The [illegible] surface coverage [illegible] of the available surface. The effect the coverage has on the [illegible] properties of the material has not been studied. If the coverage is sufficient [illegible] higher [illegible] the level of [illegible] between the [illegible] would be expected to [illegible] effect on the [illegible] in [illegible].

Faraday Discussions

PERSPECTIVE

Triboelectric nanogenerators as new energy technology and self-powered sensors – Principles, problems and perspectives

Zhong Lin Wang*[a,b]

Received 17th August 2014, Accepted 8th September 2014
DOI: 10.1039/c4fd00159a

Triboelectrification is one of the most common effects in our daily life, but it is usually taken as a negative effect with very limited positive applications. Here, we invented a triboelectric nanogenerator (TENG) based on organic materials that is used to convert mechanical energy into electricity. The TENG is based on the conjunction of triboelectrification and electrostatic induction, and it utilizes the most common materials available in our daily life, such as papers, fabrics, PTFE, PDMS, Al, PVC *etc.* In this short review, we first introduce the four most fundamental modes of TENG, based on which a range of applications have been demonstrated. The area power density reaches 1200 W m^{-2}, volume density reaches 490 kW m^{-3}, and an energy conversion efficiency of ~50–85% has been demonstrated. The TENG can be applied to harvest all kinds of mechanical energy that is available in our daily life, such as human motion, walking, vibration, mechanical triggering, rotation energy, wind, a moving automobile, flowing water, rain drops, tide and ocean waves. Therefore, it is a new paradigm for energy harvesting. Furthermore, TENG can be a sensor that directly converts a mechanical triggering into a self-generated electric signal for detection of motion, vibration, mechanical stimuli, physical touching, and biological movement. After a summary of TENG for micro-scale energy harvesting, mega-scale energy harvesting, and self-powered systems, we will present a set of questions that need to be discussed and explored for applications of the TENG. Lastly, since the energy conversion efficiencies for each mode can be different although the materials are the same, depending on the triggering conditions and design geometry. But one common factor that determines the performance of all the TENGs is the charge density on the two surfaces, the saturation value of which may independent of the triggering configurations of the TENG. Therefore, the triboelectric charge density or the relative charge density in reference to a standard material (such as polytetrafluoroethylene (PTFE)) can be taken as a measuring matrix for characterizing the performance of the material for the TENG.

[a]*School of Materials Science and Engineering, Georgia Institute of Technology, Atlanta, Georgia 30332-0245, USA. E-mail: zlwang@gatech.edu*

[b]*Beijing Institute of Nanoenergy and Nanosystems, Chinese Academy of Sciences, Beijing, China*

1. Introduction

A drastic change occurring in the world in the last two decades is the fast development of portable and personal electronics, which is now even shifting toward wearable electronics. A unique character of this technology trend is the huge increase in the number of electronic devices/systems, in a total of billions to trillions, each of which requires a mobile power source. The most traditional approach is to use batteries to power these electronics, which is likely to face a few possible consequences. First, a battery has a limited lifetime, replacing the battery for each component becomes a huge task, especially, how do we know it is out of power, when should we replace the battery, who will replace the battery, *etc.* This is not a problem if the number of batteries is limited, but the situation drastically changes if the number of batteries to be replaced becomes huge. Secondly, with the use of huge amount of batteries, recycling of these batteries become a major task, because widely distributed batteries will inevitably cause environmental issues if the chemicals for making the batteries leak out.

The trend in the development of portable electronics is toward low power consumption, which makes it possible to use the energy harvested from the working environment of the device to power directly the device, forming a trend of self-powered systems[1] for application in ultrasensitive chemical and bimolecular sensors, nanorobotics, micro-electromechanical systems, remote and mobile environmental sensors, homeland security and even portable/wearable personal electronics. New technologies that can harvest energy from the environment as sustainable self-sufficient micro/nano-power sources are the newly emerging field of nanoenergy, which is concerned with the application of nanomaterials and nanotechnology for harvesting energy to power micro/nano-systems.[2] The objective of this paper is to give a review of the fundamentals of the triboelectric nanogenerator (TENG)[3] and its updated progress and potential applications as a new energy technology and in self-powered active sensors.[4]

2. Fundamental principle modes of triboelectric nanogenerators

The triboelectric effect is the phenomenon that a material becomes electrically charged after it contacts a different material through friction, and has been known for thousands of years. Although this is one of the most frequently experienced effects that each and every one of us inevitably sees every day, the mechanism behind triboelectrification is still being studied, possibly with debate. It is generally believed that after two different materials come into contact, a chemical bond is formed between some parts of the two surfaces, called adhesion, and charges move from one material to the other to equalize their electrochemical potential. The transferred charges can be electrons or may be ions/molecules. When separated, some of the bonded atoms have a tendency to keep extra electrons, and some a tendency to give them away, possibly producing triboelectric charges on the surfaces. The presence of triboelectric charges on dielectric surfaces can be a force for driving electrons in the electrode to flow in order to balance the electric potential drop created. Based on such a principle, we have invented four different modes of TENGs, as elaborated in the following.

2.1 Vertical contact-separation mode

We use the simplest design of TENG as an example (Fig. 1a).[5,6] Two dissimilar dielectric films face with each other, and there are electrodes deposited on the top and the bottom surfaces of the stacked structure. A physical contact between the two dielectric films creates oppositely charged surfaces. Once the two surfaces are separated by a small gap by the lifting of an external force, a potential drop is created. If the two electrodes are electrically connected by a load, free electrons in one electrode would flow to the other electrode to build an opposite potential in order to balance the electrostatic field. Once the gap is closed, the triboelectric-charge-created potential disappears and the electrons flow back.[7]

2.2 Lateral sliding mode

The structure to start with is the same as that for the vertical contact-separation mode. When two dielectric films are in contact, a relative sliding in parallel to the surface also creates triboelectric charges on the two surfaces (Fig. 1b).[8,9] A lateral polarization is thus introduced along the sliding direction, which drives the electrons on the top and bottom electrodes to flow in order to fully balance the field created by the triboelectric charges. A periodic sliding apart and closing generates an AC output. This is the sliding mode TENG. The sliding can be a planar motion, a cylindrical rotation,[10] or disc rotation.[11] Related theoretical studies have been carried out for understanding the basic mode and grating structured TENG.[12,13]

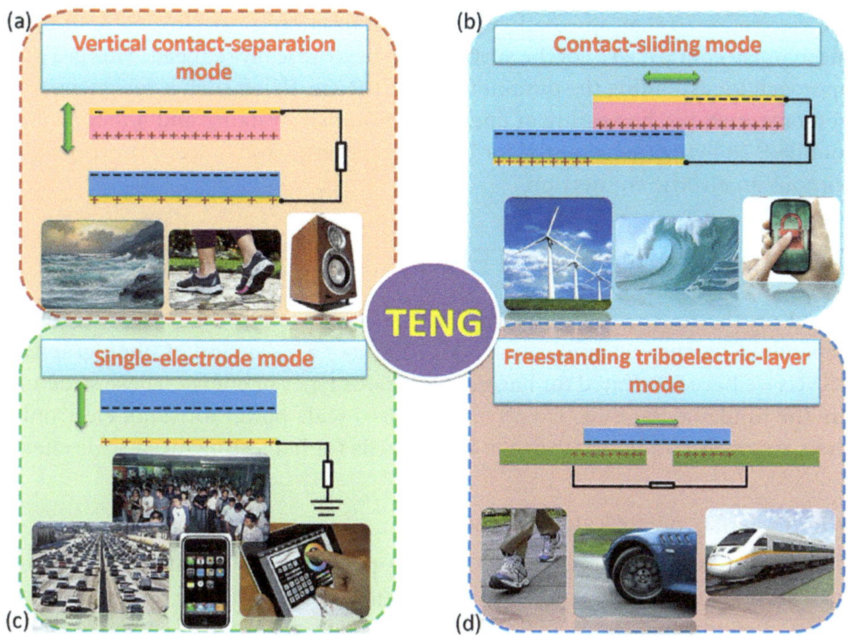

Fig. 1 The four fundamental modes of triboelectric nanogenerators: (a) vertical contact-separation mode; (b) in-plane contact-sliding mode; (c) single-electrode mode; and (d) freestanding triboelectric-layer mode.

2.3 Single-electrode mode

The two modes introduced in Sections 2.1 and 2.2 have two electrodes interconnected by a load. Such TENGs can freely move so that they can work for mobile cases. In some cases, the object that is part of the TENG cannot be electrically connected to the load because it is a mobile object, such as a human walking on a floor. In order to harvest energy from such a case, we introduced a single electrode TENG, in which the electrode on the bottom part of the TENG is grounded (Fig. 1c). If the size of the TENG is finite, an approaching or departing of the top object from the bottom one would change the local electrical field distribution, so that there are electron exchanges between the bottom electrode and the ground to maintain the potential change of the electrode. This energy harvesting strategy can be in both contact-separation mode[14] and contact-sliding mode.[15,16]

2.4 Freestanding triboelectric-layer mode

In nature, a moving object is naturally charged due to its contact with air or other object, such as our shoes walking on floors that are usually charged. The charges remain on the surface for hours and contact or friction is unnecessary within this period of time because the charge density reaches a maximum. If we make a pair of symmetric electrodes underneath a dielectric layer, and the size of the electrodes and the gap distance between the two are of the same order as the size of the moving object, the object's approach to and/or departure from the electrodes creates an asymmetric charge distribution in the media, which causes the electrons to flow between the two electrodes to balance the local potential distribution (Fig. 1d).[17] The oscillation of the electrons between the paired electrodes produces power. The moving object does not have to touch directly the top dielectric layer of the electrodes so that, in rotation mode, free rotation is possible without direct mechanical contact; wear of the surfaces can be drastically reduced. This is a good approach for extending the durability of the TENGs. Using such a design, we have demonstrated the harvesting of energy from human walking and a mobile automobile,[13] showing the potential for harvesting energy from a freely moving object without an electric connection.

3. Applications of TENGs

Based on the four modes illustrated above, we have fabricated various TENGs depending on specific applications. Fig. 2 shows a collection of photographs of TENGs we have fabricated for harvesting various types of energy. These structures are the fundamental units for providing micro-scale power for small electronics, their assembly and integration can be the basis for harvesting mega-scale energy.

3.1 TENGs as micro-scale power source

The first goal of developing TENGs is to power small electronics for sensor networks. Using the four modes demonstrated above and their combinations, a range of energy harvesting methods has been demonstrated from cases such as body motion,[18] fabrics,[19,20] vibrations from human walking,[21] hand pressing,[22,23] shoe insole,[24,25] vibration of a string or tree branch,[26] vibration of a machine,[27,28] elastic energy in sponge structure,[29] sound wave in air[30] and in water.[31]

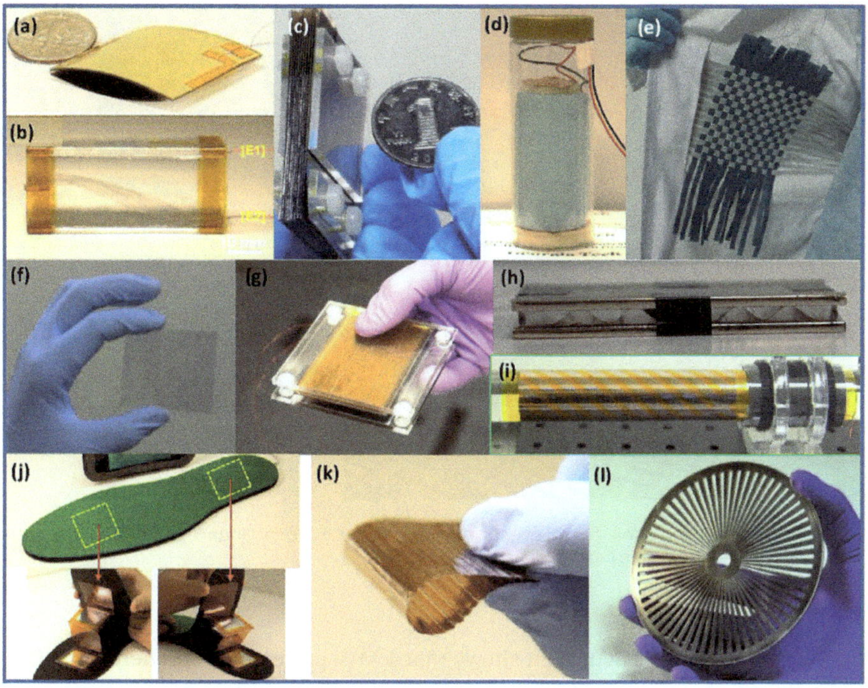

Fig. 2 Typical photographs of some triboelectric nanogenerators fabricated for harvesting: (a) finger tapping energy; (b) air-flow/wind energy; (c) relative in-plane sliding energy; (d) enclosed cage for harvesting oscillating/disturbing energy in water or mechanical vibration; (e) fabric for harvesting body motion energy; (f) transparent TENG for harvesting energy in touch pad; (g) foot/hand pressing energy; (h) water impact energy; (i) cylindrical rotation energy; (j) shoe insole for walking energy; (k) flexible grating structure for harvesting sliding energy; and (l) disc shape rotation energy.

The mechanical action can be either linear or in rotation mode. By designing a grating structure, we can lower effectively the output voltage and greatly increase the output current, which is required for high efficiency of charging of an energy storage unit. An energy conversion efficiency of 50% has been demonstrated for a sliding mode TENG[32] and 24% for a rotation based TENG,[33] and the output power density reaches as high as 1200 W m^{-2}. Such a power output is high enough to power small electronics, establishing a solid foundation for self-powered systems.

3.2 TENG as macro-scale power source

By using the contact-electrification between a liquid and a solid surface as well as the packaged TENGs,[34] we have shown that TENGs can be used for harvesting energy from flowing water in river, rain drops,[35,36] tide and ocean waves. Using the four fundamental modes presented above, we have successfully harvested the kinetic energy such as the up and down fluctuation of a water surface,[37] water waves, water streams and the impact of water to the shore.[38] Such structures can be used to harvest water energy from creeks without building a dam. Looking into the future, by constructing a three-dimensional network of the TENGs, such as a 3D fishing net, as schematically shown in Fig. 3, if the output of each unit is 1 mW

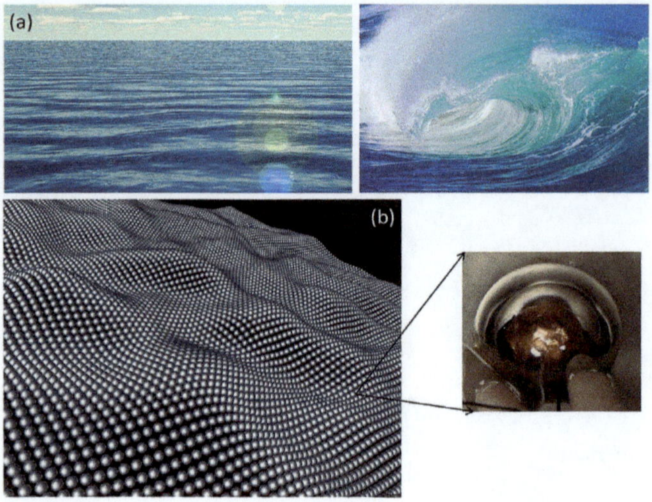

Fig. 3 A proposed idea for harvesting water energy from ocean using an integrated assembly of small unit TENGs in a 3D network.

on average,[39,40] our calculation shows that 1 MW power can be generated using 1 km square of surface area in ocean, possibly provides a feasible blue energy for large-scale energy needs of the world in the near future.

Using the contact-separation mode, a mechanical resonance of a polymer film between two dielectrics can be used to harvest wind energy.[41] If we can build a wall and solve the problem of surface wetting, we can effectively harvest energy from wild wind.

3.3 TENG as self-powered sensors

TENG is a device that converts a mechanical trigger into an electric output signal (voltage and current), which means that it can be used to direct sensing a dynamic mechanical action without applying a power unit to the device. This is a self-powering sensor. We have applied TENGs for a range of sensing types such as finger touch,[42–44] vibration detection,[19] tracking of a moving object (location, velocity and acceleration),[45–47] fine displacement in MEMS,[48] rotation sensors[49] and even chemical sensors.[50]

4. Problems to be discussed

Since their first demonstration in 2012,[3] TENGs have experienced a very rapid development both in fundamental understanding and technological improvements. As toward the future applications, there are a number of issues and problems need to be addressed, as listed in the following.

4.1 Fundamental mechanism of contact-electrification

Although triboelectrification is a phenomenon known to everyone, the basic mechanism of why two materials become charged once in physical contact remain to be investigated extensively. It has been debated that the charging is due to electron transfer as a result of energy level misalignment, ion transfer and/or materials species transfer. We have used AFM to study the charge density that a surface can be charged to, how long does it take for the charges to dissipate on an insulator surface (which can be hours), and how fast the charge diffuses on the surface.[51] We also used a conductive AFM tip to manipulate the sign of charges to be delivered to a surface by applying a preset bias, so that the surface charge can be positive, null or negative.[52] More detailed measurements have to be made in order to fully understand the electrostatic charging on surfaces.

4.2 Quantitative understanding the surface charge density

A key factor that dictates the performance of the TENGs is the surface charge density, which can be taken as a standard to characterize the matrix of performance of a material for a TENG. Quantitative techniques have to be developed to measure accurately the surface charge density. It is unclear how the surface structures, such as roughness, dielectric properties and the presence of nanoparticles/nanowires, would affect the magnitude of the charge density. More fundamental studies are required.

4.3 Measurement matrix – a standard for calibrating the performance of a material for a TENG

We have presented four modes of TENGs. The design and operation conditions of these TENGs are vastly different and it is important to find a parameter that can be used to quantify the performance of the TENG, similar to defining the efficiency for a solar cell, and a *ZT* factor for thermoelectrics. The energy conversion efficiencies for each mode can be different although the materials are the same, depending on the triggering conditions and design geometry. But one common factor that determines the performance of all the TENGs is the charge density on the two surfaces, the saturation value of which may independent of the triggering configurations of the TENG. Therefore, the triboelectric charge density or the relative charge density in reference to a standard material (such as polytetrafluoroethylene (PTFE)), can be taken as a measuring matrix for characterizing the performance of the material for the TENG.

4.4 Choice of materials

Although all materials exhibit triboelectricity, finding the right paired materials can give maximum output. Although there is guidance from the triboelectric series, it is a qualitative indication on the capability of a material to gain or lose an electron but it lacks a quantitative standard of calibration. Choice of materials, surface structure configurations and patterning, surface functionalization and more need to be studied. By introducing nanomaterials and possibly nanocomposites, we can optimize the mechanical, dielectric and surface properties of materials to receive the maximum power output.

4.5 Power management

A typical characteristic of the TENGs is high output voltage but low output current, but general electronics require a regulated power of a few volts. Approaches have to be developed to lower the output voltage without sacrificing the output power. This can be done by designing TENGs with smaller size or thickness, but the entire package can integrate many units to increase the output current. As for rotation based TENGs with a periodic AC output, a transformer can be applied to lower the voltage, but finding a right match is required to maximize the efficiency of power transfer.

4.6 Durability and stability

Since the triboelectrification is a result of two materials in physical contact, especially in contact-sliding mode, the durability and stability of the TENGs must be continuously improved. We have tested a disc based rotation TENG for 10 million cycles in contact mode;[32] no degradation in output was observed. This may be good for low frequency applications. For practical application at high frequency, finding and using materials with a high durability is important.

4.7 Packaging

Since the TENG is based on the surface charging effect, its performance is largely affected by the environment such as humidity and surface adsorption layers. Packaging materials and technologies are required to protect the device from contamination or liquid infiltration but without reducing too much of the efficiency and output power. This is because the TENG is a device that converts mechanical energy into electricity; preserving its flexibility and elasticity is important for improving the energy conversion efficiency. Such packaging is different from conventional packaging because of the presence of mechanical triggering.

4.8 Energy storage

Energy harvesting from the environment is subject to the variation of the environment, which is time dependent, unstable and sometime unpredictable, but the power required to drive electronics is regulated with a fixed input voltage and power. It is important to store the generated energy in a battery or capacitor, so that it can be used to power a device in a regulated manner. Between the power generation unit and a storage unit, a power management circuit is required to maximize the efficiency of power storage; the integration of the three forms a self-charging power pack.[53]

4.9 Hybridized cell for harvesting all types of energies

The working environment of a sensor can vary and the energy available in its working environment can also vary, such as solar, thermal and mechanical. It is important to harvest all types of energy that are available for the sensor. Hybrid cells that simultaneously or individually harvest solar, thermal and/or mechanical energy have been developed,[54–57] but the performance and coupling among them are unsuitable, because the output characteristics of each type of energy harvester are drastically different; for example, solar is a constant DC signal, while a TENG provides a pulsed AC signal.

4.10 Hybridization with a traditional electromagnetic induction generator

Electromagnetic induction was first discovered by Faraday in 1831, and later becomes the dominant mechanism used to convert mechanical energy into electricity. The generator requires a strong magnet, a metal coin and a rotator. The electromagnetic generator (EMG) usually is heavy, fairly costly and has a large volume. EMG's output characteristic is low voltage but high output current. In contrast, TENGs use organic materials so that they are light, smaller and cost-effective. The output characteristics of TENGs are high voltage but low current. The disadvantage of the TENG is a likely lower durability. The power performances of the two approaches are quite compatible and each has its own unique advantages, but the two can be complementary according to the application concerned.[58,59]

5. Perspectives

The discovery of the triboelectric nanogenerator (TENG) is a major milestone in the field of converting mechanical energy into electricity for building self-powered systems. It offers a completely new paradigm for harvesting mechanical energy effectively using organic and inorganic materials. An energy conversion efficiency of 50% and a total energy conversion efficiency of 85% have been demonstrated, and an output power density of 1200 W m^{-2} has been realized. The TENG can not only serve as a micro-scale power source for mobile and portable electronics, but it also has the potential to harvest water energy from ocean and wind, opening a new field of *blue energy*, which is a new chapter beyond green energy. Fig. 4 shows a proposed technology development road map for nanogenerators and its commercial applications. Furthermore, TENGs can serve as a self-powered sensor for sensing mechanical triggering, pressure, muscle stretching and more, which may become commercially available sooner than energy harvesters.

The discovery of TENGs opens a new field for materials scientists and chemists for using organic nanogenerators to convert mechanical energy at a high

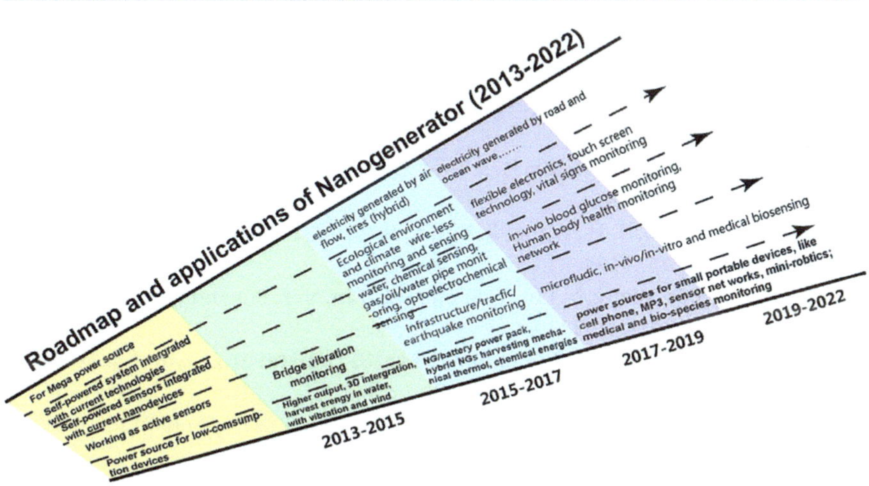

Fig. 4 Proposed technology roadmap for nanogenerators (2012).

efficiency, a so called organic nanogenerator, which is a disruptive technology for energy. We anticipate that much more enhancement of the output power density will be demonstrated in the next few years. We anticipate a worldwide study of TENG in the next few years, and soon some industrial products and applications will be demonstrated.

Acknowledgements

Research was supported by BES DOE, NSF, Airforce, Samsung, SKKU (Korea), MANA NIMS (Japan), and the Knowledge Innovation Program of the Chinese Academy of Sciences (KJCX2-YW-M13), the Hightower Chair foundation, and the "thousands talents" program for pioneer researcher and his innovation team, China. I thank my group members and my collaborators for their contributions to the work reviewed here, especially to those ones: Fengru Fan, Guang Zhu, Sihong Wang, Ya Yang, Zong-Hong Lin, Long Lin, Simiao Niu, Jun Chen, Yusheng Zhou, Weiqing Yang, Gang Cheng, Jin Yang, Hulin Zhang, Yannan Xie, Peng Bai, Cao-feng Pan, Qingshen Jing, Ying Liu, Yuanjie Su, Chi Zhang, Changbao Han, Wei Tang, Aifang Yu, Zhong-Qun Tian, Mengxiao Chen, Limin Zhang, and Weiming Du.

References

1 Z. L. Wang, *Sci. Am.*, 2008, **298**, 82.
2 Z. L. Wang and W. Z. Wu, *Angew. Chem.*, 2012, **51**, 11700.
3 F. R. Fan, Z. Q. Tian and Z. L. Wang, *Nano Energy*, 2012, **1**, 328.
4 Z. L. Wang, *ACS Nano*, 2013, **7**, 9533.
5 G. Zhu, C. F. Pan, W. X. Guo, C. Y. Chen, Y. S. Zhou, R. M. Yu and Z. L. Wang, *Nano Lett.*, 2012, **12**, 4960.
6 S. H. Wang, L. Lin and Z. L. Wang, *Nano Lett.*, 2012, **12**, 6339.
7 S. M. Niu, S. H. Wang, L. Lin, Y. Liu, Y. S. Zhou, Y. F. Hu and Z. L. Wang, *Energy Environ. Sci.*, 2013, **6**, 3576.
8 S. H. Wang, L. Lin, Y. N. Xie, Q. S. Jing, S. M. Niu and Z. L. Wang, *Nano Lett.*, 2013, **13**, 2226.
9 G. Zhu, J. Chen, Y. Liu, P. Bai, Y. S. Zhou, Q. S. Jing, C. F. Pan and Z. L. Wang, *Nano Lett.*, 2013, **13**, 2282.
10 Q. S. Jing, G. Zhu, P. Bai, Y. N. Xie, J. Chen, R. P. S. Han and Z. L. Wang, *ACS Nano*, 2013, **7**, 10424.
11 L. Lin, S. H. Wang, Y. N. Xie, Q. S. Jing, S. M. Niu, Y. F. Hu and Z. L. Wang, *Nano Lett.*, 2013, **13**, 2916.
12 S. M. Niu, Y. Liu, S. H. Wang, L. Lin, Y. S. Zhou, Y. F. Hu and Z. L. Wang, *Adv. Mater.*, 2013, **25**, 6184.
13 S. M. Niu, S. H. Wang, Y. Liu, Y. S. Zhou, L. Lin, Y. F. Hu and Z. L. Wang, *Energy Environ. Sci.*, 2014, **7**, 2339.
14 Y. Yang, Y. S. Zhou, H. L. Zhang, Y. Liu, S. Lee and Z. L. Wang, *Adv. Mater.*, 2013, **25**, 6594.
15 Y. Yang, H. L. Zhang, J. Chen, Q. S. Jing, Y. S. Zhou, X. N. Wen and Z. L. Wang, *ACS Nano*, 2013, **7**, 7342.
16 S. M. Niu, Y. Liu, S. H. Wang, L. Lin, Y. S. Zhou, Y. F. Hu and Z. L. Wang, *Adv. Funct. Mater.*, 2014, **24**, 3332.

17 S. H. Wang, Y. N. Xie, S. M. Niu, L. Lin and Z. L. Wang, *Adv. Mater.*, 2014, **26**, 2818.

18 W. Q. Yang, J. Chen, X. N. Wen, Q. S. Jing, J. Yang, Y. J. Su, G. Zhu, W. Z. Wu and Z. L. Wang, *ACS Appl. Mater. Interfaces*, 2014, **6**, 7479.

19 J. W. Zhong, Y. Zhang, Q. Z. Zhong, Q. Y. Hu, B. Hu, Z. L. Wang and J. Zhou, *ACS Nano*, 2014, **8**, 6273.

20 T. Zhou, C. Zhang, C. B. Han, F. R. Fan, W. Tang and Z. L. Wang, *ACS Appl. Mater. Interfaces*, 2014, **6**, 14695–14701.

21 W. Q. Yang, J. Chen, G. Zhu, J. Yang, P. Bai, Y. J. Su, Q. S. Jing and Z. L. Wang, *ACS Nano*, 2013, **7**, 11317.

22 X. S. Zhang, M. D. Han, R. X. Wang, F. Y. Zhu, Z. H. Li, W. Wang and H. X. Zhang, *Nano Lett.*, 2013, **13**, 1168.

23 S. Kim, M. K. Gupta, K. Y. Lee, A. Sohn, T. Y. Kim, K. S Shin, D. Kim, S. K. Kim, K. H. Lee, H. J. Shin, D. W. Kim and S. W. Kim, *Adv. Mater.*, 2014, **26**, 3778.

24 G. Zhu, P. Bai, J. Chen and Z. L. Wang, *Nano Energy*, 2013, **2**, 688.

25 B. Meng, W. Tang, X. S. Zhang, M. D. Han, W. Liu and H. X. Zhang, *Nano Energy*, 2013, **2**, 1101.

26 J. Yang, Y. Yang, J. Chen, H. L. Zhang, W. Q. Yang, P. Bai, Y. J. Su and Z. L. Wang, *Adv. Energy Mater.*, 2014, **4**, 1301322.

27 W. Q. Yang, J. Chen, Q. S. Jing, J. Yang, X. N. Wen, Y. J. Su, G. Zhu, P. Bai and Z. L. Wang, *Adv. Funct. Mater.*, 2014, **24**, 4090.

28 J. Chen, G. Zhu, W. Q. Yang, Q. S. Jin, P. Bai, Y. Yang, T. C. Hou and Z. L. Wang, *Adv. Mater.*, 2013, **25**, 6094.

29 K. Y. Lee, J. Chun, J. H. Lee, K. N. Kim, N. R. Kang, J. Y. Kim, M. H. Kim, K. S. Shin, M. K. Gupta, J. M. Baik and S. W. Kim, *Adv. Mater.*, 2014, **26**, 5037.

30 J. Yang, J. Chen, Y. Liu, W. Q. Yang, Y. J. Su and Z. L. Wang, *ACS Nano*, 2014, **8**, 2649.

31 A. F. Yu, M. Song, Y. Zhang, Y. Zhang, L. B. Chen, J. Y. Zhai and Z. L. Wang, *Nano Res.*, 2014, DOI: 10.1007/s12274-014-0559-z.

32 G. Zhu, Y. S. Zhou, P. Bai, X. S. Meng, Q. S. Jing, J. Chen and Z. L. Wang, *Adv. Mater.*, 2014, **26**, 3788.

33 G. Zhu, J. Chen, T. J. Zhang, Q. S. Jing and Z. L. Wang, *Nat. Commun.*, 2014, **5**, 3456.

34 Z. H. Lin, G. Cheng, L. Lin, S. Lee and Z. L. Wang, *Angew. Chem., Int. Ed.*, 2013, **52**, 12545.

35 Z. H. Lin, G. Cheng, S. Lee and Z. L. Wang, *Adv. Mater.*, 2014, **26**, 4690.

36 Z. H. Lin, G. Cheng, W. Z. Wu, K. C. Pradel and Z. L. Wang, *ACS Nano*, 2014, **8**, 64.

37 G. Zhu, Y. J. Su, P. Bai, J. Chen, Q. S. Jing, W. Q. Yang and Z. L. Wang, *ACS Nano*, 2014, **8**, 6031–6037.

38 X. N. Wen, W. Q. Yang, Q. S. Jing and Z. L. Wang, *ACS Nano*, 2014, **8**, 7405.

39 Y. F. Hu, J. Yang, Q. S. Jing, S. M. Niu, W. Z. Wu and Z. L. Wang, *ACS Nano*, 2013, **7**, 10424.

40 Y. Yang, H. L. Zhang, R. Y. Liu, X. N. Wen, T. C. Hou and Z. L. Wang, *Adv. Energy Mater.*, 2013, **3**, 1563.

41 Y. Yang, G. Zhu, H. L. Zhang, J. Chen, X. D. Zhong, Z. H. Lin, Y. J. Su, P. Bai, X. N. Wen and Z. L. Wang, *ACS Nano*, 2013, **7**, 9461.

42 Y. Yang, H. L. Zhang, X. D. Zhong, F. Yi, R. M. Yu, Y. Zhang and Z. L. Wang, *ACS Appl. Mater. Interfaces*, 2014, **6**, 3680.

43 Y. Yang, H. L. Zhang, Z. H. Lin, Y. S. Zhou, Q. S. Jing, Y. J. Su, J. Yang, J. Chen, C. G. Hu and Z. L. Wang, *ACS Nano*, 2013, **7**, 9213.
44 B. Meng, W. Tang, Z. H. Too, X. S. Zhang, M. D. Han, W. Liu and H. X. Zhang, *Energy Environ. Sci.*, 2013, **6**, 3235.
45 M. X. Chen, X. Y. Li, L. Lin, W. M. Du, X. Han, J. Zhu, C. F. Pan and Z. L. Wang, *Adv. Funct. Mater.*, 2014, **24**, 5059–5066.
46 C. B. Han, C. Zhang, X. H. Li, L. M. Zhang, T. Zhou, W. G. Hu and Z. L. Wang, *Nano Energy*, 2014, **9**, 325–333.
47 Y. J. Su, G. Zhu, W. Q. Yang, J. Yang, J. Chen, Q. S. Jing, Z. M. wu, Y. D. Jiang and Z. L. Wang, *ACS Nano*, 2014, **8**, 3843.
48 Y. S. Zhou, G. Zhu, S. M. Niu, Y. Liu, P. Bai, Q. S. Jing and Z. L. Wang, *Adv. Mater.*, 2014, **26**, 1719.
49 Q. S. Jing, G. Zhu, W. Z. Wu, P. Bai, Y. N. Xie, R. P. S. Han and Z. L. Wang, *Nano Energy*, **10**, 305–312.
50 Z. H. Lin, G. Zhu, Y. S. Zhou, Y. Yang, P. Bai, J. Chen and Z. L. Wang, *Angew. Chem., Int. Ed.*, 2013, **52**, 5065.
51 Y. S. Zhou, Y. Liu, G. Zhu, Z. H. Lin, C. F. Pan, Q. S. Jing and Z. L. Wang, *Nano Lett.*, 2013, **13**, 2771.
52 Y. S. Zhou, S. H. Wang, Y. Yang, G. Zhu, S. M. Niu, Z. H. Lin, Y. Liu and Z. L. Wang, *Nano Lett.*, 2014, **14**, 1567.
53 S. H. Wang, Z. H. Lin, S. M. Niu, L. Lin, Y. N. Xie and Z. L. Wang, *ACS Nano*, 2013, **7**, 11263.
54 S. Lee, S. H. Bae, L. Lin, S. Ahn, C. Park, S. N. Cha, Y. J. Park, H. Chang, S. W. Kim and Z. L. Wang, *Nano Energy*, 2013, **2**, 817.
55 Y. Yang, H. L. Zhang, S. Lee, T. C. Hou and Z. L. Wang, *Energy Environ. Sci.*, 2013, **6**, 1744.
56 Y. Yang, H. L. Zhang, Z. H. Lin, Y. Liu, J. Chen, Z. Y. Lin, Y. S. Zhou, C. P. Wong and Z. L. Wang, *Energy Environ. Sci.*, 2013, **6**, 2429.
57 Y. Yang, H. L. Zhang, S. Lee, D. Kim, W. Hwang and Z. L. Wang, *Nano Lett.*, 2013, **13**, 803.
58 C. Zhang, W. Tang, C. B. Han, F. R. Fan and Z. L. Wang, *Adv. Mater.*, 2014, **26**, 3580.
59 F. R. Fan, W. Tang, Y. Yao, J. J. Luo, C. Zhang and Z. L. Wang, *Nanotechnology*, 2014, **25**, 135402.

Faraday Discussions

PAPER

Next generation materials for energy chemistry: concluding remarks

Leroy Cronin*

Received 20th January 2015, Accepted 27th January 2015
DOI: 10.1039/c5fd00009b

Materials chemistry and electrochemistry have long been very active fields of research accelerated by chance discoveries, societal needs, and the ever expanding constituency of researchers around the globe. Advances in synthesis, characterisation techniques, and interdisciplinary research have caused an almost exponential growth in research where applications have been driving the focus – especially in energy related research. This is important because despite the vast increase in accessible fossil fuel reserves, expanded partly by fracking for oil and gas, we must surely stop using fossil fuels as soon as possible. Even if we burn a fraction of these reserves, then climate change leading to a rise in average global temperatures greater than 2 °C will be unstoppable, threatening not just society, but the entire global ecosystem on planet Earth. Life as we know it will surely decline. The key realisation is intimidating. Two thirds of our current fossil fuel reserves are effectively unusable if we want to preserve our climate, no matter how low the oil price. So how can the next generation of materials help solve these problems? Well, some of the most important needs in materials for energy relate not only to conversion and new energy systems, but also to the ability to store energy at high densities. Also, the need for scalable systems that are durable cannot be over emphasised and these stringent requirements will surely allow scientists to creatively rise to the challenge. At *Faraday Discussion* 176, I observed such a transformation, where discussions not only about scientific principles but real world applications and practical implementation were combined.

Faraday Discussions have a great tradition, but *Faraday Discussion* 176 marked a special development as the first time that a *Faraday Discussion* has been held in Asia; the venue was the University of Xiamen in China. Also, it was fitting that the topic, next generation energy materials, was discussed in a country that has recently become the largest producer and consumer of energy in the world, and is also rolling out the largest renewable energy program. However, many serious issues must be tackled beyond the science, not only economic, political, social and environmental. The role of science today has not changed. Beyond

School of Chemistry, University of Glasgow, University Avenue, Glasgow, G12 8QQ, UK. E-mail: lee.cronin@glasgow.ac.uk; Web: http://www.croninlab.com

fundamental developments, we must use our current technologies to develop next generation technologies, established by the current energy system. There is some urgency. This next generation of energy systems has to cope with increasing population, climate change, the need for food and water, as well as higher living standards. There are many questions. How can we create scalable energy systems? Which blend of systems (wind, tidal, solar, nuclear) could help the energy mix? How are we going to produce carbon-dense fuels? If we keep using carbon, how can we move to a renewable carbon cycle that removes and converts carbon dioxide at least at the rate it is produced? What are the material challenges? How can we encourage the next generation of researchers?

In my opinion, one of the greatest challenges is how do we ensure that research starts at the fundamental end of the spectrum, leading to application, rather than application first. The gradual shift in publication trends to high volume and citations is worrying. This is driven by rewarding publication volume by research funding, which in turn serves to distort the system. This in turn reduces scholarship and forces bright researchers to instead focus on the incremental. This should be avoided at all cost. Since we all understand there is a problem, and we all take an active role in the system to varying degrees, it is up to us to change the status quo for the better by striving to do the best science driven by curiosity, big challenges, and chance discoveries. Researchers need to be willing to take risks, but all too often this risk is seen through the lens of a failed career, and this should be highly discouraged. By celebrating the process of discovery and encouraging researchers to explore the unknown, I think there is hope that we can reverse this trend. Where better to discuss fundamentals to application in next generation materials for energy than at a university that has one of the top chemistry departments in China, and one of the top 100 chemistry departments worldwide, Xiamen.

This *Faraday Discussion* had three themes, electrochemical conversion and energy storage systems, solar cells and photocatalytic systems, and energy-related catalytic and other materials. Whilst I will not give a lecture-by-lecture summary (this can be gained by reading this volume), I will try and make some observations about the current status of the field and the current perspectives and challenges for the future. The Introductory Lecturer (Peidong Yang, University of California Berkeley) provided an overview of the state-of-the-art in inorganic materials, showing how control on the nanoscale allows the design of architectures with vast potential for a range of relevant applications *e.g.* solar cells. Yang described how nano-materials science has developed rapidly over the last decade with the ability to direct the formation and faceting of nanoparticles, nanocrystals and nanorods/wires.[1] Three levels of control are needed: assembly, spatial and temporal organisation. Only temporal as well as spatial organisation will yield functional systems that can exploit the nanoscale non-equilibrium organisation.[2] This requires that we understand the control parameters for the development of electrochemical materials and appreciate that as such, non-equilibrium systems are developed by the structural and compositional complexity increase. Maier proposed[3] that the distance from equilibrium and the complexity of the system can be understood if one distinguishes between the intrinsic (*in situ*) and extrinsic (*ex situ*) parameters. The control of defects, morphology and topology are also vital in defining the properties of the system and materials science can be viewed in terms of size and complexity, see Fig. 1. These ideas provoked discussion about the nature of defects and pondering on the effects on a system if random defects could be 'programmed' intentionally into a material.

The topic then shifted to battery electrodes in electrochemical circuits, where many complex issues of stability, diffusion, and composition are highly relevant, and also in fuel cells where diffusion is vital. In the case of SOFC cathode materials, substitution reactions can be used to improve performance,[4] for example where calcium was partly substituted with strontium in $Ca_3Co_4O_{9+\delta}$, and this may be likely due to an increase of the volume of the rock salt layers in which the conduction process takes place. The role of structural studies was discussed in understanding not only the overall structure, but also local environments, which is key to developing new materials. Griffin explained in detail how solid state NMR spectroscopy[5] could be used to study the local environments of anions in supercapacitor electrodes and to quantify changes in the populations of adsorbed species during charging. Exploring the size of the anions absorbed within the carbon micropores of the electrode as a function of concentration, potential, and polarisation also revealed much about the mechanism, including the relationship with anion size. The role of size was highlighted again in the investigation of nickel hexacyanoferrate (NiHCFe) as a cathode material.[6] This is a very interesting material in both aqueous and organic electrolytes due to its ease of synthesis, use of earth abundant ingredients, and framework. It is hoped that such systems will lead to long life and high performance. The recent developments of new techniques increases dramatically the amount of useful information that can be gained from electrochemical systems such as batteries and supercapacitors. For instance the use of *in situ* liquid cell TEM allows the imaging of electrode interfaces in liquids with high spatial resolution, and under operation.[7] The imaging of a gold electrode–electrolyte interface with lithium during charging and discharging using an electrochemical liquid was discussed, including control experiments and how to interpret the data. Also both *in situ* and *ex situ* SEM[8] were discussed in the context of exploring the formation of lithium dissolution and dendrite formation.

Moving to fuel cells and electrolysers, there was much discussion about the development of new electrode materials, which is currently seen as a big barrier for the wide adoption of fuel cells and electrolysers. In electrolysers, understanding the oxygen evolution reaction (OER) is important, and probing the surface chemistry of Ru oxide[9] in acidic and alkaline environments revealed that

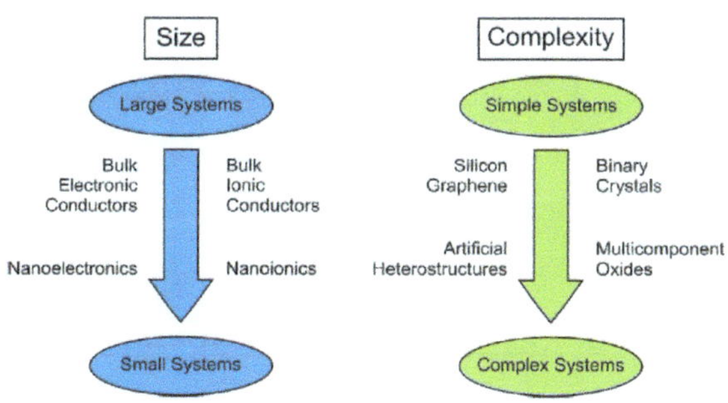

Fig. 1 Materials research in terms of size and complexity; reproduced from ref. 3.

the kinetics of the OER are almost entirely controlled by the stability of the Ru surface atoms. This is because during the OER, surface defects are created and the oxide is only active for OER if it is itself unstable. For the hydrogen evolution reaction (HER) and oxygen reduction reaction (ORR), non-precious metal catalysts suitable for acidic electrolytes are vital. Several solutions were discussed including the use of graphene supported iron-based nanoparticles encapsulated in a nitrogen-doped carbon.[10] Also, Co oxide (Co_3O_4) nanotubes are shown to act as efficient water oxidation catalysts when driven with a visible light sensitizer (pH 7).[11] The role of inorganic membranes was also discussed for the designing of solar-fuel type devices. Nickel compounds were also found to be effective for the OER when small nanoparticles of nickel oxide and nickel hydroxide were investigated.[12] The catalytic activity was comparable to that of the best mixed metal oxide catalysts under the same conditions, whereas plate-like WO_3 photoanodes[13] were discussed as being good for photoelectrochemical (PEC) water oxidation. In my own lecture I mentioned our efforts to develop a new electrolyser system based upon electron-coupled proton buffers,[14] and focused on a system using a polyoxometalate redox mediator, see Fig. 2.[15] In our new electrolyser system, oxygen and hydrogen production can be both spatially and temporally separated. In the first step, the polyoxometalate buffer is both reduced and protonated whilst oxygen is evolved and in the second step, on addition of the reduced POM to a catalyst, hydrogen is produced up to 30 times faster than in conventional PEME (proton exchange membrane electrolysers) systems.

Next, the discussion moved to a more bio-inspired system with a direct comparison of photoelectrochemical (PEC) water oxidation[16] activity between a photosystem II (PSII)-functionalised photoanode and a synthetic nanocomposite photoanode. These systems are extremely complex, requiring supramolecular control, and this was also shown for fully synthetic systems for light driven water oxidation.[17] The discussion moved to consider the oxidative stability of such systems and also suitable candidates for constructing photoanodes for light driven water oxidation[18] and water splitting in dye sensitized photoelectrochemical devices. An open question at this point is how to develop modules for oxygen evolution with those for hydrogen evolution in a simple way that can efficiently allow different catalytic 'modules' to be swapped into the system. Designing catalyst complexes that can withstand many millions of

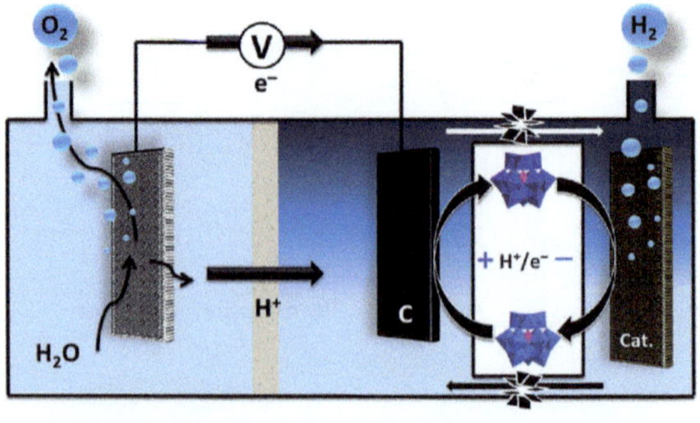

Fig. 2 Schematic of our polyoxometalate-based electrolyser system; reproduced from ref. 15.

turnovers is possible, but long term issues regarding stability and possible oxidative damage need to be explored. Biology has very efficient repair and reset mechanisms and it is not clear to what extent that artificial systems will need similar levels of maintenance.

Advances in solid-state dye sensitized solar cells was one of the highlights, with much discussion about the jumps in efficiency and scope for further control and enhancement, especially with hybrid organic–inorganic metal halide perovskites[19] and the use of double blocking layers.[20] DSSCs are already making a big impact with the first commercial systems being produced, but the leap in efficiency that allows the cost per watt to decrease is now moving toward a level where wider adoption is possible. This is because of the extraordinary leap in the solar-to-electric power conversion efficiency of devices using $CH_3NH_3PbI_3$ as a sensitizer, which has increased from 3 to 17.9%, as described by Grätzel[19] and others.[21] Cell operation of the hybrid perovskite solar cells is also an important dimension,[22] with a liquid phase deposition (LPD) method of a TiO_2 nanolayer for carbon-based perovskite solar cells. In a range of other developments, materials for solid state lighting were discussed, including which target structures to use to develop phosphors.[23] Solid state lights not only require less energy, but have longer lifetimes and could be easier to recycle. Thermoelectric materials based upon silicon were also an interesting topic.[24] This is because silicon nanoparticle based materials show especially low thermal conductivities due to their high number of interfaces, which increase the observed phonon scattering.

Approaching the final chapter of the meeting, we discussed the catalysis of small molecules, for example the reaction of CO on $Fe(OH)_x$/Pt nanoparticles.[25] This demonstrated that $Fe^{(III)}$–OH–Pt stepped sites readily react with CO adsorbed nearby to directly yield CO_2 and simultaneously produce coordinatively unsaturated iron sites for O_2 activation. There was much discussion about the role of precious metals like Pt, and how Pt-based electrocatalysts for the oxygen reduction reaction (ORR) hinder the practical application of proton exchange membrane fuel cells (PEMFCs).[26] In this respect it is critical to replace platinum with non-noble metal electrocatalysts. A number of options now exist, and this issue is one of the most crucial if the issue of scale is to be addressed. For example, carbonised metalloporphyrins are a good candidate for electrocatalysis, as is the more accessible hemin, where heat-treated nanoporous hemin activity can be comparable to that of commercial Pt/C in alkaline solution. Also, the study of CO adsorption at tetrahexahedral (THH) Pt nanocrystals (NCs)[27] was explored, which showed that CO adsorbs dominantly on the (100) sites of THH Pt NCs at low CO coverage.

Finally, the perspective lecture of Wang was extremely exciting, showing how triboelectrification could have literally thousands of applications.[28] By invention of a triboelectric nanogenerator using an elastic plastic material, mechanical energy could be converted to electrical energy. This could have applications literally everywhere and the lecture highlighted almost every conceivable option! I conclude by thanking the University of Xiamen and the organising committee Zhong-Qun Tian, Kazunari Domen, Clare Grey, Galen Stucky, Dongyuan Zhao and Nanfeng Zheng for organising such a wonderful and stimulating meeting. I can confidently predict that this will be the start of many extremely important *Faraday Discussions* in China and Asia as a whole.

References

1 C. Liu, N. Dasgupta and P. Yang, *Chem. Mater.*, 2014, **26**, 415–422.

2 V. Sans, S. Glatzel, F. J. Douglas, D. A. Maclaren, A. Lapkin and L. Cronin, *Chem. Sci.*, 2014, **5**, 1153–1157.

3 J. Maier, *Faraday Discuss.*, 2014, DOI: 10.1039/C4FD00135D.

4 Y. Hu, V. Thoréton, C. Pirovano, E. Capoen, C. Bogicevic, N. Nuns, A.-S. Mamede, G. Dezanneau and R. Vannier, *Faraday Discuss.*, 2014, DOI: 10.1039/C4FD00129J.

5 J. M. Griffin, A. C. Forse, H. Wang, N. M. Trease, P.-L. Taberna, P. Simon and C. P. Grey, *Faraday Discuss.*, 2014, DOI: 10.1039/C4FD00138A.

6 H.-W. Lee, M. Pasta, R. Y. Wang, R. Ruffo and Y. Cui, *Faraday Discuss.*, 2014, DOI: 10.1039/C4FD00147H.

7 Z. Zeng, W.-I. Liang, Y.-H. Chu and H. Zheng, *Faraday Discuss.*, 2014, DOI: 10.1039/C4FD00145A.

8 W. Li, H. Zheng, G. Chu, F. Luo, J. Zheng, D. Xiao, X. Li, L. Gu, H. Li, X. Wei, Q. Chen and L. Chen, *Faraday Discuss.*, 2014, DOI: 10.1039/C4FD00124A.

9 S. H. Chang, J. G. Connell, N. Danilovic, R. Subbaraman, K.-C. Chang, V. R. Stamenkovic and N. M. Markovic, *Faraday Discuss.*, 2014, DOI: 10.1039/C4FD00134F.

10 J. Wang, G. Wang, S. Miao, J. Li and X. Bao, *Faraday Discuss.*, 2014, DOI: 10.1039/C4FD00123K.

11 G. Yuan, A. Agiral, N. Pellet, W. Kim and H. Frei, *Faraday Discuss.*, 2014, DOI: 10.1039/C4FD00150H.

12 L.-A. Stern and X. Hu, *Faraday Discuss.*, 2014, DOI: 10.1039/C4FD00120F.

13 N. Wang, J. Zhu, X. Zheng, F. Xiong, B. Huang, J. Shi and C. Li, *Faraday Discuss.*, 2014, DOI: 10.1039/C4FD00139G.

14 M. D. Symes and L. Cronin, *Nat. Chem.*, 2013, **5**, 403–409.

15 B. Rausch, M. D. Symes, G. Chisholm and L. Cronin, *Science*, 2014, **345**, 1326–1330.

16 Y. Lai, M. Kato, D. Mersch and E. Reisner, *Faraday Discuss.*, 2014, DOI: 10.1039/C4FD00059E.

17 Y. Gao, L. Duan, Z. Yu, X. Ding and L. Sun, *Faraday Discuss.*, 2014, DOI: 10.1039/C4FD00127C.

18 H. Urabe, T. Hisatomi, T. Minegishi, J. Kubota and K. Domen, *Faraday Discuss.*, 2014, DOI: 10.1039/C4FD00122B.

19 H. A. Harms, N. Tétreault, N. Pellet, M. Bensimon and M. Grätzel, *Faraday Discuss.*, 2014, DOI: 10.1039/C4FD00160E.

20 H.-W. Kang, J.-W. Lee and N.-G. Park, *Faraday Discuss.*, 2014, DOI: 10.1039/C4FD00126E.

21 Y. Zhao, A. M. Nardes and K. Zhu, *Faraday Discuss.*, 2014, DOI: 10.1039/C4FD00128A.

22 H. Chen, Z. Wei, K. Yan, Y. Yi, J. Wang and S. Yang, *Faraday Discuss.*, 2014, DOI: 10.1039/C4FD00155A.

23 J. Brgoch, K. Hasz, K. A. Denault, C. K. H. Borg, A. A. Mikhailovsky and R. Seshadri, *Faraday Discuss.*, 2014, DOI: 10.1039/C4FD00125G.

24 S. P. Ashby, J. A. Thomas, J. García-Cañadas, G. Min, J. Corps, A. V. Powell, H. Xu, W. Shen and Y. Chao, *Faraday Discuss.*, 2014, DOI: 10.1039/C4FD00109E.
25 Y. Zhao, G. Chen, N. Zheng and G. Fu, *Faraday Discuss.*, 2014, DOI: 10.1039/C4FD00144C.
26 Y. Xie, C. Tang, Z. Hao, Y. Lv, R. Yang, X. Wei, W. Deng, A. Wang, B. Yi and Y. Song, *Faraday Discuss.*, 2014, DOI: 10.1039/C4FD00121D.
27 H.-X. Liu, N. Tian, J.-Y. Ye, B.-A. Lu, J. Ren, Z.-C. Huangfu, Z.-Y. Zhou and S.-G. Sun, *Faraday Discuss.*, 2014, DOI: 10.1039/C4FD00136B.
28 Z. L. Wang, *Faraday Discuss.*, 2014, DOI: 10.1039/C4FD00159A.

Poster titles

Nickel electrode for urea electrooxidation: nanowire array structure and catalytic reaction pathway, **D. Cao**, *Harbin Engineering University, China*

Corannulene-based materials for organic field-effect transistors and organics solar cells, **X. Y. Cao,** *Xiamen University, China*

Macroscopic assemblies derived from graphene oxide for thermal management and supercapacitor, **C. Chen, Q. Kong, X. Li, Z. Liu, Q. Zhang, D. Su and R. Cai,** *Institute of Coal Chemistry, Chinese Academy of Sciences, China*

Synchrotron X-ray techniques in energy material researches, **G. Chen, X. Liu, G. Wang, H. Geng and X. Zang,** *Shanghai Institute of Applied Physics, CAS, China*

Hybrid perovskite materials: from epitaxial assembly of nanoparticles to plasmon enhanced solar cells, **L. Chen, B.-W. Mao *et al.***, *Xiamen University, China*

Competitive activation of methane and O_2 on Pd and Pt surfaces, **M. Chen, X. Weng, H. Ren and H. Wan,** *Xiamen University, China*

Oxygen reduction reaction on Au@Pt nanoparticles and Au(100) electrodes, **Y. L. Zheng, Y. Yao, D. Mei and Y. Chen,** *University of Science and Technology of China, China*

Catalytic hydrogen generation with low-cost magnetically recyclable nanocatalysts, **Y. Chen, H. Guo, A. Lu, X. Liu, L. Wang and D. Peng,** *Xiamen University, China*

Improving electrochemical performance of $0.4Li_2\ MnO_3$-$0.6LiNi_{1/3}Co_{1/3}Mn_{1/3}O_2$ cathode materials for lithium secondary batteries with coating $LiNi_{0.5}Mn_{1.5}O_4$, **Y. Chen, K. Xie, Y. Pan and C. Zheng,** *National University of Defense Technology, China*

New solution-processed diketopyrrolopyrrole based donor for small molecule solar cell, **Y. Chen, A. Tang, C. Zhan and J. Yao,** *Institute of Chemistry, Chinese Academy of Sciences, China*

Ethanol photoelectrochemical splitting: a more energy efficient route for H_2 evolution, **Y. Chen, M. Metzler and T. Jacob,** *University of Ulm, Germany*

Aqueous processed CdTe–CdS bulk heterojunction solar cells, **Z. Chen, H. Zhang and B. Yang,** *Jilin University, China*

Functionalization of graphene with polyhedral oligomeric silsesquioxane (POSS) for supercapacitance application, **W. S. Chin, S. Y. Tang, B. B. Huang and H. Chi,** *National University of Singapore, Singapore*

Li-biphenyl-1,2-dimethoxyethane solution: electrical properties and applications in Li-ion batteries, **G. Chu, F. Luo, B. Liu, H. Lu, S. Wang, N. Liu, W. Li, J. Peng, H. Zheng, S. Zhang, H. Li, X. Huang and L. Chen,** *Institute of Physics, Chinese Academy of Sciences, China*

Efficient $CH_3NH_3PbI_3$-based planar solar cells with magnetron sputtered nickle oxide, **J. Cui, F. Meng, F. Huang and M. Wang,** *Huazhong University of Science and Technology, China*

Research progress for the foamed metal in the preparation and application, **Z. Dai and K. Ji,** *Nanjing University of Aeronautics and Astronautics, China*

Mechanical modulation of electrosorption and electrocatalysis, **Q. Deng, M. Smetanin and J. Weissmüller,** *Technische Universität Hamburg-Harburg, Germany*

Enhanced visible-light-driven hydrogen production from water with the cocatalysts made of earth-abundant elements, **P. Du, Z. Yan, Z. Sun and X. Liu,** *University of Science and Technology of China, China*

Dip-coated gold nanoparticle electrodes for aqueous-solution-processed large-area solar cells, **X. Du, Z. Chen and B. Yang,** *State Key Laboratory of Supramolecular Structure and Materials, Jilin University, China*

Si-nanoparticles encapsulated in hollow carbon fibre for lithium storage with large capacity and high reversibility, **S. Fang, L. Shen, Z. Tong, H. Zheng and X. Zhang,** *Nanjing University of Aeronautics and Astronautics, China*

Two-dimensional CoS nanosheets used for high performance organic–inorganic hybrid solar cells, **X. Fang, T. Song, R. Liu and B. Sun,** *Soochow University, China*

Fundamental issues of $Li\text{-}O_2$ and $Na\text{-}O_2$ batteries, **X. Guo and N. Zhao,** *Shanghai Institute of Ceramics Chinese Academy of Sciences, China*

Enhanced electrocatalysis of alkyne-functionalized ruthenium NPs for oxygen reduction reaction, **G. He and B. Long,** *Guangxi University, China*

Printable cheap counter electrodes for high-performance dye-sensitized solar cells, **J. He and T. Chen,** *Department of Physics, the Chinese University of Hong Kong, Hong Kong*

Planar heterojunction perovskite/PCBM solar cells with different purity of lead iodide, **D. Huang, Z. Xu, S. Zhao and Y. Li,** *Beijing Jiaotong University, China*

Enhanced performance of polymer solar cells by solvent-treated PEDOT:PSS, **D. Huang, Z. Xu, S. Zhao and Y. Li,** *Beijing Jiaotong University, China*

In situ Raman spectroscopic study of the potential decay mechanism of lithium-rich cathode material $Li[Ni_{0.2}0Li_{0.20}Mn_{0.60}]O_2$ in LIB, **J. Huang, B. Li, B. Liu, B. Liu, B. Ren and J. Zhao,** *Department of Chemistry, Xiamen University, China*

Thermochemistry of the LnI_3 and $Ln\text{–}LnI_3$ systems, **D. Ivanov, A. Dunaev, D. Sergeev, L. Kudin and M. Butman,** *Ivanovo State University of Chemistry and Technology, Russian Federation*

High-efficiency aqueous-processed hybrid solar cells with an enormous Herschel infrared contribution, **G. Jin, H. Sun and B. Yang,** *Northeast Normal University, China*

Self-oxidation of nanoporous NiCoMn alloy for high-performance pseudocapacitors, **S. Zhang and J. Kang,** *Tianjin Polytechnic University, China*

Bifunctional metal-organic framework for a highly robust single-component photo-hydrogenevolving device, **D. Kim, D. R. Whang and S. Y. Park,** *Seoul National University, South Korea*

Advanced electron transport layers for high-efficiency organic solar cell devices, **J. Kim, L. Quan, H. Kim, Y. H. Jang, Y. J. Jang and D. Kim,** *Ewha Womans University, South Korea*

AuNP@PANI@Pt–transition metal core–shell nanostructured electrocatalysts with enhanced activity for oxygen reduction reaction, **J. Lee and D. Kim,** *Ewha Womans University, South Korea*

Ag and Ag–Cu nanoclusters as a photosensitizer for the plasmonic solar cell, **W. Li, F. Chen and H. Wang,** *State Key Laboratory of Solidification Processing, Northwestern Polytechnical University, China*

All-solution-fabrication of an autonomous surface-plasmonic photosynthetic system towards hydrogen production, **D. Liu, B. Wu, S. Mubeen, G. Stucky and M. Moskovits,** *University of California Santa Barbara, United States*

Lanthanide ion doped $NaYF_4$ films with solar energy down/up conversion grown by electrodeposition, **R. Liu, L. Tian, H. Wang and X. Wu,** *Zhejiang University, China*

Electrodeposition and photoelectrochemical properties of Ag_3PO_4, **R. Liu, Q. Lu, K. Chen and W. Pan,** *Zhejiang University, China*

Nanostructured copper oxide electrodeposited from copper(II) complexes as an active catalyst for electrocatalytic oxygen evolution reaction, **X. Liu, P. Du and Z. Sun,** *University of Science and Technology of China, China*

Preparation of TiO_2 mesocrystalline microrods by electrochemical oxidation for photocatalytic application, **L. Long and H. Yu,** *University of Science & Technology of China, China*

Research on high capacity silicon anode for lithium ion battery, **F. Luo, G. Chu, J. Zheng, J. Li, H. Li, X. Huang and L. Chen,** *Institute of Physics, Chinese Academy of Sciences, China*

Solution transformation of Cu_2O into $CuInS_2$ for solar water splitting, **J. Luo, S. D. Tilley and M. Graetzel,** *Ecole Polytechnique Federale de Lausanne, Switzerland*

Perovskite solar cells with Ag–TiO_2 layer for spectrum adjusting and plasmon-enhanced absorption, **S. Ma, Z. Chen, L. Xiao, B. Qu and Q. Gong,** *Peking University, China*

A highly efficient mesoscopic solar cell based on $CH_3NH_3PbI_{3-x}Cl_x$ fabricated *via* sequential solution deposition, **Y. Ma, L. Zheng, Y. Chung, S. Chu, L. Xiao, Z. Chen, S. Wang, B. Qu, Q. Gong, Z. Wu and X. Hou,** *Peking University, China*

Crumpled graphene encapsulation optimized electrode material for high-energy lithium storage, **L. Mai, Q. Wei, X. Tian, Y. Zhao, X. Liu and L. Xu,** *Wuhan University of Technology, China*

Multiscale quantum mechanics/electromagnetics simulations for emerging energy materials, **L. Meng,** *College of Chemistry and Chemical Engineering, Xiamen University, China*

Bulk nanoporous gold with a hierarchical microstructure, **Z. Qi, M. Ritter and J. Weissmüller,** *Hamburg University of Technology, Germany*

One-step hydrothermal synthesis of high dispersity nanocrystallines of Fe_2O_3/polypyrrole/graphene composites, **S. Ren, S. Ma, Y. Yang and C. Hao,** *Dalian University of Technology, China*

Preparation of single-crystalline platinum nanowires with high aspect ratio, **Z. Shen, K. Miyabayashi and M. Miyake,** *Japan Advanced Institute of Science and Technology, Japan*

Interfacial effects of glassy carbon electrode on electrodeposition and catalytic durability of platinum nanoparticles, **K. Shi, Y. Yang, W. Wang, H. Zhang and Z. Xie,** *Department of Chemistry Xiamen University, China*

A discuss to the energy density limit of porous carbon based supercapacitors, **Z. Shi, Y. Xuewen, W. Jing and W. Chengyang,** *College of Materials Science and Engineering, Tianjin Polytechnic University, China*

Interpenetrated inorganic hybrid for efficiency enhancement of PbS quantum dot solar cells, **F. Tan, S. Qu, Q. Jiang, F. Li, G. Yue, S. Li, W. Zhang, C. Chen and Z. Wang,** *Henan University, China*

Template-free synthesis of hierarchical Mo-based oxides microflowers with improved lithium storage capability, **S. Tang and J. Liu,** *School of Materials Science and Engineering, Central South University, China*

Efficient and reproducible perovskite solar cells fabricated with a ternary precursor, **B. Wang and T. Chen,** *The Chinese University of Hong Kong, Hong Kong*

Influence of carbon surface properties on the oxygen reduction and Li_2O_2 growth, **H. Wang, K. Xie and C. Zheng,** *Department of Material Science and Engineering, National University of Defensive Technology, China*

Self-rechargeable transparent battery, **J. Wang, X. W. Lou and X. W. Sun,** *Nanyang Technological University, China*

Identifying Li^+ ion transport properties of $Li_{1+x}Al_xTi_{2-x}(PO_4)_3$ electrolyte at wide temperature range, **S. Wang, L. Ben, H. Li and C. Liquan,** *Institute of Physics Chinese Academy of Sciences, China*

Synthesis and performance of $0.4NiO.0.6Li_2Ti_{1-x}Mn_xO3$(x=0, 0.3, 0.5, 0.8) *via* sol gel method, **X. Wang, S. Zhao, S. Zhao and J. Wu,** *Department of Applied Chemistry, School of Environment and Chemical Engineering, Tianjin Polytechnic, University, China*

Numerical simulation of perovskite solar cells with inorganic HTM by using wxAMPS, **Y. Wang, Z. Xia, X. Du, Y. Liu and H. Zhou,** *Peking University, China*

Development of highly selective photocatalysts for CO_2 reduction with H_2O: a case study of TiO_2 promoted by noble metal and MgO cocatalysts, **Y. Wang, S. Xie, Q. Zhang, Y. Wang, G. Liu and W. Deng,** *Xiamen University, China*

An Ir(III)–Pt(II) supramolecular device for highly efficient visible light driven hydrogen production from water: beyond molecules, **D. R. Whang and S. Y. Park,** *Seoul National university, South Korea*

The facile synthesis of flower-like microspheres $CoMn_2O_4$ spinel material with superior supercapacitor properties, **J. Wu, L. Ren, L. Diu, X. Wang and M. Zhi,** *Harbin Institute of Technology Shenzhen Graduate School, China*

The effect of samarium doped ceria coating on the performance of $Li_{1.2}Ni_{0.13}$ $Co_{0.13}Mn_{0.54}O_2$ cathode material for lithium-ion battery, **J. Wu, F. He, C. Du, X. Wang and X. Zhang,** *Harbin Institute of Technology Shenzhen Graduate School, China*

Decoupling interrelated parameters for designing high performance thermoelectric materials, **C. Xiao and Y. Xie,** *University of Science & Technology of China, China*

Copper oxide nanomaterials for electrochemical reduction of carbon dioxide and the deposition of amorphous carbon, **J. Xie, Y. Huang and H. Yu,** *University of Science and Technology of China, China*

How supersaturation affects the surface structure of micro/nanocrystallites, **Z. Xie, W. Chen, H. Lin, Z. Jiang and Q. Kuang,** *State Key Laboratory of Physical Chemistry of Solid Surfaces, Xiamen University, China*

The preparation and characterization of perovskite thin films by dual-source vapor deposition, **H. Xu, R. Xu, Y. Yuan, W. Wang, S. Ge, L. Wang and F. Xu,** *Department of Materials Science and Engineering, Shanghai University, China*

Application of Mo-Blue as electron-coupled proton buffer (ECPB) for water splitting, **W. Xuan, H. Miras, D.-L. Long and L. Cronin,** *University of Glasgow, United Kingdom*

Hydrodeoxygenation of phenol over noble metal–Ni_2P/SiO_2, **Y. Li, X. Yang, N. Zhang, J. Zheng and B. Chen,** *Department of Chemical and Biochemical Engineering, College of Chemistry and Chemical Engineering, Xiamen University, China*

Nanoporous covalent organic polymers (COPs) for CO_2 capture, **C. T. Yavuz, J. Byun and H. Patel,** *Korea Advanced Institute of Technology and Science, South Korea*

In situ growth of ZIF-67 on carbon paper with enhanced bifunctional activity toward ORR and OER in alkaline electrolyte, **F. Yin, H. Wang, W. Shen, D. Sui and B. Chen,** *Beijing University of Chemical Technology, China*

Co_3O_4/C porous nanocomposite derived from MOFs: synthesis and bifunctional activities towards oxygen reduction/evolution reactions in an alkaline electrolyte, **F. Yin, X. He and R. Shi,** *Beijing University of Chemical Technology, China*

Tailored photocatalytic activity of ZnO/Au hybrid films coupled with thermoresponsive polymers, **M. Yoon, J. Lee and D. H. Kim,** *Ewha Womans University, South Korea*

Ni–Co layered double hydroxide (LDHs)/RGO as enhanced electrode for supercapacitor with ultrahigh capacitance, **M. Luo, C. You, S. Liang, J. Liang and L. Li,** *Ningxia University, China*

Sodium storage and pseudocapacitive charge in textured $Li_4Ti_5O_{12}$ thin films, **P. Yu, C. Li, X. Guo and X. Liu,** *Shanghai Institute of Microsystem And Information Technology, Chinese Academy of Sciences, China*

Ultrahigh-rate and high-density lithium-ion capacitors through hybriding nitrogen-enriched hierarchical porous carbon cathode with prelithiated microcrystalline graphite anode, **X. Yu, Z. Huang and F. Kang,** *Tsinghua University, China*

Au–TiO_2 junction with SPR effect for water oxidation, **C. Zhan, S. Wang, C. Li and Z. Tian,** *Collaborative Innovation Center of Chemistry for Energy Materials, Xiamen University, China*

Effect of temperature on the electrical characteristics of hybrid perovskite solar cells, **H. Zhang, X. Qiao, H. Yuang and M. Wang,** *Huazhong University of Science and Technology, China*

Preparation of nanoporous graphene sheet with encapsulated iron oxide showing enhanced electrochemical performance, **J. Zhang, Y. Yang, J. Yan and S. Guo,** *Shanghai Jiaotong University, China*

High-performance flexible broadband photodetector based on organolead halide perovskite, **X. Zhang and Y. Xie,** *University of Science and Technology of China, China*

Facile synthesis of $CuInCaS_2$ quantum dot nanoparticles for bilayer-sensitized solar cells, **J. Zhao, J. Zhang, P. Wang, Z. Liu, L. Wei and Y. Mai,** *Shijiazhuang Tiedao University, China*

Time dependent wavepacket diffusion method and its applications in organic semiconductors, **Y. Zhao, L. Han and X. Zhong,** *Xiamen University, China*

CH_3NH_3Cl-assisted growth of $CH_3NH_3PbX_3$ *via* Thermal decomposition, **Y. Zhao and K. Zhu,** *Shanghai Jiao Tong University, China*

Plantinum nanoparticles deposited mesoporous TiO_2 nanocrystals with enhanced photocatalytic, activities for solar-to-fuels conversion, **Y. Zhao, Z. Xiong, H. Wang, J. Zhang and C. Zheng,** *Huazhong University of Science & Technology, China*

Conducting porous Mo_2C as catalyst with superior performance for hydrogen evolution, **Y. Zhao, S. Chen, A. Monda, K. Kretschmer and G. Wang,** *University of Technology, Sydney, Australia*

Nanowires for energy conversion, storage, biosensing and cell interface, **G. Zheng,** *Fudan University, China*

In situ TEM study on lithium transportation in Li_2O_2, **H. Zheng, D. Xiao, D. Xie, P. Lu, Y. Wang, L. Gu, H. Li and Z. Shan,** *Institute of Physics, The Chinese Academy of Sciences, China*

A hydrophobic hole transporting oligothiophene for planar perovskite solar cells with improved stability, **L. Zheng, Y. Chung, Y. Ma, L. Zhangi, L. Xiao, Z. Chen, S. Wang, B. Qu and Q. Gong,** *School of Physics, Peking University, China*

Small sulfur molecules intercalation in expanded graphite for Li–S battery cathode, **S. Zheng, P. Han, Z. Han and J. Yang,** *University of Shanghai for Science and Technology, China*

Rational design of orderly-packed electrodes for high power and long cycle-life lithium ion batteries, **Y. Zhong and Z. Zhou,** *Nankai University, China*

Sb nanoparticles decorated N-rich carbon nanosheets as anode materials for sodium ion batteries with superior rate capability and long cycling stability, **X. Zhou and Z. Zhou,** *Nankai University, China*

Segregation and chemical ordering of AuPd nanoalloy under reactive condition, **B. Zhu, H. Guesmi and Y. Gao,** *Shanghai Institute of Applied Physics, Chinese Academy of Sciences, China*

How to get a conversion reaction reversible? Lithium and sodium storage in MoS_2 nanodots, **C. Zhu, X. Mu, P. Van Aken, Y. Yu and J. Maier,** *Max Planck Institute for solid state research, Germany*

Liquid fuel cells with polymer fiber membrane and non-precious metal catalysts, **X. Yang, Y. Liu, Z. Zhou and S. Sun,** *Collaborative Innovation Center of Chemistry for Energy Materials, Xiamen University, Xiamen, Fujian, China*

Electrospun SnO_2–graphene composite nanofibers as high performance anode for lithium ion batteries, **H. Guo, Y. Wu, J. Lin and J. Wang,** *Pen-Tung Sah Institute of Micro-Nano Science and Technology, Xiamen University, China*

Thermoelectric research at nanoscale, **J.-C. Zheng,** *Department of Physics, Xiamen University, China*

Efficient supported bimetallic nanocatalysts for selective hydrogenolysis of carbon–oxygen bonds, **Y. Yuan, W. Li, L. Ye and X. Duan,** *Xiamen University, China*

Solid state NMR study of cathode materials Na_2FePO_4F for sodium ion batteries, **X. Hou, G. Zhong, Q. Li, Z. Liu and Y. Yang,** *State Key Lab of Physical Chemistry of Solid Surfaces and Department of Chemistry, Xiamen University, China*

Evolution, separation and characterization of the active species of Ti doped $NaAlH_4$ for hydrogen storage, **R. Xiong, G. Sang, X. Yan and G. Zhang,** *China Academy of Engineering Physics, Mianyang, China*

Supercapacitive performance improvement of hydrous RuO_2 by soft-template modification, **C.-C. Hu, I.-L. Chen and C.-T. Hsu,** *Department of Chemical Engineering, National Tsing Hua University, Taiwan*

Bipolar interface derived membrane self-humidification in polyelectrolyte membrane fuel cell, **S. Peng, X. Xu, P.-C. Sui, S. Lu, Y. Xiang and N. Djilali,** *Beihang University, China*

Atomically engineered materials for solar energy chemistry, **M. Yu and H. G. Yang,** *East China University of Science & Technology, Shanghai, China*

Development of counter electrode materials for dye-sensitized solar cells, **P. Liu and H. G. Yang,** *East China University of Science & Technology, Shanghai, China*

A comprehensive study on the capacity fading mechanism of rechargeable lithium/sulfer–graphene oxide battery, **J. Zhu, E. J. Cairns, J. Guo, X. Feng *et al.*,** *Shanghai Institute of Microsystem and Information Technology, Chinese Academy of Sciences, China*

In situ STXM and APXPS study of block copolymer electrolyte in Li-Ion battery, **R. Chang, W. Hudson, M. Grass and Z. Liu,** *Shanghai Institute of Microsystem and Information Technology, Chinese Academy of Sciences, China*

Tuning the energy band-gap of crystalline gallium oxide to enhance photocatalytic water splitting: mixed phase junctions, **M.-G. Ju, W. Liang, Y. Zhao and C. Li,** *Xiamen University, China*

Carbon nanofibers as a conductive additive in $LiFePO_4$ cathode material for rechargeable lithium ion batteries, **A. Aitbekova *et al.*,** *Nazarbayev University, Kazakhstan*

Selective transformation of CO_2 to methanol by modified Cu–ZnO based catalysts, **F. Liao, X. Hong and E. Tsang,** *University of Oxford, UK*

In situ probe at interfaces using shell-isolated nanoparticle-enhanced Raman spectroscopy, **J.-F. Li, T. Wandlowski and Z. Tian,** *Xiamen University, China*

Porous conducting membranes and their application in vanadium flow batteries, **X. Li, H. Zhang and H. Zhang,** *Dalian Institute of Chemical Physics, China*

Key materials for lithium/sulfer soft package battery with advanced energy density, **H. Zhang and H. Zhang,** *Dalian Institute of Chemical Physics, China*

Nanocarbon based Pd–Pt electrocatalyst for methanol oxidation, **M.-E. Oh and H.-C. Choi,** *Chonnam National University, Korea*

Graphene oxide supported ZnO photocatalyst for methylene blue (MB) degredation, **J.-H. Jeong and H.-C. Choi,** *Chonnam National University, Korea*

Photocatalytic efficiency in presence of GO–TiO_2 nanocomposites, **E.-H. Kim and H.-C. Choi,** *Chonnam National University, Korea*

Pt modified electrode for electrocatalytic oxidation of hydrazine, **J.-D. Kim and H.-C. Choi,** *Chonnam National University, Korea*

The Skinner Prize for the best poster was jointly awarded to Yuxia Chen of the Institute of Chemistry, Chinese Academy of Sciences, China, for their poster on new solution-processed diketopyrrolopyrrole based donors for small molecule solar cells and Zhen Qi of Hamburg University of Technology, Germany, for their poster on bulk nanoporous gold with a hierarchical microstructure.

List of participants

Aisulu Aitbekova, *Nazarbayev University, Kazakhstan*
Hua Bai, *Xiamen University, China*
Xinhe Bao, *Chinese Academy of Sciences Dalian Compound, China*
Alisa Becker, *Royal Society of Chemistry, United Kingdom*
Shuping Bi, *Nanjing University, China*
Deirdre Black, *Royal Society of Chemistry, United Kingdom*
Jeremie Brilliet, *Shanghai Silicate Research Institute, China*
Jeehye Byun, *KAIST, Korea*
Jiabai Cai, *Minnan Normal University, China*
Xiao-Yu Cao, *Xiamen University, China*
Rui Chang, *Shanghai Institute of Microsystem and Information Technology, China*
Yimin Chao, *University of East Anglia, United Kingdom*
Gang Chen, *Shanghai Institute of Applied Physics, China*
Hongwei Chen, *Shanghai Institute of Microsystem and Information Technology, China*
Jianlin Chen, *Changsha University of Science and Technology, China*
Liang Chen, *Xiamen University, China*
Mingshu Chen, *Xiamen University, China*
Yanxia Chen, *University of Science and Technology of China, China*
Yan-Xin Chen, *Universität Ulm, Germany*
Yuanzhi Chen, *Xiamen University, China*
Yufang Chen, *National University of Defense Technology, China*
Yuxia Chen, *Lanzhou Institute of Chemical Physics, China*
Zeng Chen, *Henan University, China*
Zhaolai Chen, *Jilin University, China*
Wee Shong Chin, *National University of Singapore, Singapore*
Geng Chu, *Chinese Academy of Sciences Institute of Physics, China*
Lee Cronin, *Glasgow University, United Kingdom*
Jin Cui, *Huazhong University of Science and Technology, China*
Dehui Deng, *Chinese Academy of Sciences Dalian Compound, China*
Qibo Deng, *Technische Universität Hamburg-Harburg, Germany*
Kazunari Domen, *University of Tokyo, Japan*
Chunwei Dong, *Jilin University, China*
Hua Dong, *Xi'an Jiaotong University, China*
Quanfeng Dong, *Xiamen University, China*
Pingwu Du, *University of Science and Technology of China, China*
Xiaohang Du, *Jilin University, China*
Youxin Duan, *Lanzhou Institute of Chemical Physics, China*
Fengru Fan, *Energy Materials Chemistry Synergy Innovation Center, China*
Qinghua Fan, *South China University of Technology, China*
Shan Fang, *Nanjing University of Aeronautics and Astronautics, China*
Xiao Fang, *Suzhou University, China*
Xuefei Feng, *Shanghai Institute of Microsystem and Information Technology, China*

Heinz Frei, *Lawrence Berkeley National Laboratory*, *United States*
Gang Fu, *Xiamen University*, *China*
Jv Fu, *Sichuan University*, *China*
Heping Geng, *Shanghai Institute of Applied Physics*, *China*
Michael Graetzel, *École Polytechnique Fédérale de LausanneÉcole Polytechnique Fédérale de Lausanne*, *Switzerland*
Clare Grey, *University of Cambridge*, *United Kingdom*
John Griffin, *University of Cambridge*, *United Kingdom*
Hang Guo, *Xiamen University*, *China*
Xiangxin Guo, *Shanghai Silicate Research Institute*, *China*
Robert Hammond-Smith, *Merck Chemicals Ltd*, *United Kingdom*
Hongwei Han, *Huazhong University of Science and Technology*, *China*
Wei Han, *South China University of Technology*, *China*
Guoqiang He, *Guangxi University*, *China*
Jian He, *Chinese University of Hong Kong*, *Hong Kong, China*
Xiaobo He, *Beijing University of Chemical Technology*, *China*
Takashi Hisatomi, *University of Tokyo*, *Japan*
Rudolf Holze, *Technische Universitaet Chemnitz*, *Germany*
Xiang Hong, *Amperex Technology Limited*, *China*
Chun-Tsung Hsu, *National Tsing Hua University*, *Chinese Taipei*
Xile Hu, *École Polytechnique Fédérale de Lausanne*, *Switzerland*
Di Huang, *Beijing Jiaotong University*, *China*
Jiale Huang, *Xiamen University*, *China*
Jingxin Huang, *Xiamen University*, *China*
Yan-Feng Huang, *Tianjin Polytechnic University*, *China*
Yunhua Huang, *Mikrouna Shanghai Mechanical and Electrical Technology*, *China*
Hua Huo, *Harbin Institute of Technology*, *China*
Graham Hutchings, *Cardiff University*, *United Kingdom*
Dmitry Ivanov, *Ivanovo State University of Chemistry and Technology*, *Russia*
Keju Ji, *Nanjing University of Aeronautics and Astronautics*, *China*
Yun-Bao Jiang, *Xiamen University*, *China*
Gan Jin, *Northeast Normal University*, *China*
Zheng Jincheng, *Pen-Tung Sah Education and Scientific Research Foundation*, *China*
Wang Jiong, *Nanjing University*, *China*
Minggang Ju, *University of Science and Technology of China*, *China*
Ji Hyun Jung, *Chonnam National University*, *Korea*
Firoz Babu Kadumudi, *University of Ulsan*, *Korea*
Jianli Kang, *Tianjin Polytechnic University*, *China*
Ryoji Kanno, *Tokyo Institute of Technology*, *Japan*
Dong Ha Kim, *Ewha Womans University*, *Korea*
Dongha Kim, *Seoul National University*, *Korea*
Eun Hye Kim, *Chonnam National University*, *Korea*
Ji Dang Kim, *Chonnam National University*, *Korea*
Jihyeon Kim, *Ewha Womans University*, *Korea*
Wenping Kong, *Henan University*, *China*
Hyun-Wook Lee, *Stanford University*, *United States*
Ji-Eun Lee, *Ewha Womans University*, *Korea*
Zhang Lei, *Strasbourg University*, *France*
Can Li, *Chinese Academy of Sciences Dalian Compound*, *China*
Fei Li, *Dalian University of Technology*, *China*
Hong Li, *Chinese Academy of Sciences Institute of Physics*, *China*
Jiajie Li, *Harbin Institute of Technology*, *China*
Jun Li, *Xiamen University*, *China*

Liang Li, *University of Science and Technology Beijing*, *China*
Minghua Li, *University of Science and Technology Beijing*, *China*
Qi Li, *Xiamen University*, *China*
Qidong Li, *South China University of Technology*, *China*
Shengjun Li, *Henan University*, *China*
Shiyuan Li, *Mikrouna Shanghai Mechanical and Electrical Technology*, *China*
Weiyin Li, *Northwestern Polytechnical University*, *China*
Wenjun Li, *Chinese Academy of Sciences Institute of Physics*, *China*
Xianfeng Li, *Dalian Institute of Chemical Physics*, *China*
Xianggao Li, *Tianjin University*, *China*
Yu Li, *Peking University*, *China*
Wanzhen Liang, *Xiamen University*, *China*
Fenglin Liao, *University of Oxford*, *United Kingdom*
Changjian Lin, *Xiamen University*, *China*
Kaiwen Lin, *Jiangxi Normal University of Science and Technology*, *China*
Bo Liu, *Xiamen University*, *China*
Deyu Liu, *University of California, Santa Barbara*, *United States*
Guokun Liu, *Xiamen University*, *China*
Hong Liu, *Shanghai Jiaotong University*, *China*
Jun Liu, *Central South University*, *China*
Pengfei Liu, *East China University Of Science and Technology*, *China*
Qian Liu, *New Energy Technology*, *China*
Run Liu, *Zhejiang University*, *China*
Xiang Liu, *University of Science and Technology of China*, *China*
Xin Liu, *Shanghai Institute of Applied Physics*, *China*
Xingang Liu, *Sichuan University*, *China*
Xinmei Liu, *Guangxi University*, *China*
Binli Long, *Guangxi University*, *China*
Lulu Long, *University of Science and Technology of China*, *China*
Xinghui Long, *Shanghai Institute of Microsystem and Information Technology*, *China*
Peijie Lu, *Central South University*, *China*
Qinyan Lu, *Zhejiang University*, *China*
Sarah Lu, *Royal Society of Chemistry*, *China*
Fei Luo, *Chinese Academy of Sciences Institute of Physics*, *China*
Jingshan Luo, *École Polytechnique Fédérale de Lausanne*, *Switzerland*
Mandy Ma, *Royal Society of Chemistry*, *China*
Shengbo Ma, *Peking University*, *China*
Yingzhuang Ma, *Peking University*, *China*
Liqiang Mai, *Wuhan University of Technology*, *China*
Joachim Maier, *Max Planck Institute for Solid State Research*, *Germany*
Lingyi Meng, *Xiamen University*, *China*
Jonathan Midgley, *Royal Society of Chemistry*, *United States*
Shouli Ming, *Jiangxi Normal University of Science and Technology*, *China*
Heather Montgomery, *Royal Society of Chemistry*, *United Kingdom*
Andrew Mount, *University of Edinburgh*, *United Kingdom*
Myeong Eun, Oh, *Chonnam National University*, *Korea*
Richard Oldroyd, *Elkington & Fife LLP*, *United Kingdom*
Fuping Pan, *Lanzhou Institute of Chemical Physics*, *China*
Wei Pan, *Zhejiang University*, *China*
Xiulian Pan, *Dalian Institute of Chemical Physics*, *China*
Nam-Gyu Park, *Sungkyunkwan University*, *Korea*
Hongmei Peng, *Royal Society of Chemistry*, *China*
Zhen Qi, *Hamburg University of Technology*, *Germany*

Erwin Reisner, *University of Cambridge*, *United Kingdom*
Bin Ren, *Xiamen University*, *China*
Suzhen Ren, *Dalian University of Technology*, *China*
Yuanhua Sang, *Shandong University*, *China*
Ram Seshadri, *University of California, Santa Barbara*, *United States*
Xuyi Shan, *Institute of Metal Research, Chinese Academy of Sciences*, *China*
Pei Kang Shen, *Sun Yat-Sen University*, *China*
Zhongrong Shen, *Japan Advanced Institute of Science and Technology*, *Japan*
Kang Shi, *Xiamen University*, *China*
Zhiqiang Shi, *Tianjin Polytechnic University*, *China*
Bo Song, *Harbin Institute of Technology*, *China*
Guanqun Song, *Royal Society of Chemistry*, *China*
Yu-Fei Song, *Beijing University of Chemical Technology*, *China*
Yujiang Song, *Dalian Institute of Chemical Physics*, *China*
Galen Stucky, *University of California, Santa Barbara*, *United States*
Licheng Sun, *Dalian University of Technology*, *China*
Mengna Sun, *Tianjin University*, *China*
Shi-Gang Sun, *Xiamen University*, *China*
Xuhui Sun, *Suzhou University*, *China*
Furui Tan, *School of Physics & Electronics*, *China*
Shasha Tang, *Central South University*, *China*
Sarah Thomas, *Royal Society of Chemistry*, *United Kingdom*
Linlin Tian, *Zhejiang University*, *China*
Wei(Wayne) Tian, *Royal Society of Chemistry*, *China*
Zhongqun Tian, *Xiamen University*, *China*
Rose-Noelle Vannier, *École Nationale Supérieure de Chimie de Lille*, *France*
Baohua Wang, *Chinese University of Hong Kong, Hong Kong*, *China*
Chengxiu Wang, *China University Of Petroleum*, *China*
Gang Wang, *China University Of Petroleum*, *China*
Geng Wang, *Shanghai Institute of Applied Physics*, *China*
Guoxiong Wang, *Dalian Institute of Chemical Physics*, *China*
Hao Wang, *Beijing University of Chemical Technology*, *China*
Hui Wang, *National University of Defense Technology*, *China*
Huiqiong Wang, *Xiamen University*, *China*
Jinmin Wang, *Shanghai Second Polytechnic University*, *China*
Lijuan Wang, *Royal Society of Chemistry*, *China*
Shaofei Wang, *Chinese Academy of Sciences Institute of Physics*, *China*
Tian-Yan Wang, *New Energy Technology*, *China*
Xiaoqing Wang, *Tianjin Polytechnic University*, *China*
Yan Wang, *Peking University*, *China*
Ye Wang, *Xiamen University*, *China*
Zhi Peng Wang, *Jiangxi Normal University of Science and Technology*, *China*
Zhonglin Wang, *Georgia Institute of Technology*, *United States*
Chaochao Wei, *Henan University*, *China*
Jing Wei, *Peking University*, *China*
Xiaoru Wen, *Xiamen University*, *China*
Dong Ryeol Whang, *Seoul National University*, *Korea*
Junwei Wu, *Harbin Institute of Technology*, *China*
Xiaowen Wu, *Zhejiang University*, *China*
Zhi Wu, *Xiamen University*, *China*
Jun Xi, *Xi'an Jiaotong University*, *China*
Bin Xia, *Xi'an Jiaotong University*, *China*
Xinyuan Xia, *Shandong Normal University*, *China*

Zhonggao Xia, *Peking University*, China
Yan Xiang, *Beijing University of Aeronautics and Astronautics*, China
Chong Xiao, *University of Science and Technology of China*, China
Lixin Xiao, *Peking University*, China
Yin Xiao, *Tianjin University*, China
Jiafang Xie, *University of Science and Technology of China*, China
Kai Xie, *National University of Defense Technology*, China
Zhaoxiong Xie, *Xiamen University*, China
Renjin Xiong, *China Academy of Engineering Physics*, China
SHIZHAO Xiong, *National University of Defense Technology*, China
Haitao Xu, *Shanghai University*, China
Shoubin Xu, *China Academy of Engineering Physics*, China
Weijing Xu, *East China Normal University*, China
Weimin Xuan, *University of Glasgow*, United Kingdom
Tingting Yan, *Shanghai University*, China
Bo Yang, *Guilin University of Electronic Technology*, China
Long Yang, *Huaqiao University*, China
Peidong Yang, *University of California*, United States
Qian Yang, *Central South University*, China
Shihe Yang, *Hong Kong University of Science and Technology, Hong Kong*, China
Xiaodong Yang, *Xiamen University*, China
Xiaoli Yang, *Donghua University*, China
Xin Yang, *Xiamen University*, China
Xudong Yang, *Shanghai Jiaotong University*, China
Yang Yang, *East China Normal University*, China
Zhi Yang, *Wenzhou University*, China
Zixiang Yang, *Chinese Academy of Sciences Institute of Qinghai Salt Lake*, China
Linmin Ye, *Xiamen University*, China
Meidan Ye, *Xiamen University*, China
Na Ye, *Wenzhou University*, China
Fengxiang Yin, *Beijing University of Chemical Technology*, China
Ying Yin, *Harbin Institute of Technology*, China
Minji Yoon, *Ewha Womans University*, Korea
Chunqin You, *Ningxia University*, China
Mingquan Yu, *East China University Of Science and Technology*, China
Pengfei Yu, *Shanghai Institute of Microsystem and Information Technology*, China
Xiaoliang Yu, *Tsinghua University*, China
Youzhu Yuan, *Xiamen University*, China
Gentian Yue, *Henan University*, China
Hongying Zang, *Northeast Normal University*, China
Chao Zhan, *Xiamen University*, China
Chuhong Zhang, *Sichuan University*, China
Danfei Zhang, *Peking University*, China
Dengsong Zhang, *Shanghai University*, China
Fei Zhang, *Tianjin University*, China
Hua Zhang, *Huazhong University of Science and Technology*, China
Jian Zhang, *Guilin University of Electronic Technology*, China
Jianan Zhang, *Shanghai Jiaotong University*, China
Liwu Zhang, *University of Cambridge*, United Kingdom
Nan Zhang, *Xiamen University*, China
Tiansheng Zhang, *Huzhou Teachers College*, China
Xiaodong Zhang, *University of Science and Technology of China*, China
Xiaonan Zhang, *Shanghai Institute of Applied Physics*, China

Dongyuan Zhao, *Fudan University*, *China*
Jinjin Zhao, *Shijiazhuang Tiedao University*, *China*
Ning Zhao, *Shanghai Silicate Research Institute*, *China*
Wei Zhao, *Xi'an HuaYan Optoelectronic Technology*, *China*
Yi Zhao, *Xiamen University*, *China*
Yicheng Zhao, *Peking University*, *China*
Yin Zhao, *Shanghai University*, *China*
Yixin Zhao, *Shanghai Jiaotong University*, *China*
Yongchun Zhao, *Huazhong University of Science and Technology*, *China*
Yufei Zhao, *University of Technology, Sydney*, *Australia*
Shijie Zhen, *Jiangxi Normal University of Science and Technology*, *China*
Dajiang Zheng, *Xiamen University*, *China*
Gengfeng Zheng, *Fudan University*, *China*
Hao Zheng, *Chinese Academy of Sciences Institute of Physics*, *China*
Lingling Zheng, *Peking University*, *China*
Nanfeng Zheng, *Xiamen University*, *China*
Shiyou Zheng, *University of Shanghai for Science and Technology*, *China*
Guiming Zhong, *Xiamen University*, *China*
Jie Zhong, *Huazhong University of Science and Technology*, *China*
Yiren Zhong, *Nankai University*, *China*
Feng Zhou, *Shanghai Jiaotong University*, *China*
Xianlong Zhou, *Nankai University*, *China*
Zhao-Hui Zhou, *Xiamen University*, *China*
Beien Zhu, *Shanghai Institute of Applied Physics*, *China*
Changbao Zhu, *Max Planck Institute for Solid State Research*, *Germany*
Kai Zhu, *National Renewable Energy Laboratory*, *United States*
Ruimin Zhu, *Shandong University*, *China*

